Cyber-Physical Laboratories in Engineering and Science Education

Michael E. Auer • Abul K.M. Azad
Arthur Edwards • Ton de Jong
Editors

Cyber-Physical Laboratories in Engineering and Science Education

Springer

Editors
Michael E. Auer
Carinthia University of Applied Sciences
Villach, Austria

Abul K.M. Azad
Northern Illinois University
DeKalb, IL, USA

Arthur Edwards
University of Colima
Colima, Mexico

Ton de Jong
University of Twente
Enschede, The Netherlands

ISBN 978-3-030-08329-8 ISBN 978-3-319-76935-6 (eBook)
https://doi.org/10.1007/978-3-319-76935-6

Softcover re-print of the Hardcover 1st edition 2018

Printed on acid-free paper

This Springer imprint is published by the registered company Springer International Publishing AG part of Springer Nature.
The registered company address is: Gewerbestrasse 11, 6330 Cham, Switzerland

Foreword

Cyber-physical laboratories were but a theoretical paradigm until they first became a reality around the turn of the century, when technological advances in the areas of hardware, software, networking, and control made the first rudimentary laboratories possible. Since then, the accelerated evolution of the technologies required by cyber-physical labs has substantially expanded their versatility and applicability to the degree that their use in the educational realm is expanding monumentally. Today, almost all definitions of cyber-physical laboratories, although some experts may disagree on some discreet points, involve either monitoring, controlling, or twinning an object in the physical world by means of software algorithms which permit the dynamic interaction between said object and the real world, maintained through either cabled or wireless communications to computer-based resources. Also, digital twins and simulations are widely used in the online laboratory field.

Of course, this implies that major advantages of cyber-physical laboratories are that they are scalable, often shared resources that are not constrained by spatial-temporal considerations.

Adequate laboratory experience at a time and place convenient for students has always been a major challenge for science, engineering, and technology educators. This applies to both traditional laboratory courses, where classes are scheduled only for a specified time period when students attend a laboratory class located within a laboratory of an academic institution, and distance learning programs which, in the great majority of existing Internet-based distance learning programs, lack any significant laboratory-based courses.

In the case of traditional laboratories, in many cases, they do not adequately compensate for the mixed ability level of students, and the allocated time for carrying out activities is many times insufficient for all students to complete their tasks satisfactorily to gain the sufficient experience they need to internalize often complex processes and internalize them. Also, in some cases, students want or feel a need to perform additional experiments beyond their assigned tasks. It is difficult to accommodate any extra experimentation because universities often lack resources to keep their laboratories open. Additionally, laboratory facilities are often inaccessible to the students of other departments within the same institution because of their geographical location. Ironically, too much laboratory equipment lies idle during most of their usable lifetime.

Although cyber-physical laboratories provide important advantages, they can be very difficult to implementbecause these facilities involve the areas of instrumentation, computer interfacing, health and safety, video streaming, data collection and management, web application development, database management, network security, learning management systems, pedagogical design, and course management. The cyber-physical remote or virtual laboratory, either as replacement of or supplement to traditional laboratories, must be able to address the above difficulties before they can be effectively integrated into learning environments.

Cyber-physical laboratories, however, offer valuable benefits in that properly managed, they can allow for their full integration into distance-learning or blended learning programs, which can potentially make them extremely scalable, affording easy access when integrated into online learning systems. Additionally, but equally important, cyber-physical laboratories provide the opportunity for greater collaboration at more affordable costs among universities and research centers by providing both researchers and students access to a wide collection of shared experimental resources by sharing costs and reducing the duplicity, which often occurs when institutions purchase the same, often expensive equipment individually.

Another very important consideration is that cyber-physical laboratories have been shown to be equally or more effective than some more direct forms of instruction and at least as effective as traditional physical laboratories. However, this has been shown to be true only when online guidance provides students resources as part of an integrated learning system. This guidance can be provided using a variety of forms ranging from providing students with tools for inquiry (such as a scratchpad for creating hypotheses), adding augmentations to the lab, or embedding it in background information. Research is now progressing to determine what kind of guidance is necessary for students to better learn from specific kinds of laboratories.

Recognizing the benefits cyber-physical laboratories can potentially offer, there has been an increased interest and effort toward applying or developing relevant technologies and how to most effectively implement them, as well as how to identify their effectivity insofar as student learning and educational outcomes are concerned. However, there are various factors that influence the development of remote laboratories, including the nature of the input(s) and output(s) of the experiments, the speed of operation, data collection restrictions, the need for video and audio feedback, data presentation, security safety requirements, scalability, and interfacing with other similar systems. In the case of virtual laboratories, a specific development aspect is the level of required fidelity, with at its extreme virtual reality laboratories that fully mimic the real laboratory (except for the olfactory aspects).

Considering the abovementioned factors, each of the current developments in this area is unique, and there is currently little room for further integration with other systems or for expanding different experiments for local, regional, and global collaboration. To address these factors, a number of issues need to be investigated to develop modular, effective, versatile, cost effective, user friendly, and sustainable remote and virtual laboratory systems that can deliver its true potential in the national and global arena, which will allow individual researchers develop their own modular system with a level of creativity and innovation, while at the same time ensuring continued growth by separating the responsibility for creating online labs from the responsibility for overseeing the students who use them. This feature is critical for scaling the number of users of a particular laboratory experiment and for expanding the development of new laboratories.

Part I of this volume, "State of the Art and Future Developments in Cyber-Physical Laboratory Architectures," introduces the reader to several system architectures that have proven successful in many online laboratory settings. The first online laboratory developments were reported in the late 1990s. Since then the emergence of new technologies has influenced the design structure of these developments and has allowed remote laboratories to have new features and capabilities.

This section will include chapters describing the state-of-the-art structure of remote laboratories as well as ongoing and potential future development. Authors are encouraged to include sufficient detail to enable an informed decision as to which approach best fits your needs. These chapters will describe the technologies used along with pedagogical issues to keep in mind while designing the architecture. The section will also provide a comparative picture of various technologies and developments. In addition, there will be an effort to report the standardization outcomes that are conducted by professional organizations to streamline online laboratory development.

Part II of this book, "Pedagogy of Cyber-Physical Experimentation," discusses the pedagogical questions that come along with the introduction of virtual and remote laboratories in the curriculum. Pedagogical questions concern, for example, the amount of freedom to hand over to students but also the type of guidance provided to students and the fading of this guidance over time, the differentiation of the lab experience for students with differing prior knowledge and/or inquiry skills, and how to shape students' collaboration when learning through an online lab, etc. This section offers a unique collection of chapters each describing one of the world's five most widely used ecosystems for online labs for science education. In these chapters, the latest developments of these ecosystems are presented, including the design and development of integrated student guidance, the online measuring and interpretation of student activities as a basis for providing students with adaptive feedback, (teacher) authoring facilities, accessibility of online labs for students, and the use of advanced learning scenarios such as collaborative learning and learning by modelling.

Finally, Part III is titled "Cyber-Physical Laboratories: Best Practices and Case Studies." This section highlights a number of remote laboratory case studies, covering a range of application areas that can be considered as representative best practices. There is a total of six chapters highlighting remote laboratories for life science experiments, automation engineering, hardware in the loop systems, integration of augmented reality and haptic devices, heat transfer experiments, additive manufacturing, and utilization of mobile devices for remote laboratories. The contributions provide an insight from a different perspective and each discussion leads the reader to understand the rationale behind the approaches taken and obtain further information of interest. Almost all the chapters in this section report the developments in engineering, technology, and physics topics.

It is our sincere hope that by reading the valuable contributions to this book, you will gain a greater insight as to the many considerations persons wishing to develop and implement cyber-physical laboratories must take into consideration, by reflecting upon the actual thoughts and experiences of some of the foremost developers and practitioners in this important and quickly evolving area. It is our further hope that any knowledge gained by our experiences serve to motivate you to become still more informed and motivated to join us in providing more valuable experimental and experiential tools to induce, motivate, and help students gain practical knowledge about real-world principles and phenomena.

Carinthia University of Applied Sciences, Villach, Austria Michael E. Auer
Northern Illinois University, DeKalb, IL, USA Abul K.M. Azad
University of Colima, Colima, Mexico Arthur Edwards
University of Twente, Enschede, The Netherlands Ton de Jong

Contents

About the Editors

Michael E. Auer received his Ing. degree (1971) and his Ph.D. degree (1975) from the Dresden University of Technology. His working field was the analysis and design of high-speed integrated microelectronic circuits. From 1974 to 1991, he was an assistant professor at the faculties "Electrical Engineering" and "Informatics" of this university. His research at this time was related to high-speed digital circuit simulations, design systems and platforms for the design and simulation of microelectronic circuits, and real-time and network programming in UNIX environments.

From 1991 to 1995, he was head of the software department F+O Electronic Systems GmbH, Heidelberg. His research there was related to real-time and network programming, embedded control systems, programming in C, C++, PERL, as well as system and network administration of heterogeneous networks with UNIX, VMS, and Windows workstations.

In 1995, Michael Auer was appointed Professor of Electrical Engineering at the Carinthia University of Applied Sciences, Villach, Austria, and built up the teaching domain "Fundamentals of Electrical Engineering and Circuit Design."

Michael Auer has also a teaching position for "Microelectronics" at the University of Klagenfurt, Austria, and works as a visiting professor at some universities worldwide.

His current research is directed to technology-enhanced learning and remote working environments.

Besides being a co-chair or member of the program committees of several international conferences and workshops, he is especially involved as founder and general chair of the annual international conferences "Interactive Collaborative Learning" (ICL) and "Remote Engineering and Virtual Instrumentation" (REV).

In 2009, Michael Auer was appointed as a member of the Advisory Board of the European Learning Industry Group (ELIG). Furthermore, he is chair of the Advisory Board of the International E-Learning Association (IELA).

Michael Auer is Managing Editor of the OnlineJournals.ORG platform with a number of open access journals in the fields of "Online Engineering," "Emerging Technologies in Learning", Mobile Technologies," and "Engineering Pedagogy."

Michael Auer is Founder, President, and CEO of the "International Association of Online Engineering" (IAOE), a nongovernmental organization that promotes the vision of new engineering working environments worldwide. From 2009 to 2016, he was President of the "International Society for Engineering Education" (IGIP). In 2015, he was elected as President of the International federation of Engineering Education Societies (IFEES) for the term 2016–2018.

Abul K.M. Azad is a Professor in the Technology Department of Northern Illinois University, USA. He has a Ph.D. in Control and Systems Engineering and M.Sc. and B.Sc. in Electronics Engineering. He has been in academics for 25+ years, and his research interests include remote laboratories, mechatronic systems, mobile robotics, and educational research. In these areas, Dr. Azad has over 115 refereed journal and conference papers as well as 5 edited books. So far, he has attracted around $2.6 M of research and development grants from various national and international funding agencies. He is a member of the editorial board for a number of professional journals as well as an Editor-in-Chief of the International Journal of Online Engineering. Dr. Azad is active with remote laboratory field and is the President of the Global Online Laboratory Consortium (GOLC) as well as the Vice-President of the International Association of Online Engineering (IAOE). He is also active with few other professional organizations like IEEE, IET, ASEE, ISA, and CLAWAR Association, and served as Chair and Co-Chairs of numerous conferences and workshops. He was a program evaluator for the ABET and is active in evaluating research and development projects for various national and international funding agencies in the USA, Europe, and Australia.

Arthur Edwards holds a master's degree in education from the University of Houston (1985). He has collaborated at the University of Colima, Mexico, for 29 years as lecturer/researcher, where he has been instrumental in the area of curricula and instruction. He is a co-founder of the College of Foreign Languages, the University English Program, and the Self Access Centers of this institution. As head of the Self Access Center, he developed an interest in Computer-Assisted Language Learning (CALL) and moved to the College of Telematics to follow up this line of research, where he is currently a senior tenured researcher. During this first period of his career, he authored two English textbooks published by the University of Colima Press.

In 1999, he was assigned to the College of Telematics full time, where he developed additional interests in eLearning and other related topics. He was awarded funding in 1999 to follow up his project of eLearning by the Ministry of Scientific Research of the Mexican government, being the first project approved for financing of the College of Telematics.

Over the last decade, Arthur Edwards has been integrated into the Mobile Computing workgroup, where he has collaborated on a series of nationally and internationally funded research programs in the area of ad hoc networking (primarily vehicular ad hoc networks) and remote mobile self-organizing robotics.

During this time, he has participated in the publication of approximately 50 scientific articles, 30 book chapters, and 6 books. He has also participated internationally as editor in four journals (two related to technology and two related to sustainability in education). Arthur Edwards has also participated in various national and international organizations, where he has evaluated research projects, publications, conferences, etc.

Ton de Jong holds a chair in Instructional Technology at the University of Twente, the Netherlands. He has specialized in inquiry learning and collaborative learning (mainly in science domains) supported by technology. He was coordinator of several EU projects and several national projects, including the ZAP project in which interactive games/simulations for psychology were developed. ZAPs commercial licences now go over 80,000 in number. He was coordinator of the 7th framework Go-Lab project on learning with online laboratories in science and currently is coordinator of its H2020 follow-up project Next-Lab (see www.golabz.eu). He published over 200 journal articles and book chapters, was an associate editor for the *Journal of Engineering Education* and for *Instructional Science*, and currently is on the editorial board of eight journals. He has published papers in *Science* on inquiry learning with computer simulations (2006), design environments (2013), and virtual laboratories (2013). He is AERA fellow and was elected member of the Academia Europaea in 2014. He is dean of the master program Educational Science and Technology at the University of Twente. For more info see: http://users.edte.utwente.nl/jong/Index.htm

Abbreviations

AABB	Axis Aligned Bounding Boxes
AD	Automation Device
ADDIE	Analysis, Design, Development, Implementation, and Evaluation
ANN	Artificial Neural Network
API	Application Protocol Interface
AR	Augmented Reality
AWS	Amazon Web Services
BKT	Bayesian Knowledge Tracing
CAD	Computer-Aided Design
CGI	Common Gateway Interface
CMS	Content Management System
CPPS	Cyber-Physical Production System
CPS	Cyber-Physical System
CPU	Central Processing Unit
CSS	Cascading Style Sheets
CV	Computer Vision
DAQ	Data Acquisition
DMZ	Demilitarized Zone
DV	Dependent Variable
EA	Evolutionary Algorithm
FIFO	First In First Out
FREVO	Framework for Evolutionary Design
FSM	Finite State Machine
GBVL	Game-Based Virtual Learning
GCM	Gesture Control Module
Go-Lab	Global Online Science Labs for Inquiry Learning at School
GUI	Graphical User Interface
HMD	Head Mounted Display
HMI	Human Machine Interface
HTTP	Hypertext Transfer Protocol
ICT	Information and Communication Technologies

IIoT	Industrial Internet of Things
iLab	Interactive Lab
ILS	Inquiry Learning Space
IMS	Instructional Management System
IMS-CP	IMS Content Packing
Inq-ITS	Inquiry Intelligent Tutoring System
IoT	Internet of Things
ISA	iLab Shared Architecture
IV	Independent Variable
JSON	JavaScript Object Notation
LaaS	Laboratory as a Service
LiaaS	Lab Server Infrastructure as a Service
LiLa	Library of Labs
LMS	Learning Management System
LTI	Learning Tools Interoperability
MOOC	Massive Open Online Course
MOOL	Massive Open Online Lab
MQTT	Message Queue Telemetry Transport
NGSS	Next Generation Science Standards
NNGA	Neural Network Genetic Algorithm
NUI	Natural User Interface
OBBB	Object Oriented Bounding Boxes
OECD	Organisation for Economic Co-operation and Development
Olab	Online Labs
OTAP	Over-the-Air Programming
PDOM	Parallel Document Object Model
PhET	Physics Education Technology
PhET-iO	Interoperable PhET Simulations
PISA	Program for International Student Assessment
PLE	Personal Learning Environment
RAL	Remote Access Laboratory
RAMI	Reference Architectural Model Industry
RCE	Remote Code Editor
REST	REpresentational State Transfer
RFC	Request for Comment
RFID	Radio Frequency Identification
RL	Remote Laboratory
RLMS	Remote Laboratory Management System
RSDL	RESTful Service Description Language
RT Lab	Remote Lab
RT-WSN Lab	Remote Triggered Wireless Sensor Network Lab
SAR	Search and Rescue
SCADA	Supervisory Control and Data Acquisition
SCORM	Shareable Content Object Reference Model
SLAM	Simultaneous Localization and Mapping

SOAP	Simple Object Access Protocol
TCP	Transmission Control Protocol
UAV	Unmanned Aerial Vehicle
UI	User Interface
VD	Virtual Device
VE	Virtual Environment
VIS	Viewable Image System
VISIR	Virtual Instrument Systems in Reality
VL	Virtual Laboratory
VLCAP	Virtual Labs Collaborative Accessibility Platform
VNC	Virtual Network Computing
VR	Virtual Reality
W3C	World Wide Web Consortium
WDG	WOAS Device Gateway
WISE	Web-Based Inquiry Science Environment
WOAS	Web-Oriented Automation System
WPG	WOAS Protocol Gateway
WSDL	Web Services Description Language
WSN	Wireless Sensor Network

Part I
State of the Art and Future Developments in Online Laboratory Architectures

Introduction

Ian Grout

Today, we consider the use of the Internet as an everyday activity and routinely expect access to a rich set of resources which are presented to us in audio, visual and even tactile forms that suit our particular wishes or needs. Access to resources which are interesting, of a high technical quality, beneficial to the individual and easy to access, have a high quality of service (QoS) and are typically available continuously (on a "24/7" basis) is required. To reach the current situation that provides online services with these attributes, a great amount of work has been undertaken within higher education research and industry globally, led by individuals who have visions of what can be achieved and why they should be achieved. In higher education, one particular vision has been to widen access to engineering and scientific laboratory resources using online and remote access by embracing the positive power that the Internet can provide. Over the last number of years, the development of the online laboratory has evolved from an interesting engineering or computer science exercise where a primary question was "How can we use the Internet to remotely access our experiments and form an online laboratory?" to "How can we maximise the potential of our online laboratory?". The considerations and focus for practitioners in the field are evolving from a purely but interesting engineering, or computer science, technical challenge to a more end-user requirement challenge. This requires the laboratory developers to embrace new perspectives and challenges whilst maintaining or enhancing the technical foundations that underpin any laboratory infrastructure. Since the initial work undertaken in online laboratory design and development, a wealth of ideas, information and experiences have been collated. In addition, the number of laboratory providers and users has expanded so that now online experimentation is an integral part of many higher education programmes. Given that each laboratory resource developer has a particular set of aims and ideas of how the laboratory can be developed and used, each laboratory may have a different "look and feel", as well as different available resources from teaching materials through to the physical laboratory infrastructure itself. These would be based on a number of different developer requirements including availability of suitable electronic hardware and software, access to experiments, the ability to access the Internet for specific requirements and developer knowledge and experience, along with end-user requirements and needs. To provide a right balance between what set of outcomes would be desired, what would need to be created and what would be possible is a challenging task which is a problem to solve that has multiple dimensions.

This section, "State-of-the-Art and Future Developments in Online Laboratory Architectures", provides an insight into current developments in online laboratories. It is aimed to consider the current status in the field, end-user requirements and future directions in online laboratory development. The section consists of contributions from practitioners in the field and their insights into how these laboratories are designed, developed, deployed and can evolve. Each contribution provides an insight from a different perspective, and each discussion leads the reader to understand the rationale behind the approaches taken and obtain further information of interest.

In the contribution "Online Laboratory Architectures and Technical Considerations", developments in online laboratories are provided. A background into online laboratory development is initially provided along with examples of existing laboratory arrangements such as the "iLab Shared Architecture" from the Massachusetts Institute of Technology (MIT) and "WebLab-Deusto". Ad hoc solutions and Remote Laboratory Management Systems are introduced, along with frameworks and tools that are available and can be used to simplify the development and deployment of online laboratories. The contribution concludes with trends towards the creation of a common online laboratory architecture.

In the contribution "The WebLab-Deusto Remote Laboratory Management System Architecture: Achieving Scalability, Interoperability, and Federation of Remote Experimentation", WebLab-Deusto, an open-source Remote Laboratory Management System (RLMS), is introduced and discussed. It allows access to remote experiments and developers to share their own laboratories. This work is the result of collaborators, mainly from the University of Deusto in Portugal, working since 2004 in the field of remote/online engineering. The contribution provides a useful background history to the WebLab-Deusto project and the results obtained. The laboratory structure is shown, reasons why decisions in the laboratory development were made discussed, and its uses and the future directions such as the "LabsLand" spin-off activity provided.

In the contribution "Deploying Large-Scale Online Labs with Smart Devices", the upcoming challenges in moving remote experimentation from small-scale deployment to very large-scale deployment are considered. This is referred to as "Massive Open Online Labs (MOOLs)". Sharing resources whilst minimizing or even cancelling the waiting time to access a particular resource is a major challenge to support the end-user experience. This requires the resource provider to revisit both the pedagogical and technical methodologies of online laboratory implementation, sharing and deployment. The concept of the "Smart Device" model, which follows the "Laboratory as a Service (LaaS) scheme", is introduced and attempts to describe the physical laboratory equipment, its digital components and interfaces as a unique entity.

In the contribution "Augmented Reality and Natural User Interface Applications for Remote Laboratories", two key areas of focus to potentially enhance the end-user experience are considered. Firstly, augmented reality (AR), which the potential to create rich user interfaces, where users can view and interact with virtual objects, is considered. The discussion considers the use of AR and the use of virtual objects in remote laboratories as an alternative, immersive user interface. Secondly, natural user interfaces (NUIs), mechanisms to take input from users without using a fixed position or dimensionally restricted input, are considered. By using the natural movement of the user, a computer can be controlled without the use of objects such as the computer mouse and keyboard. Typically, a NUI incorporates some form of computer vision. By considering AR and NUIs, new and exciting ways in which a remote laboratory can be interacted with may be developed and deployed.

In the contribution "Designing Cyber-Physical Systems with Evolutionary Algorithms", cyber-physical systems (CPSs) are considered, and the need for suitable design tools is discussed. As the degree of interaction among CPSs increases, this can lead to unpredictable and partially unexpected behaviour. Such potentially unwanted behaviour must be addressed in the design process. Hence, CPS design must be supported with suitable design methods and tools. Whilst a number of methods and tools that support CPS design already exist, there is no comprehensive toolset available. This chapter presents a proposal for a common CPS design toolset that combines existing and emerging tools to design, simulate, evaluate and deploy solutions for complex, real-world problems. A case study of swarms of unmanned aerial vehicles (UAVs) is presented as part of this discussion.

Chapter 1
Online Laboratory Architectures and Technical Considerations

Danilo Garbi Zutin

Abstract While a traditional, hands-on laboratory experience may be ideal, it is not always feasible. The costs associated with providing laboratory resources to students and the logistics of providing access can be prohibitive. This is particularly the case with laboratories that utilise limited resources or with students who may be performing their coursework remotely. In such cases, an Online Laboratory a laboratory that students can control remotely via the Internet can provide students with a valuable practical experience that is complementary to available hands-on laboratories. At the beginning, Online Laboratories were developed as ad hoc solutions and, as such, were designed for a very specific purpose and were not intended to be adapted or generalised. Ad hoc implementations of Online Laboratories are likely to neglect important aspects and requirements of an Online Laboratory system, such as scalability (ability to cope with a growing number of users and laboratories to manage). Furthermore, sharing Online Laboratories was also not a trivial issue. This chapter will present an overview of the main technical developments, software architecture models and access schemes used to deploy Online Laboratories.

Keywords Online Laboratories · Service-oriented architectures · Cloud computing · Web services · E-learning · Remote systems · Peer-to-peer networks

1.1 Introduction

The development of online laboratories has undergone major changes since the first systems were introduced almost 15 years ago. In the beginning, online laboratories were developed as ad hoc solutions, usually designed for a very specific purpose, and were not intended to be adapted or generalised, making online laboratories a

---

D. G. Zutin (✉)
Carinthia University of Applied Sciences, Villach, Austria
e-mail: dgzutin@ieee.org

M. E. Auer et al. (eds.), *Cyber-Physical Laboratories in Engineering and Science Education*, https://doi.org/10.1007/978-3-319-76935-6_1

closed, self-contained system. Fortunately the community soon realised that this strategy did not favour the large-scale use and scalability of online laboratory systems in formal education and research. The initial attempts to address some of these issues began with the development of the first Remote Laboratory Management Systems (RLMS), such as the iLab Shared Architecture (Harward et al. 2008) and WebLab-Deusto (Ordua et al. 2011), that took place mainly during the last decade. RLMSs grouped common functionalities around a single framework.

It is beyond the scope of this chapter to discuss the pedagogical setups that favour the use of Online Laboratories; however, one of its advantages, as pointed out by Cooper (2005), is the improved access for disabled students and the possibility to better combine experimentation with distance education programmes. In fact, the last one has been recognised by some authors as one of the driving forces that pushed the development of Online Laboratory systems (Feisel and Rosa 2005).

1.2 Online Laboratories and Architectures

Online Laboratories are computer applications that allow students, teachers and researchers to conduct experimentation from a remote location. These experiments can be of any kind and from any domain (e.g. Physics, Chemistry, Electronics, etc.).

Access to an Online Laboratory is usually delivered via a client application that can run in a Web browser, standalone or even in an embedded system. If this client interacts with a remote server (in this work referred to as a Lab Server) and controls a piece of equipment where the experiment data is measured, the Online Laboratory is referred to as being a Remote Laboratory. If the Online Laboratory delivers data generated as the result of a simulation, it is classified as a Virtual Laboratory. In this sense, remote and virtual laboratories are subsets of Online Laboratories.

This section will provide an overview of the software architectures commonly used to deliver Online Laboratories.

1.2.1 Ad Hoc Online Laboratory Architectures

In the beginning, these Online Laboratories were developed mainly following an ad hoc approach as depicted in Fig. 1.1. This means these solutions were designed for a very specific purpose and were not intended to be adapted or generalised. As a consequence, each Online Laboratory was a closed, self-contained system with no communication whatsoever with other entities, even if they implemented the same functionalities repeatedly. For example, in Fig. 1.1, both Laboratory 1 and Laboratory 2 implement user management, experiment data management and booking of laboratory sessions and their users cannot sign in with the same credentials to the other Online Laboratory.

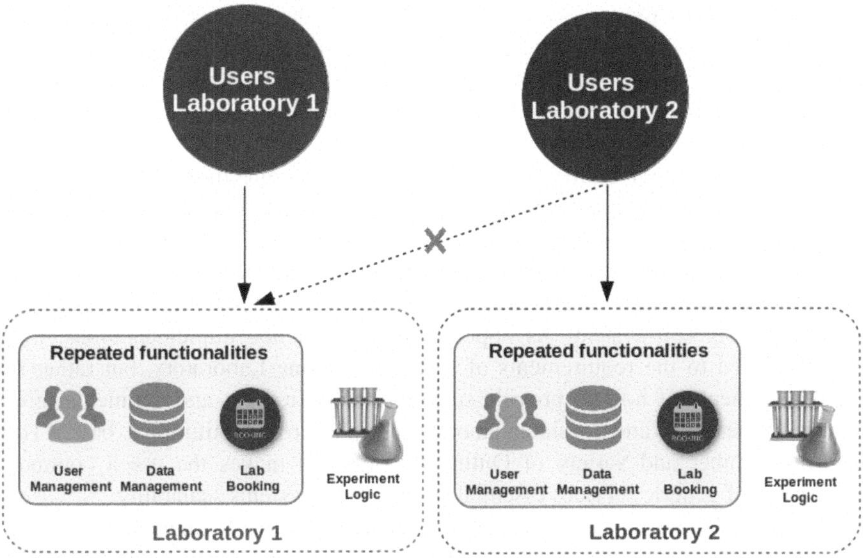

Fig. 1.1 Ad hoc implementations of Online Laboratories

1.2.2 Remote Laboratory Management Systems

When online laboratories began gaining uptake, a more structured approach was necessary to ensure the scalability of these Online Laboratory systems. The scalability of a system can be defined as its ability to efficiently handle a growing workload. In a network system, workload is mainly affected by an increasing number of users, instances and network nodes. Remote Laboratory Management Systems (RLMSs) were the first attempt to address this situation. An RLMS is a software that groups functionalities common to every online laboratory system around a single framework. Some of the initial functional requirements of these RLMSs were:

- User management, single sign-on with institution's authentication systems
- Implementation of laboratory scheduling services
- Support for a scalable federation of online laboratories
- Support single sign-on in a cross-institutional federation of online laboratories
- Support for management of experiment data (data storage and retrieval)
- Integration with Learning Management Systems (LMS)

Remote Laboratory Management Systems contributed significantly for important advancements in the field of Online Laboratories, but their main contributions were the new possibilities created for sharing access to online experiments in an efficient and scalable manner, often across institutional boundaries. By grouping the functionalities described above around a common framework, the Online Laboratory system became a very specialised component, designed to process exclusively experiment-related requests. As a consequence of the decreased complexity of Online Laboratory systems, their development was also simplified to some extent. From this point of view, an RLMS could be considered a set of services available to laboratory servers that allowed for sharing of common functionalities among a cluster of Online Laboratories.

An example of an RLMS is the iLab Shared Architecture (ISA). ISA is a Web service infrastructure developed at the Massachusetts Institute of Technology (MIT) that provides a unifying software framework for online laboratory developers (Harward et al. 2008). ISA supports the access to a globally distributed network of Online Laboratories, and their users can access these laboratories by means of a single sign-on system. As opposed to most ad hoc implementations, ISA is not tailored to the requirements of a specific Online Laboratory, but rather to the requirements of how to provide support for the framing and maintenance of laboratory sessions and to share laboratories in a cross-institutional basis. The growing number and variety of Online Laboratories makes the use a common framework of generic services essential to ensure the systems scalability.

The use of Web services was favoured mainly due to its characteristics as a technology that allows the loose coupling of the different components of the ISA. Beyond that, the architecture should support the use of a diverse number of laboratory hardware and software and should not tie client and server platforms. It should also not make any assumptions on the network policies (firewalls, proxy servers) that a user might be under. These requirements favour the use of Web services for the implementation of this architecture due to their platform independence and standardisation. According to the World Wide Web Consortium (W3C 2002), Web services provide a standard means of interoperating between different software applications, running on a variety of platforms and/or frameworks. The iLab Shared Architecture is depicted in Fig. 1.2.

ISA supports asynchronous (or batched) and synchronous (or interactive) Online Laboratories. Asynchronous labs are special types of Online Laboratories whose experiments can have their entire execution course specified before it starts. The task of performing a batched experiment can be summarised in submitting an experiment specification, executing the experiment, retrieving, analysing and displaying the results to the user. Synchronous labs, in the other hand, are those Online Laboratories whose experiments require real-time control of the laboratory equipment. In fact the terms *batched* and *interactive* were coined by the ISA developers. The support for one or the other type of laboratories guided decisions on the framework development, described in the following sessions. The ISA batched architecture follows a typical three-tier model as depicted in Fig. 1.3. The role of Web services in this three-tier architecture is to provide the interfaces between the different components.

- The client application typically runs in the browser and is a domain-specific programme that communicates via the ISA API to carry out experiment-related tasks. It must be able to parse the batched parameters and experiment results and therefore understand the schemes used to encode these messages.

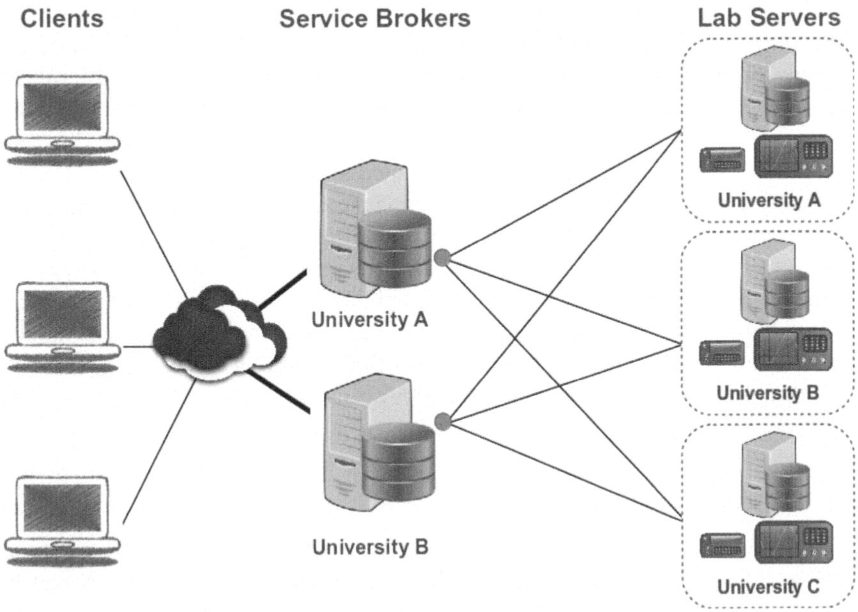

Fig. 1.2 The iLab Shared Architecture

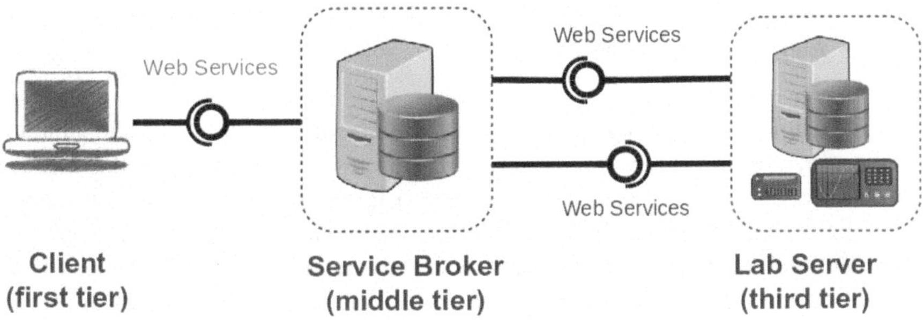

Fig. 1.3 The iLab Shared Architecture – batched architecture

- The Service Broker mediates the communication between the client application and Lab Servers and provides user management functions, for example, to assign students. It normally resides on a server at the students' institution, where all accounts are managed.
- The Lab Server is the component of the architecture responsible to run experiments. Lab Servers serve the Service Broker with experiment results. They do not implement any lab user management and know nothing about the user running an experiment. The Lab Server exposes its functionalities to the Service Broker via its Web services API and never communicate directly with the client.

The ISA API allows the different components to be loosely coupled. The cardinality between Service Brokers and lab servers can be represented by a many-to-many relationship. This means that a Service Broker can request experiment execution to several different lab servers located anywhere in the globe, and a lab server can serve requests from several different Service Brokers located at different institutions. This relationship can be represented as shown in Fig. 1.2.

RLMSs have played a major role concerning the adoption of Online Laboratories. They contributed by providing common frameworks that laboratory developers could use to build their systems upon. RLMS aggregated several common functionalities of Online Laboratories and exposed them to developers via well-defined APIs. Although this section focused on the iLab Shared Architecture, this is not the only existing RLMS. Other examples of RLMSs are WebLab-Deusto (Ordua et al. 2011) and Labshare Sahara (Lowe et al. 2009).

1.3 Frameworks and Tools

This section will introduce the reader to some frameworks and tools created to simplify the development and deployment of Online Laboratories.

1.3.1 The Experiment Dispatcher: A Tool for Ubiquitous Deployment of Labs

The Experiment Dispatcher is a software architecture used to provide ubiquitous deployment of Online Laboratories. It is a framework that provides Online Laboratory server infrastructure as a service (LIaaS) consumed by the laboratory developers to enable and/or facilitate the deployment and development of these systems (Zutin et al. 2016). This approach makes no assumption on how the access to the experiment will be delivered, since this is left for the discretion of the lab owner. For example, lab owners might decide to deliver remote experimentation using an RLMS and thereby take advantage of all benefits provided by these systems. Alternatively, the lab owner might decide not to use an RLMS, but rather interface their client application directly with the laboratory using the provided channels to relay messages between them. This new approach can complement the Smart Device specification (Salzmann et al. 2015) from the Go-Lab project (de Jong et al. 2014). The Go-Lab Smart Device paradigm aims to decouple client and server by defining interfaces between them and enabling thereby their easy integration with other third-party services. As pointed out by Salzmann et al. (2015), it originates from the RFID and sensor world, where information (metadata) is added to the device allowing user interface it to adapt itself based on the provided services. The specification makes no assumption on the inner working of the online laboratory, but rather defines the interfaces (message schema and protocol) for a client application to communicate with it.

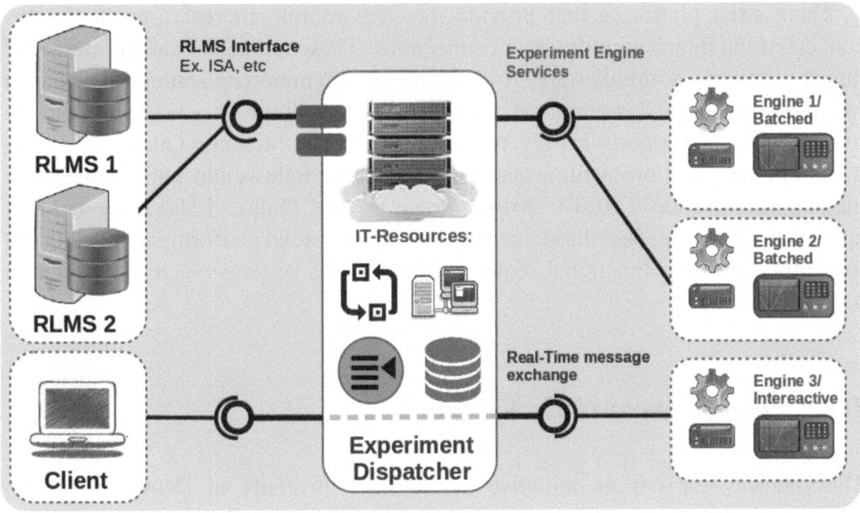

Fig. 1.4 The Experiment Dispatcher

The Experiment Dispatcher centralises functionalities commonly provided by Online Laboratory servers and allows its seamless reuse by heterogeneous Online Laboratories. Additionally, it abstracts the development of the software necessary to deliver remote experimentation. It supports experiments that run according to the batched execution model. The term batched experiment, alternatively also called *asynchronous experiment*, concerns the scheduling schema employed. Batched experiments should have their entire execution course specified before it begins. The task of performing a batched experiment can be summarised in submitting an experiment specification, executing the experiment, retrieving, analysing and displaying the results to the user. Batched laboratories are designed to run relatively fast, therefore scheduling of lab session by means of a calendar-based booking is not necessary. Batched experiments are usually queued prior to processing by the Lab Server. Examples of batched Online Laboratories are the MIT Microelectronic Device Characterization (del Alamo et al. 2003) and the University of Queensland Radioactivity Laboratory (Jona and Vondracek 2013). Batched Online Laboratories are very reliable, since the user interaction with the lab equipment is limited and highly controlled by the RLMS framework. Interactive experiments are also supported, but with limited functionalities at the present moment. The Experiment Dispatcher is depicted in Fig. 1.4.

The relaying service of the Experiment Dispatcher is an essential functionality to transfer messages to the laboratory equipment and support an ubiquitous deployment of remote experimentation. This approach allows a laboratory equipment to reside at any network, even behind NATs, as long as Internet connectivity is provided. In that sense, it shares some characteristics of a peer-to-peer network.

There exist platforms that provide message routing in real time to devices connected via intermittent Internet connections. These are the so-called Internet of things platforms or middleware. An example of a commercial solution is Amazon IoT platform (https://aws.amazon.com/iot/) that also employs a publish-subscribe mechanism and supports HTTP, WebSockets and the Message Queue Telemetry Transport (MQTT) protocol, a machine-to-machine lightweight publish/subscribe messaging transport (MQTT 2014). However, the Online Laboratory field of application is a very specialised one, and these commercial platforms are not tailored to comply with the functional requirements of most systems, such as integration with RLMSs.

1.3.2 The Gateway4Labs System

The Gateway4Labs is an initiative led by the University of Deusto, Spain, to facilitate the integration of Online Laboratories in different learning tools such as LMSs, PLEs and CMSs (Ordua et al. 2015). It attempts to provide a unifying centralised software framework to a level of interoperability between the lab management and learning management layers. As pointed out (Sancristobal Ruiz et al. 2014), there are different ways to integrate Online Laboratories with learning tools; however, most implementations are ad hoc and tailored to a specific system. The main motivation for that is the fact that several functionalities implemented by RLMSs and LMSs are duplicated (Ordua et al. 2015), such as user management and user tracking.

The Gateway4Labs system is a middleware that provides a centralised component for the integration of Online Laboratories and learning tools such as LMSs (Ordua et al. 2015). Support for different RLMSs is achieved by means of a plug-in-based architecture. A plug-in wraps the authentication mechanism with a particular system (e.g. Online Laboratories, RLMSs). Consumer tools can interact with Gateway4Labs via three different interfaces, namely, via a HTTP RESTful interface, via IMS-LTI or via OpenSocial. In the case of LTI, the Gateway4Labs middleware handles the LTI launch request and calls the Online Laboratory system requesting the launch of a client application. To launch an application, it might have to authenticate against the Online Laboratory system, if the last one requires so. According to the LTI security model, tool providers and consumers must exchange out-of-band a permanent launch link and a credentials necessary to launch the application on the remote system. LTI compatibility ensures that Gateway4Labs can be used by adopters of a large range of LMS users, assuming they implement the LTI interface.

In the context of the Go-Lab project (de Jong et al. 2014), Gateway4Labs offers the software framework that allows for the integration of third-party Online Laboratories into the Go-Lab ecosystem. The Go-Lab system uses the OpenSocial API, a specification originally developed by Google to allow third-party trusted applications to run in a container hosted by other Web applications. Gateway4Labs became an essential tool to ensure a more lightweight integration of external online laboratory systems with Go-Lab.

1.3.3 The Go-Lab Smart Device Paradigm

Go-Lab (Global Online Science Labs for Inquiry Learning at School) is a European Commission co-funded project that aims at creating a European-wide federation of Online Laboratories and the pedagogical framework to allow for their effective use by secondary school teachers to enrich classroom experience as well as the learning activities out-of-class. Go-Lab is a European collaborative project co-funded by the European Commission in the context of the Seventh Framework Programme with 18 partner organisations from 12 countries.

Go-Lab proposes a new paradigm to facilitate the reuse and sharing of laboratory equipment and user interface components (widgets) by defining laboratory equipment as smart devices. The smart device paradigm aims to decouple client and server by defining interfaces between them and enabling thereby their easy integration with other third-party services. As pointed out by Salzmann et al. (2015), it originates from the RFID and sensor world, where information (metadata) is added to the device (therefore smart device). This information can be the data range of the sensor, the measured unit, etc. This metadata contains the information necessary for a hypothetical client application to adapt itself to interact with the sensor or actuator in question.

Instead of providing a monolithic interface, smart device paradigm decomposes the functionalities of the Online Laboratory and specifies its interfaces in terms of sensors and actuators, each one with well-defined interfaces (Salzmann et al. 2015). Although not a requirement, Web browsers are the typical environment to render the user interface of the smart device, preferably with HTML5 and JavaScript, due to their cross-platform support. The Go-lab Smart Device specification is divided into different parts that define the behaviour of the smart device and how to interact with it. The specification defines the transport protocols used by the client applications to interact with the smart device, the message schema used, the schema of the metadata that carries the description of the smart device sensors and actuators and a description chosen to describe the sensors and actuator services. This is depicted in Fig. 1.5.

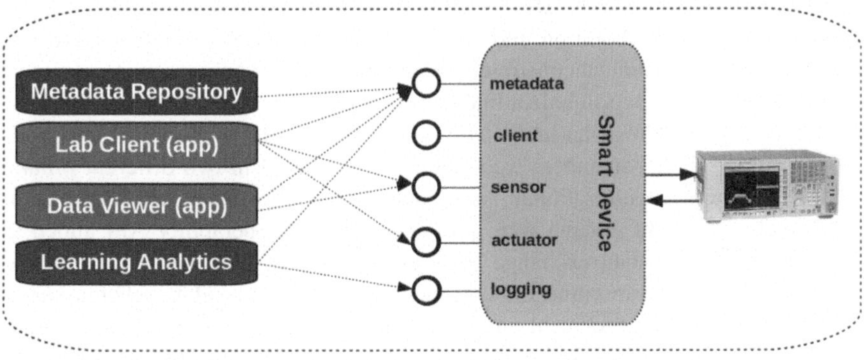

Fig. 1.5 Different clients consuming a smart device service (Salzmann et al. 2015)

The specification defines WebSocket (Fette and Melnikov 2011) as the transport protocol for the messages sent to and received from the smart device sensors and actuators. According to Salzmann et al. (2015), the decision towards WebSockets is justified by its asynchronous nature. The protocol allows for a bidirectional full-duplex communication channel. Clients can send messages to a smart device server, which can also send data to the client application asynchronously. This eliminates the need for the client application to perform long polling when the clients wants to check on the status of an asynchronous request.

The Go-Lab Smart Device specification for Online Laboratories has served as a draft for the IEEE P1876 (IEEE Networked Smart Learning Objects for Online Laboratories) working group (IEEE-SA 2015). The IEEE P1876 focuses additionally in other aspects of Online Laboratories and learning objects in general, such as defining methods for linking learning objects to design and implement smart learning environments (IEEE-SA 2015).

1.4 Conclusions and Trends Towards a Common Architecture

The advancements brought by the development of RLMSs fostered numerous cooperation initiatives between different Online Laboratory developer groups. As the community around RLMSs grew, we observed the development of segmented clusters, of adopters of a specific RLMS. If on one side it was becoming easier to share Online Laboratories among systems within the same RLMS, on the other side, sharing these systems across different RLMS was not a trivial issue, as the APIs and interfaces used to communicate different components (e.g. laboratory clients and servers, booking services, experiment data retrieval) were not compatible. Furthermore, this incompatibility was not limited to the APIs, but spanned from lower level communication layers (e.g. hardware interfaces) to higher layers of abstraction, such as the metadata schemes and the terminologies used. For example, even terms that in a first glance appear to be of trivial understanding, such as "experiment", had different interpretations among different RLMSs. As a consequence, several opportunities for collaboration and sharing of remote experimentation were left unexplored due to technical constraints. When this problem became more apparent, some initiatives, like the Global Online Laboratory Consortium (GOLC 2012), were started aiming to address the incompatibility between different systems. The work carried out by GOLC members was mainly concentrated in two different pillars, namely, technical and educational. As a result, an interoperability API (Lowe et al. 2015) and a metadata schema were proposed. The interoperability API aimed at defining an API for data exchange between different RLMSs that would enable these systems to share Online Laboratories. The GOLC metadata schema was a joint effort of GOLCs technical and education committees and aimed at creating different metadata profiles and defining the semantics between different resources in a Remote Laboratory system (Richter et al. 2012).

The numerous initiatives and frameworks presented in the previous sections are a testimonial for the great efforts spent with the development, deployment and usage of Online Laboratories. All these initiatives, such as the Global Online Laboratory Consortium, the standardisation efforts and the numerous research groups, contributed to many advancements in the field of Online Laboratories and software architectures for their deployment and federation. However, until the present moment, no trend towards a convergent software architecture is observed. Instead, RLMSs continue to follow independent paths. Current trends show that the role of RLMSs is changing. This change is mainly driven by the necessity to seamlessly include Online Laboratories into a learning activity, which is not in the scope of RLMSs. As pointed out by Sancristobal Ruiz et al. (2014), a possible solution is packing the virtual and Remote Laboratories within content packages that comply with e-learning standards such as IMS content packaging (IMS-CP) or Sharable Content Object Reference Model (SCORM), since these standards are supported by the majority of LMSs.

The IMS Global Learning Consortium also provides since 2010 the LTI (Learning Tools Interoperability) specification to integrate third-party tools (usually remotely hosted by a tool provider) into a tool consumer such as a LMS. This is the approach used with the Gateway4Labs initiative (Sect. 1.3.2). These different initiatives and projects show that RLMSs are being reduced to a set of services, instead of a Web application rendering user interfaces. For example, Colbran and Schulz (2015) pointed out with a new implementation of the iLab Shared Architecture that when the RLMS is reduced to a set of services, lab clients can reside anywhere on the Internet such as in Learning Management Systems. In this way, the developer can modify the behaviour of the RLMS in a programmatically way.

References

Colbran, S., & Schulz, M. (2015). An update to the software architecture of the iLab service broker. In *2015 12th International Conference on Remote Engineering and Virtual Instrumentation (REV)* (pp. 90–93). IEEE.

Cooper, M. (2005). Remote laboratories in teaching and learning issues impinging on widespread adoption in science and engineering education. *International Journal of Online Engineering (iJOE), 1*(1).

de Jong, T., Sotiriou, S., & Gillet, D. (2014). Innovations in stem education: The Go-Lab federation of online labs. *Smart Learning Environments, 1*(1), 1–16.

del Alamo, J. A., Chang, V., Hardison, J., Zych, D., & Hui, L. (2003). An online microelectronics device characterization laboratory with a circuit-like user interface. In *ICEE*, Valencia.

Feisel, L. D., & Rosa, A. J. (2005). The role of the laboratory in undergraduate engineering education. *Journal of Engineering Education, 94*(1), 121–130.

GOLC (2012). GOLC – global online laboratory consortium website. http://www.online-lab.org/. Last visited: 11 Feb 2015.

Harward, V., del Alamo, J., Lerman, S., Bailey, P., Carpenter, J., DeLong, K., Felknor, C., Hardison, J., Harrison, B., Jabbour, I., Long, P., Mao, T., Naamani, L., Northridge, J., Schulz, M., Talavera, D., Varadharajan, C., Wang, S., Yehia, K., Zbib, R., & Zych, D. (2008). The iLab Shared Architecture: A web services infrastructure to build communities of Internet accessible laboratories. *Proceedings of the IEEE, 96*(6), 931–950.

IEEE-SA (2015). P1876 – networked smart learning objects for online laboratories. Last visited 12 Nov 2015.

Jona, K., & Vondracek, M. (2013). A remote radioactivity experiment. *The Physics Teacher, 51*(1), 25.

Lowe, D. B., Berry, C., Murray, S., & Lindsay, E. (2009). Adapting a remote laboratory architecture to support collaboration and supervision. *International Journal of Online Engineering, 5*(S1), 51–56.

Lowe, D., Yeung, H., Tawfik, M., Sancristobal, E., Castro, M., Ordua, P., & Richter, T. (2015). Interoperating remote laboratory management systems (RLMSs) for more efficient sharing of laboratory resources. *Computer Standards & Interfaces, 43*, 21–29.

MQTT (2014). MQTT. http://mqtt.org/. Last visited: 02 Apr 2016.

Ordua, P., Irurzun, J., Rodriguez-Gil, L., Garcia-Zubia, J., Gazzola, F., & Lpez-de Ipia, D. (2011). Adding new features to new and existing remote experiments through their integration in WebLab-Deusto. *International Journal of Online Engineering (iJOE), 7*(S2).

Ordua, P., Zutin, D., Govaerts, S., Lequerica, I., Bailey, P., Sancristobal, E., Salzmann, C., Rodriguez-Gil, L., DeLong, K., Gillet, D., Castro, M., Lopez-de Ipina, D., & Garcia-Zubia, J. (2015). An extensible architecture for the integration of remote and virtual laboratories in public learning tools. In *IEEE Revista Iberoamericana de Tecnologias del Aprendizaje* (pp. 1–1).

Richter, T., Grube, P., & Zutin, D. (2012). A standardized metadata set for annotation of virtual and remote laboratories. In *2012 IEEE International Symposium on Multimedia* (pp. 451–456).

Salzmann, C., Govaerts, S., Halimi, W., & Gillet, D. (2015). The smart device specification for remote labs. In *2015 12th International Conference on Remote Engineering and Virtual Instrumentation (REV)* (pp. 199–208).

Sancristobal Ruiz, E., Pesquera Martin, A., Ordua, P., Martin, S., Gil, R., Ruiz Larrocha, E., Albert, M. J., Diaz, G., Meier, R., & Castro, M. (2014). Virtual and remote industrial laboratory: Integration in learning management systems. *IEEE Industrial Electronics Magazine, 8*(4), 45–58.

Zutin, D. G., Auer, M., Ordua, P., & Kreiter, C. (2016). Online lab infrastructure as a service: A new paradigm to simplify the development and deployment of online labs. In *2016 13th International Conference on Remote Engineering and Virtual Instrumentation (REV)* (pp. 208–214).

Danilo Garbi Zutin was Senior Researcher (2007–2017) at the Carinthia University of Applied Sciences, Villach, Austria

Chapter 2
The WebLab-Deusto Remote Laboratory Management System Architecture: Achieving Scalability, Interoperability, and Federation of Remote Experimentation

Pablo Orduña, Javier Garcia-Zubia, Luis Rodriguez-Gil, Ignacio Angulo, Unai Hernandez-Jayo, Olga Dziabenko, and Diego López-de-Ipiña

Abstract WebLab-Deusto is an open-source Remote Laboratory Management System (RLMS). On top of it, one can develop and manage remote laboratories and share them with other institutions. This chapter describes the architecture and features of the system, as well as a nontechnical view of other aspects such as how to share laboratories in the context of WebLab-Deusto, different institutions using WebLab-Deusto for their remote laboratories, research projects where it has been used, and sustainability plans.

Keywords Remote laboratories · Online education · Remote laboratory management systems

2.1 Introduction

WebLab-Deusto[1] is an open-source RLMS (Remote Laboratory Management System) developed mainly at the University of Deusto since 2004. As a RLMS, it provides some of the shared features of most remote laboratories, such as

[1] http://weblab.deusto.es

P. Orduña (✉)
DeustoTech-Deusto Foundation and LabsLand, Bilbao, Spain
e-mail: pablo.orduna@deusto.es

J. Garcia-Zubia
Faculty of Engineering, University of Deusto, Bilbao, Spain
e-mail: zubia@deusto.es

L. Rodriguez-Gil · I. Angulo · U. Hernandez-Jayo · O. Dziabenko · D. López-de-Ipiña
DeustoTech-Deusto Foundation, Bilbao, Spain
e-mail: luis.rodriguezgil@deusto.es; ignacio.angulo@deusto.es; unai.hernandez@deusto.es; olga.dziabenko@deusto.es; dipina@deusto.es

M. E. Auer et al. (eds.), *Cyber-Physical Laboratories in Engineering and Science Education*, https://doi.org/10.1007/978-3-319-76935-6_2

authentication, authorization, administration tools, scheduling, analytics, or laboratory sharing. Therefore, if a laboratory is developed on top of WebLab-Deusto (see Sect. 2.2), the laboratory developers do not need to deal with any of those issues: they are provided for them. Furthermore, as new versions of the system are released, the developer can simply upgrade the system without changes in the laboratory and automatically benefit from the new features built into the new versions.

During these years, WebLab-Deusto has been used for the development of remote laboratories in different institutions (see Sect. 2.4), and it is being tested in other universities. The architecture and most of the code have been deeply changed through different iterations during the last years while keeping backward compatibility for existing laboratories. One of the key features of WebLab-Deusto is the sharing model (Sect. 2.3), which has been extended to be integrated in different learning tools (such as LMS or CMS; see Sect. 2.5). The final goal of this sharing architecture is to aim a sustainable model for remote laboratories (see more information in Sect. 2.7).

The remainder of the chapter is structured as follows: Sect. 2.2 explains the technical details of WebLab-Deusto. This section is oriented to a rather technical audience and mainly covers architecture and software topics, while the rest of the sections are intended for technical and nontechnical audience. Section 2.3 explains how any laboratory can be shared among different universities or schools using WebLab-Deusto in a transparent way for the users and administrators. Section 2.4 covers some examples of WebLab-Deusto deployments in different countries. Section 2.5 explains the integration of WebLab-Deusto in different learning tools (including LMS and PLE). Section 2.6 explains research projects where WebLab-Deusto has been used. And finally Sect. 2.7 covers the spin-off of WebLab-Deusto, called LabsLand. Section 2.8 summarizes the chapter.

2.2 WebLab-Deusto RLMS

When creating a remote laboratory, a remote laboratory developer (a person or group who wants to create a remote laboratory) needs to deal with the management tools (e.g., authentication, authorization, storing logs, viewing logs, administrative tools, etc.) and with the particularities of the remote laboratory (e.g., interfacing with the hardware, developing the user interface, etc.).

WebLab-Deusto is a Remote Laboratory Management System (RLMS). This means that it is a set of software tools and libraries that let remote laboratory developers to focus only on the development of the remote laboratory itself and let the RLMS manage the rest (by managing authentication, groups, authorization, scheduling, integration in learning tools, analytics, or sharing).

To this end, WebLab-Deusto provides an architecture (see Fig. 2.1), where the clients interact with WebLab-Deusto for requesting access to a laboratory, and the laboratory is only contacted when the current user has permission to use the laboratory at a particular time slot (managed internally by a queue of users).

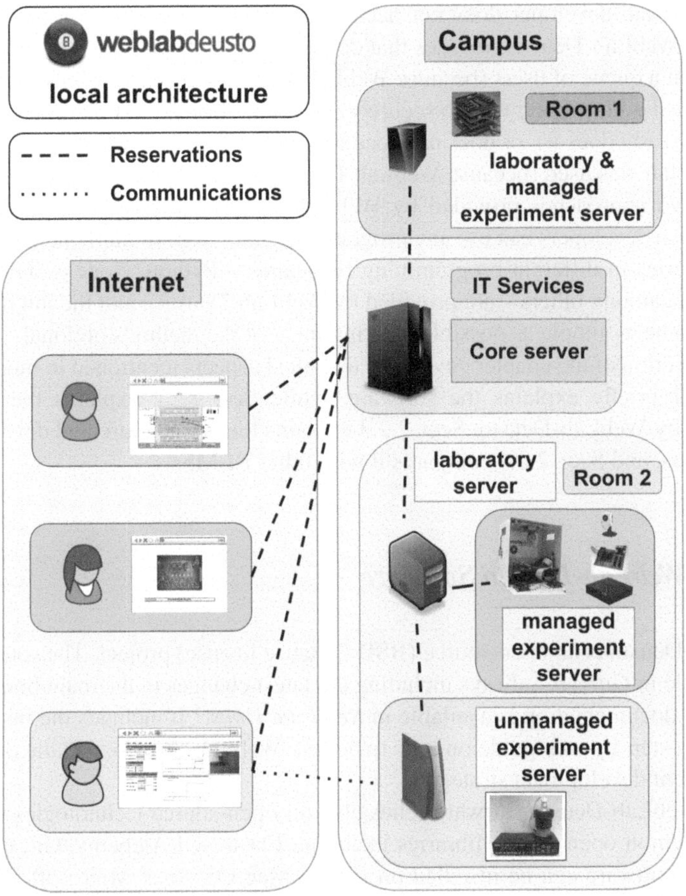

Fig. 2.1 WebLab-Deusto local architecture

Depending on the approach selected by the remote laboratory developer (*managed* or *unmanaged*, explained in more detail in Sect. 2.2.3), the communications with the laboratory will be performed through the WebLab-Deusto main server (which will store and forward the user requests) or directly with the final laboratory.

This way, remote laboratory developers can develop their laboratories in any programming language or web framework, and by using certain interfaces, WebLab-Deusto automates the rest of the processes. For example, if someone wants to develop a laboratory using a Raspberry Pi (an inexpensive – $35 – single-board computer) controlling a set of electronic devices, the remote lab developer can develop the software as they prefer (in Python, PHP, or other software frameworks) running in the Raspberry Pi and a web interface for the users. Then, by providing some HTTP interfaces and sharing a couple of secret messages with a WebLab-Deusto server (deployed in the same Raspberry Pi or in a regular computer), the remote lab developer does not need to implement a connection to a Moodle (because WebLab-Deusto provides that connection; see Sect. 2.5), does not need to implement a queue of users (because WebLab-Deusto already provides it), does not need to deal with sharing the lab securely with other institutions (because WebLab-Deusto already does it), or does not need to deal with writing analytics software on when the lab was used (because WebLab-Deusto already does it).

The two approaches provided by WebLab-Deusto are programming language agnostic, so developers can use any programming language to implement them (and even libraries in different programming languages – Python, Node.js, Java, .NET, C++, or C, among others – are provided by WebLab-Deusto); and the interfaces are defined to be as simple as possible, making most of the methods optional.

This section of the chapter covers the technical aspects mentioned in more detail: Sect. 2.2.1 briefly explains the software, while Sect. 2.2.2 explains the features provided by WebLab-Deusto, Sect. 2.2.3 explains the architecture and development approaches, and Sect. 2.2.4 compares it with other RLMSs.

2.2.1 WebLab-Deusto Software

WebLab-Deusto is an open-source (BSD 2-clause license) project. The source code is available in *GitHub*,[2] always including the latest changes in the main branch. The complete documentation is available in *Read the Docs*.[3] It includes the installation process, a step-by-step guide on how to deploy WebLab-Deusto, and the details on the different development strategies.

The WebLab-Deusto software relies also on open-source technologies (Python, several Python open-source libraries including Flask-SQLAlchemy, Git, etc.), and its server software can be installed on Linux, Mac OS X, or Microsoft Windows. The client is a regular web application, so WebLab-Deusto itself does not impose any special requirement to the client other than a modern web browser (working on mobile devices, tablets, and regular desktops). However, each laboratory can use different technologies, so if the laboratory uses a technology that does not work in certain devices (e.g., Adobe Flash, which does not work in mobile devices or tablets), those limitations will be applied. All the laboratories developed by the WebLab-Deusto team only require a web browser (Garcia-Zubia et al. 2008), and the team has contributed to other remote laboratories for HTML5 adoption (as in the case of VISIR[4]).

---

[2]https://github.com/weblabdeusto/weblabdeusto

[3]https://weblabdeusto.readthedocs.io

[4]https://github.com/JohanZackrisson/visir_html5

2.2.2 *WebLab-Deusto Advanced Features*

This section describes briefly each of the main features of WebLab-Deusto. It just aims to give a more narrow idea on what each of the features mean, but more documentation is available in the WebLab-Deusto documentation site. An important feature (sharing laboratories) is missing in this section since it has a dedicated section for it (Sect. 2.3).

2.2.2.1 Security

WebLab-Deusto has been designed to be secure. At user level, as stated in Sect. 2.2.2.7, it supports different authentication mechanisms, most of which enable that the passwords are not stored in the WebLab-Deusto server, and in some, the password does not even go ever through the WebLab-Deusto servers.

Additionally, all the communications between client and server (and between servers when sharing laboratories) are managed through HTTP, and therefore they can be secured using HTTPS. The only exception is the unmanaged laboratories (Sect. 2.2.3.2), where the communications are managed by the remote laboratory developer, but in the case of the University of Deusto, all the unmanaged laboratories use HTTPS. This also solves any problems with firewalls and proxies (Garcia-Zubia et al. 2009), and given that all the software provided by WebLab-Deusto relies on HTML5, it can run on any web browser (even in mobile phones).

At laboratory level, the WebLab-Deusto architecture offers the managed approach (Sect. 2.2.3.1), where the software developed by the remote lab developer is isolated in a local network at the university level. This avoids common issues generated by remote lab developers who are experts on the hardware side but may not know several pitfalls when dealing with security. Given that all the communications are forced to work through an API on the client side and on the server side, and WebLab-Deusto stores and proxies all those communications, the vulnerability window is considerably smaller than in other architectures.

2.2.2.2 Interoperability

WebLab-Deusto supports interoperability with external tools at two levels:

- Other Remote Laboratory Management Systems (supporting bidirectional integration with iLab laboratories and Universidad Nacional de Rosario laboratories through the federation mechanisms Orduña et al. 2013)
- Learning tools (including a wide range of learning management systems) as explained in more detail in Sect. 2.5.

Furthermore, WebLab-Deusto relies on different open-source technologies and supports deploying the server in Linux, Microsoft Windows, and Mac OS X.

2.2.2.3 Learning Analytics

When a student uses a laboratory, WebLab-Deusto stores who accessed and when and, in the case of the managed laboratories, also all the commands and files that were sent during the session.

This enables WebLab-Deusto to report this information to both the administrators and the instructors of the groups. In particular, instructors can see (Fig. 2.2) both the global trend of the class and also the usage done by particular students.

Furthermore, as detailed in Orduña et al. (2014), if a particular laboratory requests files to be submitted (e.g., typically code or binary files to be programmed in a device), WebLab-Deusto automatically compares which previous uses of the lab in class were done with the same files and had not been previously used by an instructor or administrator. This way, it can discover potential plagiarism among students, reporting the non-explicit social network in class.

Uses

Property	Value
Users:	45
Total uses	1595 uses (35.44 uses per user)
Total time	2429894.35 seconds - over 28 days; 53997.65 seconds per user

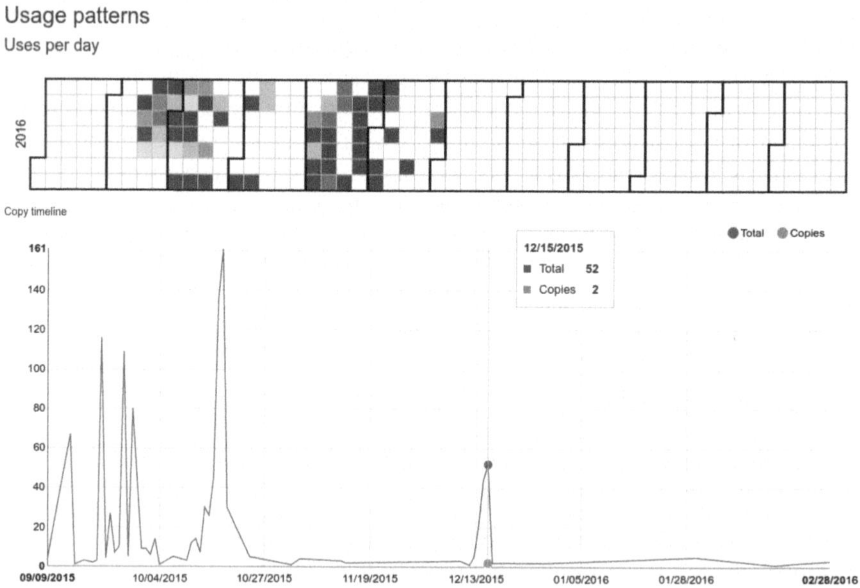

Fig. 2.2 Parts of the learning analytics panel of WebLab-Deusto for a group of students

2.2.2.4 Scalability and Scheduling

WebLab-Deusto has been designed to be scalable at different levels: user, laboratory, and sharing.

At *user level*, it supports different deployment options so as to be able to scale horizontally (adding more servers) and vertically (using more powerful servers) to deal with higher levels of usage of concurrent students. It supports both lightweight deployments (using SQLite Sect. 2.2.2.6) and scalable deployments (using Redis and MySQL). Furthermore, it comes with tools for measuring a particular deployment, by simulating students using different strategies, so it is possible to establish what policies are better for a particular server. In Orduña (2013) it is analyzed how the software behaves with different loads of simulated users (up to 150).

At *laboratory level*, WebLab-Deusto supports load balancing. WebLab-Deusto supports an internal queue of users that aim to access a particular experiment. This way, WebLab-Deusto guarantees that only one user can access a laboratory at a time, and the rest of the users are waiting in that queue. The architecture has also been designed to manage multiple *experiment servers* (the experiment-dependent software) through a different *laboratory server* (a software component of WebLab-Deusto that performs regular checks on the labs and manages part of the communications with the laboratories), so it is possible to increase the number of *laboratory servers* without adding more processing load to the main servers.

However, when a laboratory is going to be accessed by several students, a common solution is to provide multiple copies of the same laboratory. For example, in the University of Deusto, in electronics laboratories more than one copy has been provided (see Fig. 2.3). WebLab-Deusto can be configured, so the students will be randomly going to one or other copy of the laboratory, and the rest will still wait in a shared queue. The local scheduling mechanisms are detailed in Orduña and García-Zubia (2011).

Finally, at *sharing level* WebLab-Deusto supports federated load balance: if two institutions have a copy of a laboratory, they can balance automatically the load of users of both institutions among the different copies in each system, without the need to register students from one side to the other. This is covered in Sect. 2.3 of this chapter.

2.2.2.5 Hybrid Laboratories

Remote labs let users access real, physical equipment through the Internet, while virtual labs let users access simulations (Ma and Nickerson 2006). Traditionally, remote labs have been designed to mimic a hands-on experience. However, there are ongoing research efforts to design new models of laboratories which provide additional features and advantages. One of such models is hybrid labs (Rodriguez-Gil et al. 2016).

A hybrid lab mixes virtual and real components to leverage some of the advantages of both. Though those depend on the specific lab, examples of such

Fig. 2.3 Example of local load balance of users among copies of the same laboratory in the University of Deusto

Fig. 2.4 FPGA-water tank hybrid lab architecture

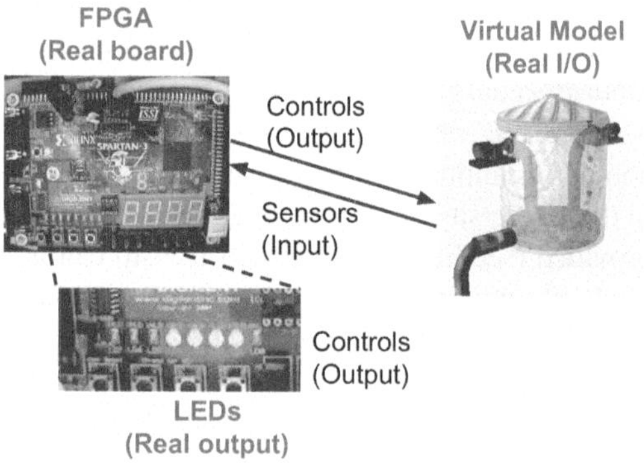

advantages could be cost-effectiveness, higher realism than purely-virtual labs, and features such as gamification or virtual environments.

The extensible and fully web-based architecture of WebLab-Deusto allows and facilitates the development of such labs. An example is the hybrid FPGA laboratory described in Rodriguez-Gil et al. (2016) and depicted in Fig. 2.4. Through it, students can program a real, physical FPGA which is used to control a virtual model of an industrial water tank. The virtual model has both sensors (water level and temperature) and actuators (two water pumps), which interact bidirectionally with the real, physical board. Thus, students program in a realistic environment (the board they program is real) but with a lower cost than purchasing and maintaining a real industrial water tank model would entail.

2.2.2.6 Embedded Deployments

WebLab-Deusto is a very light system, which does not require much memory or processing power. Indeed, we have successfully deployed the whole system even in Raspberry Pi 1 devices and measured the results (Angulo et al. 2013). On this ARM device, with only 256 MB RAM, could manage different amounts of concurrent users. So as to measure this, WebLab-Deusto comes with tools for simulating concurrent students in different environments.

However, the amount of time increased fastly per student, so it is not recommended to deploy the whole system in such an inexpensive device (around $35). The typical deployment would include the general layers of WebLab-Deusto in a regular server (which can be a normal computer, depending on the expected load of concurrent users), and the experiments then can be deployed in constrained devices such as Raspberry Pi.

2.2.2.7 Authentication and Authorization

WebLab-Deusto supports different authentication mechanisms. The simplest (and default) one is storing in the database the username and a salted hash of the password (the standard secure procedure when working with passwords: instead of storing the password, it stores a hash of the password, so if someone gets access to the database, the person can't figure out what was the original password).

However, in many contexts this is inconvenient, since this leads to creating yet another account for students and maintain and remember the passwords. So as to provide a more integrated solution, WebLab-Deusto also supports LDAP, which is a protocol used internally in many universities for user management. This way, the users can have the same credentials as in the university, and no password will be stored in the WebLab-Deusto server.

In addition to these two approaches, WebLab-Deusto provides support for other protocols such as OpenID and OAuth 2.0 (used with Facebook) as well as an API for developing third-party protocols. For example, certain universities count with other types of authentication (e.g., based on encrypted cookies), so with this API, it becomes possible to add support for those mechanisms.

Regarding authorization, WebLab-Deusto provides tools for adding users to groups and granting access to certain laboratories to each group or individual. However, it also supports delegating it to remote other tools such as LMS as explained in Sect. 2.5.

2.2.3 Managed and Unmanaged Labs

A remote laboratory consists of two parts: a client and a server. Depending on the technology used, both can be very isolated or not. For example, in many web frameworks, there is no such distinction, while in many occasions, the remote lab developer might want to have a clearly separated set of technologies (such as a JavaScript client that only performs some interactions with the server).

As shown in Fig. 2.1, a WebLab-Deusto server can be managing both *managed* and *unmanaged* laboratories. The distinction between the two approaches is the following:

- A *managed* laboratory (Sect. 2.2.3.1) is a laboratory where the communications between the client and the server side are managed by WebLab-Deusto. Therefore, the remote lab developer must implement a client (using a JavaScript library provided by WebLab-Deusto) and a server (using any programming language, either using the libraries provided – for Node.js, Python, Java, .NET, C++, or C – or implementing the required protocol by WebLab-Deusto). WebLab-Deusto will be in charge of showing the client when it is necessary and to communicate both (i.e., when the client calls a function `sendCommand(message)`, the server will receive that message).
- An *unmanaged* laboratory (Sect. 2.2.3.2) is a laboratory where WebLab-Deusto is totally unaware of the communications between the client and the server. Typically this means that the remote laboratory developer provides a complete web application (managing all the communications) and implements a RESTful interface that WebLab-Deusto calls. This way, WebLab-Deusto will call certain methods to state "a new user, called Michael and identified by this token, comes now" or "the current user left or has timed out," and the web application will be responsible of making this happen.

There are advantages and disadvantages in both approaches, explained in each section, and that is the reason for both approaches coexisting and being supported.

2.2.3.1 Managed Labs

As represented in Fig. 2.1, in the managed approach, all the users interact only with the WebLab-Deusto "core" servers and never with the particular experiment servers. A single "core" server can be in charge of dozens of managed and unmanaged laboratories at the same time. In the case of the "managed" laboratories, all the commands will be sent through these servers, which will forward the commands to each laboratory. The system has been designed to minimize the latency added to each command submitted (Orduña 2013).

The *managed* approach is simpler for certain developers that might not be familiar with web technologies. Writing an HTML + JavaScript code that calls simple functions for submitting commands and not dealing with tokens, authentication, or communications can be easier.

Additionally, all the communications are managed by the WebLab-Deusto servers, which also add some simplicity:

- If the administrator adds support for HTTPS to provide encryption, all the managed laboratories automatically support it too.

- Only a web server (WebLab-Deusto) requires to be deployed. Once running, all the managed laboratories can be in a private network without dealing with ports, firewalls, configuring web servers, etc.
- Since the access from the Internet is totally restricted, each managed laboratory will never have more load of users than the one defined by the configuration of WebLab-Deusto (e.g., only one user at a time or some users if it requires some collaboration). This makes the approach ideal for constrained devices, since no security needs to be managed at experiment server level.
- Everything is automatically stored in the database, and it is available to instructors and administrators.

2.2.3.2 Unmanaged Labs

As represented in Fig. 2.1, in the unmanaged approach, all the users interact with the WebLab-Deusto "core" servers but also with the experiment servers, using whatever protocol is desired by the remote lab developer. WebLab-Deusto is unaware of these communications, so no constraint is established.

The *unmanaged* approach is better for web developers who use web frameworks and do not want to limit to a simple client and server. The fact of not managing communications enables developers to use any type of protocol (such as WebSockets) without constraining to the libraries of WebLab-Deusto. Additionally, the approach is more scalable since, if there are more laboratories, the requests for one laboratory and for the other are not managed by the same core servers.

However, the flow is also more complicated, since it requires implementing a set of features such as:

- Receiving a message of a user coming (with some metadata of the user and a token for identifying).
- Rejecting users who do not have a valid token.
- Receiving a message that the user session is expired and therefore the user must be rejected in the next request.
- Tracking if the user is active or not and notifying WebLab-Deusto when requested (so if the user left, WebLab-Deusto can assign the laboratory to the next student in the queue).

In this case, examples for Python and PHP are provided in the documentation site, as well as definitions of the particular methods and parameters.

2.2.4 Comparison with the State of the Art

The concept of RLMS, through different names, is commonly used for almost a decade. The main RLMSs available are MIT iLabs[5] (Harward et al. 2008a,b),

[5]http://ilab.mit.edu

RemLabNet[6] (Schauer et al. 2016), and Labshare Sahara[7] (Lowe et al. 2012a,b). Other systems (such as LiLa Richter et al. 2011) are not really comparable as they are an index of existing publicly available resources.

Both MIT iLabs and Labshare Sahara and WebLab-Deusto are open source and publicly available, while RemLabNet is still not available for usage by third parties. MIT iLabs and WebLab-Deusto are the only ones supporting federation, while MIT iLabs does not support federated load balance or transitive federation. WebLab-Deusto and Labshare Sahara are the only ones supporting local load balance when it comes to scheduling. MIT iLabs and Labshare Sahara are the only ones supporting calendar-based booking (and in particular Labshare Sahara supports a very interesting and complex mechanism for supporting queueing and scheduling at the same time Lowe and Orou 2012). The approach taken by MIT iLabs to support LMS is a joint effort with WebLab-Deusto (as described in Sect. 2.5), and it has also been extended to RemLabNet for its support in the Go-Lab project (Sect. 2.6).

2.3 Sharing Laboratories

WebLab-Deusto supports *federating laboratories* (Orduña et al. 2013): this means that one WebLab-Deusto deployment can share its laboratories with other instances, as well as consume them.

The way it works is the following (Fig. 2.5): a *University A* has an *Experiment 1*, and *University B* has *Experiment 2* and *Experiment 3*. University A can create a particular type of user (*federated user*), which represents another institution (*University B* in this case). Then, it can share with that particular user a set of laboratories (in this case, *Experiment 1*). From this moment, *University B* has access to *Experiment 1*, so it can be treated just as any other local experiment. *University B* might let only one group of *University B* students access this laboratory. In any case, *University A* is unaware of who has permission at *University B*: the relations are in a university-to-university basis, never in a university-to-other-students basis. This simplifies the mechanism, since each university manages its own students, groups, and permissions, and in addition to that, they just have access to some more laboratories from other universities.

The way this sharing is performed can also be configured in different ways. For example, *University A* might decide that their own students will advance faster in the queue, so they will use the laboratories faster. This way, if one student is using the *Experiment 1*, and a student of *University B* comes into the queue and then a student from *University A* comes in, this student will go first in the queue. WebLab-Deusto enables this type of policies, but it is up to the particular universities to define them.

[6] http://www.remlabnet.eu

[7] http://labshare-sahara.sf.net

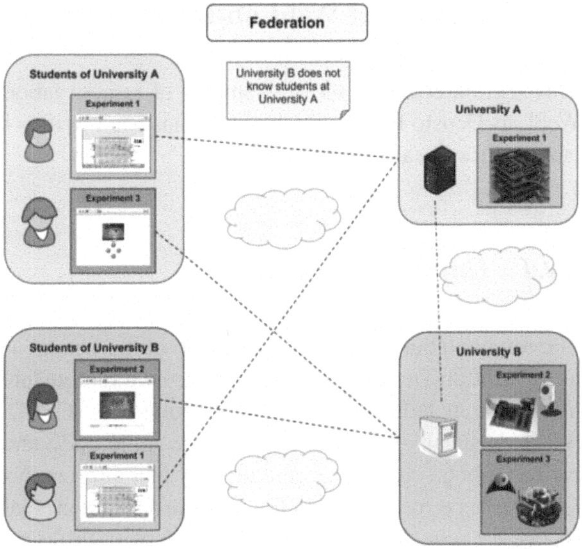

Fig. 2.5 WebLab-Deusto federation

Also, in the case of *University B* sharing laboratories with *University A*, *University B* can select whether to share all the laboratories or only few of them to *University A*.

Two key features of the federation model of WebLab-Deusto (Orduña 2013) are that it supports:

1. *Transitive federation*: if *University A* shares a laboratory with *University B*, then *University B* can re-share that laboratory with *University C*. At every moment, when a student comes, *University A* will be aware of the student coming from *University C*. This enables complex chains (Orduña et al. 2013) of sharing laboratories where *University C* could be a secondary school of the area of *University B* which would otherwise less likely be aware of the laboratory.
2. *Federated load balance*: if *University A* and *University B* have the same laboratory (which happens, e.g., in the VISIR laboratory Gustavsson et al. 2007, available in several countries), then they can balance the load of users between both copies of the laboratory. This way, students of *University A* requesting the laboratory would always be redirected to the local laboratory. But, if there is a queue, a meta-queue is formed in both institutions, and whichever laboratory is available earlier will be used. On top of this queue, the same rules explained above (e.g., priority of local students) are maintained.

The combination of both features is also possible, so if *University A* and *University B* had the same laboratory, any of them could still share it with *University C*.

2.4 Examples of WebLab-Deusto Deployments

This section covers a set of examples of remote laboratories developed using the WebLab-Deusto RLMS. Each section has been written by the authors of each of the laboratories in each institution.

2.4.1 University of Deusto

The remote laboratories research group of the University of Deusto has extensively used WebLab-Deusto for developing its own remote laboratories, with over 100,000 uses. This includes programmable laboratories (CPLDs, FPGAs, Microchip PICs, ARM devices) – where students write some code and send it to the device – as well as robots, electronics, and biology laboratories. A demo of the currently active laboratories is available.[8] In Fig. 2.6, the Archimedes laboratory is presented, where secondary school students can drop balls with different mass to tubes and measure the results (Garcia-Zubia et al. 2015).

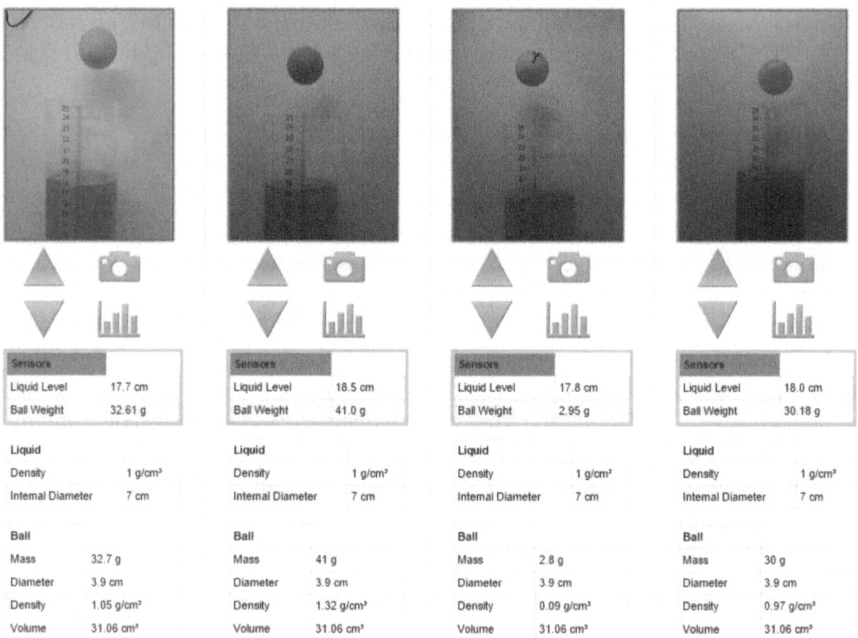

Fig. 2.6 Archimedes remote laboratory (Garcia-Zubia et al. 2015)

[8] https://weblab.deusto.es/weblab/

2.4.2 STUBA: Process Control Remote Laboratories[9]

Institute of Information Engineering, Automation and Mathematics (IAM) at STU in Bratislava is one of the main Slovakian educational institutions in the field of automation and process control. IAM has been active in the development of remote laboratories for almost a decade. Most of the developed labs are designed for the applications of control theory and practice. This also influences the nature of the experiments provided as remote labs. Unlike the labs from the fields of electronics and robotics, the IAM labs mostly use the training models of technological processes and systems, which exhibit an internal dynamics in a continuous manner. These are, e.g., processes with transfer and accumulation of heat or mass. This fact brings new requirement on server-side hardware since it must be able to sense the analog signals and control the actuators also in the analog fashion.

In the past, each remote laboratory at IAM was designed as the ad hoc solution, managing its own resources, as well as the user access mechanism. In 2013 the institute has adopted the Remote Laboratory Management System (RLMS) WebLab-Deusto[10] (Kalúz et al. 2013). Approximately in that time, a new type of architecture for development of remote laboratories has been developed (Kalúz et al. 2015) and used for implementation of several experiments for automatic control. Developed laboratories contain the following experiments:

- DC-motor (Fig. 2.7) – this experiment exhibits the dynamical behavior of first-order mechanical system;
- Thermo-optical device (Fig. 2.8) – provides a multi-input-multi-output system with coupled states. This system contains thermal and optical channel to control;
- Air heat exchanger – designed as air turbine with heating element is the system with two inputs and two outputs;
- System of cascaded liquid storage tanks – is a very popular educational device, which represents the dynamical system of second order.

The graphical user interface (GUI) of laboratories is fully customizable, and it is designed as a workspace with draggable windows. It contains several types of predefined components, which are main control panel located at top of Web page; tables with inputs, outputs, and variables located at fixed position in left-hand side of interface; a set of charts with visualization of signals and variables; window with selection of control algorithms; a set of video streams from remote video devices; window for download of measure data; and event-logging window.

These laboratories have been used in the education process for more than 3 years as the supplementary tools for practical exercises. They have been directly incorporated into curricula of two courses, namely, in the *Integrated Control in Process Industries* and *Theory of Automatic Control*. Students have to handle several

[9]Section contributed by Martin Kaluz (STUBA – Slovak University of Technology in Bratislava).
[10]http://weblab.chtf.stuba.sk/

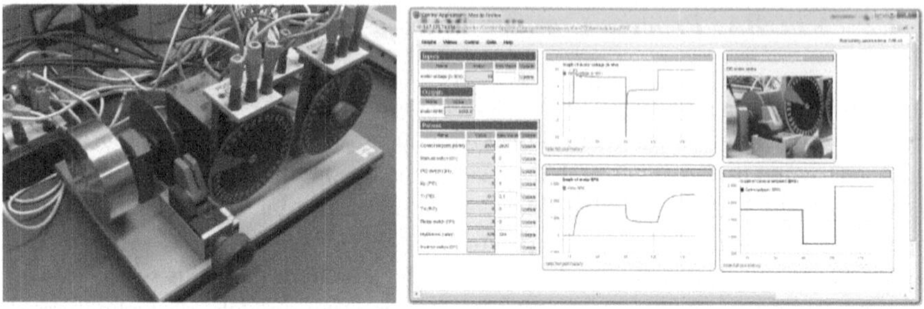

Fig. 2.7 DC-motor remote laboratory

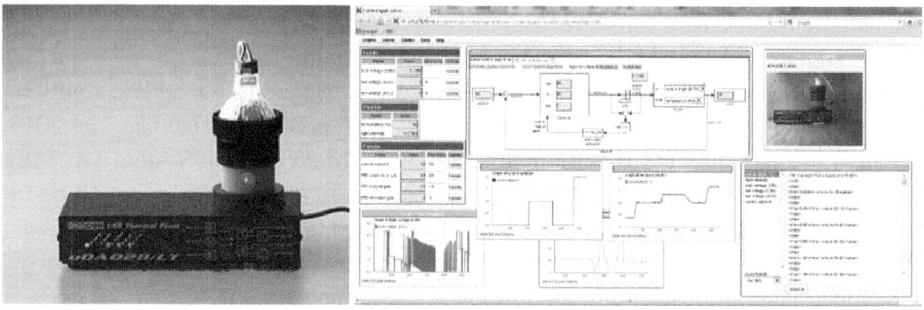

Fig. 2.8 Thermo-optical device remote laboratory

tasks during their practical work. These are the measurement of step responses of laboratory system in order to obtain the mathematical model, experimental identification of model's parameters, a design of appropriate controller (usually the PID), and evaluation of control quality.

During the period when the laboratories were used in teaching, there have been 2179 laboratory sessions by students. The DC-motor laboratory has been used 897 times and thermo-optical device laboratory 927 times. Other laboratories are not the direct part of curricula, but they have been still used 373 times, mostly during the students' projects.

2.4.3 Control Systems Remote Laboratories in Flipped Classroom Context[11]

In 2012, the Ecole des Mines school of engineering (Nantes, France) and the University of Los Andes' school of engineering (Bogotá, Colombia) started a collaboration around the implementation of control system remote laboratories

[11] Section contributed by Michael Canu and Mauricio Duque (University of los Andes, Colombia) and Philippe Chevrel and Ismael Meja (Ecole des Mines des Nantes, France).

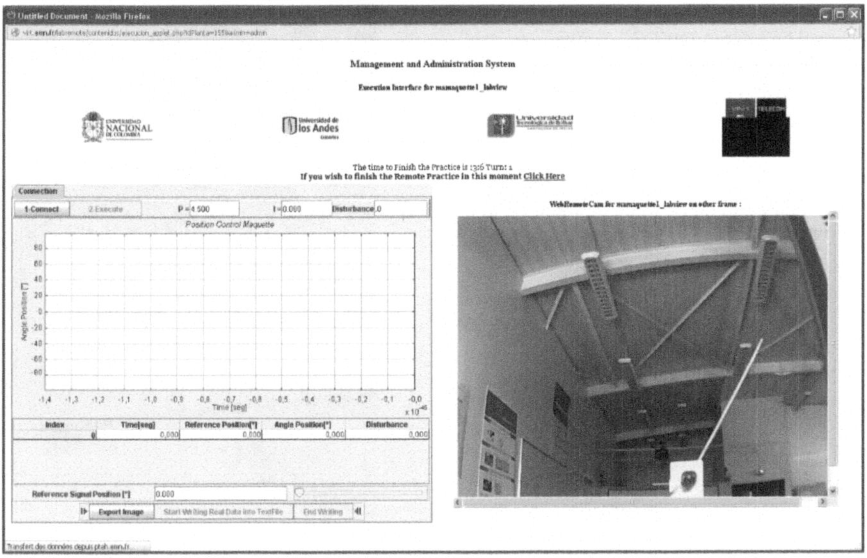

Fig. 2.9 Remote laboratory user interface

within the WebLab-Deusto initiative. The web interface (Fig. 2.9), LMS integration and back-end interfaces developments were conducted in collaboration between the three entities during more than 3 years in order to reach an effective managing interface from heterogeneous technologies infrastructures (Barrios et al. 2013b). While in the French context, the laboratory was proposed to students on a traditional course modality during 3 years, in the Colombian university, a flipped classroom (or inverted classroom) modality was used after 2 years of traditional course. Three control system plants were developed for the Colombian context: two axle position control systems (Barrios et al. 2013a,b) and a "ping-pong ball" control system (Caro and Quijano 2011). The last investigations we have conducted on this kind of laboratories were about the comparison between two modalities, real lab vs remote lab (Barrios et al. 2013a), and the interpretation of our results which contrast to some of international research results (Canu and Duque 2015).

2.4.4 UNED: VISIR in SPOC and MOOC[12]

DIEEC-UNED (Electrical and Computer Engineering Department, National University for Distance Education) has been working actively with a modified version of VISIR (see Fig. 2.10) and WebLab-Deusto[13] since February 2016. WebLab-Deusto

[12]Section contributed by Felix Garcia-Loro (UNED DIEEC, Madrid, Spain).

[13]http://weblab.ieec.uned.es

Fig. 2.10 VISIR system adapted by UNED-DIEEC, used through WebLab-Deusto

allows management VISIR operation in expertise courses and SPOCs (Small Private Online Courses).

During the 1st semester of the academic year 2016–2017, over 150 students from an engineering degree subject ("Fundamentals of Electronics Engineering") will access VISIR remote electronic lab through WebLab-Deusto. Furthermore, VISIR and WebLab-Deusto will be part of a redesigned massive open online course (MOOC) titled "Circuits Fundamentals and Applied Electronics" (Macho et al. 2016; García et al. 2014), 2017. This MOOC course had just had over 7,000 enrolments in the last 3 years.

In the course of these months of activity with VISIR and WebLab-Deusto, over 750 accesses were performed. These students have accessed VISIR lab as federated users from LMS (Learning Management System) platforms (Ruiz et al. 2014) by means of the WebLab-Deusto federation API through a custom system developed at UNED that uses the internal authentication mechanism. The rest of accesses are the stem from sessions aimed at experiments design by teachers/instructors and administrative tasks.

Besides the authentication mechanisms, a key factor for integrating WebLab-Deusto at DIEEC-UNED is that WebLab-Deusto includes the necessary support for monitoring the requests/responses between remote lab VISIR and final user. Before WebLab-Deusto, users' activity in VISIR was almost invisible for administrators. WebLab-Deusto allows monitoring all the users' interactions within the lab. This feature makes feasible the development of educational self-assessment tools and grading tools for VISIR.

2.4.5 Other Examples

In addition to the examples presented in this section, WebLab-Deusto has been used in other institutions, such as in the Pontifical Catholic University of Sao Paulo in Brazil and[14] University of Niš (Milošević et al. 2016), among others.

2.5 Integration in Learning Tools: gateway4labs

The strengths of a RLMS are the management and development of remote laboratories and making it easy to integrate them in external systems. This last point is particularly important since RLMSs must support its integration in systems where instructors or teachers put the learning resources, including Learning Management Systems (LMSs) and Personal Learning Environments (PLEs).

In this line, in 2012, the WebLab-Deusto team started a project originally called lms4labs (then gateway4labs/Go-Lab Smart Gateway Orduña et al. 2015) which aimed to be an open-source tool that would help the remote laboratories community to address this issue, by providing a simple protocol to connect virtual and remote laboratories to different types of learning tools (including PLEs and LMSs). As depicted in Fig. 2.11, gateway4labs is a middleware that supports three protocols:

- IMS LTI (Learning Tools Interoperability): a standard that supports interoperability between different learning tools in a secure way. In the context of gateway4labs, it is possible to create credentials for each teacher or groups of teachers, and they can use them in their courses in their LMS for accessing each particular laboratory.
- OpenSocial: a standard used in Graasp,[15] partially developed as part of the Go-Lab project[16] (de Jong et al. 2013, 2014; Gillet et al. 2013).
- HTTP: a custom simple protocol using HTTP and JSON, used for those circumstances where there is no support for IMS LTI or OpenSocial

Thanks to this component, any of the laboratory systems on the right side (including WebLab-Deusto and therefore any remote laboratory developed using WebLab-Deusto) can be consumed by any of the tools on the left side. In the case of IMS LTI tools (which include Moodle, Sakai, Blackboard, or Open edX), the integration is secure, so the laboratories do not need to be publicly available if the remote laboratory developer does not need to. In the case of OpenSocial, the remote laboratory developer must enable that the particular laboratory becomes an open educational resource, and therefore anyone will be able to use it without any authentication.

[14]http://weblabduino.pucsp.br/weblab/

[15]http://graasp.eu

[16]http://www.go-lab-project.eu

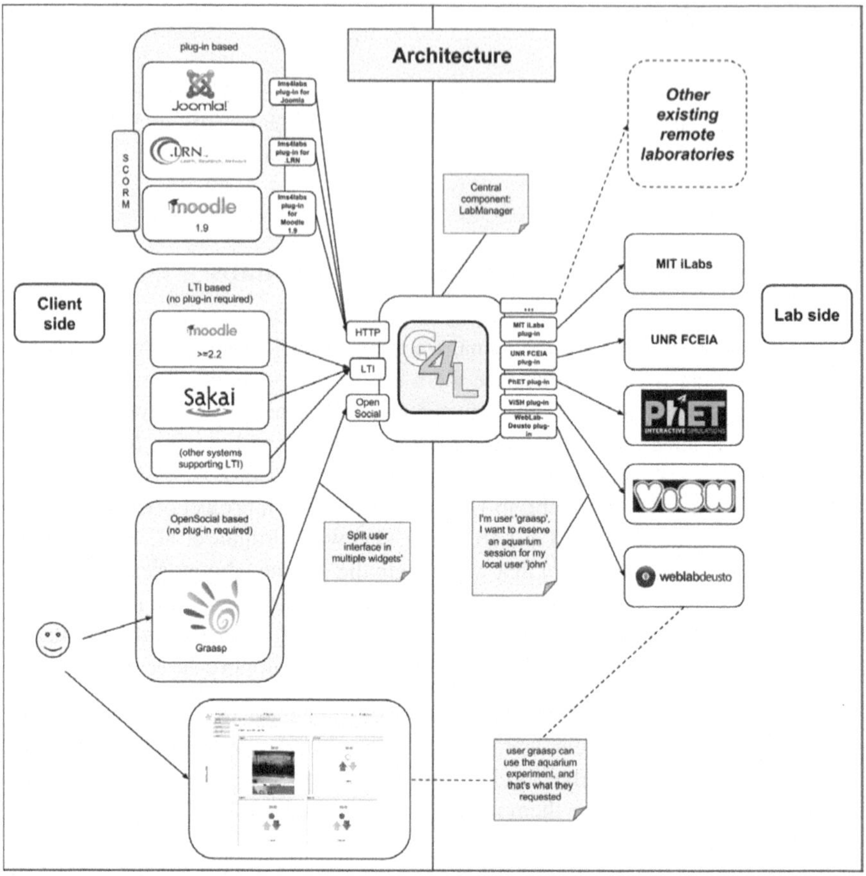

Fig. 2.11 gateway4labs/Go-Lab Smart Gateway

The advantage of using gateway4labs for laboratories is that, if the instructor is using a LMS (such as Moodle or Sakai), then there is no need to register the students and groups in WebLab-Deusto. This way, the analytics tools of WebLab-Deusto are not available, and the authentication and authorization features of WebLab-Deusto are not used (which is not a problem since LMS provide these features). And the management of the laboratories (queueing, federating, administration tools) is still made by the WebLab-Deusto instance.

2.6 WebLab-Deusto in Research Projects

WebLab-Deusto research team has designed and deployed the WebLab-Deusto platform and some remote experiments during the last more than 10 years. The first design of WebLab-Deusto was an intrusive program written in C to control and program remotely a programmable device. It was tested during the CLEI Conference in 2003 in La Paz (Bolivia). This section is devoted to describe the projects in which the WebLab-Deusto is involved and the tasks assigned to it. This set of projects and tasks describes the potentiality of the WebLab-Deusto platform in the present and in the future.

Under the name of WebLab-Deusto, different products and research have been obtained and applied in different national and international granted projects. There are two different products in WebLab-Deusto: the remote experiments and the Remote Laboratory Management System, WebLab-Deusto RLMS. The second product is the most important, but during the last 10 years, more than 20 remote experiments have been designed in and for supported projects: FPGA, CPLD, PIC, ARM, Archimedes principle, robotics, aquarium sensoring, etc. In 2005 the Rex-Net (Alves et al. 2005) was the first European project that included WebLab-Deusto experiments.

The WebLab-Deusto RLMS also allows users and partners to connect, to offer, to control, and to track remote experiments, designed or not by the partner or by other designers. This is the main result of the WebLab-Deusto research team.

In some projects, all the design experiments were offered through the WebLab-Deusto RLMS, being this platform the portal of the project. In OLAREX,[17] WebLab-Deusto was used by the National Polytechnic Museum of Bulgaria, by the Urdaneta School in Bilbao, and by other partners, schools, and institutions to offer remote experiments. In iCo-op project[18] WebLab-Deusto deployed a set of experiments in Georgia under the WebLab-Deusto RLMS, and the same situation occurred in Serbia with the NeReLa project.[19] In VISIR+ (Alves et al. 2016) the UNR (Argentina) is using the WebLab-Deusto platform to offer the students its VISIR remote lab. All these projects were granted by the EU within the Erasmus, Tempus, Leonardo, and Alfa programs.

WebLab-Deusto RLMS can be also used to integrate remote experiments into any LMS (Moodle, Sakai, Claroline, etc.) in a transparent (plug and play) way (as mentioned in Sect. 2.5). Using the WebLab-Deusto as a gateway, any institution can connect a remote experiment as an additional resource of the LMS, and under this tool the remote experiment can be accessed, tracked, etc. This was the main task of WebLab-Deusto in sLabs project (Sancristobal et al. 2012), granted by the Spanish Government in 2009; the same application was given in ePRAGMATIC project,[20] granted by EU in 2010 and in "Building an ecology of online labs" project,[21] granted by NSF (USA) in 2011. Currently, WebLab-Deusto is being used by different educational institutions to implement remote labs and remote experiments.

[17] http://www.europamedia.org/projects/olarex

[18] http://www.ico-op.eu/

[19] http://projects.tempus.ac.rs/en/project/855

[20] http://www.e-pragmatic.eu/

[21] https://www.nsf.gov/awardsearch/showAward?AWD_ID=1132813

The Go-Lab[22] project was granted by the EU in the FP7 and finished in 2016. One of the objectives of the project was to gather the remote experiments available in the world and offer them through a common platform[23] after an agreement with the owners of the experiments. The gateway4labs/Go-Lab Smart Gateway (as explained in Sect. 2.5) was designed for this task. Usually a remote experiment is included in a specific portal, but Go-Lab would like to offer this remote experiment through its own portal. When a user clicks on a remote experiment that is offered by the Go-Lab portal but it is stored in another portal, gateway4labs will make this link transparent for the user and the Go-Lab portal. gateway4Labs has been used to connect hundreds of labs to be used for thousands of teachers. At this moment, a new project called NextLab has been approved by EU in the H2020 program to continue with the main objectives of Go-Lab.

Currently, WebLab-Deusto RLMS is part of the PILAR project, granted by EU in the Erasmus+ program in 2016 (2016-1-ES01-KA203-025327), and it will be responsible of the federation of all the VISIR remote labs deployed in Europe. Federation is the highest level of integration of remote labs; under this approach, when a user creates a VISIR remote experiment, he/she will not know where it is going to be executed, in Austria, Spain, Portugal, or Sweden? This approach improves the scalability and the load balance, and it assures the availability of the lab if one of the VISIR is broken or not available. In the same line, in the proposal of the EOLES 2 project, WebLab-Deusto is expected to create a federation of remote labs designed and deployed in Europe and Maghreb. Under the EOLES federation, the different countries will be able to design and implement new engineering degrees.

Summing up, WebLab-Deusto RLMS can be used as a platform to integrate and offer remote experiments and remote labs or as a platform to help and facilitate another platform in integrating different remote experiments and remote labs.

2.7 LabsLand: A Spin-Off of WebLab-Deusto Aiming for Sustainability of the Service

The WebLab-Deusto team has started a spin-off called LabsLand[24] (Orduña et al. 2016). The goal of LabsLand is to provide a platform (Parker et al. 2016; Chase 2015) where different types of entities (schools, universities, research centers) can share their laboratories (using WebLab-Deusto or other systems) either free or for some price. The new platform uses the vision of the federation approach of WebLab-Deusto, and its aim is to create an ecosystem of providers and consumers of remote laboratories, as well as content providers.

[22]http://www.go-lab-project.eu/

[23]http://www.golabz.eu

[24]https://labsland.com

The platform will continuously be tracking what laboratories are available and when, so it will be able to tell consumers what laboratories have actually been working during the last months. This way, providers of expensive equipment will be able to share the cost with other consumers, and consumers will know which providers they can trust and how reliable they are.

While in the beginning LabsLand counts with own equipment, most of the laboratories available in LabsLand will be property of the provider, and LabsLand will only manage the connections. The first pilots of LabsLand are running at the time of this writing.

2.8 Conclusions

WebLab-Deusto is a Remote Laboratory Management System (RLMS) that enables remote laboratory developers to create and manage laboratories focused on the particularities of the laboratories (and not on the management layers that can be managed by WebLab-Deusto). WebLab-Deusto provides a flexible architecture to this end: enabling developers coming from different backgrounds (e.g., those familiar with web technologies and those who are not) to create laboratories with the different approaches (managed and unmanaged) and benefit from all the features of WebLab-Deusto (authentication, authorization, analytics, scheduling, or integration in different learning tools).

A key feature of WebLab-Deusto is the way it shares laboratories, in a university-to-university basis rather than on a university-to-student basis. This flexible design enables a simpler customization by the administrators of each entity, since they see as local the external resources, and therefore they can easily assign who can access what remotely and locally.

On top of this experience, the WebLab-Deusto team has created LabsLand, a spin-off that provides a common marketplace for accesses to remote laboratories. As more entities join LabsLand for sharing their laboratories, more value will all the LabsLand network in an exponential trend as defined by the Reed's law.

References

Alves, G. R., Ferreira, J. M., Muller, D., Erbe, H. H., Hine, N., Alves, J., Pereira, C. E., Chiang, L., Herrera, O., & Sucar, E. (2005). Remote experimentation network-yielding an inter-university peer-to-peer e-service. In *2005 IEEE Conference on Emerging Technologies and Factory Automation* (Vol. 2, pp. 8–pp). IEEE.

Alves, G. R., Fidalgo, A., Marques, A., Viegas, C., Felgueiras, M. C., Costa, R., Lima, N., Castro, M., Diaz-Orueta, G., San Cristobal Ruiz, E., Garcia-Loro, F., Garcia-Zubia, J., Hernandez-Jayo, U., Kulesza, W., Gustavsson, I., Pester, A., Zutin, D., Schlichting, L., Ferreira, G., de Bona, D., da Silva, J. B., Alves, J. B., Biléssimo, S., Pavani, A., Lima, D., Temporao, G., Marchisio, S.,

Concari, S., Lerro, F., Fernandez, R., Paz, H., Soria, F., Almeida, N., de Olieveira, V., Pozzo, M. I., & Dobboletta, E. (2016). Spreading remote lab usage. In *CISPEE 2016*. IEEE.

Angulo, I., García-Zubia, J., Orduña, P., & Dziabenko, O. (2013). Addressing low cost remote laboratories through federation protocols: Fish tank remote laboratory. In *Global Engineering Education Conference (EDUCON), 2013 IEEE* (pp. 757–762). IEEE.

Barrios, A., Duque, M., Canu, M., Luis Villa, J., Chevrel, P., Grisales, V. H., Prieto, F., & Panche, S. (2013a). Academic evaluation protocol for monitoring modalities of use at an automatic control laboratory: Local vs. remote. *International Journal of Engineering Education, 29*(6), 1551–1563.

Barrios, A., Panche, S., Duque, M., Grisales, V. H., Prieto, F., Villa, J. L., Chevrel, P., & Canu, M. (2013b). A multi-user remote academic laboratory system. *Computers & Education, 62*, 111–122.

Canu, M., & Duque, M. (2015). Laboratorios remotos: ¿qué interés pedagógico? Una formación de calidad en ingeniería, para el futuro.

Caro, C., & Quijano, N. (2011). Low cost experiment for control systems. In *Robotics Symposium, 2011 IEEE IX Latin American and IEEE Colombian Conference on Automatic Control and Industry Applications (LARC)* (pp. 1–6). IEEE.

Chase, R. (2015). *Peers Inc: How people and platforms are inventing the collaborative economy and reinventing capitalism*. New York: PublicAffairs.

de Jong, T., Linn, M. C., & Zacharia, Z. C. (2013). Physical and virtual laboratories in science and engineering education. *Science, 340*(6130), 305–308.

de Jong, T., Sotiriou, S., & Gillet, D. (2014). Innovations in stem education: The go-lab federation of online labs. *Smart Learning Environments, 1*(1), 1.

García, F., Díaz, G., Tawfik, M., Martín, S., Sancristobal, E., & Castro, M. (2014). A practice-based mooc for learning electronics. In *2014 IEEE Global Engineering Education Conference (EDUCON)* (pp. 969–974). IEEE.

Garcia-Zubia, J., López-de Ipiña, D., & Orduña, P. (2008). Mobile devices and remote labs in engineering education. In *2008 Eighth IEEE International Conference on Advanced Learning Technologies* (pp. 620–622). IEEE.

Garcia-Zubia, J., Orduña, P., Lopez-de Ipina, D., & Alves, G. (2009). Addressing software impact in the design of remote laboratories. *IEEE Transactions on Industrial Electronics, 56*(12), 4757–4767.

Garcia-Zubia, J., Angulo, I., Martnez-Pieper, G., López-De-Ipiña, D., Hernández, U., Orduña, P., Dziabenko, O., Rodrguez-Gil, L., Riesen, SANv., Anjewierden, A., Kamp, E. T., & de Jong, T. (2015). Archimedes remote lab for secondary schools. In *2015 3rd Experiment International Conference, exp.at'15* (pp. 60–64), Azores. https://doi.org/10.1109/EXPAT.2015.7463215, 00000.

Gillet, D., de Jong, T., Sotirou, S., & Salzmann, C. (2013). Personalised learning spaces and federated online labs for stem education at school. In *Global Engineering Education Conference (EDUCON), 2013 IEEE* (pp. 769–773). IEEE.

Gustavsson, I., Zackrisson, J., Håkansson, L., Claesson, I., & Lagö, T. (2007). The visir project–an open source software initiative for distributed online laboratories. In *Proceedings of the REV 2007 Conference*, Porto.

Harward, V. J., del Alamo JA, et al. (2008a). Building an ecology of online labs. In Proceeding of the International Conference on Interactive Collaborative Learning – ICL2008.

Harward VJ, et al. (2008b). The ilab shared architecture: A web services infrastructure to build communities of internet accessible laboratories. *Proceedings of the IEEE, 96*(6), 931–950.

Kalúz, M., García-Zubía, J., Orduña, P., Fikar, M., & Čirka, Ľ. (2013). Sharing control laboratories by remote laboratory management system WebLab-Deusto. In S. Dormido (Ed.), *Proceedings of 10th IFAC Symposium on Advances in Control Education,* University of Sheffield, International Federation of Automatic Control, Sheffield, Advances in Control Education (Vol. 10, pp. 345–350). https://doi.org/10.3182/20130828-3-UK-2039.00048

Kalúz, M., García-Zubía, J., Fikar, M., & Čirka, L'. (2015). A flexible and configurable architecture for automatic control remote laboratories. *IEEE Transactions on Learning Technologies, 8*(3), 299–310. https://doi.org/10.1109/TLT.2015.2389251

Lowe, D., & Orou, N. (2012). Interdependence of booking and queuing in remote laboratory scheduling. In *Remote Engineering Virtual Instrumentation 2012, REV 2012*.

Lowe, D., Machet, T., & Kostulski, T. (2012a). UTS remote labs, Labshare, and the Sahara architecture. Using Remote Labs in Education: Two Little Ducks in Remote Experimentation (p. 403).

Lowe, D., de la Villefromoy, M., Jona, K., & Yeoh, L. (2012b). Remote laboratories: Uncovering the true costs. In *2012 9th International Conference on Remote Engineering and Virtual Instrumentation (REV)* (pp. 1–6). IEEE.

Ma, J., & Nickerson, J. V. (2006). Hands-on, simulated, and remote laboratories: A comparative literature review. *ACM Computing Surveys (CSUR), 38*(3), 7.

Macho, A., Sancristobal, E., Rodriguez-Artacho, M., & Castro, M., et al. (2016). Remote laboratories for electronics and new steps in learning process integration. In *2016 13th International Conference on Remote Engineering and Virtual Instrumentation (REV)* (pp. 112–117). IEEE.

Milošević, M., Milošević, D., Dimopoulos, C., & Katzis, K. (2016). Security challenges in delivery of remote experiments. In *XXII Skup TRENDOVI RAZVOJA*.

Orduña, P. (2013). Transitive and scalable federation model for remote laboratories. Ph.D. thesis, University of Deusto.

Orduña, P., & García-Zubia, J. (2011). Scheduling schemes among internet laboratories ecosystems. In *Proceeding of Eighth International Conference on Remote Engineering and Virtual Instrumentation (REV11)* (pp. 1–6).

Orduña, P., Lerro, F., Bailey, P., Marchisio, S., DeLong, K., Perreta, E., Dziabenko, O., Angulo, I., López-de Ipiña, D., & Garcia-Zubia, J. (2013). Exploring complex remote laboratory ecosystems through interoperable federation chains. In *2013 IEEE Global Engineering Education Conference (EDUCON)* (pp 1200–1208). IEEE.

Orduña, P., Almeida, A., Ros, S., López-de Ipiña, D., & Zubía, J. G. (2014). Leveraging non-explicit social communities for learning analytics in mobile remote laboratories. *J UCS, 20*(15), 2043–2053.

Orduña, P., Zutin, D. G., Govaerts, S., Zorrozua, I. L., Bailey, P. H., Sancristobal, E., Salzmann, C., Rodriguez-Gil, L., DeLong, K., Gillet, D., et al. (2015). An extensible architecture for the integration of remote and virtual laboratories in public learning tools. *IEEE Revista Iberoamericana de Tecnologias del Aprendizaje, 10*(4), 223–233.

Orduña, P., Rodriguez-Gil, L., Garcia-Zubia, J., Angulo, I., Hernandez, U., & Azcuenaga, E. (2016). Labsland: A sharing economy platform to promote educational remote laboratories maintainability, sustainability and adoption. In *Frontiers in Education Conference FIE2016*.

Parker, G., Van Alstyne, M. W., & Choudary, S. P. (2016). Platform revolution: How networked markets are transforming the economy and how to make them work for you. 1st ed. New York: W. W. Norton & Company.

Richter, T., Boehringer, D., & Jeschke, S. (2011). Lila: A European project on networked experiments. *Automation, Communication and Cybernetics in Science and Engineering, 2009/2010*, 307–317.

Rodriguez-Gil, L., Garcia-Zubia, J., Orduña, P., & Lopez-de Ipiña, D. (2016). Towards new multiplatform hybrid online laboratory models. *IEEE Transactions on Learning Technologies, 10*(3), 318–330.

Ruiz, E. S., Martin, A. P., Orduna, P., Martin, S., Gil, R., Larrocha, E. R., Albert, M. J., Diaz, G., Meier, R., & Castro, M. (2014). Virtual and remote industrial laboratory: Integration in learning management systems. *IEEE Industrial Electronics Magazine, 8*(4), 45–58.

Sancristobal, E., Martin, S., Gil, R., Orduna, P., Tawfik, M., Pesquera, A., Diaz, G., Colmenar, A., Garcia-Zubia, J., & Castro, M. (2012). State of art, initiatives and new challenges for virtual and remote labs. In *2012 IEEE 12th International Conference on Advanced Learning Technologies* (pp. 714–715). IEEE.

Schauer, F., Krbecek, M., Beno, P., Gerza, M., Palka, L., Spilakov, P., & Tkac, L. (2016). Remlab-net III – federated remote laboratory management system for university and secondary schools. In *2016 13th International Conference on Remote Engineering and Virtual Instrumentation (REV)* (pp. 238–241). IEEE.

Pablo Orduña is co-founder of LabsLand and external collaborator of DeustoTech – Fundación Deusto

Javier Garcia-Zubia is Full Professor in the Faculty of Engineering of the University of Deusto, head of the WebLab-Deusto research group and cofounder of LabsLand

Luis Rodriguez-Gil is co-founder of LabsLand and researcher in DeustoTech

Ignacio Angulo is researcher and project manager in DeustoTech and cofounder of LabsLand

Unai Hernandez-Jayo is researcher and project manager in DeustoTech and cofounder of LabsLand

Olga Dziabenko is researcher and project manager in the University of Deusto

Diego López-de-Ipiña is the head of MORElab group in DeustoTech and Associate Professor of the Faculty of Engineering of the University of Deusto

Chapter 3
Deploying Large-Scale Online Labs with Smart Devices

Christophe Salzmann, Wissam Halimi, Denis Gillet, and Sten Govaerts

Abstract Deploying remote laboratories at a very large scale is one of the upcoming challenges in remote experimentation. It is referred to as Massive Open Online Labs (MOOLs). Being able to sustain a massive number of users accessing a specific resource concurrently is a complex task. The challenge is to maximize the use of the shared resource (the online lab) while minimizing or even canceling the waiting time to access the online lab such that the user feels it is dedicated to him/her. Tackling this problem requires revisiting both the pedagogical and the technical methodologies of online lab implementation, sharing, and deployment. In this chapter, we use indifferently online labs or remote labs to refer to physical labs accessible at distance through the Internet. A remote lab can also be considered as a cyber-physical system (CPS) as a combination of sensors, actuators, and embedded intelligence to fulfill given operational objectives.

Remote experimentation is becoming a mature technology, and it is usual to see institutions or platforms offering a significant number of remote labs. The model often chosen to enable online access to these labs is *Laboratory as a Service* (LaaS), where the lab is seen as a set of resources that the user can select on demand. The Smart Device model follows to the LaaS scheme and tries to describe the physical lab equipment and its digital components and interfaces as a unique entity. The Smart Device is seen through a set of services that the user can connect to. There is an ongoing effort to standardize the relationship between all the components (software, hardware, and learning environments). The aim of this standard is to ease the design, the implementation, and the usage of pedagogically oriented online laboratories as smart learning objects and their integration in learning environments and learning object repositories. The initial Smart Device model has been enriched to provide remote application developers a set of noteworthy configurations since not all combinations of sensors, actuators, and services are meaningful.

The Smart Device and other LaaS models are the cornerstone for deploying Massive Open Online Labs (MOOLs), but alone, they just provide a one-to-one

---

C. Salzmann (✉) · W. Halimi · D. Gillet · S. Govaerts
Ecole Polytechnique Fédérale de Lausanne, EPFL, Lausanne, Switzerland
e-mail: christophe.salzmann@epfl.ch

M. E. Auer et al. (eds.), *Cyber-Physical Laboratories in Engineering and Science Education*, https://doi.org/10.1007/978-3-319-76935-6_3

(1:1) access: one user accesses one real equipment at a time. Various solutions such as efficient time-sharing and resource duplication are proposed to increase the numbers of concurrent users, and a ratio of 5–10:1 is possible. The next step is to be able to handle the massive access, in the range of 50–100:1. Accommodating such a large number of concurrent users to access a real critical resource is a challenge that can be addressed by first giving priority to some users, for example, using *gamification* mechanisms. The analysis of the online usage pattern also permits a real-time adaptation of the various session parameters and session ordering. Lastly providing usage awareness is a key to help users select the best experimentation time.

This paper first provides some historical perspective and rationale to introduce the Smart Device model and its recent modifications toward completeness. Then, it proposes the required modifications in both the pedagogical and technical aspects to "traditional" remote lab in order to support the massive aspect of MOOCs. The MOOC and MOOL infrastructures are then described, especially how a Smart Device client application is integrated in a MOOC as an LTI module and how this application is able to interact with other applications or tools such as simulations.

This paper focuses on technical aspects, currently deployed, to implement such a remote lab within a MOOC infrastructure. It first covers the Smart Device specifications and its latest extensions. Then, the technical aspects such as the Smart Device implementation (server side) and the HTML5 client application are described. The components used for the deployment such as the LTI interface, the user storage, and other interactive tools are presented. The load balancer and the methods used to control the access are then depicted.

This paper provides a learning scenario for a MOOC session using the above elements, and an example is given with the control system lab deployed in edX, where more than 200 students access concurrently a farm of 30 electrical drives.

Keywords Smart Device · Massive Open Online Lab (MOOL) · Laboratory as a Service (LaaS) · Massive open online course (MOOC) · Cyber-physical systems · Remote experimentation · Standardization · edX · LTI component · WebSockets

3.1 Historical Context: Why a Smart Device

Remote experimentation started more than two decades ago. For a long time, lab owners, i.e., the persons or teams who own the physical equipment to be remotely controlled, were also the designers and developers of the software. The remote lab is mainly split in three components: the physical equipment, the client, and the server applications. Note that the server application can be a group of applications. This *closed* architecture worked and still works today but has a weakness that refrains its deployment at a large scale, namely, the strong coupling between the client, the server, and the equipment. For example, if the lab equipment is enhanced with the addition of a sensor, both the server and client applications have to be updated or rewritten. Similarly, to add another kind of lab equipment, modifications of both the client and the server are required. Sharing remote experiments between institutions requires the guest institution to use the provided client application. If, for some reasons, this client does not fit with the guest institution ecosystem, it may require a major upgrade or full rewrite of the client and possibly also the server. Such a time-consuming upgrade has to be carried out by the lab owner who may not have the needed resources (Salzmann and Gillet 2008). Quickly lab owners understood the need to agree on communication protocols and technologies. The advent of Web browsers accelerated this effort. The usage of Web browsers as the de facto environment for the client application, first as a container for embedded client running via plug-ins like Flash or Java and then without plug-in using plain HTML (Salzmann and Gillet 2013). The HTML5 norm included all the necessary components: rich user interface and communication via WebSocket to develop a client application that matches native application functionalities and UI. The client application environment being clearly identified, developing reusable and easily adaptable client applications, becomes possible provided that the interface with the server is also clearly identified.

The Smart Device specification is an effort to standardize the interface description between the client and the server such that it is complete, self-explanatory, and self-contained. These specifications were first documented in deliverable D4.1 of the European FP7 project, Go-Lab (Gillet et al. 2013; De Jong et al. 2014).

3.2 Smart Device Paradigm

The Smart Device paradigm originates from the RFID and sensor world, where one adds information to a static sensor to enhance its functionality. Thus, instead of a thermometer just returning a voltage, a sensor provides additional information such as the sensor ID, a time stamp, or a data range. Thomson (2005) defines that smart objects connected to the Internet need some or all of the following capabilities: (i) communication, (ii) sensing and actuating, (iii) reasoning and learning, (iv) identity and kind, and (v) memory and status tracking. We extended Thomson's proposition to support more complex devices that are using web-based technologies, namely, to support remote labs (Salzmann and Gillet 2008). We used this paradigm to specify on one hand the remote lab interfaces exposed on the Internet and on the other hand its internal functionalities (Salzmann and Gillet 2013). Since the Smart Device interfaces are well defined, a Smart Device becomes interoperable with other Smart Devices, external services, and client applications. Such interoperability fosters reuse of applications and external services and can provide extra functionality to any Smart Device (e.g., booking and authentication), simplifying the development of remote labs. The specification is designed to enable any client application developer to easily interface with a remote lab. Moreover, the specification of the services is machine readable, enabling the automatic generation of a skeleton ofthe client application (Halimi et al. 2017). The actual implementation of the specification, as well as the remote lab software and hardware implementation, is left to the lab owner's discretion.

The Smart Device paradigm revisits the traditional client-server architecture, on which many remote lab implementations rely. The main differences between existing implementations and the Smart Device's ones are first the complete decoupling between the server and the client and second the server representation as a set of well-defined services and functionalities that enable interoperability (Salzmann and Gillet 2013; Gillet et al. 2013; Tawfik et al. 2014). Similar approaches were proposed at the sensor/actuator level to enable the plug and play mechanism for smart electronic transducers, which provide electronic data sheets describing themselves (IEEE 2007). This paper proposes a specification that handles the interaction between clients and servers at the service level.

The decoupling removes the umbilical cord between the client and the server so that they can live their own separate lives. In a traditional client-server architecture (Chen et al. 2010), the server and client share a specification that is often uniquely used by them. On the contrary, the Smart Device paradigm defines one common specification that is shared by all Smart Devices. This reuse of a common specification and the client-server decoupling alleviates most of the problems developers are facing when the client application needs to be adapted to new OS/platforms, or if the client application is to be integrated in other environments such as learning management systems (LMS), or simply if additional features are added to the server. Furthermore, interoperability with, and reuse of, existing applications and services becomes possible when labs share a common specification.

Smart Devices mainly provide Web services to access sensors and actuators. Traditional solutions often provide a monolithic interface without the possibility to specifically access a given sensor or actuator (Salzmann and Gillet 2011). The Smart Device specification fully describes the Smart Device from a client point of view by specifying only the interfaces, not the inner working of the lab, which is left to the lab owner's discretion. The Smart Device specification is agnostic about the server-side hardware but suggests to reengineer the software components by adding "intelligence" to handle complex tasks accessible through the API.

There is no assumption regarding the communication channels for Smart Devices (Cascado et al. 2011). The Internet is the de facto choice for online labs (Salzmann and Gillet 2008; Auer et al. 2003). In addition, open Web technologies enable a broader compatibility and adoption, while proprietary technologies break the core ubiquitous access requirement.

The Smart Device may not necessarily provide a user interface (UI) but often proposes a minimal client UI. Thanks to the interoperability provided by the Smart Device specification, client applications can be developed to operate with different Smart Devices promoting reuse. Due to their ubiquity, Web browsers are the preferred environments to render the client UI. There is often a direct relation between the Smart Device sensors and actuators and the client app rendering this information. For example, an oscilloscope app renders the voltage evolution measured by a sensor of the Smart Device. In general, the Smart Device paradigm defines an ideal autonomous device which provides internal functionalities and that can be accessed through well-defined services.

The *Smart Device* model belongs to the *Laboratory as a Service* (LaaS) model where the lab is seen as a set of resources that the user can select on demand (Tawfik et al. 2014; Salzmann and Gillet 2013).

3.3 Smart Devices for Remote Labs

A generic Smart Device can already be seen as an autonomous online lab. On the other hand, it does not target a specific purpose, and therefore the expected requirements may not be satisfied. The principal aim of a remote lab is to represent its partial or full state to the client side and to enable real-time interaction. For example, it could be implemented in the form of a simple oscilloscope depicting the temporal evolution of a given sensor or a full 3D representation of the system. Interacting with the physical lab by directly controlling actuators or indirectly through a supervision stage (local controller or other logic) should also be possible. When considering remote labs, the client side that renders the server information needs also to be taken into account. Remote lab client applications are typically running in a Web browser. This specific choice of open Web technologies enables a broader compatibility and favors adaptation as well as adoption. The Smart Device paradigm enables the rethinking of such an interface into a Web 2.0 interface.

The Smart Device provides interfaces to remote labs for clients and external services through well-defined services and internal functionalities. A precise definition of these services and functionalities permits the decoupling between the client and the server. Some of these services and functionalities are meant for the client application, while others are meant for the Smart Device. The Smart Device's additional intelligence and agility mainly come from these internal functionalities. The services' and functionalities' definition enables anyone to design his/her own interface for accessing the Smart Devices for any remote lab.

A service represents, for instance, a sensor or an actuator exposed to the outside world (e.g., a client) through the API. Services are fully described through metadata, so that a client can use them without further explanation. A functionality is an internal behavior of the Smart Device. There may be communication between internal functionalities and client applications or external services through Smart Device services. While the required services are fully specified, the functionalities are only recommended, and best practice guidelines are provided.

For example, imagine an actuator service that enables the client application to set the voltage of a motor and a functionality that checks if the maximum voltage is not exceeded. The actuator service is well described by the Smart Device metadata (see Sect. 3.5.3). The internal validation is left to the lab owner's discretion, since it will be mainly ad hoc. Still, such a mechanism has to be implemented to ensure the protection of the server and the connected equipment.

The Smart Device specification (see Sect. 3.5) defines the communication and interfaces between the client and server, and sufficient information is provided to generate client applications or reuse existing client applications. Since the specification is common to many Smart Devices, client apps are not tightly coupled to one server, encouraging interoperability and reuse.

3.4 The Smart Device Architecture

The Smart Device specification provides a set of well-defined interfaces that enable communication between the remote lab, external services, and applications. Figure 3.1 illustrates a basic architecture with interaction examples that abstract the implementation of a remote lab, by providing a set of required and optional interfaces. The specification does not define the communication between the *Smart Device* and the *remote lab equipment* as shown in Fig. 3.1. The communication on the left side of Fig. 3.1 is what the Smart Device specifies, namely, the protocols and data formats of the interfaces of the Smart Device (i.e., the "metadata," "client," "sensor," "actuator," and "logging" interface shown in Fig. 3.1).

For instance, a metadata repository can retrieve the metadata of any Smart Device, index it, and provide a lab search engine. Because the interfaces are well defined, client apps can be reused among Smart Devices. For example, one Data Viewer Client or Learning Analytics Processing Client could retrieve data from any Smart Device and present it to the user. Additionally, a metadata format that describes the Smart Device, its functionalities, and its services is specified. Section 3.5 will elaborate on this metadata and each service and functionality in detail.

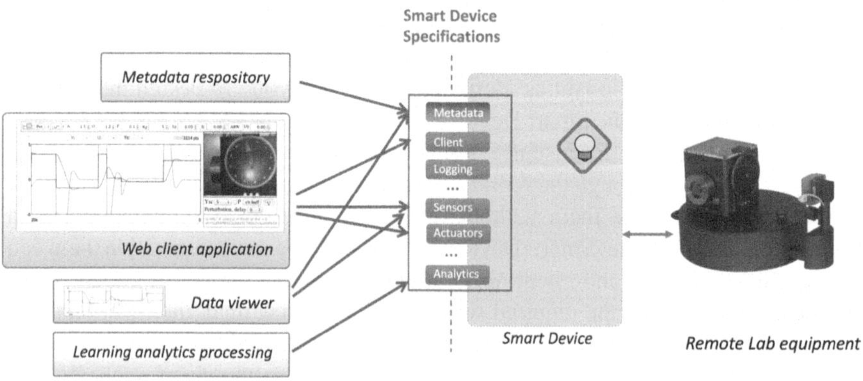

Fig. 3.1 Component diagram of different clients making use of the same common Smart Device services (arrows represent calls)

3.5 The Smart Device Specifications

This section presents selected parts of the Smart Device specifications in more detail. The complete Smart Device specifications are available at https://github.com/go-lab/smart-device-metadata/raw/master/smart-device-specification/SmartDevice specification.pdf

First, the communication protocol and the terminology used are described. Then, we will elaborate on the Smart Device well-defined services and internal functionalities.

3.5.1 Data Transfer Protocol

The goal of the Smart Device is to enable access to remote laboratories via the Internet. The targeted client application is a Web-enabled client, which can run on a tablet. We rely on open, standardized Web protocols to provide the data transfer between the Smart Device, external services, and applications to avoid dedicated plug-ins or customer lock-in. Typically, widely used candidates are HTTP and recently WebSockets. The problem with most HTTP-based Web services is that they follow a synchronous request-response schema. Hence, data can often only be "pulled" from the server, and the server cannot initiate a "push" of information to the clients. However, remote laboratory experiments often require asynchronous data transfer, e.g., a lengthy experiment should be able to push its results to the clients upon completion. HTTP solutions are often inefficient, e.g., via long polling (Loreto et al. 2011).

WebSocket specification (2011) on the other hand is asynchronous by nature and allows both pushing and pulling. This provides a bidirectional, full-duplex communication channel. Although WebSockets are a recent technology, they are supported by all modern browsers. Since WebSockets support both push and pull technologies efficiently and often with less programming effort than HTTP-based services, the Smart Device specification uses the WebSocket protocol. Only the metadata service that defines the other services (see Sect. 3.5.3) will be provided via HTTP GET to enable easy text retrieval.

3.5.2 Terminology and Concepts

The following terminology and concepts are used:

- The terms *sensors* and *actuators* reflect the traveling direction of information relative to the Smart Device. For example, a sensor enables the reading of a thermometer. An actuator enables the setting of a value, e.g., setting a motor voltage.

- Sensors and actuators can be physical (temperature sensor), virtual (computed speed derived from a position measurement), or complex, i.e., an aggregation of sensors/actuators (the front panel buttons of an oscilloscope or a 3D accelerometer).
- Both sensors and actuators can be configured; see the metadata service in Sect. 3.5.3.

3.5.3 Metadata Service

The metadata service is a required service that is at the core of the interoperability provided by the Smart Device specification. The requirements of the metadata are:

- Describe the lab (e.g., the contact person and the goals), which can be useful to allow automatic indexing by search engines (see Sect. 3.4).
- Describe the integration with external services (e.g., authentication with a booking service).
- Describe the concurrency mechanisms (e.g., are lab observations allowed, while someone is doing an experiment?).
- Describe and define the provided services (e.g., specify the service requests and response formats).
- Be easily extensible to enable adding extra services.

First, we survey different Web service description languages and highlight our choice. Afterward, the metadata design choices and the metadata for the services and how metadata can be added for additional services are described.

1. *Comparison of Web Service Description Languages*: Several options to describe Web service specifications have been surveyed with the goal not to reinvent the wheel but to use open, robust, and complete specifications. Furthermore, some specifications already allow the automatic generation of client applications. Since no Web service description languages specific to the WebSocket protocol were found, SOAP- and REST-based description languages were considered.

One of the most popular Web service description languages is WSDL,[1] which originally strongly focused on SOAP and provided support for REST since version 2.0. However, currently limited software is available for WSDL.[2] Other description languages are dedicated to RESTful services. WADL (Hadley 2009) can be considered as the REST equivalent of the original SOAP-only WSDL. RSDL[3] is

[1] Web Services Description Language (WSDL) 1.1, http://www.w3.org/TR/wsdl

[2] Web Services Description Language – Wikipedia, http://en.wikipedia.org/wiki/Web Services Description Language.

[3] RESTful Service Description Language (RSDL), http://en.wikipedia.org/wiki/RSDL

more focused on the structure of the Web service URIs, while RAML[4] relies on markdown and JSON Schema.[5]

Since all the abovementioned languages were hard to use WebSockets with, we have opted for Swagger v1.2.[6] Swagger is a JSON-based description language meant for RESTful APIs, but it was easily extensible to WebSockets while conserving all of Swagger's features. Since Swagger aims to describe Web services for both humans and computers, it strongly focuses on automatically generating user interfaces, which is one of our goals. Using JSON Schema, Swagger specifies the data format of requests and responses. Due to its large and growing list of supporting software, Swagger is growing in popularity. The specification is open, and the community is currently finalizing an updated version.

In the remainder of this section, we will elaborate on how we have applied and extended Swagger for the Smart Device specifications.

2. *Smart Device Metadata Design Choices*: Based on the requirements elicited above, the following main design choices were made:

    - *Sensor and actuator metadata service*: The metadata that describes the available sensors and actuators is provided by separate services. In this way, a developer of a simple Smart Device needs just to edit a few lines of metadata and does not need to add complex descriptions and models of actuators and sensors. The Smart Device software packages provided (see Sect. 3.7) already implement these services, so the developer can just edit this implementation, which also keeps this metadata very close to the actual sensor and actuator implementation.
    - *Service names*: Each service requires a method name, and each request and response of a service needs to pass this method name (e.g., the service for the sensor metadata is called *getSensorMetadata*). By passing this name, a WebSocket can be reused (channeled) by different services since the requests and responses can be identified by method name. Additionally, the method names are used to control access to services.

3. *General Smart Device Metadata Specification*: The official Swagger RESTful API documentation specification can be found online.[7] The Swagger specification is typically split over multiple files per service and served in the path of a REST service. Since WebSockets are not hierarchically organized in different URLs, we have opted to provide one specification file, containing the general metadata and all service-specific metadata.[8] This section will introduce the

[4]RESTful API Modeling Language (RAML), http://raml.org/

[5]JSON Schema specification – JSON Schema: core definitions and terminology json-schema-core, http://json-schema.org/latest/json-schema-core.html

[6]Swagger website: http://swagger.wordnik.com/

[7]https://github.com/wordnik/swagger-spec/blob/master/versions/1.2.md

[8]Metadata specification examples for Smart Devices are available at GitHub: https://github.com/Go-Lab/smart-device-metadata

general structure of the adapted Swagger file. However, code samples and exact field names are omitted for brevity but are available in the full specifications.[9] The metadata consists of six parts:

(a) Swagger-related metadata: Swagger requires to declare the version of Swagger and the API. The version of Swagger should not be changed by the developer.
(b) General metadata: These default Swagger fields provide information about the lab, such as the name, a short description, a contact person, and licensing information.
(c) API metadata: The root URL path of the Smart Device services is described, and all services are defined. Each service will be described from Sects. 3.5.6, 3.5.7, 3.5.8, 3.5.9, and 3.5.10.
(d) Authorization metadata: Swagger supports common REST-based authentication and authorization mechanisms, e.g., OAuth. All these mechanisms can be used in the Smart Device. For instance, in the Go-Lab booking system, we are using a token-based authorization, which can be modeled with Swagger's API key type since the booking token is a sort of temporary API key for the duration of the booking.
(e) Concurrent access metadata: We have extended Swagger to model the concurrency models of remote labs. Different concurrency schemes exist, and it is up to the lab owner to decide on an appropriate scheme. One can interact with a lab in a synchronous or asynchronous way. In a synchronous lab, the users are interacting directly with the experiment and are aware of actions of other concurrent users. When in the asynchronous mode, the user typically prepares an experiment, submits it, waits to get results back, and is not aware of other users. The rest of this metadata is for synchronous labs, since asynchronous labs can deal internally with concurrency issues. Typically, two concurrency schemes are possible: "concurrent" and "roles." Either users are allowed to use the experiment at the same time, or different user roles control the access. Each role has a name and can declare which services will be accessible for a user with that role and a mechanism to select the role.
(f) Different mechanisms have been identified to switch roles:

   - *Fixed role*: The user cannot be promoted from one role to another, e.g., the teacher can control the remote lab, but the students can only observe.
   - *Dynamic role*: The role can change during the session, e.g., a user observing can later control.
   - *Race*: When nobody is using the lab, the first user who accesses it gets the control. Subsequent users have to retry until the access is granted to one of them.

[9]The full Smart Device specification is available at https://github.com/go-lab/smart-device-metadata/raw/master/smart-device-specification/SmartDevice specification.pdf.

- *Queue*: Upon access, the user is added to a first-come-first-served waiting queue; other queuing schema can also be proposed.
- *Interruptor*: The user can abort the session of another user and take control of the Smart Device.

4. *Service Metadata Specification*: This section discusses how a service can be added as a JSON object in the API metadata on a high level (for details, refer to the full specification). Optionally, new data models need to be declared in the model section. However, we have tried to design the specification so that for simple Smart Devices, developers do not need to learn how to describe a service in Swagger. The specification provides reusable service metadata descriptions and models for the sensor, actuator, and logging services.

A new API object needs to contain the path, description, and also an optional "protocol" field that the Smart Device specification has been extended to support the WebSocket protocol. Then a list of all operations of the service is specified and its response messages that describe the error messages (relying on HTTP status codes (RFC7231 2014)). Each operation can specify the protocol method; in case of WebSockets, this is typically "Send;" and one can define the type of WebSocket: text or binary. Binary WebSockets can make the transmission of binary data much more efficient, e.g., for video streaming. Additional documentation can be provided in the "summary" and "notes" fields. Next, the service arguments and results can be configured using JSON Schema primitives[10] or the ID of a model from the model metadata section. One can also model the response format using any Internet media type (Freed et al. 2014), e.g., for a service that returns images. The service input arguments are typically represented as a data model. Simple request models are provided, but more complex models can be defined when needed. More information on adding a new service can be found in Swagger RESTful API specification (2014).

3.5.4 Sensor Metadata Service: getSensorMetadata

As mentioned, the sensor and actuator metadata are provided via separate services and not in the metadata description itself. This section elaborates on the sensor metadata.

The service is called *getSensorMetadata* and can be called like most Smart Device services with a JSON object by specifying the "method" field and an optional authentication token in case booking is required. As mentioned before this, method field enables the reuse of one WebSocket to channel multiple services. The service returns an array describing each sensor exposed to the outside world. Each sensor contains:

[10]JSON Schema specification – JSON Schema: core definitions and terminology json-schema-core, http://json-schema.org/latest/json-schema-core.html

- The ID to identify the sensor, e.g., "3D-acc."
- The full name, e.g., "3D acceleration."
- The description, e.g., "the robot arm 3D acceleration."
- The WebSocket type is "text" or "binary" (e.g., for video).
- The response type of the sensor service for the sensor defined as an Internet media type (Freed et al. 2014), e.g., a webcam sensor using JPEG compression uses image/JPEG.
- The measurement value array will contain a single value for a simple sensor like a thermometer, but for a complex sensor like an accelerometer, the array contains, for example, three elements for the X-Y-Z acceleration. Values are described with a name and unit. Since the set of possible units is almost infinite, we recommend to use the SI units (Taylor and Thompson 2008) and the SI derived units.[11] Optionally, a last measured time stamp and a range of minimum, maximum, and iteration step of the range in which the values safely operate can be added. Furthermore, for continuously measured values, the frequency at which the measurement is updated can be provided in Hertz (s^{-1}).
- The configuration parameters can be used to adjust the sensor when requesting a sensor value (see Sect. 3.5.6). Each parameter has a name and data type such as a JSON Schema primitive, array, or data model for complex parameters, e.g., to configure the video resolution.
- The access mode describes how the sensor can be accessed, e.g., some sensors can be measured once (pull), while others provide a continuous data stream (push or stream). For "push" sensors, one can specify the nominal update interval and whether the measurement frequency can be modified by the user.

Both sensors and actuators can be configured, which means that the information can be sent and received even for the sensor. For example, the image resolution of a webcam sensor can be configured. Similarly, for actuators some aspects may be set through configuration (e.g., the gain of a power amplifier could be configured), while the actual value is set through the actuator value itself (see Sect. 3.5.7). Typically, sensors and actuators are rarely configured.

Streaming video of the experiment is an essential service that a Smart Device should provide through a sensor. We recommend that such sensor treats the video image as an encoded image, for example, JPEG encoded. Using JPEG encoding results in binary data which either should be transmitted through a binary Web-Socket (recommended) or BinHex'ed prior to sending it using a textual WebSocket. If further processing is required at the client side, a pixmap (pixel array) could be used, this at the cost of being 10–90% larger in size (Furht 1995).

[11] SI derived units – Wikipedia, http://en.wikipedia.org/wiki/SI derived unit

3.5.5 *Actuator Metadata Service: getActuatorMetadata*

As mentioned, the actuator metadata is also provided via a service, named "getActuatorMetadata." The service is very similar to the sensor metadata service, so we will only discuss the difference in the service response: the input type expresses what data type can be used for a specific actuator in the actuator service. By default, this is JSON, but it can be set to any Internet media type (Freed et al. 2014). This replaces the response type of the sensor metadata service.

3.5.6 *Sensor Service: getSensorData*

The sensor and the actuator data services are at the core of the Smart Device interaction, and both are quite similar. They handle the main data exchange between clients and the Smart Device. Both services in combination with their metadata services enable developers to create apps that can adapt to different Smart Devices, enabling app reuse and interoperability. Similarly, different apps could be developed for a Smart Device. For example, for a Smart Device that provides a temperature measurement every second, one app could just update a text field, while another app could visualize the temperature evolution over time. This difference in app functionality requires absolutely no change on the Smart Device services. Furthermore, using the sensor metadata service, these two proposed apps could be made interoperable and reusable with any Smart Device.

Different sensors and actuators exist:

- Real: Represents a physical sensor, e.g., a thermometer
- Virtual: Represents a computed sensor, e.g., a speed measurement derived from a position measurement
- Complex: Represents the aggregation of sensors/actuators, e.g., buttons on the front panel of an oscilloscope

The data structure returned by a sensor or sent to an actuator may vary depending on the number of values and the measurement data structures. The data structure (see Fig. 3.2) contains three fields to enable flexible data representation. In the "valueNames" field, the names of the sensor or actuator measurement value is listed as returned by the sensor or actuator metadata services (see Sect. 3.5.4). Then, the actual data for each value is listed. Finally, the optional "lastMeasured" array contains the time stamps when a value was measured. This time stamp array should not be included when sending data to set an actuator. The data and the "lastMeasured" time stamps are listed at the same array index as the value name, as indicated by the dashed lines in Fig. 3.2. The elements in the data array can be in different formats: (1) a single value, e.g., temperature; (2) an array of values representing a set of single values over time, e.g., temperatures over the last minute; (3) aggregated values representing a sensor or actuator that returns multiple values,

Fig. 3.2 Sensor and actuator data structures

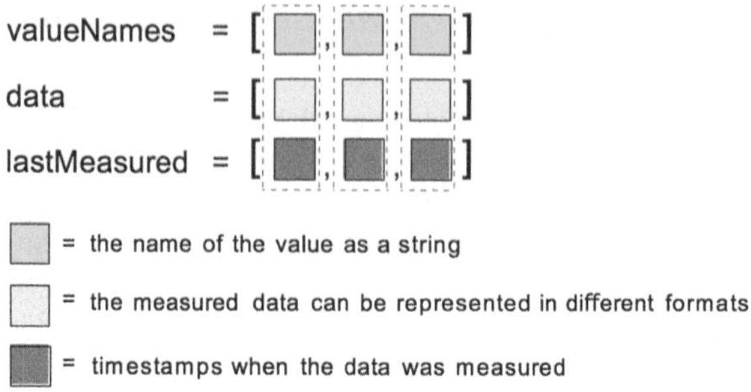

e.g., a 3D accelerometer; (4) an array of aggregated values representing a set of aggregated values over time, e.g., the 3D acceleration over the last minute; and (5) complex data structures that are used when sensors and actuators require input and output not definable with primitive variables or arrays, e.g., for complex JSON objects or binary data. This data representation was chosen, because flat array-based data can be more efficient to process than complex data structures interleaved with time stamps.

As an example of a complex data structure, a webcam can be modeled as a single value sensor that returns a compressed image, as an array of values based on the image bitmap or as a binary value with JPEG-encoded data. The choice between the three representations is up to the lab owner.

A request to the *getSensorData* service is more complex than the previous services due to possible authentication, concurrency, and configuration settings. Optionally, an access role from the concurrency role list (see Sect. 3.5.3) can be passed. If no *accessRole* is available, the Smart Device can decide the role. The Smart Device will decide whether these rights can be granted and react accordingly.

The *getSensorData* service will return the data in the above-described data format (see Fig. 3.2) together with the method name, sensor ID, and access role to foster possible WebSocket reuse. This is in case the user has the controller role. But when the user is an observer and does not have access to the measured data, the service can optionally provide extra waiting information that can be used to display how long the user has to wait and how many people are in front of her (e.g., the queue size, position, and waiting time left). Furthermore, the sensor configuration might be used (e.g., for a video sensor), if it is described in the sensor metadata. For example, this can be very useful to adapt to the client screen size and network speed by reducing the transmitted image resolution and compression (if configurable). Similarly, the data transmission pace could also be controlled. If the user temporarily needs to throttle the video stream, the client can ask the Smart Device to reduce the number of images sent per second by setting the update frequency (see Sect. 3.5.4). The sending may even be interrupted by setting the update frequency to 0. It is up to the application developer to take advantage of these features.

3.5.7 Actuator Service: *sendActuatorData*

The actuator service is very similar to the sensor service (see Sect. 3.5.6). The main difference with the sensor service is the fact that the *sendActuatorData* service allows the user to actually set the desired actuator value, meaning that the data model of Fig. 3.2 is sent in the request.

The internal functionality of the Smart Device should first validate the sent value (see V-K) prior to applying it to the actuator itself. While sensors often do not have concurrency issues, the actuator may also be controlled by another client concurrently, and its access needs to be moderated. Various schema can be implemented by the lab owner to internally manage the actuator access (see Sect. 3.5.3). In the following examples, we will assume one of the most common scenarios: a user can either control the lab or observe what others are doing. Given that the user has a controller role, the actuator may set the value and acknowledge the actuator change by returning the set values in the payload of the response. The payload is optional, and the format is not specified. As a good practice, we recommend to return the data of the actuator in the same format as the request data format. This returned actuator data in the payload could be used to update the client application UI with the actual value. The client can assume that the actuator has fulfilled the request when no errors are returned. If the actuator is currently in use, a more specific payload, detailing some information regarding the time the user has to wait prior to control the actuator can be set, similar to the example in Sect. 3.5.6.

Furthermore, a user with the "interruptor" role can abort the actuator control of current user. The way the conflict is resolved and the policy to grant this role are defined by the lab owner and/or the client application.

3.5.8 User Activity Logging Service: *getLoggingInfo*

The optional user activity logging service returns logged user actions or lab status info in the Activity Streams 1.0 JSON format.[12] The Activity Streams format is a JSON-based specification to describe a sequence of user actions with a time stamp, and it is often used in social media platforms. To retrieve a continuous stream of real-time user activities of the Smart Devices, the *getLoggingInfo* service can be called with an optional authentication token to validate access (which is recommended due to the privacy-sensitive data).

[12]The Activity Streams specification is available at http://activitystrea.ms/specs/json/1.0/

3.5.9 Client Application Service: getClients

This optional service provides links to the client applications to operate the Smart Device. The client technology is not strongly specified. It can be a simple HTML page or a packaged Web app that carries additional functionalities to enable interaction with the containing environment, for example, the Go-Lab project advocates OpenSocial gadgets (Marum 2013), since they effortlessly run on the Go-Lab ILS platform (Govaerts et al. 2013). Upon sending a request to the *getClients* service, a client app list will be returned, with for each item a type that specifies the kind of application and a URL. The current version of the Smart Device specification contains the following extensible list of types: "OpenSocial Gadget," "W3C widget," "Web page," "Java Web Start," and "desktop application."

3.5.10 Models Service: getModels

This optional service can provide several models of the physical lab (i.e., the instrumentation) and its theoretical back-scale object that students can manipulate. With a mathematical model of the experiment, a client app can be built with a local simulation. This can provide an interactive simulated version of a remote lab that can be used by students when the lab is already in use (i.e., to provide a better observer mode). Due to the wide range of existing formats to express graphical and theoretical models (e.g., VRML,[13] X3D,[14] and MathML,[15]) we do not limit the specification and leave the model language choice up to the lab owner.

3.5.11 Functionalities: Best Practices

Internal functionalities are implementation suggestions for the Smart Device. They are provided as best practices, since the implementation of these functionalities is often ad hoc and strongly related to the connected equipment.

(a) *Authentication functionality*: The Smart Device may not contain a booking system. It can make use of an external booking system. When a user reserves a lab, the booking system provides an authentication token. At the booked time, the user can connect to the Smart Device with this authentication token. The Smart Device then contacts the booking system to validate whether the user

[13] Virtual Reality Modeling Language (VRML), http://gun.teipir.gr/VRML-amgem/spec/index.html

[14] X3D, http://www.web3d.org/standards

[15] MathML, \http://www.w3.org/Math/

is currently allowed to access the Smart Device. Thus, integrating the booking service in the Smart Device requires little effort, compared to providing its own authentication and booking mechanisms.

(b) *Self and known state functionality*: The precise implementation of this recommended functionality is left to the lab owner's discretion. This functionality ensures that the remote lab is reset to a proper state after an experimentation session is completed or a system outage occurred, so that the next user can properly use it. Since remote experiments are supposed to be conducted from faraway, nobody is expected to be around the experiment to put it back in a known state. Thus, the system should be as autonomous as possible, which implies an adequate and defensive software and hardware design that is able to adapt to "any" situation. We suggest to implement the following procedures in the Smart Device: (1) automatic initialization at startup, (2) reset to a known state after the last client disconnects, and (3) potentially hardware calibration.

(c) *Security and local control*: This functionality is recommended, and its implementation is left to the lab owner's discretion. At all time the security of the server and its connected equipment must be ensured. All commands should be validated before being forwarded to the connected equipment. This step may require the addition of a local controller to track the connected equipment's state, e.g., a speed increase may need to follow a ramp before being applied to a motor. Users often try to take the system to its limits, i.e., not only the physical limit of a given sensor/actuator but also signal patterns on a sensor over time may also need to be considered. Since the actuators may be connected to the Internet, it is essential to validate all applied values and to consider potential external constraints. The lab owner should implement the following procedures in the Smart Device: (1) value validation before applying data to actuators and (2) actuator state validation to check if the command to be applied is safe.

(d) *Logging and alarms*: This functionality logs session and lab information, as well as user interactions. In case of problems, alarms may be automatically triggered by this functionality. Since a Smart Device will be typically online unattended for an extended period of time, it is essential to monitor it and have a method to perform post hoc analysis. The user action should be logged and can be made accessible via the user activity logging service (see Sect. 3.5.8). But extra information should also be logged, e.g., the system state and the environment (e.g., room temperature). Note that some sensors may be available internally to the Smart Device, but not necessarily accessible via the sensor service. We suggest to track the following information: (1) user actions, (2) the complete system state, and (3) its environment state. Additionally, by definition the Smart Device is connected to the Internet and has no knowledge of its clients. Proper action is required to prevent abuse. A firewall or a DMZ[16] may protect it from attacks. While some hostile actions may be reduced using such mechanisms, the Smart Device should add internally additional measures: (1) validate the

[16]Demilitarized zone (DMZ), http://en.wikipedia.org/wiki/DMZ (computing)

requests sent by clients, (2) throttle continuous requests of a malicious client, and (3) log all Internet connections for later analysis. If an unexpected event occurs, its potential danger should be assessed by the Smart Device, and an alarm may be triggered.

(e) *Local simulation*: When the experiment is busy or unavailable, a local simulation might be a useful alternative for waiting users. The simulation data could be read or modified through virtual sensors/actuators. A mathematical model describing the physical equipment can be made available to the client via the model service, which the client developer can use to simulate the hardware. Such simulations can require computational resources unavailable at the client. However, this computation can be done in the server side, and the results can be sent to the client using virtual sensors and actuators.

3.6 A Detailed Smart Device Example

This section illustrates how a Web client interacts with a simple Smart Device, with one sensor and one actuator. Both the Smart Device and the Web client are available on GitHub.[17] The full JSON messages are omitted for brevity, but similar examples can be found in the full specification.[17]

The first step taken by the Web client is to ask the Smart Device about its general capabilities using the metadata service. This is done with a regular HTTP GET request to http://serverIP/metadata. The Smart Device returns JSON containing the metadata (see Fig. 3.3). Then, the client requests the available sensors from the sensor metadata service. This request is performed via a WebSocket. A JSON object containing {“method”: “getSensorMetadata”} is sent to the server, upon which the Smart Device replies with another JSON object {...[“sensorID”:“discPos”, ...]} containing an array of available sensors and related information such as range, etc. The next step is to ask about available actuators with a similar request (see Fig. 3.4). The Smart Device replies that there is one actuator: a motor with “actuatorID” :“motor”, “rangeMinumum”:“-5”, and “rangeMaximum”:“5”. The client app has now enough information to build a basic UI. In this case two UI fields are present: one to display the discPos sensor value and one to set the motor actuator value.

The fields of the generated skeleton UI need to be populated with the data coming from the Smart Device. In other words, we need to tell the Smart Device to start sending measured values to the client via a WebSocket. This is done by sending the request {“method”: “getSensorData”, “sensorID”:“discPos”, ...}. The Smart Device will start pushing the measured values continuously to the client (see Fig. 3.5). The client application needs to parse the received JSON objects and update the sensor field in its UI with the received value.

[17] https://github.com/go-lab/smart-device/tree/master/Desktop/Simple-examples

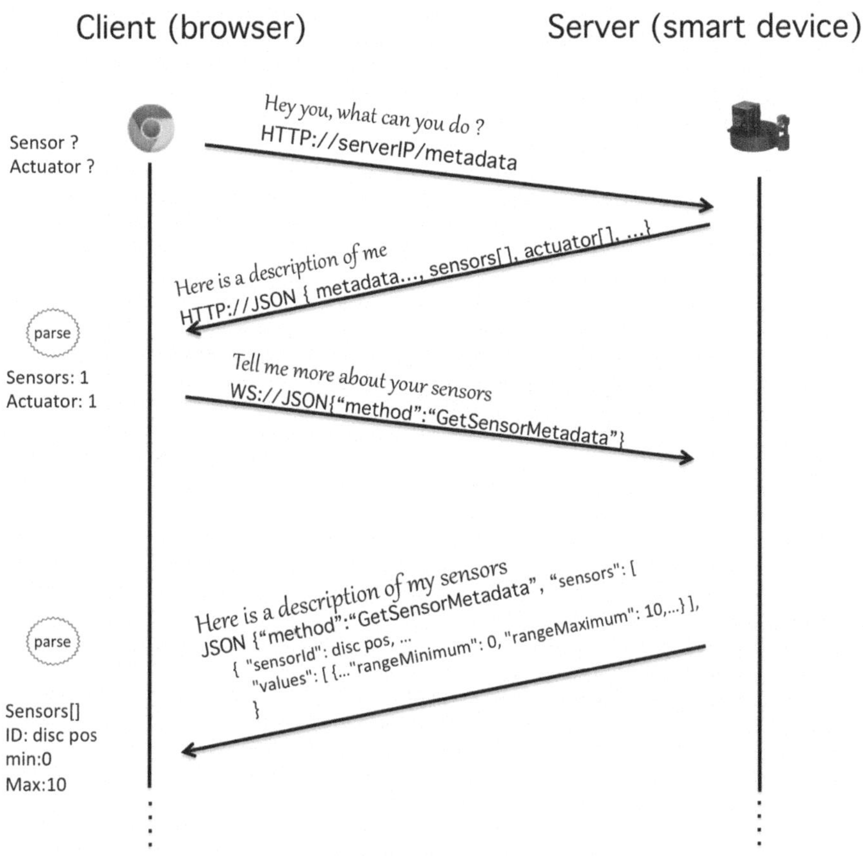

Fig. 3.3 The Web client asks the Smart Device about the available sensors

When the user modifies the actuator value in the client UI, a WebSocket request is sent to the Smart Device with the new actuator value, {"authToken":"42FE36", "method": "SendActuatorData", "actuatorID":"motor", "values": [...]}. This request carries an authentication token, which will be used by the Smart Device to verify that access to the actuator is granted to the client application (e.g., based on a lab booking of a user at a given time). To control the access to the actuator, the Smart Device will contact the booking service with the provided token. If the booking service confirms the token, the new actuator value will first be internally validated (e.g., within a specified range) and then applied to the motor. If the token is invalid or if the value is out of range, the value will not be applied to the motor, and an error message may be returned to the client application.

Upon completion of the remote experiment, the client closes the WebSocket connections. Internally, the Smart Device should go back to a known state and wait for the next user to connect, e.g., set the motor voltage to "0" to save energy.

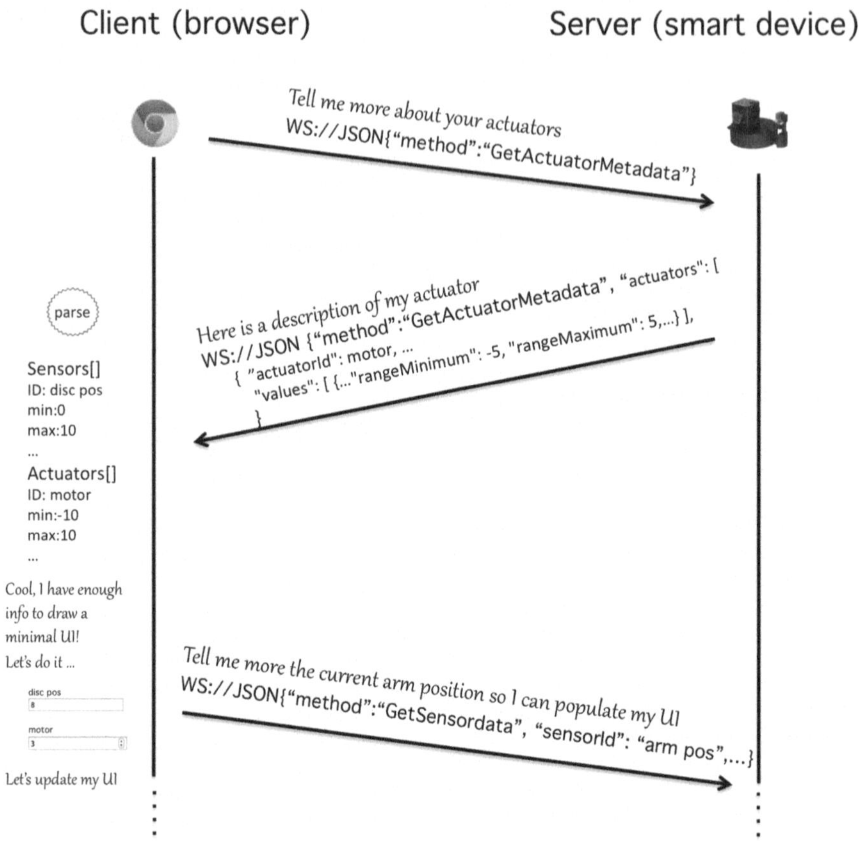

Fig. 3.4 The Web client asks the Smart Device about the available actuators

3.7 Smart Device Extensions

As illustrated in Sect. 3.5, any person with programming skills can create user interfaces to connect to a lab built according to the Smart Device specifications. It is possible with the provided APIs to personalize the user interfaces, which allow teachers to use the remote labs in different ways, according to their educational needs. This is particularly interesting for remote laboratories which are configurable to conduct different experiments corresponding to different scientific phenomena. In this context, we refer to the activity allowing the students to freely vary the parameters of control on lab equipment as an "experiment," and we refer to the combination of sensors and actuators used in an experiment as a "configuration" from a lab owner point of view. Since the APIs do not convey any information regarding the relationships and dependencies among the different components of a remote lab, the user interface programmer resorts to contacting the lab owner to have

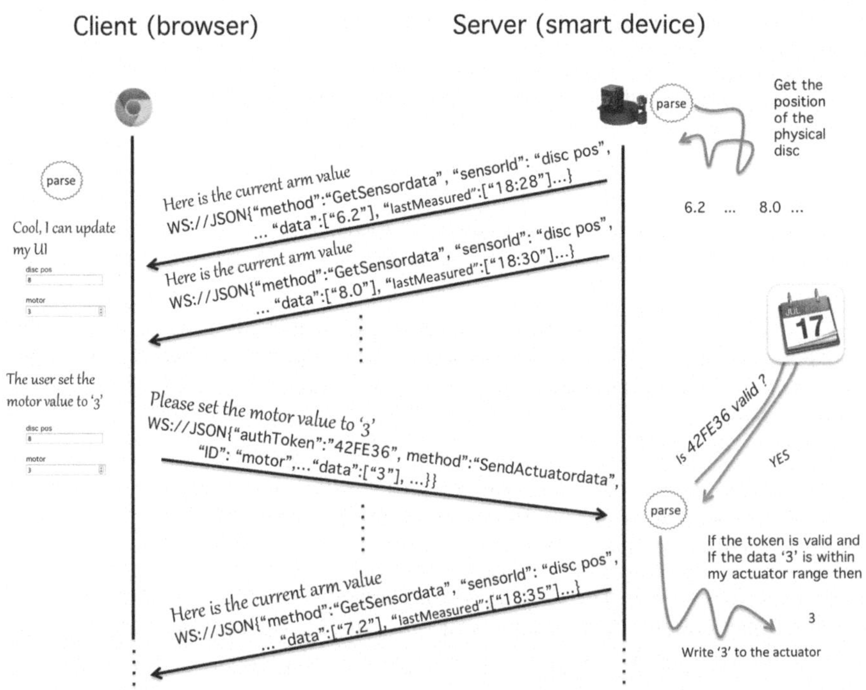

Fig. 3.5 The Smart Device pushes the measured values to the client. It also receives and validates the actuator value prior to applying it to the motor

information about what experiment(s) a considered lab implements. The aim of the proposed extension is to augment the Smart Device specifications with additional API calls, to reduce the dependence of the programmer on the lab owner in this kind of situations.

What differentiates remote laboratories from other cyber-physical systems is that they are built to fulfill an educational goal: conducting predefined experiments to reflect on certain topics. With no knowledge about the interconnections of the lab components, it is not possible to build a UI that interfaces "pedagogically meaningful" experiments. We extend the "metadata" to add a service to the API which returns the configurations or experiments supported by the remote lab, in addition to the request and response *models*. Our proposed extension is twofold:

1. Defining the *models* for an *Experiment*, *SendExperimentsRequest*, *SendExperimentRequest*, *ExperiementsMetadataResponse*, and *ExperiementMetadataResponse*
2. Defining the new api calls: *getExperiments* and *getExperiment*

Experiment model: An *Experiment* model is characterized by two fields common to all models: *id* and *properties*. The *id* characterizes the model at hand; in this case its value is *Experiment*. This *id* field gives knowledge about the format of an

Experiment JSON object for further processing. The *properties* are made up of five subfields:

- *experimentId*: Which can take any string value. The value of this field is defined by the lab provider.
- *fullName*: Which contains a non-formal name of the experiment. It can take any string value.
- *description*: A human-readable description of what the experiment is about. This field is meant to be informative for teachers, to get a high-level description of the experiment.
- *sensors*: It is an array containing a list of the sensor ids used in a particular experiment. *sensorIds* can have any string value. The string values of *sensorIds* contained in this JSON object should be corresponding to *sensorIds* defined in the metadata.
- *actuators*: It is an array containing a list of the actuator ids used in a particular experiment. *actuatorIds* can have any string value. The string values of *actuatorIds* contained in this JSON object should be corresponding to *actuatorIds* defined in the metadata.

getExperiments api: The *getExperiments* api allows the retrieval of a list of supported experiments. The *nickname* of this call is *getExperiments* which means it needs to be used when initiating a request. *summary* and *notes* fields give a high-level description of what this call does: answers with a JSON object containing the list of available experiments ids. The response of this call is formatted as an *ExperimentMetadataResponse* which will be detailed later in this section. As it can be deducted from the *properties* field, the request is formatted as a *SimpleRequest* defined in the original SD specifications. The *authorization* field designates authentication mechanisms that the remote lab is using to permit users to access the lab; if empty it means no authorization needs to be done. *responseMessages* detail the possible responses that can be received at the requester end, in case an *ExperimentMetadataResponse* cannot be received.

ExperimentRequest model: To retrieve the required *actuatorIds* and *sensorIds* for a particular experiment, an ExperimentRequest has to be sent to the Smart Device hosting the laboratory as shown hereafter. The *ExperimentRequest* should contain the *experimentId* of the desired experiment. A list of *experimentsId*s can be retrieved with the *getExperiments* call.

ExperiementMetadataResponse model: The response of an *ExperimentRequest* is an *ExperimentMetadataResponse*. The *id* of this response tells the type of JSON object to expect at the receiving end. It is formatted as to contain the *Experiment* JSON object which defines an experiment. This should be enough for an auto generator to create a UI corresponding to the required request.

In Halimi et al. (2017), we present an automatic generator of user interfaces based on the extended Smart Device specifications, to demonstrate their completeness for our purpose. A typical communication scenario between the automatic generator and the Smart Device is illustrated in Fig. 3.6.

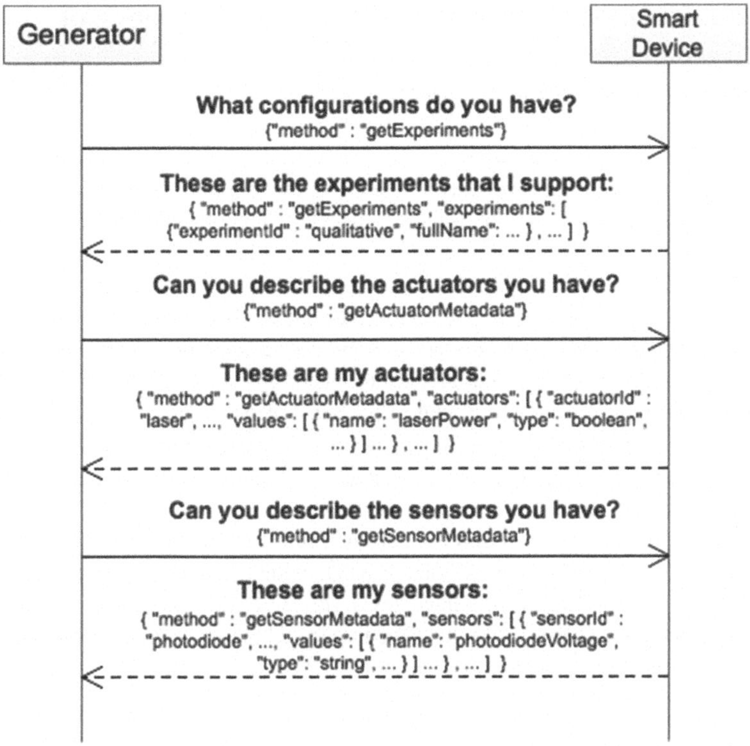

Fig. 3.6 Example interaction between the automatic UI generator and the Smart Device to build the UI

3.8 Implementation Examples

To illustrate that the Smart Device specifications are software and hardware platforms agnostic, we have implemented the specifications on various platforms and with different programming languages. These examples are publicly available on GitHub.[18] In some situations, it will not be possible to modify the server of already existing labs, e.g., due to the lack of resources. In this scenario, a Smart Gateway (Orduna et al. 2014) that lies between the client and the remote lab does the necessary translation to make the remote lab behave like a Smart Device from a user point of view. This translation is performed by the Gateway4Labs,[19] a software orchestrator that relies on plug-ins to adapt the different existing labs to the Smart Device specifications.

[18]https://github.com/go-lab/smart-device

[19]https://github.com/gateway4labs

3.9 Smart Device Integration in Learning Environment

The Smart Device specifications permit to describe a complete online laboratory that has enough resources to live its own life without necessarily reflecting the educational objectives that underlie the activities. There is an ongoing effort to streamline the relationship between all the components (software, hardware, and learning environments) in order to ease the design and implementation of pedagogically driven online laboratory activities. Traditional integration was aimed toward LMS such as MOODLE; recently the ability to integrate remote laboratories in MOOCs has been proposed (Salzmann and Gillet 2013; Salzmann et al. 2016). The Smart Device and LaaS models are the cornerstone for deploying Massive Open Online Labs (MOOLs), but alone they just provide a one-to-one (1:1) access: one user accesses one real equipment at a time. Various solutions are proposed to increase the number of concurrent users (Tawfik et al. 2014; Lowe 2014), and a ratio of 5–10:1 is possible. The next step is to be able to handle the massive access, in the range of 50–100:1.

Figure 3.7 illustrates the implementation levels for online labs. The low level refers to the online Lab as a Service (LaaS), which can be personalized at the intermediary level. The intermediary level refers to the online lab as an open educational resource (LaaR), which can be integrated in learning environments, such as edX or Graasp.

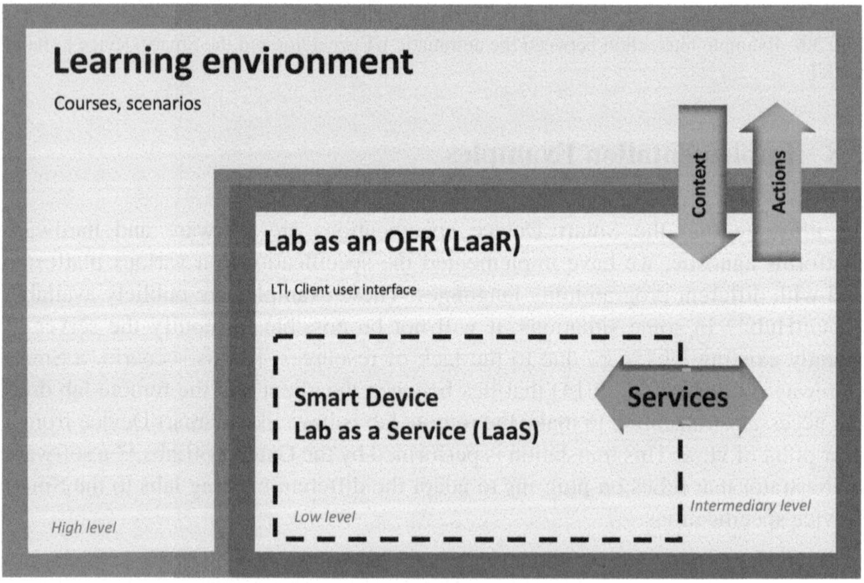

Fig. 3.7 Implementation levels for online labs

The proposed levels fit well with the required MOOLs for MOOC infrastructure. The MOOC (high level) is used to deploy a full-fledged remote laboratory made of Smart Devices (low level) integrated as LTI modules (intermediary level).

The MOOC infrastructure offers many features to support the students' learning process. To take full advantage of the MOOC paradigm, classical courses and/or hands-on laboratory sessions are to be completely reworked. Classical MOOCs consist of videos interleaved with questions and online exercises. Automatic or peer grading is provided to support self-paced user progress.

Supporting a large number of users accessing shareable resources such as video or virtual simulation is "reasonably straightforward" provided that you have the needed servers and bandwidth. To handle the same number of users accessing a critical resource (i.e., not sharable) such as a remote lab is a challenge. To tackle this challenge and efficiently integrate remote lab sessions in MOOCs, the following actions can be combined: the first one is to *rework the classical hands-on laboratory* sessions such that existing experimentation activities are split into shorter parts to ensure that the critical resource – the remote experiment – is not held for too long by a single user. Typical activities last for 30 s to a few minutes. The shorter the activity duration is, the larger the number of users handled per unit of time can be.

The second action is to *rethink how the user interacts with the remote equipment*. Is the distant equipment only to be observed? In this case, the measurements and video frames can easily be duplicated at the server side and streamed to each client. If the envisioned scenario involves acting on the equipment, a policy has to be defined if more than one user wants to act at the same time. The Smart Device specifications propose various mechanisms and policies to implement resource sharing. A controller/observer mechanism with a first-come-first-served policy is an efficient way to share a critical resource. Each user is able to observe the equipment concurrently, but only one user at a time can act on the equipment. If two or more users want to act at the same time, the requests are queued. The Smart Device proposes these mechanisms, but it is up to the client application to take advantage of them. It is especially important that the client application provides full awareness regarding the queuing/acting information to the client. This implies that waiting users are informed about the waiting time; similarly acting users are aware of the remaining time available for the current activity.

The third action is to *duplicate the critical resource* (the equipment) as a farm of remote labs to support more users.

Section 3.6 suggests additional actions to handle a larger number of concurrent users using gamification.

There exist many MOOC platforms. Currently, the two main ones are Coursera and edX. The solution presented in this paper is implemented in edX but could be implemented in other platforms provided that the needed technology is available.

3.10 MOOC Infrastructure

The complete infrastructure to support the MOOL and its integration into a MOOC is composed of the following elements (Fig. 3.8):

- One or more Smart Device servers
- An HTML client application running in a browser
- A *cgi* interface for LTI authentication, database, and other services
- An edX server

3.10.1 Smart Device Server

The Smart Device server consists of the equipment to be controlled (Salzmann and Gillet 2013). In our example, it is an electrical drive and the computer that handles (i) the connection with the equipment through sensors and actuators, (ii) the local controller, (iii) the Web and WebSocket server, and (iv) the local "intelligence" that ensures that remote users behave adequately. The latter part also interacts with the storage services and the Smart Device that acts as the load balancer.

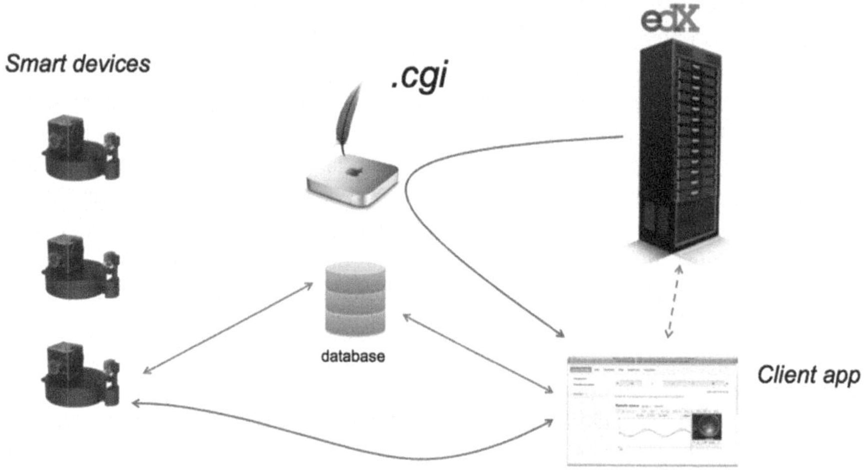

Fig. 3.8 The infrastructure with the Smart Devices, the cgi, the edX server, and the client application

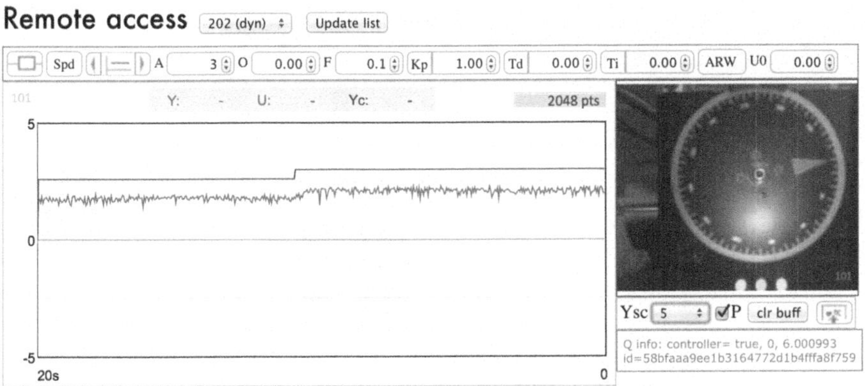

Fig. 3.9 With HTML5 client application, the control strip is made of the buttons and fields in the top part, the oscilloscope graphic on the left, the live video feedback on the right, and below video of the administrative part

3.10.2 Client Application

The client application is a pure HTML5 application that enables the user to fully interact with the Smart Device (Fig. 3.9).

It provides measurements displayed in an oscilloscope window, real-time video stream, a control strip that allows users to specify the reference or control signal, the control parameters, and the controller structure. The controller/observer modes are supported by this interface. When in controller mode, all elements are accessible; when in observer mode, the control strip is grayed out, and an indication of the waiting queue size is provided in the "administrative" part. It also provides a mean to save current measurements in the hosting environment. When in observer mode, the client application cannot save measurements and other data. This application can be used in a stand-alone mode or within a hosting environment such as a MOOC platform or an LMS.

3.10.3 The cgi Interface

The *cgi* interface is the cornerstone between edX and the external tools. When an external module is added to a MOOC session as an LTI (external) module, the *cgi* first validates the LTI encoded request containing the edX user ID and other information such as the user role, and then it servers the LTI module content that will be integrated as an iFrame in the edX page. In the proposed MOOC, the content is either a Web interface to access the Smart Device (Fig. 3.9) or one of the Sysquake tools for data analysis or controller design (Fig. 3.11).

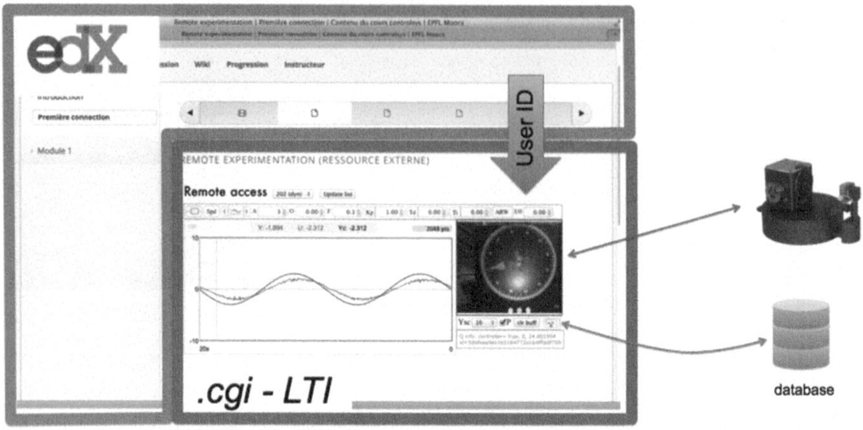

Fig. 3.10 The edX MOOC hosting an external LTI module (served by the cgi), which hosts the interface to the Smart Device. Information such as edX user ID is provided to the LTI module

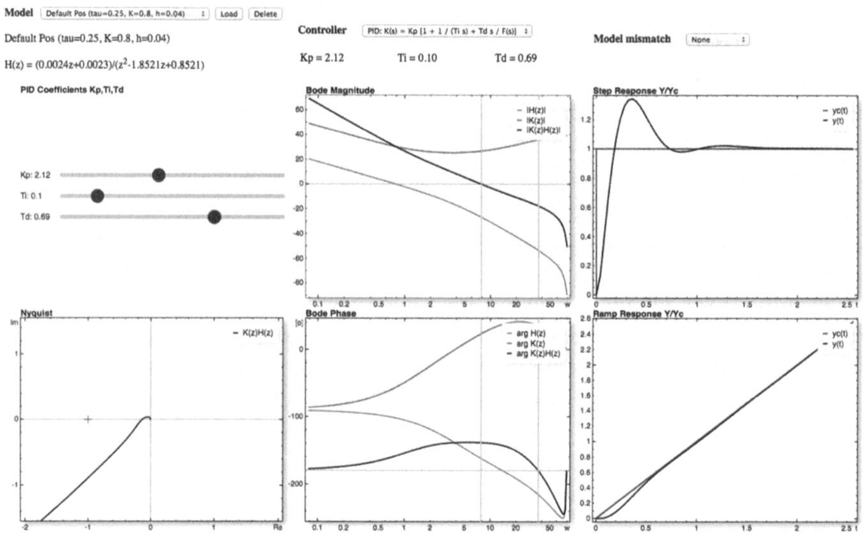

Fig. 3.11 The pure HTML5 simulation tool for PID design generated by Sysquake

The client application is integrated in the hosting environment (edX MOOC) as an LTI module (Fig. 3.10). The LTI specifications permit to exchange information such as user ID between the hosting environment and the client application. It also ensures that the provided information is not tempered by signing the transmission with a shared secret. LTI carries the advantage of being widely used. This solution proposed for *edX* could also be ported with little effort to other environments supporting LTI such as *Moodle*.

3.10.4 The edX Server

A MOOC can be hosted by the edx.org consortium; alternatively an institution can install its own instance of edX which is open source. It was decided to host the MOOC in our institution for the first run. This enables a tighter control of the MOOC and an agile development. The following iterations of the MOOC will be hosted by edx.org.

3.10.5 Additional Services

3.10.5.1 Saving

edX can remember where the student stopped the last time she/he was connected; other information such as numerical values that were entered in edX are also stored. On the other hand, the proposed LTI modules are stateless, and edX's currently implemented version of LTI does not propose a mechanism to store/retrieve data from/to edX.

A database service has been developed outside edX's infrastructure to provide a private storage to each user. The database user identity is provided by edX and passed securely to the LTI tools via the *cgi* interface. This database service is accessible only by the LTI tools. Users can save various kinds of information: measurement, model parameters, controller parameters, etc. The saved information is JSON encoded. Metadata are attached to the saved information. These metadata permit LTI tools to filter files to display to the user. At this time, the saving service is more of a database than a complete file system. For instance, users cannot retrieve the saved data or uploaded data located in their computers. An evaluation of students' needs will be conducted to see if additional features are required.

Saving information between steps is a key element to enable continuity in a MOOC session. Section 3.5 provides an example of such a session.

3.10.5.2 Data Processing

Various Web tools are proposed to process data; other tools offer simulations or controller design (Fig. 3.11). Similarly, to the Smart Device client application, these tools are encapsulated as LTI modules. These interactive tools are generated by Sysquake (2017) – a simulation engine compatible with Matlab syntax that generates pure HTML5/JavaScript applications. These lightweight tools run in the client browser and do not require any computation on the server. Since they are encapsulated in an LTI module, these tools also have access to the user private space to store or retrieve information.

3.10.5.3 Load Balancing

The wealth of users is split among the pool of Smart Devices. For the current MOOC, there are 23 identical Smart Devices. One of the Smart Devices is dynamically chosen to act as the load balancer. Each Smart Device broadcasts information about its status and the number of waiting clients. In collaboration with the Smart Devices, the load balancer guides the client to the Smart Device server that has the smallest number of clients waiting to act.

3.10.5.4 Analytics

Optimizing the user experience as well as the load balancing requires information about both the current status of the MOOL (Smart Devices) and the MOOC. Both environments are instrumented to provide live analytics. The Smart Device specifications propose such service and do not require any modification; on the other hand, the analytics recipient needs to be implemented. The live analytics processing and analysis may provide useful information to both the load balancer and the MOOC server.

3.11 Learning Scenario

The proposed infrastructure is currently exploited and validated for the first MOOC fully dedicated to hands-on activities. It is offered at the Swiss Federal Institute of Technology Lausanne (EPFL) to students of mechanical engineering, micro-engineering, as well as electrical engineering. The video parts of the MOOC describe how to use the various tools (client application, simulation, etc.), the lab infrastructure, and the experiments to be performed.

The learning scenario of each session is split into short phases (5 min each), which follows a common structure: a short video describing the experiment (with a theoretical recall if needed), a set of hands-on exercises using the remote lab clients integrated in edX, data pre- or post-processing using interactive simulation tools, a set of numerical questions to validate user's finding, and another set of open questions based on the user observations to evaluate user's understanding (Fig. 3.12).

For example, a scenario where the user has to design a controller for a given equipment could have the following sequence:

(i) Watch the introduction video, which provides a theoretical recall (e.g., loop shaping), and explain the steps to be performed in the current session.
(ii) Perform a step response on the real equipment and save the measurements. During this step, a remote connection to the Smart Device is established.

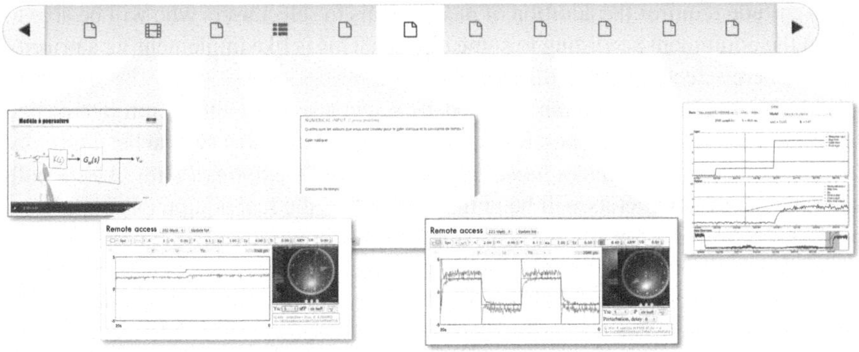

Fig. 3.12 A sequence of steps in edX

(iii) Process the measurements using the temporal fit tool (generated by Sysquake) to identify key parameters (model transfer function); save the transfer function.
(iv) Enter the transfer function parameter in edX for self-validation.
(v) Using the identified transfer function, synthetize a controller according to some specifications. With the interactive PID loop-shaping tool (Fig. 3.11), load the identified transfer function and design your controller. Simulate various cases using the provided tool. Save the controller parameters.
(vi) Test and validate the proposed controller on the real equipment through various experimentations.
(vii) Add a perturbation on the real equipment, and test that the design controller is robust to these perturbations. For example, an additional load (resistance) can be added on demand; alternatively delays can be added to the system.
(viii) Answer additional comprehension questions.

In the above scenarios, the remote experimentation is accessed in steps (i) and (v). A specific experimentation duration is defined for each task to be performed. This parameter, defined by the MOOC authors, facilitates a tight time control by setting an appropriate time slice for each experiment. The smaller the time, the larger the number of users per unit of time. Note that the current access policy permits a user to stay connected as the controller until another client connects. The other steps (ii)–(iv) do not require timing since the Sysquake Web tools used for analysis and simulation run on the client browser without live connection to the server.

3.12 Massive Access and Gamification

With the proposed actions (short activity time, controller/observer mode, FIFO queuing, and equipment duplication), a ratio of about 5–10:1 has been achieved in the control system lab MOOC. Increasing the number of users by another order of magnitude requires the addition of mechanisms to select users who will be able to act on the equipment according to some criteria. This is like implementing a priority queue where selected users will wait less time.

A gamification process can be used to select users. In this controller design scenario presented above, a possible selection criterion for step (v) can be: *how close are the controller parameters found by the user from the optimal ones?* Users with a correct set of parameters will be entitled to access the real equipment for testing, while users with demonstrably bad answers will be redirected to simulations. This simulation will highlight the problems with the proposed parameters. Similarly, such simulations could exhibit theoretical behaviors that would normally destroy the physical equipment. The proposed tools and infrastructure, especially the Sysquake JavaScript simulations, support such gamification scenarios.

Live analytics will help balance the load over time; while there is a preferred path to follow regarding the experiments order, it would thus be possible to switch experiments order dynamically according to Live analytics results. For example, if the system is heavily loaded, an experiment with a shorter duration may be proposed first.

Similarly, as the time to act on the Smart Device can be controlled, it could be reduced or increased according to the current load. This experimentation time is predefined by the MOOC teachers; having live analytics will permit a better initial duration guess.

Traditionally, reservation is the preferred method to guaranty access to a shared resource at a given time. Such a mechanism is cumbersome to manage and to explain to users; in addition it can be very time-consuming to implement. While the proposed Smart Device has the needed mechanism to support reservation, it is not used for the above rationales. Also such a mechanism is antagonist to the MOOC idea that a user should perform sessions at his/her own will.

A substitute mechanism is to inform the user about the best time, i.e., the period at which the waiting time is the smallest, to connect to the equipment on a global scale. The Web client application reports the waiting time for the current Smart Device as well as the waiting time for the other Smart Devices in its neighborhood. Only the instantaneous waiting time is provided. Providing trends over the week would permit the user to select the most appropriate experimentation time.

Some of the above mechanisms are currently implemented and tested. The final analysis will be conducted at the end of the fall semester 2016.

3.13 Initial Results

Preliminary evaluation has been conducted. The main conclusion is that students agree to wait to access a shared resource if the waiting time is provided. During the initial experimentations, the waiting time was less than 6 min, which indicated less than three users were waiting. If the waiting time information is not provided or

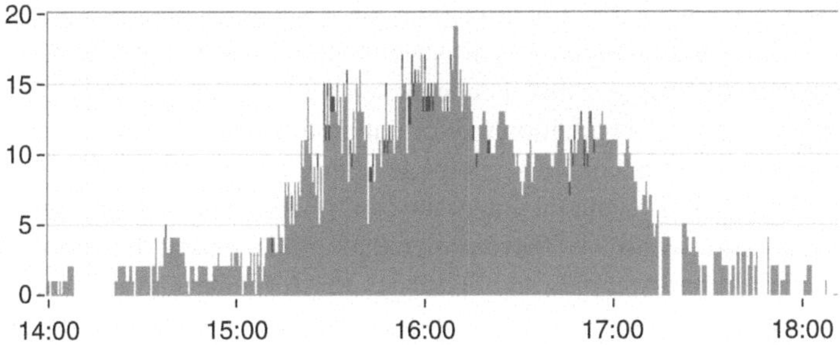

Fig. 3.13 Server loads during a planned session (15:15–17:00), 21 Smart Devices were available

incorrect, then students are not willing to wait for a "random" time and complain; we learned it the hard way.

Also, even though students could switch manually to a resource with shorter waiting time, they prefer to remain on the same resource and wait for their turn; this is probably due to the structure of the MOOC where they can perform another tasks while waiting. Switching to another task and coming back to the original remote experimentation client will put the user at the end of the waiting queue, this limitation has been quickly overcome by students who open two tabs on their browser.

Figure 3.13 shows a typical server load during a planned session with 120 students. There is a total of 21 Smart Devices. The load balancer assigns free Smart Devices to new comers. The *gray bars* represent the number of used Smart Devices at a given time. The *blue bars* represent the number of waiting users for a specific Smart Device. This picture shows that even though there are free Smart Devices (the highest *gray bar* value is 19), users prefer to wait for a given Smart Device. Also, the highest *blue bar* is 3, which represents the number of people waiting for a specific Smart Device. Outside the planned sessions, the load of the servers was rarely above 5. Figure 3.13 also shows that the available infrastructure can easily sustain the current planned MOOC sessions with a ratio of 6:1.

3.14 Conclusion

This chapter is split to two parts. The first one details the Smart Device specifications for remote experiments, and the second one presents how they are used to support Massive Open Online Lab within a MOOC. We first summarized the Smart Device paradigm and its application to remote labs. From a client or external service point of view, the Smart Device is described through well-defined services and functionalities. Services permit to access the inputs and outputs of the Smart Device, such as sensors and actuators. Functionalities refer to the provided internal behavior such as range validation for an actuator. The main goal of these specifications is to define the services and functionalities of a Smart Device using Swagger, a JSON-based description language. This specification is sufficiently detailed, thanks to the properties of Swagger that a code skeleton for the client application can be machine generated without additional information from the lab owner. Furthermore, the shared specifications enable a complete client-server decoupling by enabling interoperability, thus allowing the integration of Smart Devices in any environment, OS, or device. Additionally, we have shown that implementing the specifications is feasible by providing examples and templates for developers to get started. Some technical assumptions are made when considering the client application for remote labs. The first one implies that the client resides typically in a recent Web browser that runs on a tablet; this implies a plug-in-free solution. In addition, the means to exchange information between the client and the server is made using JSON-encoded messages that are transmitted using asynchronous WebSockets.

In the second part, we explained how a set of Smart Devices can be combined to form a MOOL that can be integrated in a learning environment. We first covered the technical aspects of this implementation using the Smart Device paradigm, HTML5 client application, LTI, interactive tools, user storage, and a load balancer. Then, we presented a way to rework classical hands-on lab session in order to accommodate a new MOOC paradigm where many users access critical resources at the same time. The MOOC infrastructure has been detailed with an example MOOC learning scenario. Finally, gamification processes combined with simulations are proposed to elect users entitled for a direct access to the real equipment. The proposed solution is currently implemented in the *Control Systems Lab* MOOC offered in the university instance of edX to more than 200 bachelor students in mechanical engineering, micro-engineering, as well as electrical engineering at EPFL. It is the first MOOC at EPFL fully dedicated to hands-on lab activities.

Acknowledgments The authors acknowledge the support provided by EPFL for the development of the control system MOOC described in this paper. The Smart Device specifications were partially developed in the context of Go-Lab (grant no. 317601) project under the ICT theme of the Seventh Framework Programme for R&D (FP7). We would like to thank Anjo Anjewierden, Lars Bollen, August'ın Caminero, Manuel Castro, German Carro, Gabriel D'ıaz, Danilo Garbi Zutin, Miguel Latorre, Irene Lequerica Zorrozua, Pablo Ordun˜a, Yves Piguet, Antonio Robles, Miguel Rodriguez-Artacho, Hamadou Saliah-Hassane, Elio San Cr'ıstobal, and Simon Schwantzer (in alphabetical order) for their input during the numerous discussions leading to the presented specifications.

References

Auer, M., Pester, A., Ursutiu, D., & Samoila, C. (2003). Distributed virtual and remote labs in engineering, In *Industrial Technology, 2003 IEEE International Conference on, vol. 2*, (pp. 1208–1213).

Cascado, D. et al. (2011). *Standards and implementation of pervasive computing applications* (pp. 135–158). Hoboken: Wiley

Chen, X., Song, G., & Zhang, Y. (2010). Virtual and remote laboratory development. *A Review, Proceedings of Earth and Space 2010, 1*(1), 3843–3852.

De Jong, T., Sotirou, S., & Gillet, D. (2014). Innovations in stem education: The go-lab federation of online labs. *Smart Learning Environments, 1*(1), 3. Retrieved from: http://www.slejournal.com/content/1/1/3.

Freed, N., Baker, M., & Hoehrmann, B. (2014). Media types, *Technical Report, 2014*. Retrieved from: http://www.iana.org/assignments/media- types/media- types.xhtml

Furht, B. (1995). A survey of multimedia compression techniques and standards. Part i: JPEG standard. *Real-Time Imaging, 1*(1), 49–67. Retrieved from: http://www.sciencedirect.com/science/article/pii/S1077201485710054.

Gillet, D., de Jong, T., Sotirou, S., & Salzmann, C. (2013). Personalised learning spaces and federated online labs for STEM education at school: Supporting teacher communities and inquiry learning. In *Proceedings of the 4th IEEE Global Engineering Education Conference (EDUCON),* (pp. 769–773).

Govaerts, S. et al. (2013). Towards an online lab portal for inquiry-based stem learning at school. In *Advances in web-based learning ICWL 2013, ser.* (Lecture Notes in Computer Science, vol. 8167, pp. 244–253). Springer: Berlin/Heidelberg

Hadley, M. J. (2009). Web application description language (WADL*), Sun Microsystems Inc., Technical Report,*. Retrieved from: http://java.net/projects/wadl/sources/svn/content/trunk/www/wadl20090202.pdf

Halimi, W., Salzmann, C., Jamkojian, H., & Gillet, D. (2017). Enabling the auto generation of user interfaces for remote laboratories. *14th International Conference on Remote Engineering and Virtual Instrumentation (REV),* New York, USA.

Hypertext transfer protocol (HTTP) status code registry (RFC7231). (2014). *IETF, Technical Report* Retrieved from: http://www.iana.org/ assignments/http-status- codes/http-status-codes.xhtml.

IEEE. (2007). IEEE standard for a smart transducer interface for sensors and actuators wireless communication protocols and transducer electronic data sheet (TEDS) formats, *IEEE Std 1451.5–2007*, 10 2007.

Loreto, S., Saint-Andre, P., Salsano, S., & Wilkins, G. (2011). Known issues and best practices for the use of long polling and streaming in bidirectional HTTP, *RFC 6202, Internet Engineering Task Force, Apr. 2011*. Retrieved from: http://www.rfc-editor.org/rfc/rfc6202.txt

Lowe, D., (2014). MOOLs: Massive open online laboratories: An analysis of scale and feasibility. *REV2014: 11th International Conference on Remote Engineering and Virtual Instrumentation*. Piscataway: (IEEE) Institute of Electrical and Electronics Engineers.

Marum, M. (2013).Opensocial 2.5.1 specification. Retrieved from: https://github.com/OpenSocial/spec

Orduna, P. et al. (2014). Generic integration of remote laboratories in public learning tools: organizational and technical challenges. In *Proceedings of the IEEE 2014 Frontiers in Education Conference,* October 2014.

Salzmann, C., & Gillet, D. (2008). From online experiments to smart devices. *International Journal of Online Engineering (iJOE), 4*(SPECIAL ISSUE: REV 2008), 50–54.

Salzmann, C., & Gillet, D. (2011). Remote labs and social media: Agile aggregation and exploitation in higher engineering education. *IEEE EDUCON Education Engineering, 2011*, 307–311.

Salzmann, C., & Gillet, D. (2013). Smart device paradigm standardization for online labs. *IEEE EDUCON Education Engineering, 2013*, 1217–1221.

Salzmann, C., Gillet, D., & Piguet. Y. (2016). MOOLs for MOOCs: A first edX scalable implementation. *13th International Conference on Remote Engineering and Virtual Instrumentation (REV),* Madrid, Spain.

Swagger RESTful API specification. (2014). Retrieved from: https://github.com/wordnik/swagger-spec/blob/master/versions/1.2.md

Sysquake. (2017). Is a product of Calerga, www.calerga.com

Tawfik, M., et al. (2014). Laboratory as a Service (LaaS): A novel paradigm for developing and implementing modular remote laboratories. *International Journal of Online Engineering, 10*(4), 13–21.

Taylor, B. N., & Thompson, A. (2008). The international system of units (SI*), National Institute of Standards and Technology,* NIST Special Publication 330.

Thompson, C. W. (2005). Smart devices and soft controllers. *IEEE Internet Computing, 9*(1), 82–85.

WebSocket specification. (2011). *rfc6455*, Retrieved from: http://tools.ietf.org/html/rfc6455

Christophe Salzmann is senior research associate and lecturer at the Ecole Polytechnique Fédérale de Lausanne (EPFL), Switzerland.

Wissam Halimi is PhD candidate at the Ecole Polytechnique Fédérale de Lausanne (EPFL), Switzerland.

Denis Gillet is MER and Head of the REACT group at the Ecole Polytechnique Fédérale de Lausanne (EPFL), Switzerland.

Sten Govaerts was post-doctoral researcher (2012–2105) at the REACT group at the Ecole Polytechnique Fédérale de Lausanne (EPFL), Switzerland.

Chapter 4
Augmented Reality and Natural User Interface Applications for Remote Laboratories

Ananda Maiti, Mark Smith, Andrew D. Maxwell, and Alexander A. Kist

Abstract Augmented reality (AR) has a great potential in creating rich user interfaces where users can view and interact with virtual objects. AR can have both passive objects and active objects. The former do not respond to interaction; the latter can be altered in their orientation, shape and position in relation to other virtual objects, for example. Remote laboratories (RLs) enable access to equipment and experiments via the Internet. This chapter focuses on the use of virtual objects in remote laboratories as an alternative, immersive user interface. Having an AR environment allows users to interact with experiments as virtual objects enabling hands-on experiences. This is made possible by using specialised natural user interface (NUI) devices. These devices can capture the natural movement of users and apply them to virtual objects. This chapter considers the role of AR and NUIs in the context of remote laboratories. It provides the context for AR and NUIs and discusses examples of systems that can be used. It demonstrates how users can interact with remote laboratories in these environments. NUI is part of the curriculum at the undergraduate level. This cyber-physical environment provides an ideal context to teach human-computer interaction (HCI). The first two sections of this chapter describe AR and using NUI in the RL environments. The last section introduces a practical example of using NUI and RL to teach HCI.

Keywords Augmented reality · E-learning · Remote laboratories · Computer vision · Virtual reality · Human-computer interface · Kinect · 3D sensors · Gesture recognition · Computer networks

A. Maiti · M. Smith · A. D. Maxwell · A. A. Kist (✉)
University of Southern Queensland, Toowoomba, Australia
e-mail: anandamaiti@live.com; mark.smith@usq.edu.au; andrew.maxwell@usq.edu.au; kist@ieee.org

M. E. Auer et al. (eds.), *Cyber-Physical Laboratories in Engineering and Science Education*, https://doi.org/10.1007/978-3-319-76935-6_4

4.1 Introduction

Augmented reality (AR) is a technology that enhances user visual and auditory perception by providing information which the users' senses cannot directly detect (Azuma 1997; Smith et al. 2016). Mostly, AR embeds additional information in video/audio streams to create a richer interactive user interface (UI). Embedded AR components can comprise of simple text or highly complex graphics and visuals. AR technology typically reacts to the changing surrounding environment, i.e. responds with AR components depending upon the visual or tracked inputs to the system.

AR components typically need to update in real time by recognising the input video frames' contents, processing it according to a predetermined logic, and then produce the AR response components. Several other factors and input components within the target environment such as image tags, room telemetry, internal GPS and inertial, magnetic tracking can also be collected and analysed to determine the AR output to produce an immersive experience for the users. Apart from objects, gestures from users themselves can be part of the environment (Oshita and Matsunaga 2010). AR is used in many areas of science and technology including computer games for recreational purposes, sports and entertainment, navigation and tourism. AR has also been used in education.

Natural user interfaces (NUIs) (Bruder et al. 2009; Liu 2010) are mechanisms to take input from users without using a fixed position (i.e. immobile) or dimensionally restricted input devices (e.g. keyboards). These are restricted in movement and only allow text or button inputs. Mice have also a limited degree of freedom. NUIs aim to take inputs from the natural movement of the users, particular gestures of their body parts. Typically, a NUI aims to incorporate computer vision processing to identify and track whole body, hand or finger movement gestures as inputs to systems.

NUIs utilising 2D cameras have been investigated for a while (Popa et al. 2015), but with recent advances incorporating depth sensors (e.g. Kinect) and other devices such as Leap Motion, it has become easier to capture complex body gestures with high accuracy. NUIs can be easily integrated with AR or virtual reality (VR) (Muller et al. 2012) in a way that users do not have to rely on conventional means for interaction (Ohta and Tamura 2014). They can directly engage with objects in AR. NUI is very common in the computer game industry and has helped to deliver very cost-effective and commonly available solutions.

AR/VR mixed reality (Müller 2009; Ohta and Tamura 2014) can run in two modes: as a *desktop mode* where the AR/VR environment and feedback are displayed on a screen and another way is using a *full AR mode* head-mounted display (HMD) to see a complete 360° view of the environment augmented with virtual objects (Barfield 2015).

It may be noted that the focus of this article is on desktop-based environments (Chang et al. 2014; Zhang et al. 2013). Comparison with an HMD-based full AR is discussed if applicable. While desktop AR has been investigated in the past, HMDs are still new, and applications regarding their use are being developed.

AR and NUI have been extensively researched. They have been applied in various ways in education and gaming industries (Jara et al. 2011; Restivo and Cardoso 2013; Wu et al. 2013). Virtual environments feature heavily within game-based learning systems (Callaghan et al. 2013). Gamification incorporates gamelike processes to create a rich learning environment that engages the user in pedagogical outcomes. Creating an ornate virtual visual experiment for the user to experience can become resource intensive. Gamification implementations require large teams over many months (Callaghan et al. 2015). Within virtual laboratories, the use of HMD (Restivo et al. 2015) and haptic devices is expanding because of their suitability to seamlessly integrate into the virtual environment. Within AR systems, providing reliable synchronisation is the most difficult process which can be addressed by using generic physics or game engines to enable smooth interactions with virtual objects. This chapter discusses techniques to use AR/VR and NUI for Remote Access Laboratories (RALs), an example of cyber-physical systems where remote experiments are controlled through the Internet by students for learning purposes. Common augmented reality applications in remote laboratories and engineering education have focused on:

- Displaying specific information (Mejías Borrero and Andújar Márquez 2012; Menezes et al. 2017; Olalde Azkorreta and Olmedo 2014; Restivo et al. 2014a, b, c; Smith et al. 2016). These applications have allowed users to view the experiments as well as additional data to explain events and the status of the experiment.
- Some applications allow interactions with virtual objects based on conventional devices such as mouse and keyboard (Andujar et al. 2011). These virtual objects represent input and output components corresponding to the experimental setup. Users can alter these objects to provide inputs for the experiments. At times, this type is combined with displaying information.

Another issue is that most remote laboratory experiments lack in providing sufficient hands-on or immersive experiences for students (Corter et al. 2004). With AR environment, the students can have greater levels of interaction with the setup of the experiment in the form of virtual objects. This enables hands-on experience to an extent, depending upon the quality of the interface of the experiment. The contents of this chapter can be summarised as follows:

- Partial issues of using AR and NUI are discussed in regard to devices such as Kinect and other wearable devices such as smart glasses. The key focus is on their use in laboratories in engineering education courses.
- Methods of implementing experiments using Kinect and Leap Motion for exchange/streaming of data for experiments are also discussed.
- Advantages and challenges of using these technologies for education in engineering laboratory courses are also discussed.

The first two sections describe augmented reality and using NUI in the RAL environments. The last part of the chapter focuses on teaching human-computer interaction (HCI) and NUI using remote laboratory technology. An example is presented where remote users log in to a Windows platform and use LabVIEW to program HCI applications. This allows the users to learn how to deal with data streams from NUI devices such as Kinect.

4.2 Augmented/Virtual Reality with NUI in RAL: Overview

An overview of AR/VR with NUI in RALs is shown in Fig. 4.1. It consists of several components involving acquiring video feedback, constructing a virtual 3D environment with the video feedback, acquiring user inputs through NUI devices, capturing gestures from the users' interactions and issuing corresponding commands to the RAL experiments. These components are described in details below.

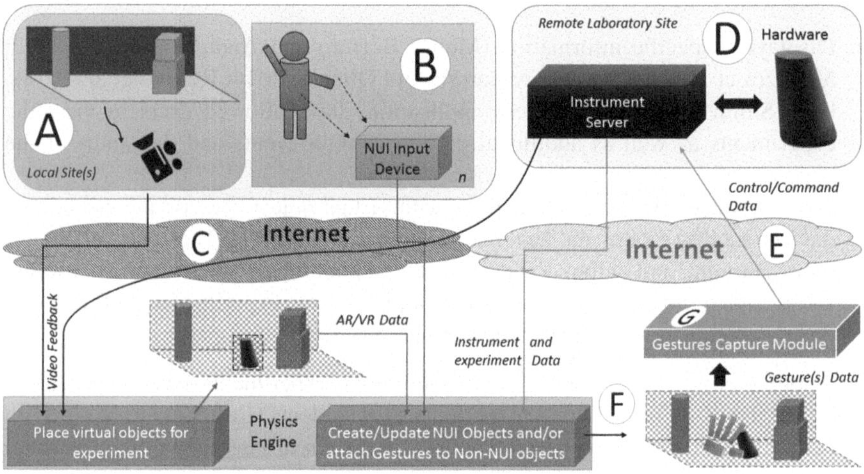

Fig. 4.1 The complete system architecture of the AR/VR environment with NUI inputs. (**a**) Video feedback from a local site. This is optional and required if using HMD. Not required for desktop. (**b**) NUI device input. The device can be Kinect, Leap Motion, etc. Optional if using simple AR. (**c**) Internet for transmission of NUI data. This is optional, required only if there are multiple users in the same experiment session. (**d**) Remote laboratory side. (**e**) Internet for transmitting the command control data and retuning experiment data to the user's site. (**f**) This is a software module which ideally runs in the local user site, but this may be implemented in cloud environment as well. (**g**) A software module to capture gestures from the physics engine generated augmented/virtual environment

4.2.1 Video Feedback (A)

If the users use a head-mounted display (HMD), the very first step is to capture the video from the local environment. If using the AR/VR in a desktop mode, there may be no need to capture the local scene. The real-world video feedback is augmented with virtual objects using a physics engine. The virtual objects are downloaded from the remote laboratory site and correspond to the experiments. In the augmented feedback, many of the real-world objects may be replaced with virtual objects to create an augmented virtual reality. In a desktop mode, the video feedback may come from the instrument server directly, which can be overlaid with information.

4.2.2 NUI Device Input (B)

In the local/user/student site, users use the NUI devices to create NUI data. Examples include Kinect (Moore et al. 2016) and Leap Motion (Weichert et al. 2013). These devices generate a constant stream of data at the minimum rate of ten frames per second. Each frame contains an array of data corresponding to the input points of the desired body part. In the case of Microsoft Kinect, the input points are up to 20 joints/bones of the user's body. In the case of Leap Motion, it is 30 points regarding bones of the fingers.

If there are multiple users participating in a collaborative session from different locations, the NUI data must be exchanged between them through the Internet. This requires a *transmitter* program to stream the data over the Internet. This also requires reducing the rate of data from the devices as only a limited number of frames can be transmitted effectively with considerable latency on the Internet.

4.2.3 Remote Laboratory Site (D)

This consists of the remote laboratory management system with the instrument server for executing commands and returning data to the user. The remote lab is connected only to the Internet.

4.2.4 Physics Engine or AR/VR Engine (G)

A physics engine or an AR/VR engine allows the developers to create virtual objects. These environments allow the developers to create complex virtual objects based on primitive geometric objects such as spheres, cylinders and boxes. The physics engine allows the objects to behave in a realistic manner as if they were in a real world with gravity and other laws of motion. Some virtual objects may be added that may not be part of the experiments, and do not have the physics engine attached to them, but to hide unwanted objects in the video feedback.

The physics engine creates a 3D environment corresponding to the video feedback. If using HMD, the environment typically consists of the video feedback. If mainly virtual objects are used, the 3D environment becomes a virtual 3D gamelike environment with little real-world data (Zhang et al. 2013). There are two types of virtual objects: the passive virtual objects corresponding to the experiment objects, e.g. a box, and the NUI virtual object which is active, i.e. all changes in the 3D environment are initiated by the NUI virtual objects, e.g. a 3D hand. In the case of HMD-based AR, the NUI virtual objects may be shown to the users. For example, a 3D hand may appear in a desktop AR to simulate the user's hand but may not appear in an HMD-based AR.

The motion or the change in orientations of the objects generate certain 'gestures' which in turn generate 'commands' that are passed to the instrument server. A Gesture Control Module (GCM) does this.

The physics engine or the AR/VR module is a software module that is ideally run in the users' computers or mobile devices. This module may be run partly as a cloud service. This module updates the users' *scene*, i.e. what the part of the 3D environment the user see (either on a desktop screen or through an HMD) when new information is available from the instrument server upon executing any command.

4.3 Part One: Augmented Reality

Augmented reality methods are employed within the Remote Access Laboratories framework to provide users with a rich immersive experience so that the disadvantage of non-proximal resource usage is counterbalanced. Providing the user with an enhanced reality such that they experience reality with computer augmentation becomes much harder to develop than virtual environments. Interacting with human sensory systems provides immersive qualities to the user, if and only if the artificial sensory information is aligned and synchronised with the real environment. In an AR, should any aspect of the computer generating feedback fail to merge in an appropriate manner with the environment, the user loses the immersive quality, and the benefits of the systems are lost. To work in this environment, the infrastructure needs of AR must be understood as described in this section.

Augmented reality systems must provide users with a sensory enhancement to the information they already have, and users must be able to interact with this enhancement. The easiest sense to manipulate is vision. Most of our work with computers is done with our eyes, and it has become our primary source of information input. Not only can our visual sense fool our perceptions of status of the real world; it is a simple matter to expand human-computer interface via enhancements to information we view.

Current AR systems rely heavily on our visual senses. The majority of research into AR furthers applications that utilise our eyesight. Augmented reality for recreational purposes including games and entertainment has a heightened level of control over the users' environment. Recent games, such as Pokemon Go™, rely on GPS coordinates and the cell phone compass orientation to define the location of the desired objects. From that point, the location dataset implements a low technological graphics rendering process to superimpose the object on the users' phone. Location accuracy is not a factor for AR applications such as these, where the virtual object is not critical to the real-world surroundings. Other AR uses, such as in the medical field (State et al. 1996), require extremely accurate alignment between real and virtual objects. Guiding surgical tools in real time cannot suffer data dropouts, frame loss, misalignment or tracking errors.

Augmented reality systems require several core components in order to provide the enhanced visual services. They depend on the level of interaction the user has with the real-world environment. For many systems, such as entertainment applications, real-world objects that appear with the video scene are part of the users' interaction. AR implementations only provide visual enhancements to simulate elements within a game. The actual environment is (technically) irrelevant.

4.3.1 Core AR Systems

For AR to comprehend and interact with the physical attributes of the real world, an understanding of the real world is required. Core components of an AR system are essential to ensure that the environment is correctly interpreted. The minimum set of services varies from installation to installation. For visual AR systems, the set can be broadly defined as video capture, object identification and tracking.

- *Video capture* of live video streams is an essential component of all AR systems. The video stream is the baseline dataset, and no understanding of the real environment is possible without quality or reliable video data. All real and virtual objects must be coordinated as they meet our eyes. Without proper processing of the video stream, numerous errors become introduced into the resultant feedback. Azuma (1997) has detailed the primary difficulties including many hardware- and software-induced issues.
- *Real object identification* (ROI) within computer vision (CV) systems is essential to interpret the images that appear in a video stream. Without the ability to interpret the scene from a video stream, AR capabilities become ineffectual. Interpretation of objects within the video scene comprises the bulk of the image processing workload. Key aspects of ROI provide the bulk of AR processing, consuming ICT resources as the detail and complexity of the dataset increase. Computer vision techniques, such as probability density functions, frame subtract and clustering, are all useful tools to extract meaningful information from the

sequence of live video frames. Identifying objects of interest with a live video stream is currently the subject of much research.

- Identifying an object is only part of the AR core components. *Object tracking* must work in conjunction with ROI. Tracking real objects provides AR systems with knowledge of the scene and the dynamic nature of attributes within the scene. Without the ability to track the identified object, there is no knowledge if the object has moved in and out of the scene or adjusted its visible surface area, for example, so the results are ineffectual. Object tracking must continue to recognise the object and the changes within its position, orientation or other parameters as the scene progresses.

4.3.2 Augmented Reality and Remote Access Laboratory Integration

Utilising AR with a Remote Access Laboratory environment creates a unique set of conditions to overcome depending upon the nature of the experiment (Cubillo et al. 2012; Odeh et al. 2013; Restivo et al. 2014a, b, c). Augmented reality services need access to the key aspects of the remote laboratory so as to interact seamlessly. Laboratory systems generally consist of the experiment equipment which is required to be operated remotely and the measurement and sensor data. The crucial requirements include the interfaces and timings.

Remote Access Laboratories are both the sources and sinks of data. Live video stream, the key aspect of any RAL systems, is provided to the AR service. Figure 4.2 shows how the live video stream is used by both the object detection and tracking systems but is also providing data back into the video stream through the virtual objects. Sensor data is important to allow the actions and measurements of the unit under test to undergo interpretation and feedback into the video feed if required. This is the primary interface between the two systems and requires careful management.

The discrepancy between the live video data appearing from the RAL experimental rig and the image overlays supplied by the AR services is finite and influences the user's immersion in the AR experience. Any delays introduced by the AR services must be kept to a minimum so as to maintain the AR synchronisation.

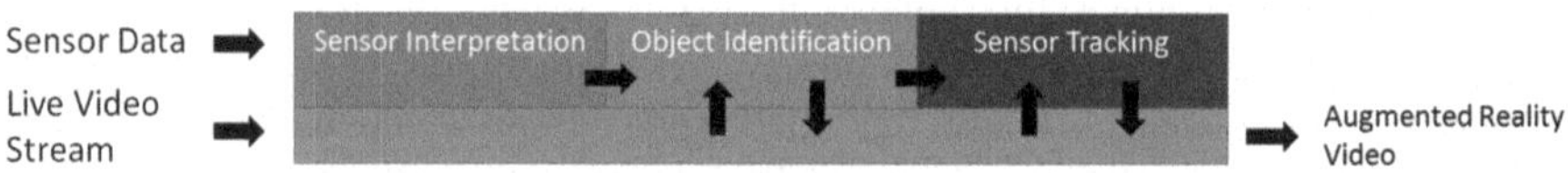

Fig. 4.2 Core components to a visual AR system

4.3.3 Augmented Reality Issues

Implementing AR for visual sensors requires overcoming two main problems: registration and timing errors.

Registration errors occur when real and virtual objects do not align or are not synchronised with each other. Figure 4.3 demonstrates a simple problem of the virtual pressure reading, not aligned with the pressure gauge (circled in red). The error occurs for several reasons such as camera calibration or position changes, optical distortion or tracking errors. The human brain is able to be deceived by what our sensors are receiving. However, only very small registration errors are needed before our brain finds the situation disconcerting and the immersive effect is lost. Consequently, registration errors have to be minimised. For AR systems where registration errors must be kept to a minimum, computer vision (CV) methods are employed. Key reference points, or fiducial markers, help to maintain stable image orientation which minimises registration errors.

Timing errors occur when delays in network traffic or delays from video processing and rendering cause the real and virtual objects to shift out of synchronisation. Like a TV broadcast where the audio delays the video, the effect completely destroys the benefits AR brings to a system. Timing problems can be minimised by coordinating the real video stream and virtual objects. Delaying key video frames until AR systems have completed processing provides simple synchronisation. Live video streams operating at 20 to 30 frames per second can incorporate virtual objects

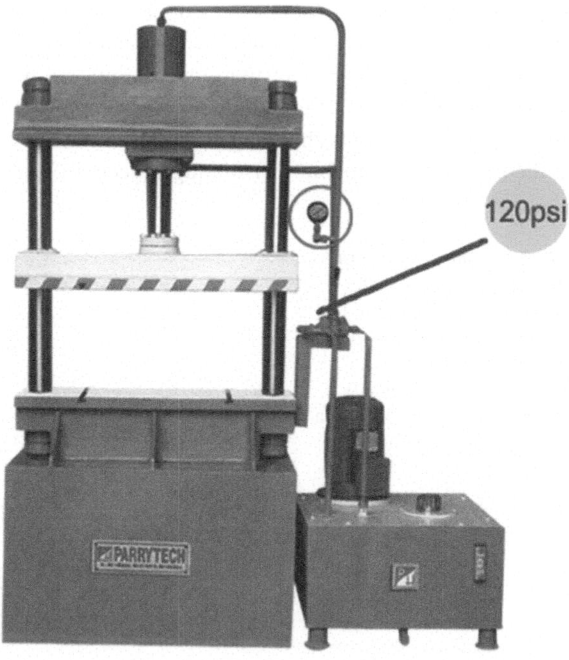

Fig. 4.3 Registration errors cause the real and virtual objects to be misaligned. The pressure dial (real) does not align with the virtual indicator

in every second to the fifth frame without the user being aware of any timing issues. Care must be taken, so that large scene changes occurring between the skipped frames do not create ghosted virtual images. Ghosted images can easily appear as a consequence of old images that belong to frame data that is now obsolete.

4.3.4 Computer Vision

Visual AR systems incorporate substantial CV processes in an attempt to understand the live video scene. Whether the AR system relies on fiducial markers (Maier and Klinker 2013) or the discovery of objects within the scene through other methods, CV models are an important aspect.

Real object identification and object tracking rely on CV models for the identification of segments within the video frames. To make AR within RAL friendly to lab creators and users alike, CV models must be capable of contending with complex images but with very little prior knowledge of the scene. Three generic CV models are leading the field as suitable candidates for AR/RAL systems, such as frame subtraction, clustering and statistical analysis.

Frame subtraction is a popular model used in current security and tracking systems and relies on assuming that objects of interest within a video scene are moving or changing their attributes from frame to frame. Figure 4.4 demonstrates the method: video frames are subtracted from a reference frame (usually the previous frame), where a changing object can be isolated. The green cross in Fig. 4.4 has moved and is the only object remaining in the resultant frame. It is assumed that the static regions of the frames simply cancel each other out. The reality is more complicated with little chance that static regions within a frame are actually static. A video camera's output rarely produces clean signals. They suffer from compression losses, poor resolution, changing lighting conditions and jitter. Frame subtraction, in theory, is quite capable of isolating objects of interest, but the reality is that it

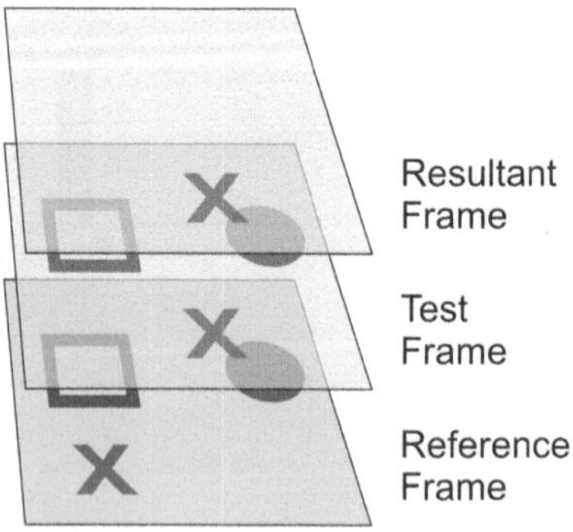

Fig. 4.4 Frame subtraction example. The frame under test is compared with the reference frame. Pixels that have not changed are removed from the scene, leaving the X as the only data in the resultant frame

needs help. It is more suited for use with other CV methods and is more often paired with statistical models, so that classified common background segments are removed from the dataset.

Clustering is a knowledge discovery and data mining technique that can be applied to vision datasets. It can be defined as segmenting heterogeneous data into subsets of homogenous data. Each pixel within a video frame holds important data about the scene. Clustering techniques catalogue a video image so that each pixel within a frame has high intra-cluster homogeny but low intercluster homogeny. Like data is grouped together. Applying clustering techniques such as DBSCAN (Ester et al. 1996) (a density-based clustering algorithm) can identify objects by classifying groups of pixels as related by common attributes and thus belonging to a cluster. Clustering is a robust method which is able to filter noise from a video frame before any other computer vision techniques are applied.

Statistical modelling of video frames has some basic operations similar to frame subtraction. There is an assumption that objects of interest are objects that some change in their attributes. Various statistical models such as the Gaussian mixture model (GMM) employ classifications which mark areas (or pixels) as part of the foreground or background. Foreground regions (or pixels) are the objects of interest and can be identified and tracked with sufficient prior knowledge. Statistical models rely on extensive training, and the more training, the better the outcomes. Training for an RAL system is not ideal, as it can be time-consuming and require technical knowledge, which is not always available.

4.3.5 Augmented Reality and Remote Access Laboratories

Laboratory experiments have long been essential for science and engineering students, reinforcing theoretical lessons through practical experiences. Remote Access Laboratories have extended the reach of the laboratory experience to any student with Internet access, which is everyone. The advantages of RAL cannot be underestimated in delivering practical curriculum outcomes. The benefits of applying AR attributes to RAL can be argued. Some RAL systems may not lend themselves to augmentation, while others may already have virtualisation as key component of the system. Older virtual instrumentation laboratories provide services that can be categorised as an augmented reality system.

The older virtualised laboratory systems demonstrate the importance of augmented enhancements for RAL environments. Virtualised environments highlight the diverse range of information and information delivery systems available, which provide the user with familiarisation and greater understanding of the operation of the equipment in their field of study. Many experiments could not translate to a remote configuration without the help of AR components. Andujar et al. (2011) developed a digital control experiment using a Xilinx FPGA Spartan-3E series circuit board. Without virtual overlays on the FPGA circuit board, the operation of the experiment would have been very sterile, with little feedback and interaction

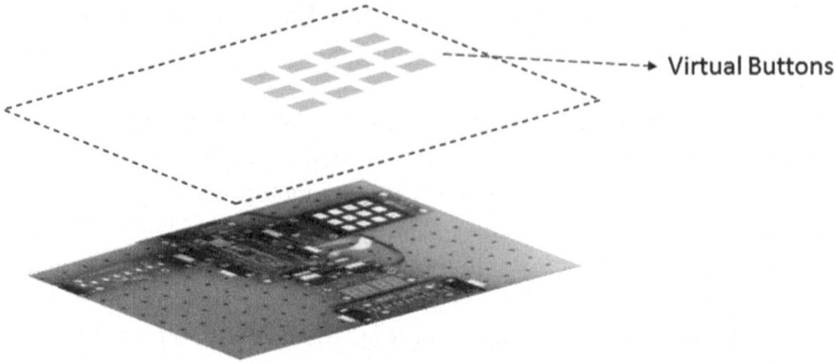

Fig. 4.5 The experiment test rig, from the remote users' point of view. The fiducial marker is not visible, and virtual interactive objects appear on the board

from the student. The conceptual structure of the system shown in Fig. 4.5 is based on the experimental rig described by Andujar and shows the layering of virtual objects over the video stream image.

Remote Access Laboratories provide solutions to many educational intuition problems, such as resource allocation, finance, expertise and maintenance. Augmented reality for RAL provides the actual users with a rich interactive environment, presenting data and experiment interaction that more closely mimics what is seen and how experiments are performed by students within the laboratory. The practical implementation of experiments, like the important configuration shown in Fig. 4.5, demonstrates the benefits AR can provide to RAL in ensuring curriculum goals are met. In addition to the curriculum goals, the implementation costs are minimal, and the overall footprint is quite small with the resources available to students 24 h a day.

4.3.6 Physics Engine

The physics engine is a software library that enables to generate graphical 3D objects and scene. The objects can have rigid bodies and can interact with each other. The movement of the objects follow the laws of physics. For example, if two objects collide, they repel each other. The physics engine provides a simulation environment.

In a desktop, AR mode, the video feedback is simply augmented with virtual objects such as text and other overlaid information. The video is always static, and the contents of the scene are always the same. The virtual objects of the scene are updated per the user's input. A desktop AR mode may not use a physics engine at all if there is no need for complex interactions between the experiment components. This is easier than a full HMD AR as the video input is directly from the remote lab site through cameras with usually static orientations.

Some common examples of 3D environment with physics engine are Unity 3D (Apostolellis et al. 2014), jMonkeyEngine (Reese and Johnson 2015) and Blender. Unity 3D supports C# and JavaScript, jMonkeyEngine uses Java, and Blender uses python to program the 3D environments. These allow creating basic 3D geometric objects such as sphere, cylinder and box. More complex object may be imported from another environment. The objects may be rotated, translated or scaled through the program or by the interaction of the objects with each other. Some objects alter the configuration of the other objects maintaining the laws of physics. In this way, an object representing the user's body or body parts such as hands and fingers using the Leap Motion can be used to interact with other virtual objects. These objects can listen to alterations to their states to generate commands for the remote experiments.

4.4 Part Two: Natural User Interfaces

This section discusses the NUI 'modules' of the AR/VR within RAL. The primary components of this module are the NUI device and then the 'users'. The 'users' operate the NUI devices to generate real-time data corresponding to their natural movement. The NUI data generated may be sent over the Internet to other users if there is a shared collaborative experiment session or if the NUI data itself is an *input* in the RAL experiment (as shown in the next part of this chapter).

Apart from real object tracking and virtual object creation within AR/VR, NUIs can also become an integral part of RAL command input system. Many experiments require operating measurement instruments that do not natively have a keyboard-type input but instead rely on their physical interface button input only. The use of these devices and activities hence involves complex hand movements to manipulate controls and physical objects and then observe the resultant behaviour of that manipulation (i.e. selection of physical controls). While implementing such experiments traditionally, hand movements are replaced with some form of automation. This results in the users rarely obtaining real hands-on experience that would otherwise be typical for an on-site laboratory. The NUI has the potential to address this particular issue. The major challenges that need to be considered when implementing NUI and AR in an RAL experiment for RAL are as follows:

- *Space requirements*: NUI-based experiments require a certain amount of physical space to operate to allow for the gestures and natural input. The actual type of the experiment will determine the space requirements needed for the relevant movement and gestures. Given that most experiments only require hand-based gestures, a smaller space such as a table or bench-type surface for creating and manipulating the virtual objects may be suitable.
- *Suitable gestures*: NUI will require a suitable gesture library for each experiment that defines what is to be recognised as a valid set of movements with respect to the experiment activity requirements.

- *Gesture control*: A suitable input device needs to be chosen for capturing the NUI data. This might consist of wearable devices, e.g. Sixense Hero, Razer Hydra and remote recognition devices (e.g. Leap Motion for hand gestures or Microsoft Kinect for full or upper body gestures).
- *Types of display*: The output of the gestures along with the graphical representations of the gestures may be displayed either on a wearable display device, e.g. smart glasses, or on a traditional computer desktop. Available display types to the user will determine if the system runs in 'desktop mode' or 'full AR' with a HMD.
- *Virtual objects*: As with any AR application, the NUI also requires virtual objects which are manipulated according to the users' input. These objects can be stored locally or downloaded during runtime.

Integrating the NUI with the AR may present the following issues:

(i) The virtual objects may be stored locally or downloaded at runtime, where the user interacts with them locally on their computing devices. If a certain gesture (e.g. turning or 'flipping/flicking' on a switch) requires actions on the remote experiment rig, then the UI will send the request to the experiment rig and then subsequently wait for a reply. Once a response from the rig has been received, the corresponding changes are then reflected in the video output device or the 3D environment scene presented to the user. This method is effective if the gestures are short and distinct, e.g. flipping a switch or pushing a button. Longer duration gestures may require a larger number of request/response interactions between the NUI and the experiment rig, potentially causing delays in updating the NUI and the displayed virtual objects. Additionally, the user may not be able to hold the hand or body positions for large intervals while waiting for a rig response to be received, thus creating a very unnatural and 'jittery/stuttered' learning experience.

(ii) The virtual object may be operated remotely where the users' inputs from the NUI are constantly streamed to the remote experiment rig. The experimental rig then determines if any gestures have been given and then makes changes to the real experimental rig as well as the virtual components. The status of the virtual rig and real experimental instrument rig are constantly streamed back to the NUI. This type of remote NUI can be used to record and process long gesture commands as the gestures are streamed to the remote location.

4.4.1 Gestures

Unlike conventional gestures (Oshita and Matsunaga 2010), activity gestures in the current context are associated with virtual objects within the physics engine or AR/VR environment. Normally when using NUI devices, inputs are not aimed at any particular object, i.e. the hands move in free space without any relation to any virtual objects. Thus, only the transitions within the state space of the users' body or body parts are analysed to obtain a gesture. However, while using a physics engine, when the objects in the scene correspond to real-world objects, the users' change of state space is consequently not as important. The more important factor is that of the transition of the virtual objects within their state space(s).

Gestures may be associated with any object in the AR/VR scene. For example, if there is a control button on a real physical instrument device, then a virtual button may have 'press' gesture which is determined by the linear transition of the object in 3D virtual space. Devices in RAL systems usually contain a definite set of interface elements. In electronics, this includes flipping switches, pressing buttons, rotating knobs and dragging wires or other connective objects. These are usually single hand/palm gestures. For mechanical and chemistry labs, two hands may be required to perform acts of lifting and moving large objects. For instance, some experiments in chemistry require very precise movements and actions such as tilting and pouring/decanting liquids into specific containers.

Gestures can be broadly classified into three categories:

- Simple gestures, where the hands/fingers present a certain orientation to cause translation, rotation or scaling of the virtual objects.
- Simple timed gestures, where a simple/complex gesture is carried out over a specific time period.
- Complex gestures, involving objects composed of multiple sub-objects or planes which behave differently with time. A complex gesture is composed of multiple simple gestures.

The data generated from input devices such as the Kinect or Leap Motion sensors are presented for processing in the form of a matrix of data. Identifying a gesture then becomes the case of checking whether the virtual object(s) had changed its X, Y and Z coordinates, rotation or shape with respect to a static limit prescribed for that particular object.

The user's body also forms a rigid object in the 3D space and is constantly updated based on the NUI device as well as interaction with other objects within the environment. Figure 4.6a, b shows 3D object representations created from data obtained from Kinect (full body scan) and the Leap Motion (palm/hand scan), respectively.

4.4.2 Gesture Libraries

The 'simple gestures' (using hands/palm scans only) are presented as follows:

- *Grab* is a multi-finger gesture, when an object has been lifted which surrounded the hands and finger. This is detected if the part of the virtual object is within the sphere of the hand, i.e. an imaginary sphere that best fits within the points of the

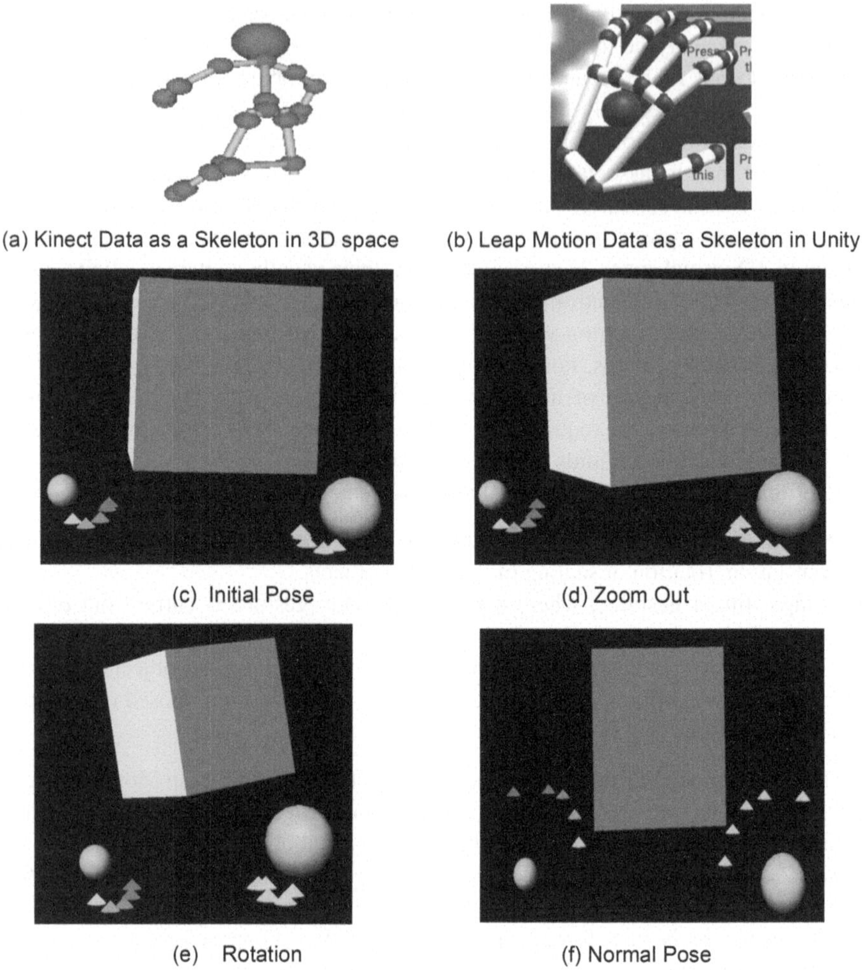

Fig. 4.6 Gestures (**a**) Kinect data as a skeleton in 3D space. (**b**) Leap motion data as a skeleton in unity. (c) Initial pose. (**d**) Zoom out. (**e**) Rotation. (**f**) Normal pose

fingers for a given hand. Grab can be detected while reshaping objects but not while displacing them. Grab can be identified as both a closed hand gesture for small objects and open handed for larger objects.

- *Rotate* is a gesture where the objects are rotated relative to the origin point ($X = 0$, $Y = 0$, $Z = 0$) or other objects in the 3D virtual space or scene.
- *Press* or *flip* can be detected by determining the linear translation of the virtual objects.

In certain experiments, a gesture can also be associated with time or be relative to its own basic components. For example:

- *Drag* is a time variant version of 'grab'. This means an object first detects a 'grab' gesture and then for a certain amount is converted to a 'drag' gesture to move that object in space.
- *Twist* is detected when elements or parts of objects, i.e. some geometric plane of the object, experience a different rate of rotation compared to other parts, i.e. geometric planes of the object.

These gestures create a corresponding command in the GCM. For example, the 'press' gesture generates a command specific for the button or input on the actual hardware device. This command is then sent to the instrument control server and executed on the real experiment hardware.

4.4.2.1 Special Gestures

There are some additional common gestures which are used for navigating and using the RAL system. These gestures are conventional, as they are not necessarily related to the physics engine or any specific action upon an object in the experiment. These special gestures are:

- *Zoom(ing)* is a gesture that allows the users to zoom a viewport around objects. The gesture is identified when two hands are closed and placed in parallel, i.e. they have the same (or nearly same) value for Z-axis. Once the two hands are in place for (example) 30 ms without any change in X-, Y- and Z-axis, the gesture is considered as started. At this stage, if the hands move away from each other, the scene is 'zoomed in'. Alternatively, if the hands move closer to each other, then the scene is zoomed out proportional to the distance between the hands. The gesture is then released when the hands are opened. The object then remains in the final state at which the gesture ended. There are particular limits to the level of zoom available corresponding to the object in the scene, and the gesture may fail to work once these limits are reached. The user places their hand at the desired position depending on whether they want to zoom in or out. If they want to zoom in they initially place their closed hands close to each other and then moves the closed hands apart. This is shown in Fig. 4.6c, d.
- *Rotation* is a gesture that allows the users to change the rotation of the scene containing the objects with respect to the camera. Like 'zoom', this gesture is identified by two closed hands placed in parallel, i.e. they have the same (or nearly same) value for Z-axis. Once the two hands are in place for (example) 30 ms without any change in X-, Y- and Z-axis, the gesture is considered to have started. At this, if the hands move back and forth with respect to each other, the scene is then rotated. Like 'zoom', the gesture ends when the hands are opened. Unlike 'zoom' however there may be no limit on rotation as the camera could revolve infinitely around the object. This is shown in Fig. 4.6c, e.
- *Panning* is another method of changing the camera's view by moving the camera position at X, Y and Z along the XY plane by keeping the Z value constant.

This is similar to the rotation but provides a different perspective as the distance between the camera and the furthest points of the object remain the same. The user cannot see the back view of a scene this way.

These gestures may not be required for HMD-based full AR/VR environments and may be implemented in different ways.

4.4.3 NUI Devices

There are several NUI devices available commercially at the time of writing. NUI device allows users to behave and interact in a natural manner. This is the main way NUI devices differ from conventional input devices such as mice and keyboards which often inhibit natural interaction. NUI devices generally return large streams of three-dimensional data. Several popular and emerging NUI devices are discussed below.

4.4.3.1 Kinect

The Microsoft Kinect is a line of motion sensing input devices from Microsoft. It is based around a webcam-style add-on USB peripheral device and enables users to naturally control and interact with software on dedicated game console hardware with a version available for use with computers. It allows human interaction without the need for direct physical control such as a keyboard, mouse or joystick. The Kinect was originally designed for XBOX 360 dedicated game console to capture and track human motion and postures. Briefly, it uses 3D image motion sensing to obtain depth information along with RGB video frame data and also incorporates a microphone to allow for voice recognition capabilities. The Kinect returns skeletal data in the form of a 2D array where each row represents a joint in the body along with corresponding data for the X, Y and Z value in 3D Cartesian space of that joint. The Kinect is a long-range HCI device and also returns data about the larger skeletal joints such as the wrist, elbows, shoulders, neck, chest, hip, knees and feet. Because of the full-body scanning nature of this device, it typically requires a larger space to operate.

4.4.3.2 Leap Motion

Leap Motion is a USB peripheral device to also directly track human motion. The Leap Motion however by design has a very limited range of about ±50 cm around the sensor. Users move their hands over the device which then tracks the joint positions. The Leap Motion can detect each bone segment position of the fingers as well as the palm position of the hands. The Leap Motion returns multiple 2D arrays of finger bones/joints. The Leap Motion also augments this with other data such as acceleration measurements of the fingers. There are also several programming language options to interface with the Leap Motion.

4.4.3.3 Sixense

The Sixense range of devices utilise a weak magnetic field in order to obtain both position and orientation data for a range of handheld paddle devices. These devices can track, in true 3D space, a controller representing the hand input of the user. These types of devices are particularly useful in that they can track more than one input device and are generally used with at least a pair of controllers. Consumer versions of the current 'Sixense STEM System' allow tracking of two handheld controllers as well as three body-worn trackers (for each leg and head tracking, for instance), providing almost full body tracking within the sensing space, being approximately a 2.5 m in diameter around the base station. Consequently, this device is more suited to larger environments than using a single piece of equipment. This device allows very precise positional tracking (in six axes; positions X, Y, and Z; as well as rotation around the same axis). Additional controls on each handheld controller allow more detailed interaction such as selection of objects or manipulation of virtual hands within the computer-generated environment. These types of devices are generally more suited to full virtual reality environments where complete positional input can be tracked allowing the teaching of the UI or haptic training for the given environment or experiment.

4.4.3.4 Other Input Devices

There are several other input devices that can be also considered. Many of these utilise a camera and lighting/illumination arrangement to triangulate controller position that may be augmented with gyro sensing within the controller to provide rotational and acceleration input. Examples of these systems are the Sony Play Station Move Controller which uses a light globe controller with integral accelerometers and a camera to discern controller (and hence hand) position and rotation within the capture space. This input can be used to control virtual appendages or provide 3D input for option selection or manipulation of computer-generated content. Other examples are accelerometer-augmented game controllers, very commonly found on the fourth- and fifth-generation dedicated game console hardware.

4.4.3.5 Other Output Devices

AR and VR features are best experienced with head-mounted displays (HMD). These are wearable glasses that provide direct visual feedback to the eyes. This form of mixed reality can also overlap into the users' personal space. The HMD concept is capable of delivering a very realistic experience as the user and can also incorporate head tracking, as well as incorporate hand position input to offer greater positional accuracy, as well as provide significantly improved immersion compared to a desktop version of the mixed reality.

One issue of the mixed reality environment in an educational setting is that there must always be an alternative offered to the UI that the does not use the HCI devices. This is because some users may not have access to the specific HCI devices and hence should not be excluded from accessing and using the experiments. Moreover, the absence of a particular device when using an experiment designed for that input would usually result in an inability to completely and properly interact with that environment. Thus, the UI must be adaptable and flexible to take inputs from already existing ubiquitous input devices such as mouse and keyboard. If it is not possible to simulate the inputs from an HCI device with mouse/keyboard for a particular experiment, then a sample set of input data must be provided to students to run the experiment.

4.4.4 Using the AR/VR with NUI in RAL

Experiment and activity design naturally becomes more complicated when using AR/VR. Not only does the hardware need to be connected to the Internet, additionally the user interface also requires significant changes. The AR/VR and NUI features typically focus on enhancing aspects the user experiences (as shown in Fig. 4.7).

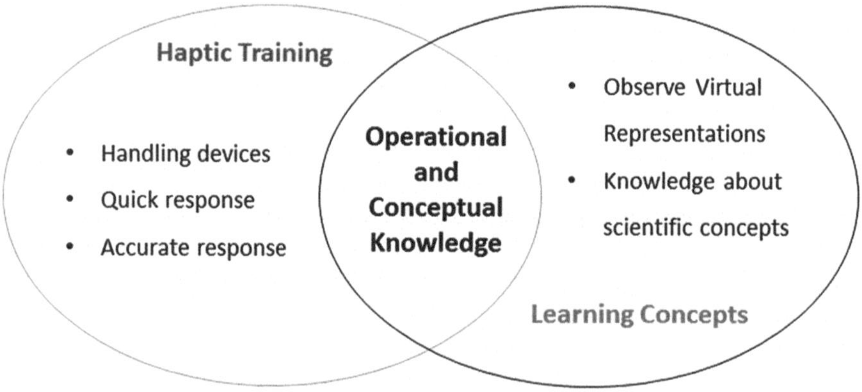

Fig. 4.7 The key focus areas of NUI with regard to RAL

4.4.4.1 Haptic Training

Haptic training is gaining the experience in creating control events, such as pressing appropriate control buttons or rotating knobs, in response to constraints such as time or position or orientation. The NUI can then facilitate training of a realistic experience for using equipment and devices.

4.4.4.2 Learning Concepts

Using AR/VR allows users to view objects of information that is otherwise hidden, such as the operations within a piston engine. In this example, when using the real hardware and a standard camera, the events and moving parts of the engine are generally not visible. A VR/AR physics engine however then allows users to view inside solid objects by simply translating the camera's position close to it or within its boundaries.

4.4.4.3 Combined Learning

There is another type of training that lies between *haptic training* and *learning concepts.* It combines both training about handling devices and learning concepts of operation of the devices, for example, as shown in Fig. 4.8, using a function generator (a test device to create simple electrical signals) an essential learning experience for electrical engineers. The users of the RAL must be capable of operating the actual device in a real laboratory/location. The video feedback, in this case, is the value shown on the screen of the real hardware device that comes from the RAL site. The video is edited to fit the 3D environment and updated in response to user interactions.

The effectiveness of the AR/VR environment depends upon the activity and the experience level. For example, connecting a wire in an electrical experiment

a b

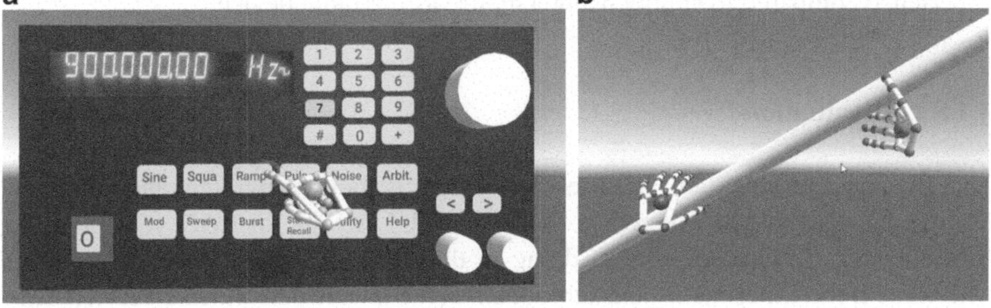

Fig. 4.8 Leap Motion and Unity 3D environment with (**a**) function generator (**b**) handling a pipe/tube

(Restivo et al. 2014a, b, c) does not contain significant educational importance; however, connecting the correct type of pipe/tube in a mechanical engineering experiment may be a critical component of the RAL activity (Chang et al. 2014).

Another issue is that of space resolution, i.e. the minimum amount of space that can be distinctly observed and interacted with when using NUI devices. The main problem is that there is a competing need between precise position detection and overall detection area. The ability to detect the positions accurately decreases as the user/users' body moves away from the device. Additionally, in AR/VR since there is no physical object to actually interact with, it is difficult for users to operate the interface based on visual cues and intuitions alone if the interface components are close to each other. For example, with respect to Fig. 4.8a, the user may accidentally press a different button than the desired one if the buttons are small and close to each other. Any NUI for RAL must consider these factors.

4.4.5 Advantages

The main aim of integrating NUI with AR/VR into the remote laboratory is to positively enhance the educational experience. This can take place in several ways.

4.4.5.1 Hands-On Experience

One of the main considered shortcomings of remote laboratories in many fields is the lack of hands-on experience (Maiti et al. 2014). The NUI can address this, allowing students to move their hands and fingers to interact with virtual objects. This allows for a very realistic haptic and educational experience. This may not be effective for an electronic experiment where the user interface may be considered simplistic and trivial, i.e. pressing control buttons, etc. which can be easily done with mouse input. However, for an experiment where moving and re-wring large components require significant and controlled hand movement, such interactions can be done with the NUI devices in a virtual environment, providing a more effective learning experience compared to simply using a mouse as the input device.

4.4.5.2 Evaluation

The biggest advantage of gesture-based NUI is both providing and allowing evaluation of hands-on experience. With the NUI AR/VR in RAL, it is possible to keep track of what the user is doing in the 3D environment and analyse it to determine if the experiments and subsequent actions were performed correctly. This can help identify issues with both haptic and conceptual learning. In the case of mouse and keyboard, with lower flexibility in interaction, there is less scope of deviation in students' interaction that is not effective towards the learning outcomes.

The interaction is very close to the optimal behaviour expected. However, the NUI with larger flexibility in interaction allows for greater amount of manoeuvring which can help the RAL system determine how efficiently the student performs and likely to perform in a real-life situation. In this manner, it is more effective than using a mouse and keyboard, as in the real world, many devices are not controlled using keyboard or mouse input.

4.4.5.3 Projecting Hidden Information

AR can allow users to see real time feedback of objects that cannot ordinarily be seen through a normal camera.

4.4.6 Disadvantages

There are some disadvantages with the NUI AR combination. The biggest shortcoming is the lack of a direct haptic feedback. There have been attempts to provide haptic feedback to students for RAL experiments (Quintas et al. 2013). These have been limited to smaller dimensions, and the devices for haptic feedback are not universal. Thus, in the AR with NUI in RAL, although the user can interact with objects directly using their hands, they are unable to feel anything (Norman 2010) or obtain haptic feedback. This means that the users may not achieve a suitable hands-on interactive experience. Within the physics engine, this also contributes to the problem of 'fake collisions'. This occurs when the hand, or the NUI virtual object, is calculated to be placed within another virtual object. For example, a user may continue to press a virtual button on a piece of equipment where the hand then moves into the 3D space of that equipment, presenting an unrealistic situation. Some experiments may neglect this as in the case of the example showed above. Others may hold the position of the NUI virtual objects at their last known positions outside the concerned RAL virtual objects.

Another issue is that at the present time, technologies to implement realistic AR/VR experiences are limited and not ubiquitous. There are limited options for acquiring NUI data, and the accuracy is not considered sufficient for this work. Additionally, devices to project the AR objects are limited and again not widespread. It is expected that with time and progress in technology, these devices will become available, allowing for the experience of students to be more accurate.

4.5 Part Three: Teaching HCI and NUI with RAL

This section presents a different approach to the AR/VR for remote laboratory as discussed in Sect. 4.1 and shown in Fig. 4.1. Instead of using an AR/VR system, a new RAL activity is presented where the learning exercise is about programming and using NUI at the local sites and controlling real hardware in a remote location. The AR/VR system is condensed into two modules using the desktop mode: a *remote laboratory location* where the video feedback originates and *local site* with a NUI device. Physics engines or augmented objects are not required for this type of RAL activities.

HCI applications have grown significantly in the last few years. Several new devices have become available as commercial products. Several engineering courses, for example, have integrated HCI devices into the teaching curriculum. As these devices become widely available, they may be used for remote laboratory input for experiments. HCI inputs can be used to control robotic equipment. This can be used in hazardous conditions where human beings would be at risk. The robotic equipment can respond to the human inputs to HCI devices like Kinect or Leap Motion in real time. Thus, there is a need for students to learn to create the programs to synchronise the human inputs and robotic movements. A case study of a RAL activity is presented where HCI devices located at personal locations are used to control equipment at the remote laboratory.

Figure 4.9 shows the system architecture of the remote laboratory experiment. The student has the HCI equipment, e.g. Kinect. They must download and run a transmitter program that uses the Microsoft API and a WebSocket to transmit the 20-point array as fast as possible. The transmission is the only one directional, from the student location to the remote laboratory location. There is a delay on the Internet, and this affects the responsiveness of the robotic arm to the human inputs. In the remote laboratory, there is a LEGO Mindstorms-based robotic arm with three actuators. The remote server runs a VirtualBox which allows the students to access a remote desktop of a virtual machine. This is transparent to students, who run LabVIEW within the remote desktop. The students must create a program including specially designed VI files to control the equipment. There is an IP camera

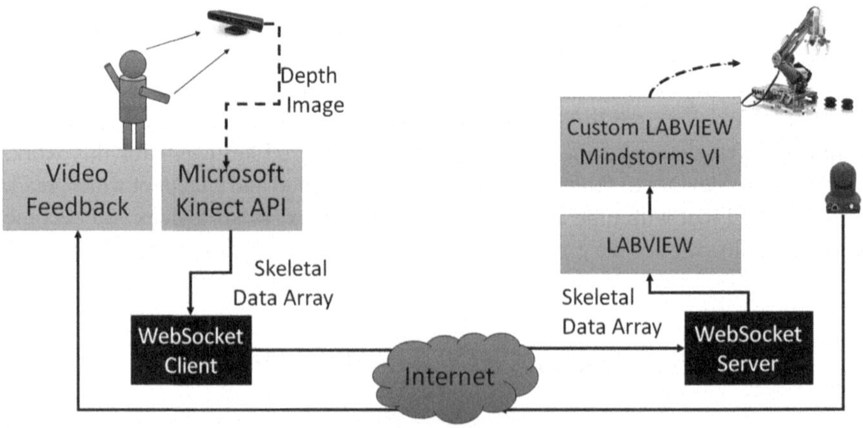

Fig. 4.9 The system architecture of the HCI, Kinect and remote desktop

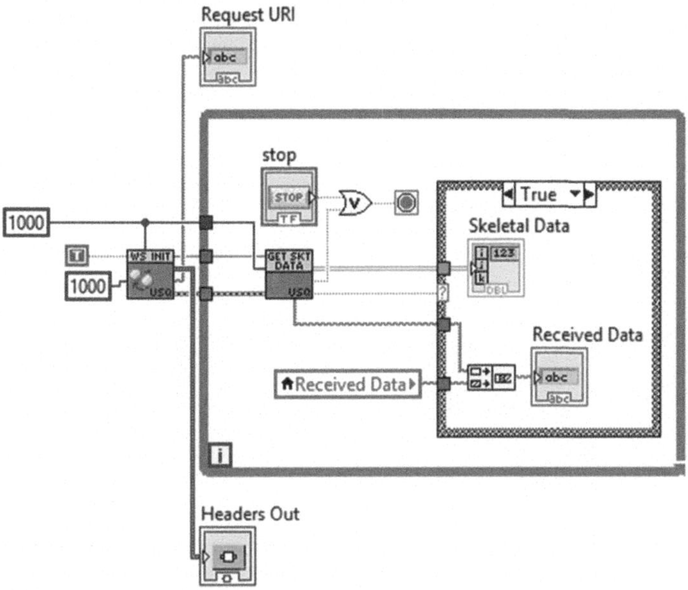

Fig. 4.10 The basic program to obtain the skeletal data

in the laboratory that allows the students to view the movement of the robotic equipment. Figure 4.10 shows the basic LabVIEW program to acquire the skeletal data from a WebSocket connection. This includes customised WebSocket server VIs that connect to the students' client and get the skeletal data.

4.5.1 Robotic Arm

The robotic arm (LEGO) is constructed with LEGO Mindstorms. There are three actuators on the robotic arm. Actuator A_1 (θ) can rotate between $0 < \theta < 450$. Actuator A_2 (ϕ) can rotate between $0 < \phi < 330$. Actuator 3 is the arm's claw. There are two sensors corresponding to the actuators 1 and 2. All actuators can also act as sensors by reading the degrees the actuator has rotated since they were last 'reset' or 'cleared'. The program created by the students should follow the general structure as shown in Fig. 4.11. Initially, the initiation step restores the robotic arm to the initial state. The sensors help in identifying the initial states. Once the initial state is reached, θ and ϕ variables in the program are set to 0, and the actuator rotation degrees are cleared. From here on, the actuator can move with the specified limit according to the input from the Kinect. The claw operates with two different hands and can close and open according to movements of the left hand of the operator.

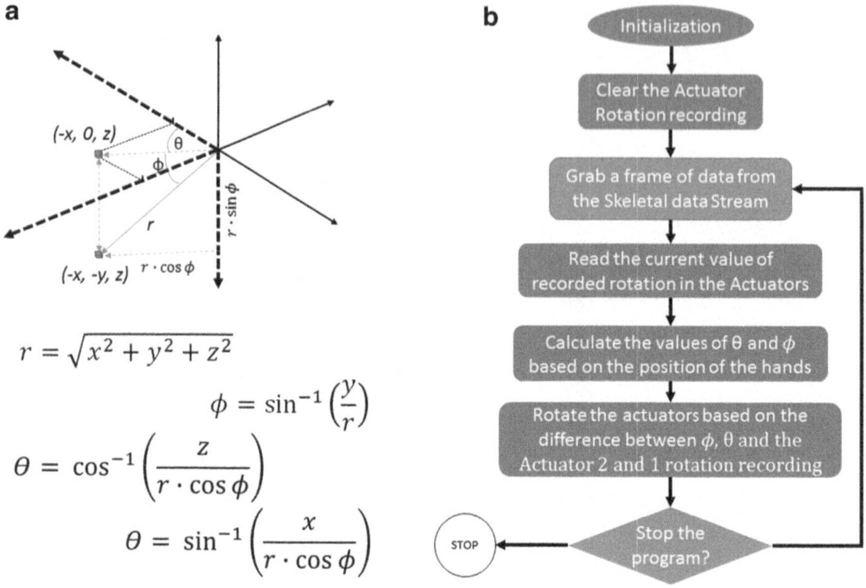

Fig. 4.11 (**a**) The formulas to calculate the value of θ, ϕ. (**b**) The flow chart

4.5.2 *Kinect Usage*

Kinect returns 20 data points with X, Y and Z values. For this experiment, the right-hand wrist and right shoulder points are used to control the values of theta, and the left-hand wrist is used for controlling the claw. The students may switch the hands and shoulder for their convenience. As shown in Fig. 4.11a, the right shoulder is considered as the origin. The right wrist W_R represents the position of the claw. The value of the θ, ϕ is calculated as the angle between YZ plane and W_R and the angle between XZ plane and W_R with respect to the origin. The values of θ and ϕ are calculated in LabVIEW using the formulas.

4.5.3 *Users' Task*

The student logs into the RAL system and gets access to the remote desktop (Melkonyan et al. 2014) with LabVIEW containing the custom VI libraries for the WebSocket and the LEGO Mindstorms. The students should follow the general flow chart as shown in Fig. 4.11b. First, initialise the robotic arm based on the sensors. The actuators' rotation recordings are cleared. Then a loop is used to grab new skeletal data from the WebSocket, and the new values of θ and ϕ are calculated. Next the change in the angles, i.e. current rotation of the actuators $\alpha(A_1)$ and $\alpha(A_2)$, is subtracted from the value of theta and ϕ. The actuator A_1 is rotated by $(\theta_t - \alpha(A_1)_t)$, and the actuator A_2 is rotated by $(\phi_t - \alpha(A_2)_t)$. This way the claw mimics the position of the right wrist corresponding to the shoulder. For the claw to open and close, the left-hand position is tested. If the left hand is above any limit, the claw is open, and if the left hand is below a certain point, the claw is closed. The claw can only move by only $\pm 90°$. The students can decide how the claw is initialised.

Once the program is created in the remote desktop, the student runs the transmitter program on their local PC and the program on the remote desktop. The student then moves their hands in their locations, and the data is transmitted to the remote desktop. The student then stops the program when they want.

4.5.4 Network Architecture

As part of a previous project (Maiti et al. 2015), a system exists that allows authenticated access from an active browser window to the remote location of the experiment. Direct connections to WebSocket endpoints are possible. As these connections are transparently relayed via an HTTP proxy, this works very reliably, also in heavily firewalled and NATed environments, particularly in environments where direct connections to remote ports are not possible. In essence, all connections appear as encrypted requests to a Web server.

There are some customised VIs that control the robot's actuators. These are based on the VIs provided by the LabVIEW Mindstorm library. The customised VIs make sure that the robotic state space remains within the limit, such that the robot does not break down. The users are restricted from using the basic VIs which are hidden as they could cause the robot to go out of control. Figure 4.12 shows the robotic setup using LEGO Mindstorms in the remote laboratory.

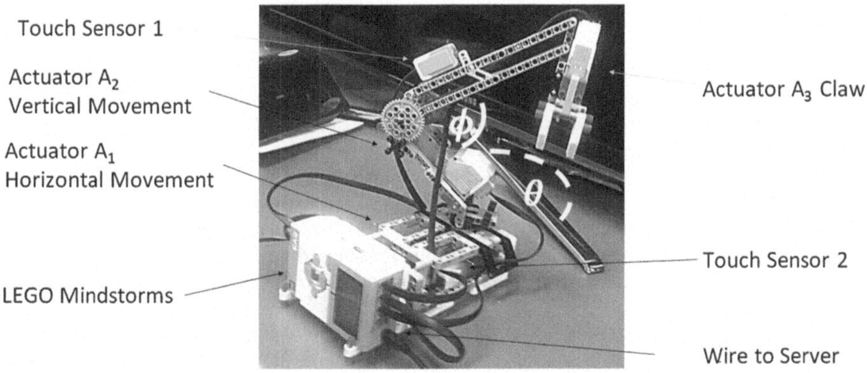

Fig. 4.12 The whole experimental setup

4.6 Conclusions

This chapter has presented the state of AR and NUI and has discussed new ideas based on their use in remote laboratories. One of the notable limitations of remote lab has been the lack of hands-on experience and training in creating and setting up experiments. NUI and AR/VR can address this shortcoming. AR/VR can also provide a mediated interface and add augment information that cannot be either seen with naked eyes. Such real-time information can help users to better identify concepts that need to be learned. Virtual objects can also be placed in the environment allowing users to experience immersive practical activities at their own (comfortable) location. Compared to a pure VR, an AR with virtual objects can look more convincing given that experiments are well designed.

NUI can allow the user a realistic experiment of handling objects and creating experiment setups inside a physics engine-driven gamelike 3D environment. Learning outcomes heavily depends on how well the interface is developed and its ability to handle interactions using the game engine. Despite the benefits of using AR/VR and NUI in RAL, it is a long way until such technologies become widely used. This is largely caused by the availability of input devices. Most students do not have access to such technology for NUI input devices. Secondly, the key limitation with most of the NUI devices discussed here is that they depend on audio-visual feedback to supplement real haptic input, and technology to create very accurate realistic haptic feedback experience is still a few years away. Also, the NUI/AR/VR needs to address new design issues while being incorporated into the RAL environment.

The last part of the chapter has presented an example of activities that can use the NUI devices for controlling real hardware in a remote location. The basic model of this experiment, replacing the remote desktop with a physics engine, can be used for other experiments and activities in remote laboratories as well.

References

Andujar, J. M., Mejias, A., & Marquez, M. A. (2011). Augmented reality for the improvement of remote laboratories: An augmented remote laboratory. *IEEE Transactions on Education, 54*(3), 492–500. https://doi.org/10.1109/TE.2010.2085047.

Apostolellis, P., Bortz, B., Peng, Mi., Polys, N., & Hoegh, A. (2014). *Poster: Exploring the integrality and separability of the Leap Motion Controller for direct manipulation 3D interaction.* Paper presented at the 3D User Interfaces (3DUI), 2014 IEEE Symposium on.

Azuma, R. T. (1997). A survey of augmented reality. *Presence, 6*(4), 355–385.

Barfield, W. (2015). *Fundamentals of wearable computers and augmented reality*. Boca Raton: CRC Press.

Bruder, G., Steinicke, F., & Hinrichs, K. H. (2009). *Arch-explore: A natural user interface for immersive architectural walkthroughs*. Paper presented at the 3D User Interfaces, 2009. 3DUI 2009. IEEE Symposium on.

Callaghan, M. J., McCusker, K., Losada, J. L., Harkin, J., & Wilson, S. (2013). Using game-based learning in virtual worlds to teach electronic and electrical engineering. *IEEE Transactions on Industrial Informatics, 9*(1), 575–584. https://doi.org/10.1109/TII.2012.2221133.

Callaghan, M., Savin, M., McShane, N., & Eguiluz, A. (2015). Mapping learning and game mechanics for serious games analysis in engineering education. *IEEE Transactions on Emerging Topics in Computing, 99*, 1–7.

Chang, Y., Aziz, El-S., Zhang, Z., Zhang, M., Esche, S., & Chassapis, C. (2014). *A platform for mechanical assembly education using the Microsoft Kinect.* Paper presented at the ASME 2014 International Mechanical Engineering Congress and Exposition.

Corter, J. E., Nickerson, J. V., Esche, S. K., & Chassapis, C. (2004). *Remote versus hands-on labs: A comparative study.* Paper presented at the Frontiers in Education, 2004. FIE 2004. 34th Annual.

Cubillo, J., Martin, S., Castro, M., & Meier, R. (2012). *Control of a remote laboratory by augmented reality.* Paper presented at the Teaching, Assessment and Learning for Engineering (TALE), 2012 IEEE International Conference on.

Ester, M., Kriegel, H.-P., Sander, J., & Xu, X. (1996). *A density-based algorithm for discovering clusters in large spatial databases with noise.* Paper presented at the Second Annual Conference on Knowledge Discovery and Data Mining (KDD-96).

Jara, C. A., Candelas, F. A., Puente, S. T., & Torres, F. (2011). Hands-on experiences of undergraduate students in automatics and robotics using a virtual and remote laboratory. *Computers & Education, 57*(4), 2451–2461.

Liu, W. (2010). *Natural user interface-next mainstream product user interface.* Paper presented at the 2010 IEEE 11th International Conference on Computer-Aided Industrial Design & Conceptual Design 1.

Maier, P., & Klinker, G. (2013). Augmented chemical reactions: 3D interaction methods for chemistry. *International Journal of Online Engineering (iJOE), 9*(S8), 80. https://doi.org/10.3991/ijoe.v9iS8.3411.

Maiti, A., Maxwell, A. D., & Kist, A. A. (2014). Features, trends and characteristics of remote access laboratory management systems. *International Journal of Online Engineering, 10*(2), 31–37.

Maiti, A., Maxwell, A. D., Kist, A. A., & Orwin, L. (2015). *Joining the game and the experiment in peer-to-peer remote laboratories for STEM education.* Paper presented at the 2015 3rd Experiment@ International Conference (exp.at'15), University of the Azores, Ponta Delgada, Portugal. http://eprints.usq.edu.au/27481/

Mejías Borrero, A., & Andújar Márquez, J. M. (2012). A pilot study of the effectiveness of augmented reality to enhance the use of remote labs in electrical engineering education. *Journal of Science Education and Technology, 21*(5), 540–557. https://doi.org/10.1007/s10956-011-9345-9.

Melkonyan, A., Gampe, A., Pontual, M., Huang, G., & Akopian, D. (2014). Facilitating remote laboratory deployments using a relay gateway server architecture. *IEEE Transactions on Industrial Electronics, 61*(1), 477–485. https://doi.org/10.1109/Tie.2013.2242416.

Menezes, P., Chouzal, F., Urbano, D., & Restivo, T. (2017). Augmented reality in engineering. In M. E. Auer, D. Guralnick, & J. Uhomoibhi (Eds.), *Interactive collaborative learning: Proceedings of the 19th ICL conference – Volume 2* (pp. 221–228). Cham: Springer International Publishing.

Moore, P., Kist, A. A., Maiti, A., & Maxwell, A. D. (2016). *Work in progress: Remote experiment control through gesture recognition.* Paper presented at the 2016 13th International Conference on Remote Engineering and Virtual Instrumentation (REV).

Müller, D. (2009). Mixed reality systems. *International Journal of Online Engineering (iJOE), 5*(S2). https://doi.org/10.3991/ijoe.v5s2.1098.

Muller, D., Chilliischi, A., & Langer, S. (2012). *Integrating immersive 3D worlds and real lab equipment for teaching mechatronics.* Paper presented at the Remote Engineering and Virtual Instrumentation (REV), 2012 9th International Conference on.

Norman, D. A. (2010). Natural user interfaces are not natural. *Interactions, 17*(3), 6–10. https://doi.org/10.1145/1744161.1744163.

Odeh, S., Abu Shanab, S., Anabtawi, M., & Hodrob, R. (2013). A remote engineering lab based on augmented reality for teaching electronics. *International Journal of Online Engineering (iJOE), 9*(S5), 61. https://doi.org/10.3991/ijoe.v9iS5.2496.

Ohta, Y., & Tamura, H. (2014). *Mixed reality: Merging real and virtual worlds*: Springer Publishing Company, Incorporated.

Olalde Azkorreta, K., & Olmedo Rodríguez, H. (2014). Augmented reality applications in the engineering environment. In P. Zaphiris & A. Ioannou (Eds.), *Learning and collaboration technologies. Technology-rich environments for learning and collaboration: First international conference, LCT 2014, held as part of HCI international 2014, Heraklion, Crete, Greece, June 22–27, 2014, proceedings, part II* (pp. 83–90). Cham: Springer International Publishing.

Oshita, M., & Matsunaga, T. (2010). Automatic learning of gesture recognition model using SOM and SVM. In G. Bebis, R. Boyle, B. Parvin, D. Koracin, R. Chung, R. Hammoud, M. Hussain, T. Kar-Han, R. Crawfis, D. Thalmann, D. Kao, & L. Avila (Eds.), *Advances in visual computing* (Vol. 6453, pp. 751–759). Berlin: Springer.

Popa, D., Gui, V., & Otesteanu, M. (2015). *Real-time finger tracking with improved performance in video sequences with motion blur.* Paper presented at the 2015 38th International Conference on Telecommunications and Signal Processing (TSP).

Quintas, M. R., Restivo, M. T., Rodrigues, J., & Ubaldo, P. (2013). Let's use haptics! *International Journal of Online Engineering.*

Reese, R., & Johnson, J. (2015). *jMonkeyEngine 3.0 game development: A practical guide.* Oakamoor: P8Tech.

Restivo, M. T., & Cardoso, A. (2013). Exploring online experimentation. *International Journal of Online Engineering, 9*, 4–6.

Restivo, M. T., Rodrigues, J., & Chouzal, M. de Fàtima. (2014a, 3-6 December). *Let's work with AR in DC circuits.* Paper presented at the 2014 International Conference on Interactive Collaborative Learning (ICL).

Restivo, M. T., Chouzal, F., Rodrigues, J., Menezes, P., Patrão, B., & Lopes, J. B. (2014b). Augmented reality in electrical fundamentals. *International Journal of Online Engineering, 10*(6), 68–72.

Restivo, M. T., Chouzal, M. d. F., Rodrigues, J., Menezes, P., Patrão, B., & Lopes, J. B. (2014c). Augmented reality in electrical fundamentals. *International Journal of Online Engineering (iJOE), 10*(6), 68. https://doi.org/10.3991/ijoe.v10i6.4030.

Restivo, M. T., Quintas, M. R., Rodrigues, J., & Cardoso, A. (2015). *Next-generation experimental lab #1.* Paper presented at the 2015 3rd Experiment International Conference (exp.at'15).

Smith, M., Maiti, A., Maxwell, A. D., & Kist, A. A. (2016). Augmented and mixed reality features and tools for remote laboratory experiments. *International Journal of Online Engineering (iJOE), 12*(07), 45. https://doi.org/10.3991/ijoe.v12i07.5851.

State, A., Livingston, M. A., Garrett, W. F., Hirota, G., Whitton, M. C., Pisano, E. D., & Fuchs, H. (1996). *Technologies for augmented reality systems (Realizing ultrasound-guided needle biopsies).* Paper presented at the 23rd annual conference on computer graphic and interactive techniques.

Weichert, F., Bachmann, D., Rudak, B., & Fisseler, D. (2013). Analysis of the accuracy and robustness of the leap motion controller. *Sensors, 13*(5), 6380.

Wu, H.-K., Lee, S. W.-Y., Chang, H.-Y., & Liang, J.-C. (2013). Current status, opportunities and challenges of augmented reality in education. *Computers & Education, 62*, 41–49. https://doi.org/10.1016/j.compedu.2012.10.024.

Zhang, Z., Zhang, M., Chang, Y., Esche, S. K., & Chassapis, C. (2013). Integration of physical devices into game-based virtual reality. *iJOE, 9*(5), 25–38.

Ananda Maiti works at the University of Southern Queensland and has a research interest in Internet of Things, computer networks and algorithms.

Mark Smith is a PhD student at the University of Southern Queensland in the field of remote laboratories and augmented reality.

Andrew D. Maxwell is a senior lecturer (Electronics and Communications Engineering) in the School of Mechanical and Electrical Engineering in the Faculty of Health, Engineering and Sciences at the University of Southern Queensland, Toowoomba, Australia.

Alexander A. Kist is an associate professor (Telecommunications) in the School of Mechanical and Electrical Engineering in the Faculty of Health, Engineering and Sciences at the University of Southern Queensland, Toowoomba, Australia.

Chapter 5
Designing Cyber-physical Systems with Evolutionary Algorithms

Melanie Schranz, Wilfried Elmenreich, and Micha Rappaport

Abstract CPSs find their application in different domains, including smart cities, Internet of Things (IoT), and Industry 4.0. The increasing degree of interaction among CPSs leads to unpredictable and partially unexpected behavior. The major steps to manage emerging behavior in CPSs are taken in the design process. Although a high number of methods and tools already exist from related disciplines (including complex system research, embedded system design, and self-organization), there is no comprehensive toolset available to address the extensive CPS design process. This chapter presents a proposal for a common CPS design toolset. It combines existing and emerging tools to design, simulate, evaluate, and deploy solutions for complex, real-world problems using evolutionary algorithms on the example of swarms of UAVs.

Keywords Cyper-physical systems · Model-based design · CPS integration · Optimization · Evolutionary algorithms · Emergent behavior

5.1 Introduction

Cyber physical systems (CPSs) are characterized by the integration of computation and networking (Martins and McCann 2017). Furthermore, they enrich embedded devices by interacting with the physical world through sensors and actuators. Other synonyms for a CPS are "networked embedded system" or "system of systems" (Schätz et al. 2015). According to the National Science Foundation's definition (Foundation 2016), CPSs "are engineered systems that are built from, and depend upon, the seamless integration of computational algorithms and physical

M. Schranz (✉) · M. Rappaport
Lakeside Labs, Klagenfurt, Austria
e-mail: schranz@lakeside-labs.com; rappaport@lakeside-labs.com

W. Elmenreich
Alpen-Adria-Universität Klagenfurt, Klagenfurt, Austria
e-mail: wilfried.elmenreich@aau.at

M. E. Auer et al. (eds.), *Cyber-Physical Laboratories in Engineering and Science Education*, https://doi.org/10.1007/978-3-319-76935-6_5

components." CPSs have the capability to link previously disjoint technical and organizational processes, including embedded systems, logistics, coordination, control, and Internet services on a local and a global scale. CPSs find their application in many domains including smart cities, Internet of Things (IoT), and Industry 4.0. A typical characteristic of CPSs is the high degree of interaction between components. Such highly networked systems become increasingly hard to design and to predict, sometimes leading to unexpected and particularly unwanted behavior (Cohen et al. 2013; Schätz et al. 2015). These conditions can lead to disrupted results with consequences in performance, efficiency, and resource consumption. Therefore, a CPS should be designed to reach its goals in a flexible, reliable, and adaptable way, considering variable environmental conditions. Researchers go even further and add more characteristics to CPSs, including scalability, resiliency, safety, security, and usability (Foundation 2016). All these characteristics show the difference between a CPS and an embedded system – as these terminologies likely lead to confusion. Moreover, CPSs will transform the way people interact with engineered systems. All of these features are going to change the drive in innovation and competition and are thus enabling breakthrough achievements in several production and service sectors like agriculture, energy, manufacturing, or transportation (Foundation 2016). Thus, CPSs change the current market incrementally and create new markets (Schätz et al. 2015; Törngren et al. 2017).

The idea of building a CPS design process comes up with many challenges that counter its construction process. Within such a design process, highly distributed and connected technologies are embedded in a multitude of CPSs. Bare CPS subsystems can come into conflicts. Therefore, abstraction layers and libraries are indispensable to enable easy development and programming of a multitude of systems. The abstraction layer should natively support connectivity and communication, thus transparently supporting highly distributed and highly interconnected setups.

Another challenge could arise as we design increasingly autonomous physical systems with various dynamics by simultaneously satisfying multiple critical constraints. Predictive engineering methodologies based on simulation and performance prediction, together with supporting iterative design refinement, will guarantee high flexibility of designed CPSs and unprecedented abilities to address multiple, dynamically varying, and critical and non-critical constraints.

Further, the combination of several CPSs in a "system of systems" gives rise to unpredictable behavior and emergent properties. Therefore, the toolset should enable unprecedented analysis capabilities for tuning the design of complex, heterogeneous swarms of CPSs. Emerging behavior is the focus and the goal of the toolset, rather than a counter effect to address. Moreover, the innovation beyond the state of the art (Bagnato et al. 2017) of this toolset is to support the design of unpredictable behavior with the help of evolutionary algorithms.

While CPSs have promising capabilities, the design process of such a networked system with emergent behavior puts a big challenge upon a toolset for CPSs. In this chapter we present a way to design such emergent systems by applying evolutionary algorithms as part of a CPS toolset.

5.1.1 Motivation

While existing methods and tools from complex system research, embedded system design, and self-organizing systems focus on individual design issues of CPSs (e.g., hardware interaction, optimization, or deployment), there is no comprehensive methodology to engineer self-organizing systems. Individual theories allow formal descriptions of different aspects of CPS design. Those theories comprise physical, technical, and organizational perspectives at different levels of detail. Considering only a set of topics to design CPSs, the complexity is increasing tremendously. This is shown with a conceptual map in Fig. 5.1. Although this list is still far from being complete, the demand for establishing design and deployment methodologies for CPSs is clearly indispensable (Lee 2008; Schätz et al. 2015). However, not all mentioned disciplines and open subjects are fully integrated in a common, generally valid system theory. Instead, individual methodologies, representations, and tools exist to address single aspects of CPS design. Nevertheless, a closed design process for CPS design is still an open issue.

5.1.2 Objectives

With a CPS design process, the following two high-level goals can be achieved: (i) The toolset should ease the process from the design to the deployment of complex, autonomous, and heterogeneous CPSs, and (ii) the whole design process should be reduced in terms of complexity and time. These goals can be split further into subgoals as follows:

- **Support the CPS design process**
  The toolset should support the CPS design process in all stages from modeling and implementation over optimization to the final prototype.
- **Provide an extensible library of reusable models**
  An initially predefined library should be provided to the modeler for describing CPSs. This library specifies models of CPS agents, environments, and goals to be reached by the CPS.
- **Provide an extensible library of swarm and evolutionary algorithms**
  A CPS design process does not just deal with descriptive modeling aspects but also needs to include a set of reusable reference algorithms. In complex tasks, the CPS behavior is typically emergent and not easy to predict. Typically, multiple

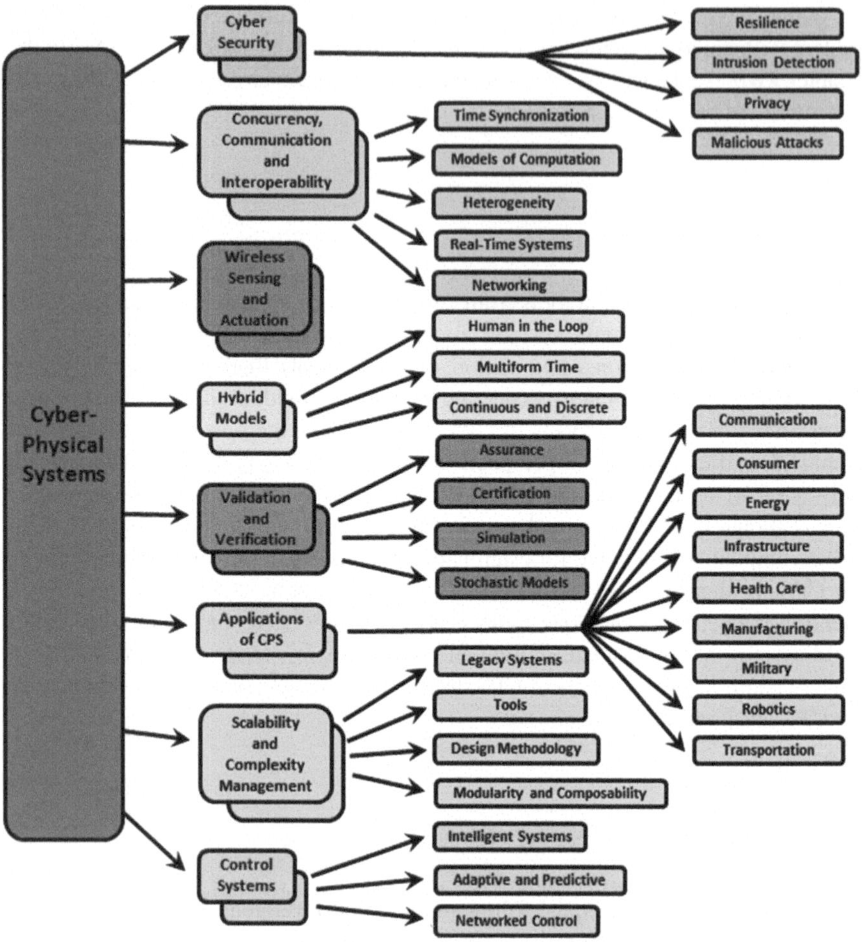

Fig. 5.1 The complexity of designing CPSs. (Adapted from Bagnato et al. 2017)

CPSs – a swarm of CPSs – collaborate to reach high-complexity goals. To solve complex tasks, swarm and self-organizational algorithms are applied and further optimized through evolutionary algorithms with respect to predefined goals. This also includes meta-heuristic design and applications, where needed.

- **Reduce complexity and time by automatic deployment**
  By reusing models and integrating existing methodologies, the CPS design process is able to reduce development and integration effort by maximizing reuse. Thus, techniques to apply an easier block-based design of CPSs are the focus. Further, methodologies to support iterative refinement of the design are needed. As the final algorithms are evaluated and optimized through the iteration process, a code generation tool is able to transform these optimized code sequences to the final piece(s) of hardware.

- **Support hardware abstraction**
  Interoperability and predictive engineering are necessary to enable estimation and prediction of the overall swarm/self-organizational behavior and performances. Further, they are necessary for a reliable integration with third-party CPSs. Concurrently, hardware abstraction issues must be managed to enable cross-platform CPS integration. Thus, a hardware abstraction layer should be introduced to isolate generated artifacts from real CPS hardware.
- **Focus on industrial needs in CPS design**
  The basis for the CPS design process are industry-driven use cases. This enables the final toolset to gather qualitative and quantitative measures of improvement to the CPS engineering process in terms of development time and involved costs.

5.2 The CPS Design Process

We propose a CPS design process combining existing and emerging tools to solve complex, real-world problems in the application area of CPSs. The proposed toolset should follow three design steps by providing (i) a reusable model library for CPS design, (ii) evolvable CPS functionality, and (iii) the deployment of prototypes. Furthermore, such a toolset is capable of partitioning the individual tasks within the CPS design process to users with a different knowledge base – the modeler, the software developer, and the engineer.

5.2.1 Roles for the CPS Design Process

As shown in Fig. 5.2, the design process is partitioned to three roles of different knowledge base.

1. **Modeler**
   The modeler is a person modeling a problem definition and all its associated parameters with a computer modeling tool. She/he has a rapid perception for complex problems and a basic understanding of programming.
2. **Software Developer**
   The software developer describes functionalities of the models provided by the modeler. Therefore, she/he uses a high-level programming language. Her/his characteristic is an analytic and structured operation.
3. **Engineer**
   The engineer has the task of deploying the final algorithms on hardware. She/he understands specific and detailed hardware characteristics as well as the deployment chain to bring code to hardware. Her/his characteristic is a structured working ability.

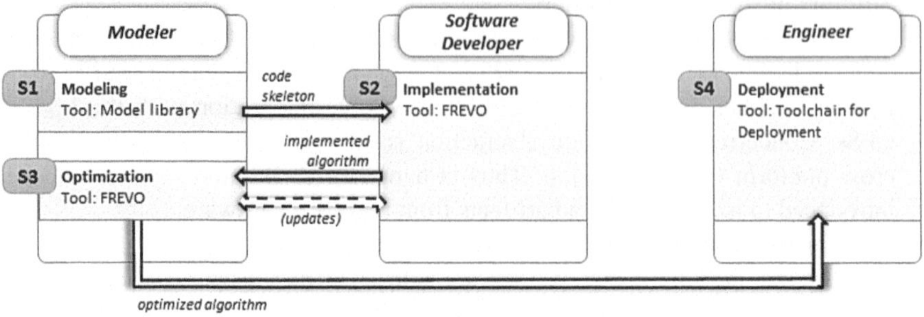

Fig. 5.2 The three roles with their individual tasks in the CPS design process

Applied to the proposed toolset in the next section, the CPS design process involves four steps, denoted with S1 to S4 in Fig. 5.2. The entire modeling is performed by the modeler in S1 using the model library (see Sect. 5.2.2.1). Out of the models, the modeling tool generates code skeletons (like classes, descriptions of the functionality, etc.), which serve as input to the software developer. In S2, the software developer implements the functionalities/algorithms and returns them to the modeler. In S3 the modeler optimizes the code with evolutionary algorithms in FREVO (see Sect. 5.2.2.2). If necessary, the modeler adds additional functionalities that are then implemented by the software developer and passed back to the modeler to be optimized again. Finally, the optimized algorithm is converted to hardware-specific code and deployed on the final piece of hardware by the engineer.

5.2.2 *Toolset for the Design of CPS*

A conceptual architecture of the toolset for the CPS design process is shown in Fig. 5.3. Designing a CPS with this approach assumes that the initial problem definition is known and understood by the modeler. This comprises knowledge about the overall CPS hardware specifications (like flight time and flight speed of UAVs), the problem that needs to be solved by the CPS, and the goal. Through three stages a CPS is created – supported by the toolset in an automated optimization process and a code generation module, which is deployment-ready for the real CPS hardware. In particular, the stages cover the following activities: (i) The models are taken out of a model library as required and the corresponding functionalities are implemented as required, (ii) these serve as input to the workbench that optimizes the algorithm until the acceptance criteria are reached, and (iii) the final algorithm can be directly deployed on the target CPS.

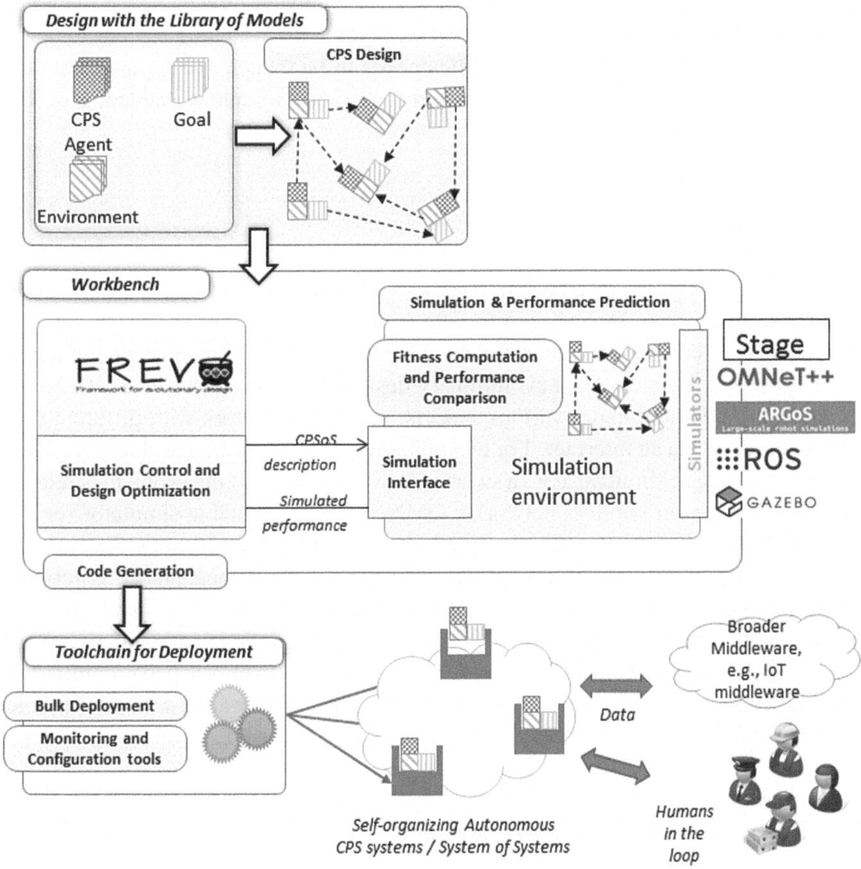

Fig. 5.3 The high-level architecture of the proposed toolset for the design, simulation, evaluation, and deployment of CPSs using evolutionary algorithms. (Adapted from Bagnato et al. 2017)

5.2.2.1 Design Using a Model Library

As proposed in Fig. 5.3, the model library consists of several models that can be customized to form the CPS. This includes models for the CPS agent, the environment, and the goal to achieve. The CPS agent library includes models for functions, behaviors, security issues, communication systems, movement models, etc. It encompasses hardware-independent models for single CPS subsystems, where it is possible to define local physical characteristics like communication technologies, sensors, actuators, and computing capabilities. Security issues can be considered by reacting with proper countermeasures to the modeled threats. Behavior routines cover interactions among a set of CPSs, e.g., by exchanging specific data. The models in swarm/self-organization are models from algorithms – including nature-inspired algorithms, like bird flocking or ant food foraging – to solve the original problem. Furthermore, in environmental models it is possible to define the environment the CPS is moving in. Finally, the goal a CPS needs to reach with all its capabilities and functionalities needs to be modeled. The model library is open for extensions and further grouping, so models could be added, e.g., for human-to-CPS interaction.

5.2.2.2 The Workbench

As the conditions for a CPS are set, the system emerges over iterations by simulation evaluation. In each iteration the simulation results are validated against the acceptance criteria and restricted by design constraints. For this functionality, we propose the framework for evolutionary design (FREVO) (Sobe et al. 2012). As a CPS can be defined under various aspects, several simulators with different foci can be included via an interface. For example, Stage[1] is used if the problem is a two-dimensional one to simulate, e.g., a swarm of ground robots. If the problem is related to a routing issue in wireless networks, OMNeT++[2] is a suitable simulator for the evaluation. Gazebo[3] is specially designed for the simulation of drones. To evaluate the simulation output, an interface is used to hand over the acceptance criteria in the form of a fitness function. This function serves as input to the next iteration for the evolution of the algorithm. Typically, several iterations of simulation need to take place, to overcome initial constraints and effectively disperse conflicting requirements from the design in the beginning. Only then an algorithm can emerge that gives the best solution satisfying the global goal of the CPS.

5.2.2.3 Code Generation for Deployment

After the design and the engineering of the algorithms for the CPSs are done, the final stage is to start the deployment by automatically transforming the generated code to hardware-specific requirements. Automatic code generation avoids errors that are quite common in manual implementation of algorithms. An automatic step for code generation and deployment further greatly reduces time and effort for software revisions.

5.3 Designing Systems by Evolution

Designing a system can be also rephrased into finding the right design for the system. While traditional design processes typically involve a top-down approach with

[1]The Stage Robot Simulator, https://github.com/rtv/Stage/, Accessed: 2017-04-25

[2]OMNeT++ – Discrete Event Simulator, https://omnetpp.org/, Accessed: 2017-04-20

[3]Gazebo, http://gazebosim.org/, Accessed: 2017-04-25

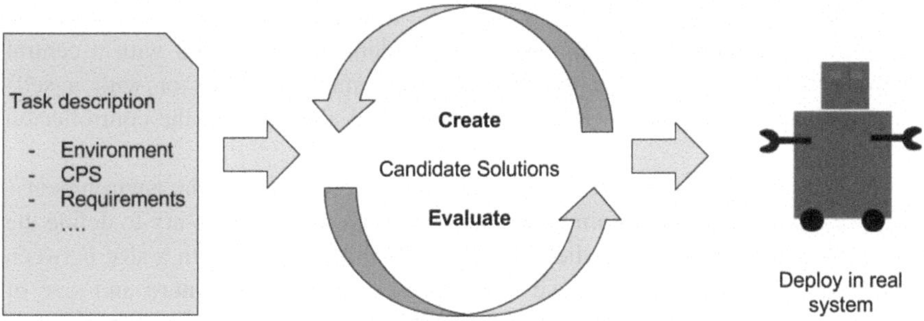

Fig. 5.4 Evolutionary design approach

hierarchically organized steps, an evolutionary design approach aims at building the target system bottom-up from many small interacting components. One of the major problems with both approaches is that, if the target system consists of highly networked interacting components, even a small change in the system can lead to significantly different outcomes. This sometimes leads to counterintuitive behavior of a system, which even (or especially) misleads domain experts in predicting the system behavior after modifying a parameter (Resnick 1997). However, an evolutionary algorithm is designed to search for a solution without a bias from a system model. Figure 5.4 shows the basic approach of such a process: a set of potential solutions is modified until a satisfying result is achieved.

In the following, we discuss the basic components which are necessary to design a CPS from an evolutionary perspective. In (Fehérvári and Elmenreich 2014), Fehervari and Elmenreich propose a design methodology and identify six parts to be addressed for an evolutionary design approach:

1. Task description: Set of requirements that the solution has to meet.
2. Target system simulation model: Describes the simulation model and the relations between the system's components and its environment.
3. Evolvable decision unit: An evolvable representation of the CPS controller.
4. Interaction interface: Describes how the decision unit interacts with its environment.
5. Search algorithm: A search algorithm responsible of finding a valid solution.
6. Fitness function: The fitness function that guides the search algorithm.

5.3.1 *Task Description*

The task description describes the problem statement for the CPS. Therefore, it contains a description of the physical environment of the system and the agents operating in the system. Furthermore, the task description encompasses intended functions, behaviors, and movement models, as described in model library in Sect. 5.2.2.1.

An example for a task of a CPS would be a number of UAVs searching for a person in an unknown environment. This problem can be solved with a central controller that integrates the information from multiple UAVs or with a self-organizing approach where we have a simple local algorithm in the controller on each vehicle.

The task description would involve the physical properties of the environment – since we said that the environment is unknown, it would be necessary to define the boundary conditions (e.g., the field to be searched is a square with a size between 1000 and 10,000 m^2). Further, we need a description of the nature and size of obstacles here. Depending on the type of problem, different physical aspects need to be modeled, for example, wind needs to be modeled in simulations with UAVs, while a simulation involving photovoltaic systems will require to model sunshine and temperature.

Another physical aspect is given by the vehicle itself. Although the selection of the type of aerial robot and determining the number of those robots applied might be part of the engineering process, we need a description for each system that is supposed to be used. For an aerial vehicle, this involves speed and battery lifetime and in a more sophisticated model also acceleration, weight, and drag.

Finally, the task description should contain the success criteria for the system. For example, we could define that the average time for the robot swarm to find the person should be minimized. In a sophisticated version of the success criteria, we could add a criterion like a maximum time to complete the task. For example, in search and rescue missions for avalanche victims, keeping the time short is critical to avoid people dying of asphyxiation. So, the task description could specify that the victims should be found as fast as possible, but at least within a time that allows rescuing a buried person before 10 min have passed.

5.3.2 *Target System Simulation Model*

The target system simulation provides a virtual test-bed for possible solutions. Therefore, an agent controller is executed in a simulation of the target system. The level of detail of the simulation is derived from the task description. For performance reasons, the simulation model should abstract over irrelevant parts while modeling the aspects that are important for the problem with sufficient detail.

Looking at the example of a search and rescue mission for a swarm, the target system simulation could be a state-of-the-art simulation for flying robots, where a model of typical weather conditions and a model for persons on the ground covered by snow have been added.

5.3.3 *Evolvable Decision Unit*

The main idea in evolutionary design for swarm systems is to find the right set of *local rules* that drive the system toward the desired global behavior. Essentially, these rules form the logic of the "brain" or controller of the individual components within the CPS system. While it might appear that these components have to be intelligent on their own, this is nonessential. Imagine a complex living being that is the result of the emergent behavior of a large number of non-intelligent cells. The idea is to encapsulate the necessary actions and responses of a unit into something, so it will result in an interplay yielding the desired results (Fehérvári 2013).

In the past decades many suitable evolvable representation models have been proposed, including evolving decision trees (Greenfield 2012), finite state machines (FSMs) (Pintér-Bartha et al. 2012), and artificial neural networks (ANNs) (Fehérvári and Elmenreich 2010a). ANNs are one of the most popular techniques used for evolving behavior in self-organizing systems (Fehérvári and Elmenreich 2014).

5.3.4 Interaction Interface

The decision unit gets sensor values as inputs and computes control decisions, e.g., via a multilayered ANNs. The *interaction interface* describes these sensors and actuators to interact with the environment. This involves the selection of type and number of sensors and actuators as well as their physical placement and the representation of transducer data. Apart from the hardware sensor configuration, the representation of data also plays a role for the feasibility in an evolved system (Fehérvári 2013). Experiments with self-organized soccer robots have shown that the same sensor configuration yields different performances for different representations of the same data (Fehérvári and Elmenreich 2010a). A similar issue exists on the actuator side. For a given set of actuators, the way how these are interfaced influences the quality of the solution that can be evolved.

Usually, there is no analytical approach applicable for planning sensors and actuators for a system so that it is guaranteed that it can be evolved well for a given purpose. However, Fehervari and Elmenreich identified the following principles to guide the system design (Fehérvári and Elmenreich 2014) for this purpose.

5.3.5 Search Algorithm

The search algorithm optimizes the decision unit models of the components according to the objective function. Therefore, the evolvable decision unit defines a design space (or search space), and the simulation of the target system assigns a fitness value to each of these design options. When the optimization problem is formulated correctly, the main task is to find the optimal solutions by some iterative mathematical solution. As figuratively expressed by Rechenberg (Rechenberg 1994), it is like finding the tallest hill in an unknown landscape. Typically the search space is by far too large to allow for exhaustive search; therefore we use meta-heuristic approaches (Yang 2008), for example, evolutionary algorithms for the search algorithm.

The search will return a possible design that fulfils the intended objective (at least according to its fitness value). In our example case, we get a control algorithm for a CPS swarm that performs a coordinated search.

5.3.6 *Fitness Function*

The fitness function (also called objective function, utility function, or cost function) is basically a numerical representation of the force that guides the search algorithm toward good solutions.

It does not necessarily have a mathematical representation such as a mathematical function but could be also a result of a continuous monitoring of mission parameters during a simulation run. For our search and rescue example, we could define a fitness function for our search mission based on the search time or success probability.

The fitness function that rewards the desired emergent behavior is usually highly problem-dependent, although there are studies available in evolutionary robotics on possible generic methods (Nelson et al. 2009). A design guide for a fitness function can be found in (Floreano and Urzelai 2000).

5.4 A Framework for Evolutionary Design, Simulation, and Evaluation

In this section we describe the framework for evolutionary design (FREVO), a software tool to design self-organizing systems. It is a general-purpose framework that optimizes a generic representation of a given problem using evolutionary methods. As it supports agent-based modeling, it is well suited to evolve the controllers for CPSs, when the controller is implemented as a generic, evolvable representation, e.g., an ANN. An iterative heuristic search is applied to find an optimized configuration of the controller for a CPS with respect to a system-level optimization measure, called fitness. The result is a controller that exhibits the local interaction rules to reach the desired global behavior of the system. The controller can be evaluated on a large scale of parameters under predefined conditions. FREVO uses a modular approach, where the distinct steps of evolutionary design are split into different components. Its graphical user interface (GUI) simplifies the design process and offers statistics and graph generation for easy evaluation of the chosen design. The main purpose of FREVO is to support the optimization process as it guides through the individual steps of the evolutionary design, whereas it requires work by the software developer to implement the modeled problem details. Besides evolving controllers for CPSs, such as in cooperative robotics or wireless sensor networks, FREVO can as well be used for other problems, such as pattern generation (Elmenreich and Fehérvári 2011) or economical simulations (Fehérvári and Elmenreich 2009). Figure 5.5 gives an overview of the FREVO architecture.

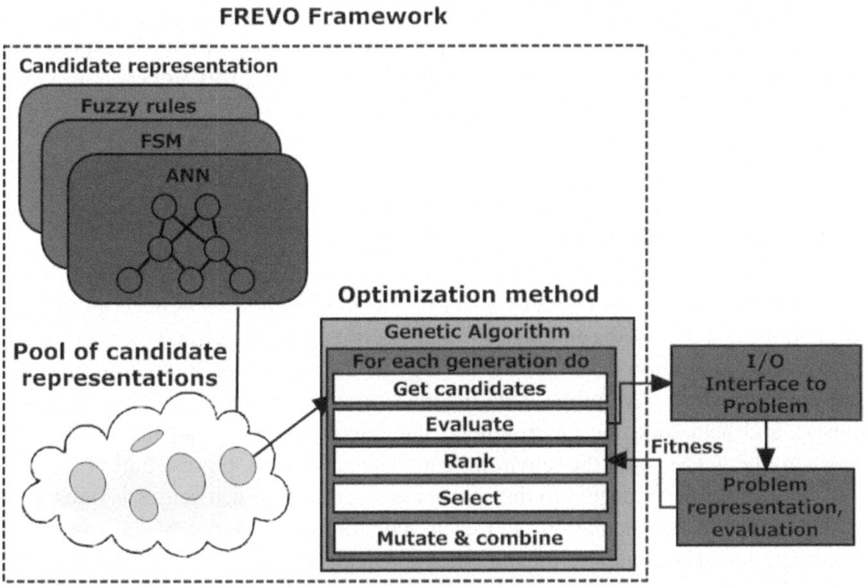

Fig. 5.5 The FREVO framework at a glance (Fehérvári and Elmenreich 2010b)

5.4.1 Architecture

FREVO's architecture is component-based, where the steps of evolutionary design are split into individual components. With this architecture it is possible to develop a single component and easily exchange individual components. Hence, different configurations can be easily evaluated to find the best-suited configuration for a given problem. Each component implements a distinct feature of the evolutionary approach. The *problem* component defines the specifics of a CPS controller, the environment, and the fitness function. The *representation* component defines how the controller of the CPS is represented. The *optimization* component defines the method for finding the optimal candidate representation. The *ranking* component defines how the candidate representations are ordered based on their performance.

FREVO is implemented in the Java programming language and makes strong use of the object-oriented programming paradigm. Each component is defined by an abstract class so that the interfaces between the components are well defined. Therefore, new components can easily be implemented requiring only the implementation of the core functionality. This step is guided by the built-in component generator that assists the software developer by generating the required code skeleton in the context of the class hierarchy. Since FREVO is published under the GNU General Public License version 3, it supports the exchange of research ideas and engineering solutions. The source code is freely available at http://frevo.sourceforge.net/.

The FREVO GUI guides a user step by step through the configuration process. This is done by picking a component for each of the tasks of the evolutionary process. One single configuration of components is called a FREVO session. Such sessions, as well as the results of the optimization process, can be exported and stored to be imported for later use.

5.4.1.1 Problem Definition

The problem definition in FREVO is an implementation of the task description, the simulation model, and the interaction interface that were described in Sect. 5.3.1. The problem definition describes the ecology of the agents, the evaluation context, and the goals of the evolution. This includes descriptions of the CPS in terms of sensors and actuators, the environment, the fitness function, and the interaction between the CPSs and the environment. The candidate representation of the controller must be connected to the sensors and actuators. Each sensor is one input to the candidate representation, and each actuator is one output of the candidate representation. The fitness of a candidate representation is usually evaluated in the phenotype (i.e., behavioral) space by calculating the objective function through a simulation run. This fitness function guides the heuristic optimization process in finding the optimal candidate representation.

There exist two types of problem definitions: The *AbstractSingleProblem* evolves a candidate representation for a CPS that can then be used in a swarm setup to cooperatively achieve a given task. In contrast, an *AbstractMultiProblem* defines a scenario where multiple representations are evolved and evaluated against each other. The first one is used for homogeneous multi-agent systems, where a single controller is evolved by absolute ranking of the fitness value. The latter one is used to derive controllers for competitive multi-agent systems. Here, the fitness of the controllers is evaluated relative to the performance of the other agents and no absolute ranking takes place. Such optimization requires a tournament algorithm to acquire a ranking of a pool of candidates. An exemplary use case is a soccer game, where two teams compete against each other (Fehérvári and Elmenreich 2010a).

To create a new problem definition, the software developer is required to implement the following parts. First, the interface between the sensor inputs and the actuator outputs and the candidate representation needs to be defined. Second, the simulation as method for evaluating the problem needs to be implemented. Third, the calculation of the fitness value according to the given performance measure needs to be implemented. These steps are guided by the predefined interfaces between the components of the system. As the other components are already existing, one can focus solely on the implementation of a new problem without the need to care about representation, optimization algorithm, or ranking components. For more complex and high-fidelity simulations, one can interface with an external simulator from the model library or model a new interface as described in Sect. 5.4.4.

5.4.1.2 Candidate Representation

The candidate representation models the internal structure of the CPS controller. It is a generic structure which is evolvable, e.g., an ANN. It represents a possible solution to the given problem and encodes the reactive behavior of the CPSs. Every representation must define the genetic operators such as mutation, crossover, and selection. For supporting the user in analyzing the representation, different output formats can be implemented.

Typically, the representation is derived from the *AbstractRepresentation* class and common among all agents. For heterogeneous multi-agent problems, one can choose the bulk representation, which can evaluate a set of candidates with distinct representations.

Currently, FREVO supports the following representations:

- *Fully meshed net*: A recurrent, fully meshed ANN with one hidden layer. During evolution the biases of the neurons as well as the connection weights are changed. Adaptive mutations are also supported.
- *Three-layered net*: A feed-forward, nonrecurrent ANN with one hidden layer. The biases of the neurons as well as the connection weights are evolved. Compared to the fully meshed ANN, it supports simpler problems and decreases the search space significantly.
- *NEAT*: An ANN where the connectivity between neurons is also evolved. This implementation is based on the NeuroEvolution of Augmenting Topologies (NEAT) method proposed by (Stanley and Miikkulainen 2002).
- *HebNet*: A recurrent, fully meshed ANN with Hebbian learning. Here the synapses are assigned a plasticity which defines their ability to learn. The plasticity and the initial weights are evolved.
- *Simple bulk representation*: A composition of multiple representations from above.

5.4.1.3 Optimization Method

The optimization method searches for the candidate representation that yields the highest fitness as defined in the problem description. It uses the genetic operators defined in the candidate representation to create new candidates in each generation to replace the worst performing candidates of the population. Through iterative heuristic search, it gradually obtains candidates with better performance. The search runs until a termination criterion, defined in the optimization method, is reached. Termination criteria are, e.g., a maximum number of generations, a maximum number of generations, or a number of generations where the fitness does not improve.

The optimization methods currently offered by FREVO are the following:

- *Random search*: A baseline comparison method, where candidates with low fitness are replaced by randomly created ones.

- *NNGA*: An evolutionary algorithm (EA) that maximizes the population diversity. It supports multiple populations and several ranking algorithms. It is based on the neural network-genetic algorithm (NNGA) described by (Elmenreich and Klingler 2007). It is suited to evolve any kind of representation.
- *GASpecies*: An EA that classifies candidates into species. Species are defined by a similarity function defined in the candidate representation. Candidates within one species share their fitness value.
- *CEA2D*: A cellular EA that arranges all candidates on a 2D torus surface. Genetic operations are performed in a local context. It features better diversity with slower convergence compared to standard EA.
- *Novelty search*: An EA that rewards behavioral diversity rather than fitness. The implementation is based on rtNEAT by Kenneth Stanley, http://nn.cs.utexas.edu/keyword?rtneat.
- *Novelty species*: An EA, where behavioral diversity across species is rewarded.

5.4.1.4 Ranking Algorithm

The ranking algorithm sorts candidate representations based on their performance, i.e., their fitness value. The ranking algorithm is also responsible for parallelization to decrease the overall simulation time of the optimization process. Two types of absolute rankings are currently implemented in FREVO:

- *Absolute ranking*: A ranking algorithm that sorts candidates by the fitness value returned from the problem component. It supports multi-threading to decrease the time needed for optimization.
- *Novelty ranking*: A ranking algorithm that sorts candidates based on their novelty in the behavioral space.

If an *AbstractMultiProblem* is evolved, pairwise comparisons between the candidates are performed to achieve a relative ranking. For this purpose, a full tournament ranking and a ranking based on the Swiss system are provided. The latter is able to provide a ranking with fewer comparisons at the cost of ranking accuracy (Elmenreich et al. 2009).

5.4.2 Graphical User Interface

The GUI of FREVO enables quick evaluation of a component of the evolutionary design process by reflecting the modular architecture underneath (see Fig. 5.6). The figure shows the FREVO GUI, where an example problem has been evolved with the *Select Problem Component* window in front. The top left *Configure Session* panel allows the user to configure a session by guiding him/her step by step through the process of selecting a problem, an optimization algorithm, a candidate representation, and a ranking method. Testing new components can therefore be

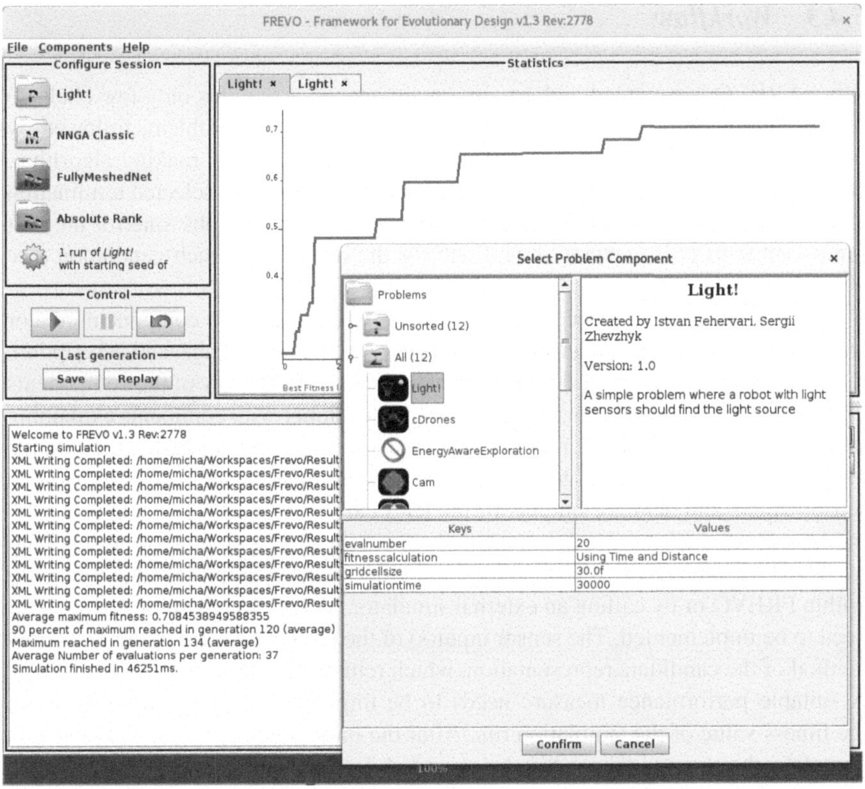

Fig. 5.6 Screen shot of the FREVO GUI showing the evolution of an example problem

done easily. For each step a new window opens, where the component can be selected and configured. Each component can have multiple parameters to fine-tune the configuration. Below there is the *Control* panel that starts, stops, or resets the optimization. The optimization process can be monitored in the *Statistics* panel to the right and the *Console* panel below. The *Statistics* panel shows, by means of graphs, how fitness and diversity evolve over the generations, whereas the *Console* panel shows the output of the active components. Once the optimization stops, it is possible to access the results from the *Last generation* panel on the left. Here it is possible to save the result or replay it for further introspection. By clicking on *Replay*, it is possible to have a closer look at each candidate representation of the last generation as well as to show the resulting behavior in simulation. The top menu offers the possibility to save and restore previous sessions and to manage components. This includes the *Component Creator*, a tool for automatic code generation of skeleton code that helps the software developer to implement new components for the model library.

5.4.3 Workflow

When FREVO is executed with existing components, it requires only few clicks to run the optimization. First, one needs to select the desired problem, followed by the optimization method, the candidate representation, and the ranking algorithm. After a click on the play button, the optimization runs until the selected termination criterion has been fulfilled. The statistics panel shows two graphs: one for the best fitness value of each generation and one for the diversity in each generation (see Fig. 5.6).

When a new problem needs to be modeled, one can use the component creator from the top menu. After selecting component type, name, package, and description, the code skeleton is created. The code is placed in a subdirectory of the components directory in FREVO and provides the software developer with comments explaining where further implementation is required (see Fig. 5.7). Also an XML file with the name of the component is generated, where configuration parameters as well as sensor inputs and actuator outputs of the CPS can be defined. The main task is to implement the *evaluateCandidate* method, where the given candidate representation needs to be evaluated. This is done by simulation, either by implementing it directly within FREVO or by calling an external simulator. Therefore environment and CPS need to be implemented. The sensor input(s) of the CPS are passed to the *getOutput* method of the candidate representation, which returns the output for the actuator(s). A suitable performance measure needs to be implemented that is used to return the fitness value of the simulation run. After the implementation of a component is complete, the component needs to be compiled. It is then automatically loaded upon the launch FREVO. Then the new component appears in the component selection window, and the workflow above can be applied. The same process also holds for creating other types of components, whereas the generated code skeleton and implementation are tailored to the specifics of the respective type.

MyProblem.java

```
package utils;

import core.AbstractSingleProblem;

public class MyProblem extends AbstractSingleProblem {

    @Override
    /** Evaluates the given representation by calculating its corresponding fitness value. A higher fitness
 means better performance.<br>
     * @param candidate The candidate solution to be evaluated.
     * @return the corresponding fitness value. */
    public double evaluateCandidate(AbstractRepresentation candidate) {
        // TODO Auto-generated method stub
        return 0;
    }

    /** Returns the achievable maximum fitness of this problem. A representations with this fitness value
 cannot be improved any further. */
    public double getMaximumFitness() {
        return Double.MAX_VALUE;
    }

}
```

Fig. 5.7 Screenshot of the source code skeleton of a newly created problem

5.4.4 Simulator

With FREVO it is possible to use external simulators that possibly offer high-fidelity simulations. If this is desired, the *evaluateCandidate* function of the problem component needs to call the simulator and pass the candidate representation, i.e., the CPS controller. Therefore, the simulator needs to implement the same representation so that FREVO is able to evolve an optimal solution. There are two possibilities for that: Compact, by code generation, or modular, by passing only the parameters of the candidate representation. For the compact approach, the code for the simulator needs to be recompiled with the newly created representation in each generation. The simulator is then directly executed from FREVO and returns the fitness value for optimization. This approach is more suitable for CPSs that offer neither file system nor network communication. For the modular approach, the simulation is implemented once with the desired candidate representation. The parameters that define the representation are then passed to the simulator in each generation, either by means of files or by network communication. The actions taken by the individual CPSs within the simulator are then fully defined by the candidate representation, whose parameters are evolved by FREVO. FREVO executes the simulator, which logs the performance measure into a log file. After the simulation finishes, one needs to compute the corresponding fitness value in the *evaluateCandidate* method.

5.5 Exemplary Use Case: Search and Rescue with Unmanned Aerial Vehicles

This section describes a use case to show the applicability and usefulness of the proposed design process that involves different tools and methodologies. These tools and methodologies are combined in the toolset for the design, simulation, evaluation, and deployment of CPSs. The use case involves a swarm of UAVs in an industry-driven, time-critical application, where complicated tasks are to be solved. This leads to complex, non-deterministic behavior that emerges from the interactions between the UAVs and is very hard to predict with traditional approaches. This is even more true if the UAVs are part of a heterogeneous network of CPSs. A typical application is the scenario of search and rescue (SAR), where the UAVs can be employed to generate a situational overview and assist first responders in finding injured persons.

5.5.1 Swarm of UAVs

Generating a situational overview of the disaster scene is an important instrument to identify dangerous sectors, e.g., areas with toxic or explosive gas leakages. Such overviews provide valuable insights to the scene, for example, through captured real-time images. Generating an overview requires the UAVs to cover vast spatial areas, which can be done, in principle, with a single UAV. However, due to the time-critical nature of the scenario, a swarm of UAVs is much better suited. The swarm can complete the task much faster, which can make a difference between life and death. Thus, especially in SAR applications, swarms are preferable over single UAVs. Additionally, the UAVs can support first responders in efficiently finding casualties or persons trapped in the disaster area. Another enhancement is achieved by extending the homogeneous swarm of UAVs to a heterogeneous swarm by adding ground robots supporting the UAVs with information from the ground. Conversely, the ground robots may order a UAV for finding an optimal path through the disaster area.

A typical swarm of UAVs is modeled as a set of autonomous agents. They are equipped with different sensors, like viewable image system (VIS) or infrared cameras, microphones, or gas sensors. After recording the data from the environment through sensors, they process the data locally and analyze it. This data is usually distributed among the UAVs to allow the other agents to optimize their local results. For communicating they can use different technologies, either through locally available infrastructure (e.g., WLAN or 4G) or by forming ad hoc networks (e.g., Bluetooth or WLAN in ad hoc mode). Ad hoc networks are often favorable as they allow the agents to form a meshed communication network to improve communication performance. Typically, each UAV is equipped with the same sensors. The swarm operates utilizing swarm behavior concepts, based on nature-inspired self-organization. Thus, the tasks that are going to be fulfilled by each UAV are not predefined at the start of the mission, but rather arise as required during the mission. Therefore, the swarm is highly adaptive and robust to changes in the environment. It acts in a dynamic way as the underlying situation requires. Moreover, in contrast to fully centralized control, such a swarm can still operate even if connectivity among the individual agents (or with a base station) is intermittent. The real swarm mission itself is defined in a central operation center in the beginning of the mission. For ease of use, the central station is equipped with suitable user interfaces to enable the operator to monitor and document the swarm mission, including the sensor data gained from the agents. The operator also has the possibility to influence the swarm behavior, e.g., to get a close-up of a certain scene. The swarm would automatically send the closest agent to the scene. In addition to the central control, members of intervention teams may directly access information collected by the swarm via wearable devices and, possibly, modify the swarm tasks, accordingly.

5.5.2 Designing a Swarm of UAVs with the Proposed Toolset

In the design of SAR tasks, as described before, a swarm of UAVs can be used to fulfill the underlying mission (Fig. 5.8). As proposed in Sect. 5.2.2, the design

Fig. 5.8 A swarm of AscTec Fireflys

process passes three steps: the model library, the workbench, and the toolchain for deployment. Further, three roles are taken into account, working on the CPS design process, namely, the modeler, the software developer, and the engineer.

1. **Modeling the CPS with the model library**
   In the first task the modeler works with the library of partially predefined models to model the SAR tasks for the swarm of UAVs. Therein, the modeler needs to define three models to describe goal, agent, and environment of the application.

   First of all, the modeler models the goal as objectives that need to be reached. In this example the goal is a SAR task, where the modeler needs to define the concrete parameters to reach this task. This could include (i) giving an overview picture of the affected scenario to the involved rescue team, (ii) finding three employees that should be still on-site, and/or (iii) providing information about the presence and the location of gas leakages.

In a next step, the UAV is modeled with its typical hardware characteristics. In Lakeside Labs[4] we use the AscTec Firefly[5] for our experimentations. Typical aspects to be modeled in this UAV include, i.a., the UAV's measure of $605 \times 665 \times 165$ mm, its flight time of 12–14 min (with payload), a flight speed of 15 m/s, and the presence of a 2.4 GHz XBee link communication. Additionally, we add a camera module[6] and a local processing unit[7] to the UAV. Further, the modeler needs to define the minimum number of the UAVs used for the swarm task, e.g., eight UAVs. If the SAR task demands heterogeneous agents, e.g., ground rovers are modeled as well. Another subsystem for the UAVs is their mission behavior. Therefore, a predefined swarm algorithm can be applied in the modeling process. As swarm algorithms are typically nature inspired, suitable examples for a SAR process include cuckoo search and ant or bee food foraging (see Wahab et al. 2015 for further information).

Another step involves modeling the environment the UAVs operate in. In a SAR task, the environment is limited by the coordinates of the boundaries of the disaster area. First of all, to simplify the environment, a 2D map can be modeled. In this model further critical way points can be marked, e.g., hazard points or areas with a high probability to find a gas leakage or threatened humans. The fitness function results from the final goal of the SAR task. For instance, the fitness function can be a combination between the number of humans found, the time it takes to find them, and the time it takes to inform the first responders with the information of their location.

A final step in this process involves the software developer. She/he uses the models provided by the modeler to give functionality to them in the form of source code. This is especially needed if the required model is not yet implemented.

2. **Applying the workbench for optimization**
   All the models constructed, adapted, and connected in the previous stage are fed into FREVO as described in Sect. 5.4. FREVO is connected with an external simulator for fitness evaluation. For this exemplary application, the choice is made on Gazebo. In each iteration the algorithm evolves and is evaluated with the simulation interface. Reaching a threshold for the fitness function interrupts this process. The output is an optimized swarm algorithm for a swarm of UAVs performing the SAR task.
3. **Using the generated code on the UAVs**
   Finally, the engineer uses the optimized code and deploys it to the final piece of hardware – in our application the UAVs. The algorithms for the SAR tasks

[4]Lakeside Labs, https://www.lakeside-labs.com/, Accessed: 2017-05-08

[5]AscTec Firefly, http://www.asctec.de/uav-uas-drohnen-flugsysteme/asctec-firefly/, Accessed: 2017-04-25

[6]Matrix Vision mvBlueFOX-MLC200wC, https://goo.gl/7Cbi85, Accessed: 2017-05-02

[7]AscTec Mastermind, http://wiki.asctec.de/display/AR/AscTec+Mastermind, Accessed: 2017-04-29

are already optimized for the task itself and the swarm of specific, pre-modeled UAVs. Thus, the final algorithm is deployed to the UAVs. As we operate a swarm with an optimized swarm intelligence, each UAV gets the same algorithm for the targeted operation. The engineer is in charge of testing the final implementation. In the case that specific hardware or functionalities were not modeled, or that additional hardware is added to the UAV, the engineer needs to make a trade-off, whether the optimized algorithm can still run on the UAVs or the design process needs to be repeated to reach the corresponding goal.

5.6 Summary and Outlook

In this chapter we addressed the design process of CPSs, especially for swarms of cooperating robots. We have identified multiple challenges in such a design process, including the need for proper abstraction layers and the difficulty to engineer properties of an emergent system. Our proposed CPS design process identified users with different expertise and working scope and proposes a model for task sharing and cooperation based on this assumption. For the actual creation of algorithms, we suggest a search approach based on an evolutionary algorithm such that the modeler of a CPS only needs to define the required property in form of a fitness function instead of sketching the actual algorithm. Besides that, we identified the need for a search algorithm, a simulation model, an evolvable decision unit, and an interface for the interaction of an agent with its physical environment using sensors and actuators. We introduced a modular tool named FREVO supporting this evolutionary design approach and providing exchangeable building blocks for different parts of the evolutionary design approach.

For the overall design approach, we propose a toolset including a model library, an evolutionary design tool, a domain-specific simulator, and a code generation and deployment tool. Hence, this toolset provides an integrated solution to design complex systems, such as swarms of CPSs with an inherent support for modeling, algorithm design, software engineering, and deployment. The toolset will be provided as open-source software and will be enhanced with a constantly growing set of features in the future.

Acknowledgements We are grateful to Andreas Kercek and Christian Raffelsberger for their critical comments on this chapter. Further we would like to thank Angelika Schauer for proofreading the text. The research leading to these results has received funding from the European Union Horizon 2020 research and innovation program under grant agreement No 731946.

References

Bagnato, A., Biró, R. K., Bonino, D., Pastrone, C., Elmenreich, W., Reiners, R., Schranz, M., & Arnautovic, E. (2017). Designing swarms of cyber-physical systems: the H2020 CPSwarm project. In *Proceedings of the ACM International Conference on Computing Frontiers*.

Cohen, I., Corman, D., Davis, J., Khurana, H., Mosterman, P. J., Prasad, V., & Stormo, L. (2013). Strategic R&D opportunities for 21st century cyber-physical systems. Technical report, National Institute of Standards and Technology – Steering Committee for Foundations in Innovation for Cyber-Physical Systems.
Elmenreich, W., & Fehérvári, I. (2011). Evolving self-organizing cellular automata based on neural network genotypes. In *Proceedings of the Fifth International Workshop on Self-Organizing Systems* (LNCS, Vol. 6557, pp. 16–25). Springer.
Elmenreich, W., & Klingler, G. (2007). Genetic evolution of a neural network for the autonomous control of a four-wheeled robot. In *Proceedings of the 6th Mexican International Conference on Artificial Intelligence – Special Session (MICAI)* (pp. 396–406).
Elmenreich, W., Ibounig, T., & Fehervari, I. (2009). Robustness versus performance in sorting and tournament algorithms. *Acta Polytecnica, 6*(5), 7–18.
Fehérvári, I. (2013). *On evolving self-organizing technical systems*. Ph.D. thesis, Institute of Networked and Embedded Systems, Alpen-Adria-Universität Klagenfurt.
Fehérvári, I., & Elmenreich, W. (2009). Towards evolving cooperative behavior with neural controllers. In *IFIP Fourth International Workshop on Self-Organizing Systems*.
Fehérvári, I., & Elmenreich, W. (2010a). Evolving neural network controllers for a team of self-organizing robots. *Journal of Robotics, 2010*(1), 1–10.
Fehérvári, I., & Elmenreich, W. (2010b). FREVO – Framework for evolutionary design. http://frevo.sourceforge.net/. Accessed 06 May 2017.
Fehérvári, I., & Elmenreich, W. (2014). Evolution as a tool to design self-organizing systems. In *Self-organizing systems* (Vol. 8221, pp. 139–144). Berlin/Heidelberg: Springer.
Floreano, D., & Urzelai, J. (2000). Evolutionary robots with on-line self-organization and behavioral fitness. *Neural Networks, 13*(4–5), 431–443.
Foundation, N. S. (2016). Cyber physical systems. https://www.nsf.gov/funding/pgm_summ.jsp?pims_id=503286. Accessed 17 Apr 2017.
Greenfield, G. (2012). A platform for evolving controllers for simulated drawing robots. In *Evolutionary and biologically inspired music, sound, art and design* (Lecture notes in computer science, Vol. 7247, pp. 108–116). Berlin/Heidelberg: Springer.
Lee, E. A. (2008). Cyber physical systems: design challenges. In *Proceedings of the 11th IEEE Symposium on Object Oriented Real-Time Distributed Computing* (pp. 363–369).
Martins, P. M., & McCann, J. A. (2017). Network-wide programming challenges in cyber-physical systems. In *Cyber-physical systems* (pp. 103–113).
Nelson, A. L., Barlow, G. J., & Doitsidis, L. (2009). Fitness functions in evolutionary robotics: A survey and analysis. *Robotics and Autonomous Systems, 57*(4), 345–370.
Pintér-Bartha, A., Sobe, A., & Elmenreich, W. (2012). Towards the light – comparing evolved neural network controllers and finite state machine controllers. In *Proceedings of the 10th International Workshop on Intelligent Solutions in Embedded Systems* (pp. 83–87). Klagenfurt.
Rechenberg, I. (1994). *Evolutionsstrategie'94*. Stuttgart: Frommann-Holzboog.
Resnick, M. (1997). *Turtles, termites, and traffic jams: Explorations in massively parallel microworlds* (Complex adaptive systems). Cambridge: MIT Press.
Schätz, B., Törngren, M., Bensalem, S., Cengarle, M. V., Pfeifer, H., McDermid, J., Passerone, R., & Sangiovanni-Vincentelli, A. L. (2015). Research agenda and recommendations for action. Technical report, CyPhERS – Cyber-Physical European Roadmap & Strategy.
Sobe, A., Fehérvári, I., & Elmenreich, W. (2012). Frevo: A tool for evolving and evaluating self-organizing systems. In *Proceedings of the 6th IEEE International Conference on Self-Adaptive and Self-Organizing Systems Workshops* (pp. 105–110).
Stanley, K. O., & Miikkulainen, R. (2002). Evolving neural networks through augmenting topologies. *Evolutionary Computation, 10*(2), 99–127.
Törngren, M., Bensalem, S., McDermid, J., Passerone, R., Pfeifer, H., Sangiovanni-Vincentelli, A., ... & Asplund, F. (2017). Characterization, analysis and recommendations for exploiting the opportunities of cyber-physical systems. Chapter in the book on *Cyber-physical systems: foundations, principles and applications*.

Wahab, M. N. A., Nefti-Meziani, S., & Atyabi, A. (2015). A comprehensive review of swarm optimization algorithms. *PLoS One, 10*(5), e0122827.
Yang, X.-S. (2008). *Nature-inspired metaheuristic algorithms*. Frome: Luniver Press.

Melanie Schranz is a Senior Researcher working in the EU H-2020 project CPSwarm in the nonprofit research company Lakeside Labs, Klagenfurt, Austria.

Wilfried Elmenreich is a Professor of Smart Grids in the Department of Networked and Embedded Systems at the Alpen-Adria-Universität Klagenfurt, Austria.

Micha Rappaport is a Researcher working in the EU H-2020 project CPSwarm in the nonprofit research company Lakeside Labs, Klagenfurt, Austria.

Part II
Pedagogy of Cyber-Physical Experimentation

Introduction by Zacharias C. Zacharia and Ton de Jong

Computer technology has revolutionized the way science experimentation is implemented nowadays in schools. Recent research has revealed a number of unique affordances of cyber-physical laboratories, which improve the learning experience of students at both a cognitive and an affective level. For instance, the use of virtual labs with affordances that could not be provided through physical labs (e.g., provision of representations of conceptual and abstract objects, such as vectors and particles) was found to positively influence science learning. On the other hand, when cyber-physical laboratories are placed within certain pedagogical contexts, especially one involving inquiry, students appear to face a series of problems that relate primarily to addressing the requirements set by the pedagogy. For example, in an inquiry-oriented context, students have difficulty in formulating hypotheses and designing informative and sound experiments. As a result of these difficulties, researchers have placed great emphasis on designing learning environments in which the use of cyber-physical laboratories is accompanied by a series of supportive resources, such as appropriate learning materials (i.e., materials that comply with the pedagogical framing at hand) and adequate guidance tools (e.g., scaffolds, prompts, heuristics). The idea behind such learning environments is to enable the student to have a smooth and productive learning path.

In this section, we deliberately selected work that reports and reflects upon the current, major online learning platforms that support experimentation through virtual and/or remote labs. All of these platforms are of high technical quality and have shown evidence of benefiting learners through their virtual labs, learning materials, and guidance tools. All of these platforms have large-scale usage. In reading each of the chapters in this section, you get a sense of how much the computer and software technology have progressed. You also get a good picture of the potentialities afforded by current online learning platforms. Therefore, there are take-home messages for educators, researchers, designers/developers (e.g., the industry), and practitioners. In particular, each of these individuals could find information in the chapters in this section that points to the new perspectives and challenges of their field. The examples/cases presented in each chapter could also be helpful in this respect. The overall goal is to provide an up-to-date overview of recently developed online learning platforms and to point out the added value that each one could bring to science learning. The section consists of five contributions from researchers who are at the forefront of online learning platform development. In this way, we provide insight into how recent, high-tech platforms are designed, developed, and implemented for learning purposes.

In the contribution, "Advances in PhET Interactive Simulations: Interoperable and Accessible," the authors describe how the PhET Interactive Simulations (phet.colorado.edu) advanced over the years and discuss how recent initiatives, PhET-iO and accessible PhET sims, have affected the design and development of the PhET simulations. PhET-iO focuses on increasing the interoperability and the level of customization of the PhET simulations, as well as supporting the inclusion of the PhET simulations in interactive e-textbooks and virtual lab notebooks. Moreover, the authors discuss how PhET-iO data could be used for the development of performance tasks. In the case of the second initiative, accessible PhET sims, the authors discuss how PhET infrastructure was changed to ensure that all students, including students with disabilities, could have access to PhET simulations.

In the contribution, "Designing Virtual Laboratories to Foster Knowledge Integration: Buoyancy and Density," the authors discuss the iterative process followed for developing an online instructional unit featuring virtual laboratory activities as developed within the WISE platform (wise.berkeley.edu). Specifically, the authors report on a learning activity that focused on investigating how mass and volume relate to the phenomenon of buoyancy. In this context, they evaluated the added value of the activity and the virtual laboratory according to three criteria: enactment of meaningful experiments, proper interpretation of evidence, and discovery of new ideas. Finally, the authors discuss the added value of such an iterative process and how it can affect practice.

In the contribution, "Scaffolding Students' Online Data Interpretation During Inquiry with Inq-ITS," the authors present an Inquiry Intelligent Tutoring System (Inq-ITS, www.inqits.com) that includes a variety of interactive simulations and virtual labs for different domains in physical, life, and earth science. They also present and discuss two affordances of the Inq-ITS. The first aims at supporting teachers with inquiry assessment by providing automatic, formative data, and the second aims at helping students enact inquiry by providing real-time, personalized guidance. Additionally, they have put Inq-ITS to the test. In particular, the authors have examined how scaffolds within Inq-ITS could help students learn skills related to data interpretation and warranting claims. Overall, this work provides a framework for the assessment and scaffolding of these practices.

In the contribution, "Providing Pedagogical Support for Collaborative Development of Virtual and Remote Labs: Amrita VLCAP," the authors present an eLearning platform, Amrita VLCAP (www.olabs.edu.in and vlab.amrita.edu), which is based on a multi-tier architecture that supports collaborative development of cyber-physical materials and carries a number of affordances, such as publishing in various online and print formats, security, auditing, and access controls. The design of the Virtual Labs Collaboration and Accessibility Platform (VLCAP) also supports the use of open technologies, provides templates for structuring the pedagogical framing of a learning activity, carries multilingual functionality, and supports sharing virtual labs from multiple geographic locations and securely accessing remote equipment. In showing the potential of the VLCAP for hosting virtual and/or remote labs, the authors present two cases of hosted ICT projects, namely, the *Online Labs (OLabs) for school education and Virtual Labs for higher education* project and the *Remote Triggered Wireless Sensor Network Lab* (RT-WSN Lab).

Finally, in the contribution, "Model-Based Inquiry in Computer-Supported Learning Environments: The Case of Go-Lab," the authors discuss how model-based inquiry in computer-supported environments could be enacted. To do so, they use the Go-Lab platform (www.golabz.eu), which includes inquiry-based environments for learning (i.e., Inquiry Learning Spaces), virtual and remote laboratories, and scaffolds that support inquiry learning processes, as an example of a learning platform in which model-based inquiry could be implemented. Specifically, the authors present three examples of virtual laboratories with modeling and simulation affordances from the Go-Lab sharing platform, to demonstrate how Go-Lab learning materials, labs, and tools could be used for the enactment of model-based inquiry.

Chapter 6
Advances in PhET Interactive Simulations: Interoperable and Accessible

Emily B. Moore and Katherine K. Perkins

Abstract Over more than a decade, the PhET Interactive Simulations project has created a suite of interactive simulations (sims) that support learning of science and mathematics content through exploration and discovery. Here we describe the state of the art in interactive science simulations, historical innovations that enabled this state, and current initiatives to advance the field.

Recently, the PhET project has engaged in two initiatives, PhET-iO and accessible PhET sims. PhET-iO increases the interoperability of sims and supports increased customization such as selection of available controls and starting conditions of the sim. PhET-iO also supports expanded integration of sims into interactive e-textbooks and virtual lab notebooks. Access to backend data streams from PhET-iO allows for the development of rich performance tasks suitable for innovative assessments that measure the learning of science practices and allow for adaptive feedback. These capabilities create new ways to positively influence science pedagogy and create targeted and adaptive learning environments for students.

Accessible PhET sims are addressing the need to ensure that all students, including students with disabilities, are allowed equitable access to high-quality learning experiences. Creating accessible interactive learning tools requires the development of new infrastructure to support communication between the simulations and assistive devices. PhET's efforts in accessibility include new features that enable sim use by students with mobility or vision impairments or learning disabilities. These features include keyboard navigation, auditory descriptions, and sonification. These features support students with disabilities and provide new opportunities for all students to engage with science content.

Keywords Interactive learning · Engagement · Representation · Science classroom · Configuration · e-Textbook · Assessment · Disabilities · Collaborative learning · Multimodal · HTML5

E. B. Moore (✉) · K. K. Perkins
Department of Physics, University of Colorado Boulder, Boulder, CO, USA
e-mail: emily.moore@colorado.edu; katherine.perkins@colorado.edu

M. E. Auer et al. (eds.), *Cyber-Physical Laboratories in Engineering and Science Education*, https://doi.org/10.1007/978-3-319-76935-6_6

6.1 Introduction

Digital interactive learning tools have the capacity to transform the teaching and learning of science in today's classrooms. Through technological advances, new capabilities are emerging for digital interactive learning tools to become increasingly customized, integrated, and accessible. These advances are resulting in new opportunities for technology-enhanced learning to support rich learning experiences for students. In this chapter, we set the stage for introducing the latest advances in PhET simulations by first introducing an example simulation and, as exemplified in this simulation, highlighting the project's design goals for the full suite of PhET simulations. We then look back upon the evolution of the PhET project, noting the prior innovations that provided the foundation on which we are advancing the capabilities of these learning tools. With this historical perspective in mind, we then introduce the latest advances in PhET simulations, PhET-iO, and accessible PhET simulations. We share the challenges that each of these advances addresses and the enhanced capabilities provided and project into the horizon new possibilities and contexts for technology-enhanced learning to support the teaching and learning of science and mathematics. Throughout this chapter, we focus on the conceptual evolution of the project and the capabilities enabled by this evolution rather than the research questions and data that supported – and at times propelled – this evolution. For example, we describe what broad goals the PhET project aims to meet, what philosophical approaches led to the state of the art of the simulations, and through these advances what challenges and opportunities are now within sight.

6.1.1 PhET Interactive Simulations

The PhET Interactive Simulations project includes a suite of over 160 interactive simulations (or "sims") for the teaching and learning of topics in science and mathematics. PhET sims are used around the world and across age groups from primary school to university. Each sim is available at no cost from the PhET website (http://phet.colorado.edu), can be used online or downloaded for offline use, and is openly licensed. For each sim we have developed materials to support teacher use (Moore et al. 2014), including Teacher Tips and PhET-created and teacher-submitted classroom activities. Many sims have an associated video primer to quickly orient teachers to sim features and to provide suggestions for scenarios in the sims that teachers may want to incorporate into their lessons.

As an introduction to the state of the art of PhET sims, we will highlight one particular sim, *Forces and Motion: Basics*, and through exemplar features in this sim, we provide a description of the goals that guide the day-to-day design and development decisions that result in PhET sims.

6.1.2 Introduction to Forces and Motion: Basics

The *Forces and Motion: Basics* sim can be used to support student learning of topics related to forces and motion, including net force, friction, and acceleration. This sim is used in classrooms from middle grades to early university level, with students from age 10 to adult.

6.1.2.1 Net Force Screen

Upon startup, the sim lists all four screens available: Net Force, Motion, Friction, and Acceleration. The Net Force screen (Fig. 6.1, upper panel) opens with a cart filled with candy in the center of the screen. Each side of the cart has a rope attached, allowing the cart to be pulled to the left or right. Below each rope is a set of puller-people: one large puller, one medium puller, and two small pullers. Each puller can be moved up to the rope above and be placed on one of four rope positions (Fig. 6.1, lower panel). By selecting a large "Go!" button located just below the candy cart, the puller-people will begin pulling, and the cart will move either to the left or to the right, or it will stay in place, depending on the net force applied by the pullers.

In addition to the interactive objects in the sim, there are also pedagogically useful representations that appear when objects are interacted with. When pullers are moved onto the ropes, a vector representation appears above the candy cart to indicate the amount of force the pullers can apply to that side. Additional options in the upper right side of the screen allow for viewing of other representations: numerical values for the forces being applied by the pullers and a sum of forces vector that indicates the net force all pullers on the ropes can apply to the candy cart.

6.1.2.2 Motion Screen

The Motion screen (Fig. 6.2, upper panel) opens with a crate on a skateboard, next to a pusher-person. At the bottom of the screen are different objects that can be stacked onto the skateboard, including a second crate, a refrigerator, a girl, a man, a trash can, and a mystery box. The pusher-person can apply a constant or an instantaneous force through direct interaction with the pusher-person or through interaction with buttons or the slider at the bottom of the screen. As the pusher-person applies force to the object(s) on the skateboard (Fig. 6.2, lower panel), a vector representation appears indicating the magnitude and direction of this force. As with the Net Force screen, additional representations can be selected, showing the numerical value of the applied force, the mass of each object, and the speed of the skateboard when in motion.

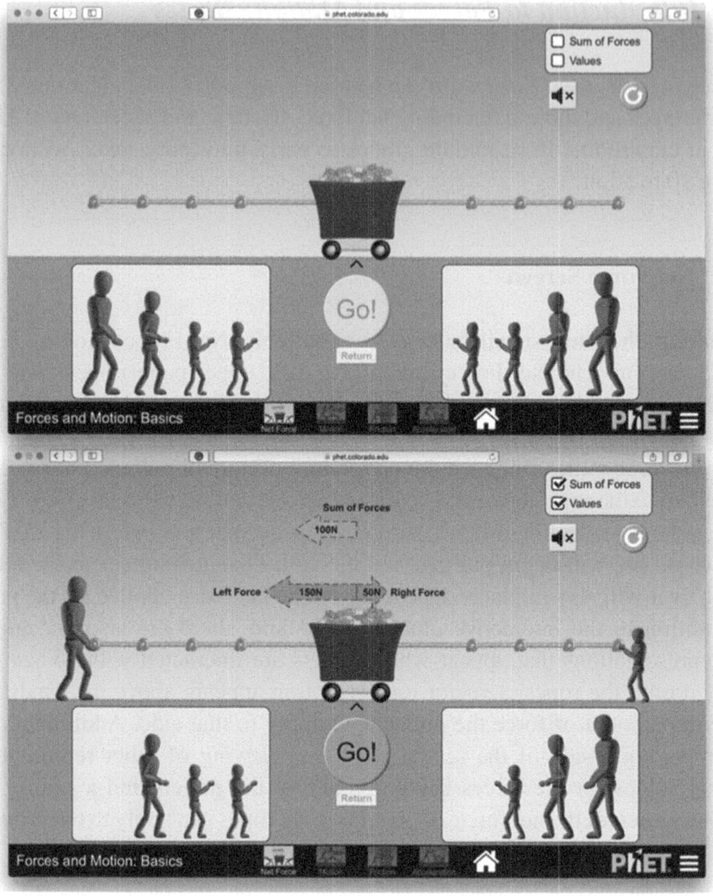

Fig. 6.1 (Upper panel) Screenshot of Net Force screen upon startup. (Lower panel) Screenshot of Net Force screen with puller-people placed on the ropes. The force vectors for each side of the cart and the sum of forces vector are visible

The two other screens available in this sim include the Friction screen and the Acceleration screen. These two screens are similar in layout and interaction to the Motion screen, with a new stackable object (large glass of water), representation (acceleration indicator), and control (friction slider) that support experimentation of friction and acceleration concepts.

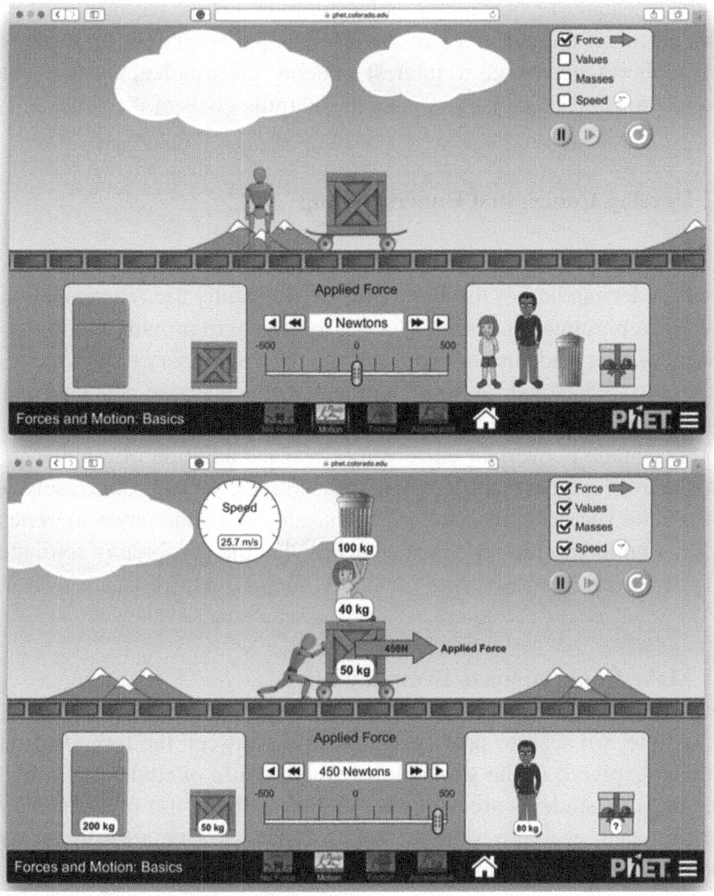

Fig. 6.2 (Upper panel) Screenshot of Motion screen upon startup. (Lower panel) Screenshot of Motion screen with a crate, girl, and trash can on the moving skateboard. The force value, mass values, and speed of the skateboard are visible

6.1.3 Addressing PhET's Goals with Forces and Motion: Basics

6.1.3.1 Engage in Scientific Exploration

Each PhET sim has the goal of engaging students in scientific exploration. In the Net Force screen, the interactive objects were chosen and designed to pique students' curiosity about moving the candy cart. The implicit goal of moving the candy cart encourages student exploration, while the presence of discrete force appliers (the large, medium, and small puller-people) supports students in setting up controlled (and easily repeatable) experiments as they explore. Thus, the resources provided to students through the sim (e.g., discrete force appliers, target locations to place the force appliers) are chosen to interest students and simultaneously enable them to notice relationships necessary to meet the learning goals of the sim.

6.1.3.2 Develop Conceptual Understanding

Beyond exploration, it is also important that students are supported in developing a conceptual understanding of the sim topic. We frequently use representations, such as the vector representations in the Net Force screen, to provide a bridge between the physical system students are exploring and the disciplinary representations used by scientists to understand the topic. In the Net Force screen, students can utilize the on-screen physical system (cart, rope, and puller-people) to make sense of the vector representations overlaid on-screen (e.g., each addition of a puller-person to the right-side rope increases the length of the right-side vector). Conversely, students can also use the vector representations to make sense of the physical system (e.g., when adding pullers to the rope in such a way that the net force is zero, the candy cart will not move).

6.1.3.3 Make Connections to Everyday Life

Where possible, we seek to provide connections between the topic and representations being explored in the sim and the everyday life of students. In the Motion screen of the sim, students are exploring motion with the use of a skateboard, and students can stack everyday objects (e.g., a refrigerator or a trash can) onto the skateboard. By providing opportunities for students to explore everyday objects within the sim, we can support students in recognizing science as a tool for understanding their world.

6.1.3.4 View Science as Accessible and Enjoyable

We aim to support students in viewing science as a discipline that any person can be a part of, and that can be a lot of fun, by providing science resources that support positive experiences. Our goal is that after using the *Forces and Motion: Basics* sim, students have enjoyed their own exploration, furthered their understanding of forces and motion, recognized connections between their investigations of the sim and that of the world around them, and are inspired to continue learning and enjoying science.

6.2 Innovations of PhET Interactive Simulations

The PhET project was started in 2002, during a time of tremendous change and growth in the creation and use of educational technology. Early philosophical perspectives became codified into the three innovations we describe below, each resulting in a set of approaches, practices, and/or resources that set the infrastructure on which the current advances rest.

6.2.1 Design: Flexible and Scaffolded

PhET sims are each designed to support a variety of teaching practices (Wieman et al. 2010; Hensberry et al. 2013; Moore et al. 2013, 2014). The sims can be used as part of lecture demonstrations, labs, in-class individual or group activities, online classes, and homework assignments. This flexibility is made possible by the open-ended design of the sims. There is no single preferred learning pathway through a sim; rather there are multiple learning pathways that can address different learning goals in different sequences. The absence of explicit instructions allows for teachers to embed the sims into their curriculum in many ways and to craft learning experiences with sims that align with their pedagogical goals, context, and practices. For examples of activities teachers have created with the *Forces and Motion: Basics* sim, see teacher-submitted activities under "For Teachers" at https://phet.colorado.edu/en/simulation/forces-and-motion-basics.

Though there are multiple learning pathways through a sim, there are some that more efficiently support the sim's primary learning goals. To highlight these pathways, PhET sims are designed to scaffold student understanding without the use of explicit instructions. We achieve this scaffolding using multiple design strategies collectively referred to as *implicit scaffolding* (Paul et al. 2012; Podolefsky et al. 2013). Implicit scaffolding includes strategies for selecting the scope of each sim and sim screen, supporting students in immediately engaging with each screen through interaction, and enabling continued engagement for sense making and understanding of the sim's learning goals.

When designing the *Forces and Motion: Basics* sim, we selected a set of 3–5 learning goals for each screen. For example, learning goals for the Net Force Screen include: (1) identifying when forces on an object are balanced or unbalanced, (2) predicting how the net force on an object will affect its subsequent motion, and (3) determining the sum of multiple forces on an object and the net force on that object. This set of learning goals is broad enough for a screen design that supports exploration of multiple relationships while being narrow enough for a screen design to implicitly support achievement of all the learning goals.

Each sim screen, upon startup, is designed to implicitly highlight a pedagogically useful starting interaction, for example, moving the candy cart in the Net Force screen of *Forces and Motion: Basics*. In this example, the presence of the large colorful cart in a central location with ropes attached for pulling is intended to cue students that getting the cart to move is a useful task. Providing this visual cue for a starting interaction can also serve to indicate to students that using the sim involves self-directed interaction, rather than passive observation (Moore et al. 2013). We design this initial cued interaction to result in the encounter of pedagogically useful relationships and representations, supporting sim use that leads into a process of experimenting and sense making. It is important to note that each screen is designed to only cue students to make certain useful interactions and it does not require them to start with these interactions. Implicit scaffolding highlights pathways that can be followed or not. As students continue interacting, the available objects (such as the differently sized people-pullers) and representations (such as the net force vector) are designed to support students in experimentation and sense making.

The approach to design the sims without embedded explicit instructions was a departure from the norm in the development of digital educational resources. Rather than creating simulations as a component of a particular curriculum, with a narrowly defined set of contexts for use, the PhET project created highly flexible sims that can be used across a wide range of student age groups, classroom contexts, and teaching practices while also supporting students to engage productively with the sims without explicit instructions (Perkins et al. 2014).

6.2.2 Dissemination: Open Licensing and Broad Compatibility

The PhET project adopted a dissemination strategy that involves an open licensing policy, with no-cost access, broad device compatibility, and online and offline use. Open licensing supports hassle-free no-cost use by teachers and students, which aligns with our belief that high-quality educational resources should be available as free public resources. In addition, PhET's licensing allows use and modification (with attribution) by third-party vendors such as textbook publishers. PhET sims are developed for broad device compatibility and can be run on desktops, laptops, tablets, and mobile devices using multiple operating systems and Internet browsers. The sims can be accessed and run online (no download required) or downloaded for offline use and distribution. The combination of these dissemination approaches has resulted in broad uptake of the sims by schools, teachers, students, and third-party vendors around the world. Additionally, these dissemination approaches influence the infrastructure decisions made by the project, and all design and development decisions are made while keeping in mind the opportunities and constraints inherent when working within a project with a focus on broad dissemination.

6.2.3 *Diversity: Translation*

As international use of the PhET sims increased, we began to seek out ways to further support learners from diverse backgrounds. The PhET project developed a translation tool (Adams et al. 2012) which allows volunteers around the world to translate the PhET sims and the PhET website into their local language or dialect. Translated versions of sims are available from the PhET website. Because of the translation tool and the efforts of many volunteer translators, the PhET sims are available in 87 languages and the full website is now available in 40 languages. While in many countries students are encouraged or required to learn the English language, we did not want English language skills to serve as a barrier to sim use. Developing this translation capability supported worldwide use of the sims.

Through the development and implementation of these innovations in design, dissemination, and diversity, the PhET project has become a leader in interactive simulation design, and PhET sims have become ubiquitous in science classrooms around the world.

6.3 New Advances in PhET Interactive Simulations

In 2013, the PhET project transitioned from developing sims in Java and Flash to developing sims in HTML5 – the new development standard in online educational resources. This transition provided a unique opportunity to build into the PhET sim infrastructure capabilities that were not possible in Java or Flash, build upon the expertise the project team had gained over the previous decade, and advance new opportunities for innovations with PhET sims. Through two initiatives, PhET-iO and accessible PhET sims, we are focused on increasing interoperability and accessibility of the sims.

6.3.1 *PhET-iO: Interoperable PhET Simulations*

Education is experiencing vast changes, including the rise of online delivery of educational resources (Beetham and Sharpe 2013), the emergence of adaptive and personalized learning environments, the demand for more engaging and interactive learning environments, and an emphasis on measuring learning progress. The learning goals themselves are shifting in science (e.g., in the United States, the Next Generation Science Standards (NGSS Lead States 2013) and in school more broadly (Trilling and Fadel 2009)), with a focus on deeper learning of content through engagement in science practices, critical thinking, and problem-solving. While interactive sims provide rich opportunities to engage students in science practices and to develop deep conceptual understanding in an online environment, the original Java and Flash PhET sims lacked certain capabilities, limiting their ability to address emerging needs in the changing education landscape. For example, the original Java and Flash sims offered no customization or configurability, no access to information about student interactions with the sim, and no communication between sims and the digital environment in which they are used.

With PhET-iO sims, we empower instructional designers with a new set of interoperable capabilities, eliminating these limitations and enabling many new pedagogical opportunities. The new capabilities include customization and configuration, integration, and real-time data.

6.3.1.1 Customization and Configuration

With PhET-iO sims, the configuration and the starting conditions of sims can be customized. For example, instructional designers can hide or show controls, hide or show any visual element, change labels, fix slider values, pre-configure a scenario, disable actions, or limit which sim screens are available. These options give the instructional designer significant ability to alter the implicit scaffolding within the sims, supporting greater alignment between the sim configuration and specific learning or assessment activities – whether embedded in an e-textbook, used as a virtual lab, or designed as a homework problem.

6.3.1.2 Integration

With PhET-iO sims, you can create an integrated learning environment where a customized sim is surrounded (or wrapped) by other instructional design elements – e.g., prompts, tables, graphs, and buttons – and can communicate with these elements. Each PhET-iO sim uses a versatile application protocol interface (API) that specifies how the software code of the instructional "wrapper" interacts with the sim to enable a range of functionality. For instance, you can load or save the sim state, record data into a table, take a screenshot of the sim, and monitor achievement of a goal. These capabilities can also be combined to create innovative interactive learning or assessment activities.

6.3.1.3 Real-Time Data

PhET-iO sims include multiple data streams that fully capture student usage and can be used in diverse ways – from real-time monitoring of student performance, to driving adaptive learning environments, to providing summaries of student use for teachers. The three data streams are (1) the event stream which logs every user interaction (button press, object dragging, slider setting, etc.) and any resulting change in the sim, (2) the state stream which logs the entire state of the sim every time the state changes, and (3) the input stream which logs the mouse or touch history (locations, clicks, holds, drags, etc.). The instructional designer decides which data streams to enable, monitor, or save to a server and how to employ that data in their learning or assessment environment.

6.3.1.4 Pedagogical Scenarios Enabled by PhET-iO

With the ability to mix and match any of these features, there is significant new advancement in the flexibility and opportunity to improve existing educational products and to create innovative new learning and assessment environments. Below, we look at how these features can be leveraged within three different learning environments: an e-textbook enhanced with an interactive learning experience, an online activity engaging students in science practices, and a homework or assessment task.

6.3.1.4.1 Enhancing the e-Textbook

While substantial research highlights the improved learning that accompanies active learning environments, passive content delivery is still pervasive. By embedding a customized PhET-iO sim, e-textbook authors can provide a targeted interactive experience that is specifically aligned to a concept the moment it is discussed. For instance, consider a passage where an e-textbook explains "When an object at rest is experiencing an applied force, the force of friction will exactly counter the applied force until the moment the object starts moving." The author could use the customization features of the *Forces and Motion: Basics* sim to create a highly constrained version of the sim to focus student interactions on experimenting with this one idea. The sim shown in Fig. 6.3 has been customized to include only the Friction screen with friction set to its default (moderate) value, speed and force values displayed, one object (the crate) to interact with, control panel hidden, and the background clouds and mountains removed. Student interaction is constrained to applying a force, and when applying an increasing force, they see the opposing force of friction exactly cancels until the crate starts to move. An example of this highly customized sim version can be found from https://phet-io.colorado.edu/examples/etextbook_friction. The addition of customized PhET-iO sims can be a powerful tool to help students interpret a complex idea in an e-textbook.

6.3.1.4.2 Engaging Students in Science Practices

In the United States, the Next Generation Science Standards (NGSS Lead States 2013) have elevated the goal to authentically engage students in science practices, such as planning and carrying out investigations, analyzing and interpreting data, and constructing explanations. Currently, sim-based lessons often include a printed activity sheet and a facilitating teacher, both providing thoughtful prompts and

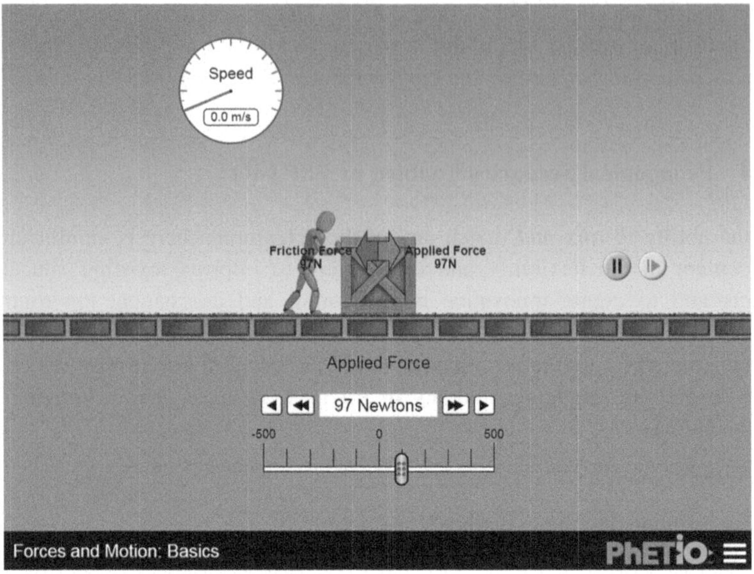

© PhET Interactive Simulations

Fig. 6.3 Screenshot of Friction screen customized using PhET-iO capabilities. The pusher-person and crate are on a surface with moderate friction, the force slider is available for interaction, and the speedometer readout is visible. Example sim available at https://phet-io.colorado.edu/examples/etextbook_friction

structures designed to engage students in science practices while the sim serves as their exploratory environment (Moore et al. 2014). PhET-iO sims enable new opportunities to achieve similar goals in integrated and fully online learning environments.

The PhET-iO sims can be used with a variety of instructional approaches with varying features and degrees of scaffolding – from specific models such as the 5E instructional model (Bybee et al. 2006) to instructional designs that share general features of inquiry-based learning (Pedaste et al. 2015). This flexibility is enabled by the use of the implicit scaffolding in PhET sims together with the capabilities of PhET-iO, which supports a wide range of scaffolding and pedagogical approaches.

In this example (Fig. 6.4), students use the Net Force screen of the *Forces and Motion: Basics* sim as an introductory activity toward achieving the Next Generation Science Standard (NGSS Lead States 2013) "Plan an investigation to provide evidence that the change in an object's motion depends on the sum of the forces on the object and the mass of the object." The instruction moves the student through five stages – predict, explore, experiment, reflect, and apply. The activity is available from https://phet-io.colorado.edu/examples/student_investigation_netforce.

Making significant use of the PhET-iO customization and communication capabilities, the sim configuration can be aligned directly with the tasks students are asked to engage in. For instance, when on the Predict tab of the activity, the sim

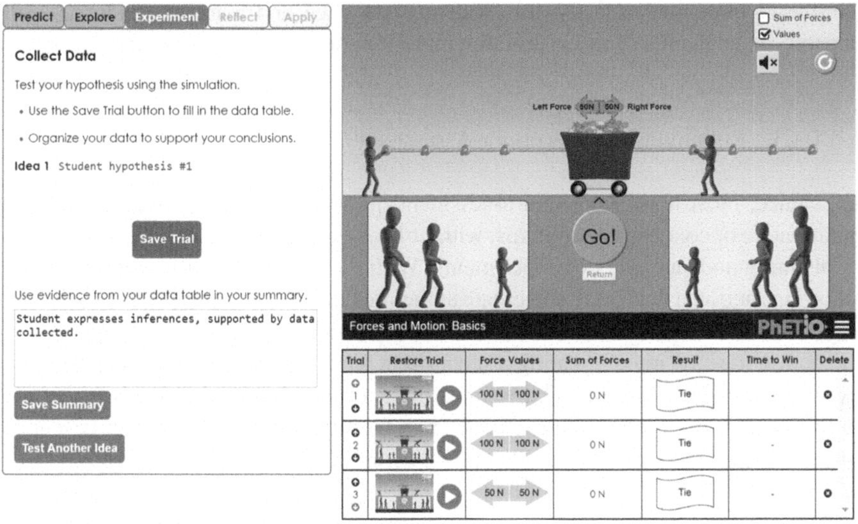

Fig. 6.4 Instructional wrapper for the *Force and Motion: Basics* sim. The "Experiment" tab prompts students to develop a hypothesis and organize their data. The "Save Trial" button collects data from the embedded sim into the table below. Example simulation available at https://phet-io.colorado.edu/examples/student_investigation_netforce

includes only preset scenarios for students to evaluate, with interactions disabled. On the Explore tab, students are invited to "play with the simulation." They are given access to all of the sim controls and readouts and are asked to describe what they notice about how force affects the cart motion. When on the Experiment tab, students are asked to articulate specific ideas they want to test and then test each idea with the sim. Using the capabilities of the PhET-iO API, the wrapper in the Experiment tab enables students to collect data into a table and organize that data with the simple push of a button. The Reflect tab asks students to generalize, writing a set of rules to predict which team wins. They continue to have access to the sim to collect more data into a table, as needed. The Apply tab presents students with four scenarios and asks them to use their rules to predict the outcome and support their reasoning. This tab uses PhET-iO to configure each scenario and to disable sim interaction until the student completes their predictions. With PhET-iO sims, *all* interactions with the sim can be logged, capturing the details of the students' sim explorations and providing the opportunity to analyze their informal exploration practices.

These same PhET-iO capabilities can be leveraged within many existing instructional environments. For instance, this same sim can be integrated within the Go-Lab project's inquiry learning spaces which scaffold students through an inquiry cycle (de Jong et al. 2014), and, indeed, the Go-Lab team has successfully used the PhET-iO API to integrate PhET sims into their environment. Importantly, for any sim-based integrated activity, the student experience and learning outcomes will depend on the details of the instructional tasks, sequencing, and interface design.

6.3.1.4.3 Rethinking Homework or Assessment Tasks

In science, most homework and assessment questions engage students in solving quantitative or conceptual problems, with little engagement in science practices such as planning and carrying out experiments. With the customizability of the PhET-iO sims, instructional designers can create authentic tasks requiring experimentation. In Fig. 6.5, the Motion screen has been altered to provide only four objects and remove the option to show mass, creating a scenario in which students can be challenged to experiment to determine the relative mass of the objects from lightest to heaviest and to justify their reasoning with evidence from the sim. This activity can be extended further, asking the students to collect data using the sim and a timekeeping device to determine the mass of the gift box.

The sim's data streams provide the opportunity to monitor a student's interaction with the sim and examine student problem-solving abilities. For instance, the data streams capture a student's investigative strategies over the course of their sim use and can be analyzed for a student's use of (or lack of) a control-of-variables

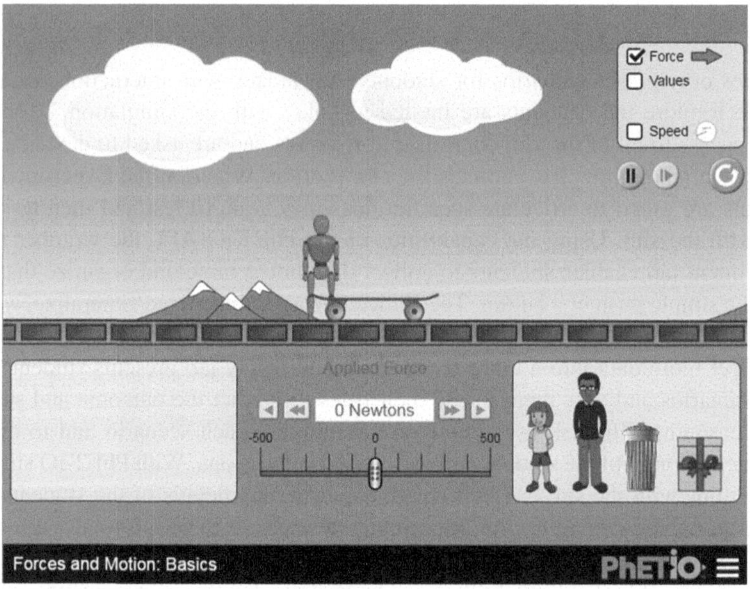

© PhET Interactive Simulations

Fig. 6.5 Screenshot of Motion screen customized for the challenge "Experiment to determine the four objects relative mass from lightest to heaviest" using PhET-iO capabilities. Four objects are provided, with the option to view mass removed

strategy. Does the student repeat a similar series of events with each object, allowing comparison? Does the student simplify the scenario or do they create complex scenarios of multiple objects and varying forces? Does the student use the measurement tools available, showing the speedometer and its values? Indeed, in examining this type of interaction data from a different simulation, Käser et al. (2017) found that exploration strategies can be distinguished and can be a significant predictor of learning outcomes. PhET-iO sims present emerging opportunities for personalized learning environments using learning analytics to leverage these data streams for real-time monitoring, providing analysis to inform teachers' classroom facilitation or to provide feedback to students in the form of new prompts or questions.

6.3.1.5 The Future of PhET-iO

PhET-iO began with a working group of educators, researchers, instructional designers, and company representatives gathering to define the new capabilities needed to enable next-generation, simulation-based learning and assessment environments. While many of these capabilities are now realized with PhET-iO, and simulations can be integrated in a wide variety of educational resources, significant work remains. Advances in techniques are needed for analyzing, visualizing, and making use of the fine process data that captures student interaction with the simulations. With these new capabilities, advances in research and methodologies are needed – in both classroom-based and online learning environments – to inform effective instructional wrapper and task design as well as effective facilitation structures. We will continue to explore this frontier together with our research partners, with the immediate next steps of outfitting more simulations with PhET-iO capabilities, advancing our understanding of sim-based science assessment, and supporting the use of these simulations and capabilities in the broader educational market. We invite others to join this endeavor. Additional information is available at https://phet-io.colorado.edu.

6.3.2 Accessible PhET Sims

Students with disabilities are currently not able to participate in science to the same extent as their nondisabled peers, with a low percentage of students with disabilities found in the sciences at the primary, secondary, and career levels (Moon et al. 2012; National Center for Education Statistics 2011; Stevens et al. 2015). For example, in the United States, 13% of the general population have disabilities, while only 4% of graduate students have disabilities (National Center for Science and Engineering Statistics 2015). A contributing factor to the absence of students with disabilities in science is the lack of high-quality accessible science learning resources.

The current paradigm in interactive simulations involves the design, development, and implementation of browser-based HTML5 applications with highly visual dynamic interfaces and the use of mouse, trackpad, and touchscreen for student input. This recent movement toward implementation in HTML5 and design for touchscreen tablet devices has significantly increased the cross-platform compatibility of these science resources but is insufficient to support access for many students with disabilities. Truly accessible interactive simulations require the expansion of interactions to include more input and output modalities, to match the true span of human diversity in perception and correspondence.

The accessible PhET sims initiative has focused on the addition of three new modalities for sim input and output – keyboard navigation, auditory descriptions, and sonification. You can experience the latest accessible sim prototypes and access video demonstrations of accessibility features at http://phet.colorado.edu/en/accessibility.

6.3.2.1 Keyboard Navigation

Keyboard navigation allows users to interact with the sims using only the keyboard (Schreep and Jani 2005). For example, in the Net Force screen of *Forces and Motion: Basics*, a student can press the Tab Key to navigate to different groupings of interactive features (e.g., the left group of puller-people). By pressing the Enter Key, the student enters the grouping and can then go on to select an individual puller and place them at a knot of their choice on the left-side rope. Once the student has added their chosen puller-people to their chosen rope knots, pressing the Tab Key will access the "Go" button. Once on the "Go" button, pressing the Enter Key selects "Go" and will initiate puller-people applying force to the rope.

Keyboard navigation benefits students with mobility impairments who are unable to easily use a mouse, trackpad, or touchscreen device. In addition, the implementation of keyboard navigation supports other alternative input devices, such as *switch devices* (where all keypresses are made through the use of a single physical button) and *sip-and-puff devices* (where all keypresses are made through inhaling and exhaling through a straw-like assistive device).

6.3.2.2 Auditory Descriptions

Auditory descriptions allow users to hear output from the sim through text-based descriptions and read through assistive software called *screen readers* (Massof 2003; "NVDA," n.d.). These descriptions provide nonvisual access to the sims and include real-time description of all interactive features, updates of changes in the sim as the user interacts, and a continuously updated summary of the overall state of the sim. When using the Net Force screen using a screen reader, upon opening the screen the student hears the name of the sim screen. Using keyboard presses, the student can use their screen reader to navigate through a Scene Summary, which provides a brief text description of the different on-screen features (e.g., "A cart with a rope attached to either side," "Two groups of puller-people," etc.) and the current state of each object. Students can also navigate to each interactive element (e.g., the puller-people, the "Go" button, the checkbox options for views). At each interactive element, the student can hear a description of the object (e.g., "Large Puller," "Medium Puller," "Button: Go! Initiates Pullers"). Upon selecting an object, the student hears any available options (such as how to move a puller-person to the rope) and any on-screen changes to the sim. For example, if a student has added a single large puller-person to the left side of the rope and selected the "Go" button, the screen reader would then read out a description of the motion of the cart to the left.

Auditory descriptions benefit students who use screen readers to access digital information, students who are blind or visually impaired, as well as students with certain learning disabilities.

6.3.2.3 Sonification

Sonification is the use of nonspeech sound to convey information (Kramer et al. 2010). While many PhET sims have sound effects emphasizing particular interactions or outcomes, sonification is the use of sound to convey information about an underlying model or data set. A classic example of sonification of data is a Geiger counter, which is a physical instrument that conveys the level of nearby radiation through a change in the rate of a series of clicking sounds. In the Net Force screen, sonification can be used as a nonvisual way to convey the magnitude and direction of the vector arrows as puller-people are added or removed from the ropes, as well as the speed and direction of the cart when in motion.

This information can also be determined visually, or provided in text description, but a sound mapping of the movement of the cart provides a new modality to access this information that can benefit students who are blind or visually impaired and also benefit students without any visual impairments to utilize a new input channel (audio) to convey subtleties of the acceleration.

While web development standards exist to guide the development of accessible web pages, interactive simulations are not structured like most web pages. To implement the features described above in PhET sims, we have developed an innovative accessible structure called a *Parallel Document Object Model* (PDOM) that couples with each sim. The PDOM updates as the state of the sim changes and provides the necessary communication between the sim and input/output devices (such as keyboards, screen readers, and speakers). The PDOM is an advance in infrastructure that supports the addition of new accessibility features as the accessible PhET sims initiative progresses and can be used as a model for the implementation of accessibility features by other interactive simulation development groups.

6.3.2.4 Pedagogical Scenarios Enabled by Accessible PhET Sims

Accessible PhET sims have the potential to transform the classroom experience for students with disabilities, their teachers, and their nondisabled peers. For example, here are two scenarios highlighting the innovative learning experiences accessible PhET sims could support.

6.3.2.4.1 Collaborative Science Inquiry

Kenzie, Iris, and Anouk are using the PhET sim *Forces and Motion: Basics* to explore net force as part of an in-class activity. Their teacher has asked them to explore the Net Force screen and discuss how the sum of forces vector arrow can help them predict the direction of the cart's motion. The three students share Anouk's computer, which has screen reader software that Anouk can control with her keyboard. Anouk is also wearing bone conduction headphones, which rest behind her ears and allow her to simultaneously hear the screen reader and the people in the room around her.

The students each take turns controlling the computer. Kenzie and Iris use a mouse, and Anouk uses the keyboard. Kenzie places all the left-side pullers on the left-side rope and clicks "Go." Anouk hears a description of all Kenzie's actions as he interacts with the sim, as well as the resulting changes to the sum of forces vector. When Kenzie clicks "Go," Anouk hears a description of the cart's motion. The students discuss the relationships between the size of the puller-people on the rope and the length and direction of the net force vector. Anouk would like to experiment with what happens when you put all the puller-people on the ropes, four pullers on the left and four pullers on the right. Using her keyboard, she navigates to each puller and puts them each on a rope, listening to the description of the change to the sum of forces vector as she goes, and then navigates to the "Go" button. Kenzie and Iris can see the changes in the sim as she does this. Before selecting the "Go" button, each student makes a prediction about the motion of the cart. Anouk selects "Go" and hears a description of the cart's motion from her headphones, and along with Kenzie and Iris, she announces who had the correct prediction – Iris!

Notice how, in this example, the student who is blind is able to seamlessly participate in the learning activity with her sighted peers. She could utilize her screen reader software and bone conduction headphones to follow along with auditory descriptions as her peers experimented, and as she experimented her peers could follow along visually. Rather than resulting in an impediment to her inclusion, the sim provided an opportunity for all of the students in the group to engage in science inquiry together. Also of note, the teacher did not have to provide an alternative activity for the student who was blind. With the appropriate assistive technology and the accessible PhET sim, no alternative activity was needed.

Other groups in this same classroom could include a student with mobility impairments using a keyboard rather than a mouse to interact with the sim and a student with a learning disability that benefits from having verbal description of visual information. In each case, the accessibility features of the sim could be used so that all students in the group could participate fully in collaborative science inquiry.

6.3.2.4.2 Multimodal Science Learning

Marie is sitting at the kitchen table with her mother's laptop, completing a homework activity for her middle school science class. In this homework activity, students are to use the PhET sim *Forces and Motion: Basics*. One of the questions prompts Marie to compare two conditions in the Net Force screen, one in which there is one large puller on one side of the cart and one small puller on the other side of the cart, and a second condition with the same setup but the small puller is replaced with a medium puller. Marie is to write three observations she makes when comparing these two conditions.

First, Marie takes a few minutes to explore the Net Force screen, adding and removing puller-people, determining how to make the cart move to the left and to the right with the puller-people, and turning on the sum of forces vectors and seeing how adding and removing puller-people cause the force vectors to change. Marie then sets up the first condition in the homework activity, one large puller on one side and one small puller on the other side. She observes that the force vector arrow on the large puller side is larger than the force vector arrow on the small puller side. She selects "Go" and observes the cart move in the direction of the large puller. Marie also has sound "on" and hears a tone representing the speed of the cart increase in pitch as the cart speeds up. Marie then sets up the second scenario, by replacing the small puller with a medium puller. She selects "Go" and again observes the cart move in the direction of the large puller and hears the tone representing the speed of the cart increase in pitch more slowly. Marie writes down her three observations: (1) the force vector for the small puller is smaller than the force vector for the medium puller, (2) the sum of forces vector is larger for the condition with large puller versus small puller than in the scenario with large puller versus medium puller, and (3) the cart increases in speed faster in the condition with large puller versus small puller than in the condition with the large puller versus medium puller. Marie then goes on to explore how to get the cart to increase in speed the quickest and to make the tone change in pitch the fastest.

In this scenario, Marie does not have a disability that we are aware of, but still makes use of the sim's sonification feature (the change in tone correlated to the change in speed of the cart). While exploring the sim, she observes the visual representations like the force vector arrows. The addition of sound provided a new modality to cue a potential relationship between the applied force by the pullers and the speed of the cart when in motion. While this relationship was represented visually by the motion of the cart across the screen, the use of sonification provided a cue using a different modality for Marie to consider. This new modality complemented the visual modality and provided a richer, more immersive learning experience than the visual modality provided alone.

6.3.2.5 The Future of Accessible PhET Sims

The accessible PhET sims initiative started with the design and development of three input and output modalities (keyboard navigation, auditory description, and sonification) for a subset of PhET sims (Moore et al. 2016; Smith et al. 2016). We will continue to refine our understanding of effective design and implementation of these features and implement these features in an expanding set of PhET sims. We will continue to share our knowledge of effective design and software infrastructure with the science education community. We will also explore new input and output modalities, with the goal of supporting all students to engage in science inquiry with PhET sims. To find the most up-to-date information about the progress and findings of the accessible PhET sims initiative, see http://phet.colorado.edu/en/accessibility.

6.4 Conclusions and Future

Since 2002, PhET has been innovating in the design and development of educational simulations. In this chapter, we highlighted some of the historical innovations in design, dissemination, and increasing diversity that we believe have contributed to the broad uptake and now ubiquitous use of PhET simulations around the world. Over the course of more than a decade, the PhET project has refined a set of goals that guide the creation of each PhET simulation. These prior innovations and project goals laid the groundwork for continued advances in educational simulations, and the transition from developing Java and Flash sims to developing HTML5 sims created the opportunity to achieve advances through new initiatives. Through the PhET-iO and accessible PhET sims initiatives, the PhET project continues to advance access and opportunity for students to engage in science inquiry with interactive simulations.

Looking forward, as part of the PhET-iO initiative, we will continue to partner with the broader science education, publishing, and assessment communities to provide opportunities for new and highly effective learning materials and assessments. As PhET-iO enables unprecedented customization, integration, and data collection, there are many open questions for the science education community to investigate to determine contexts, supports, and scenarios that are most effective in enabling student learning. As the publishing industry evolves to focus more on the use and distribution of interactive digital resources, there will be an increasing need to understand how to best utilize the sims data collection capabilities to optimize online environments for learners. The combination of highly customizable sims and rich data collection capabilities enables new approaches to assessment, requiring advances in the analysis of student choices during pursuit of a goal – which is quite a different challenge than analysis of student choices when answering more traditional assessment questions.

The work of the accessible PhET sims initiative will continue research and development efforts to explore new input and output modalities and update new sims with accessibility features, including keyboard navigation, auditory descriptions, and sonification, to support increased accessibility for students with disabilities. New input and output modalities currently being explored include the use of speech input (the ability to speak commands to the sim) and haptic feedback output (force feedback, such as vibration). Through all of these modalities, we seek to broaden understanding of effective accessible design for interactive simulations.

Continuing these advances is challenging work, requiring innovations in software development, interface design, and pedagogy as new learning contexts are envisioned and created. Through the new features developed from the PhET-iO and accessible PhET sims initiatives, the PhET project will continue to advance the capabilities of interactive HTML5 simulations, enabling new pedagogical practices for the technology-enhanced classroom and increasing engaging and effective learning opportunities for students.

References

Adams, W. K., Alhadlaq, H., Malley, C. V., Perkins, K. K., Olson, J., Alshaya, F., Alabdulkareem, S., & Wieman, C. E. (2012). Making on-line science course materials easily translatable and accessible worldwide: Challenges and solutions. *Journal of Science Education and Technology, 21*(1), 1–10.

Beetham, H., & Sharpe, R. (2013). *Rethinking pedagogy for a digital age: Designing for 21st century learning*. New York: Routledge.

Bybee, R. W., Taylor, J. A., Gardner, A., Van Scotter, P., Powell, J. C., Westbrook, A., & Landes, N. (2006). *The BSCS 5E instructional model: Origins and effectiveness* (Vol. 5, pp. 88–98). Colorado Springs: BSCS.

de Jong, T., Sotiriou, S., & Gillet, D. (2014). Innovations in STEM education: The go-lab federation of online labs. *Smart Learning Environments, 1*(1), 3.

Hensberry, K. K. R., Paul, A., Moore, E. B., Podolefsky, N. S., & Perkins, K. K. (2013). PhET interactive simulations: New tools to achieve common core mathematics standards. In D. Polly (Ed.), *Common core mathematics standards and implementing digital technologies* (pp. 147–167). Hershey: IGI Global.

Käser, T., Hallinen, N. R., & Schwartz, D.L. (2017). Modeling exploration strategies to predict student performance within a learning environment and beyond. *Proceedings of the Seventh International Conference on Learning* Analytics & Knowledge. New York: ACM.

Kramer, G., Walker, B., Bonebright, T., Cook, P., Flowers, J. H., Miner, N., & Neuhoff, J. (2010). *Sonification report: Status of the field and research agenda*http://www.icad.org/websiteV2.0/References/nsf.html. Accessed 4 June 2016.

Massof, R. W. (2003). Auditory assistive devices for the blind. *Proceedings of the 2003 International conference on Auditory Display* (pp. 271–274).

Moon, N. W., Todd, R. L., Morton, D. L., & Ivey, E. (2012). *Accommodating students with disabilities in science, technology, engineering, and mathematics (STEM): Findings from research and practice for middle grades through university education*. Atlanta: Center for Assistive Technology and Environmental Access, Georgia Institute of Technology.

Moore, E. B., Herzog, T. A., & Perkins, K. K. (2013). Interactive simulations as implicit support for guided-inquiry. *Chemistry Education Research and Practice, 14*(3), 257–268.

Moore, E. B., Chamberlain, J. M., Parson, R., & Perkins, K. K. (2014). PhET interactive simulations: Transformative tools for teaching chemistry. *Journal of Chemical Education, 91*(8), 1191–1197.

Moore, E. B., Smith, T. L., & Randall, E. (2016). Exploring the relationship between implicit scaffolding and inclusive design in interactive science simulations. In *International Conference on Universal Access in Human-Computer Interaction* (pp. 112–123). Cham: Springer International Publishing.

National Center for Education Statistics. (2011). Grade 8 national results. http://nationsreportcard.gov/science_2011/g8_nat.aspx?subtab_id=Tab_6&tab_id=tab2#chart. Accessed 6 Aug 2014.

National Center for Science and Engineering Statistics. (2015). *Women, minorities, and persons with disabilities in science and engineering: 2015*. Special Report NSF 15–311. Arlington. http://www.nsf.gov/statistics/wmpd/. Accessed 10 Dec 2016.

NGSS Lead States. (2013). *Next generation science standards: For states, by states*. Washington, DC: National Academies Press.

NVDA. (n.d.). NVDA. Retrieved from http://www.nvaccess.org. Accessed 10 Dec 2016.

Paul, A., Podolefsky, N.S., & Perkins, K.K. (2012). Guiding without feeling guided: Implicit scaffolding through interactive simulation design. *Proceedings of the 2012 Physics Education Research Conference, 1513* (pp. 302–305).

Pedaste, M., Mäeots, M., Siiman, L. A., De Jong, T., Van Riesen, S. A., Kamp, E. T., Manoli, C. C., Zacharia, Z. C., & Tsourlidaki, E. (2015). Phases of inquiry-based learning: Definitions and the inquiry cycle. *Educational Research Review, 14*, 47–61.

Perkins, K.K., Moore, E.B., & Chasteen, S.V. (2014). Examining the use of PhET interactive simulations in US college and high school classrooms. In *Proceedings of the 2014 Physics Education Research Conference* (pp. 207–210). Minneapolis.

Podolefsky, N.S., Moore, E.B., & Perkins, K.K. (2013). *Implicit scaffolding in interactive simulations: Design strategies to support multiple educational goals*. arXiv preprint arXiv:1306.6544.

Schreep, M., & Jani, R. (2005). Efficient keyboard support in web-pages. *Assistive Technologies Research Series, 16*, 504–508.

Smith, T. L., Lewis, C., & Moore, E. B. (2016). A balloon, a sweater, and a wall: Developing design strategies for accessible user experiences with a science simulation. In *International Conference on Universal Access in Human-Computer Interaction* (pp. 147–158). Cham: Springer International Publishing.

Stevens, J. J., Schulte, A. C., Elliott, S. N., Nese, J. F. T., & Tindal, G. (2015). Growth and gaps in mathematics achievement of students with and without disabilities on a statewide achievement test. *Journal of School Psychology, 53*(1), 45–62.

Trilling, B., & Fadel, C. (2009). *21st century skills: Learning for life in our times*. San Francisco: Wiley.

Wieman, C. E., Adams, W. K., Loeblein, P., & Perkins, K. K. (2010). Teaching physics using PhET simulations. *The Physics Teacher, 48*(4), 225–227.

Emily B. Moore is Director of Research and Accessibility at the PhET Interactive Simulations project at University of Colorado Boulder.

Katherine K. Perkins is Director of the PhET Interactive Simulations project at University of Colorado Boulder, and a faculty member in Physics Education Research.

Chapter 7
Designing Virtual Laboratories to Foster Knowledge Integration: Buoyancy and Density

Jonathan M. Vitale and Marcia C. Linn

Abstract In this chapter, we report upon the iterative development of an online instructional unit featuring virtual laboratory activities that target the physical science concepts of density and buoyancy. We introduce a virtual laboratory activity that was designed to facilitate exploration of the relationship of mass and volume to buoyancy. We evaluate the virtual laboratory by measuring the extent to which it fosters meaningful *experimentation*, appropriate *interpretation* of evidence, and *discovery* of new ideas. In the first revision, we simplified the exploratory tools. This revision supported better interpretation of evidence related to a specific claim, but limiting potential for discovery of new ideas. In the second revision, we introduced an intuitive graph-based interface that allowed students to specify and rapidly test properties of virtual materials (i.e., mass and volume). This revision facilitated meaningful exploration of students' ideas, thereby supporting both valid interpretations of evidence related to false claims and discovery of new ideas. We discuss the role that virtual laboratories can play in the design of all laboratory activities by tracking student strategies and offering opportunities to easily test new features.

Keywords Knowledge integration · Authentic inquiry · Reflection · Discovery · Platform · Design-based research

7.1 Introduction

In this chapter, we report upon the iterative development of an online instructional unit that enables exploration of physical science concepts of density and buoyancy. This unit is designed to support students as they conduct experiments in a virtual laboratory and use the evidence to explore the complex relationship between mass,

J. M. Vitale (✉) · M. C. Linn
Graduate School of Education, University of California, Berkeley, CA, USA
e-mail: jvitale@philasd.org

M. E. Auer et al. (eds.), *Cyber-Physical Laboratories in Engineering and Science Education*, https://doi.org/10.1007/978-3-319-76935-6_7

volume, and buoyancy. We take advantage of the affordance of virtual laboratories to log data in an effort to analyze students' exploration strategies and redesign tools. Over the course of three design iterations of virtual laboratory activities, we demonstrate how the analysis of student artifacts enables designs that support increasingly meaningful exploration of important disciplinary concepts.

Although students often demonstrate general patterns to their exploratory strategies (e.g., attempting to replicate positive outcomes rather than disconfirm claims, Tschirgi 1980), exploratory strategies are highly influenced by both the disciplinary content and the tools provided for an investigation (de Jong and Van Joolingen 1998; McElhaney and Linn 2011). Student exploration can be guided by intuitive ideas about the topic (e.g., "heavier objects sink"), grounded in everyday experiences (Perkins and Grotzer 2005). Likewise, the structure of tools and scaffolds can orient students to critical variables in an investigation or place focus on superficial features (Goldstone and Wilensky 2008). Designing learning activities around laboratory investigation requires attention to how learning materials take advantage of student ideas and enable meaningful exploration of diverse ideas.

We draw on the knowledge integration framework to design tools that help students build upon their ideas and integrate new ideas (Linn and Eylon 2011). We employ design-based research methods to investigate the effectiveness of the laboratory environment. Specifically, we use logs of student progress to reveal design weaknesses and revise the unit to better guide students to perform critical tests of their own ideas, critique their ideas in light of new evidence, and reflect upon new ideas to form a normative, coherent understanding of the studied phenomena (Linn and Eylon 2011). We compare design iterations by evaluating how well the virtual laboratory supported experimentation around a specific claim, interpretation of evidence, and discovery of new ideas.

7.2 Rationale

We designed the density unit to emphasize science practices and sustained inquiry advocated in contemporary science standards (NGSS Lead States 2013). This emphasis has motivated research on optimal ways to design and guide inquiry projects. Ideally projects will engage students in authentic practices of scientists such as asking questions, modeling, and analyzing data (Kozma et al. 2000). Yet, many science class laboratory activities are structured to avoid the challenges associated with authentic inquiry (Holbrook and Kolodner 2000). In addition some teachers worry that inquiry projects neglect important information or use instructional time inefficiently (Anderson et al. 1998; Mayer 2004). However, with appropriate guidance, research shows that inquiry activities can support both conceptual and epistemic development effectively (Hmelo-Silver et al. 2007; McNeill et al. 2006; Wilson et al. 2010).

Guidance for online inquiry-based laboratory activities must strike the right balance between providing too much structure and making tasks too vague (Koedinger and Aleven 2007). When designing a new virtual laboratory and inquiry unit, research can provide initial conjectures about how students will respond, while design-based research methods may be employed to explore unexpected patterns.

7.2.1 Students' Exploration Strategies

Students bring diverse exploration strategies to science investigations. Commonly, students have consistent difficulties generating and adapting hypotheses, designing good experiments, interpreting data, and planning sequences of investigations (de Jong and Van Joolingen 1998). Moreover, researchers have documented that students fail to control variables (Dunbar and Klahr 1989), test sufficiently diverse cases (Schauble et al. 1991), or disconfirm certain hypotheses (Dunbar 1993). In many topic areas, students and adults conduct tests to confirm their hypothesis, but fail to consider possible disconfirming cases. For example, Wason (1960) found that when trying to predict a rule underlying a numerical pattern (e.g., 2, 4, 6) by testing their own sequences, adults rarely generate counterfactual patterns that critically test their conjectured rules. More generally, such "confirmation bias" is a common feature of empirical testing, particularly when conducted by children and non-experts (de Jong and Van Joolingen 1998). Other studies suggest that many reasoners fail to test boundary conditions or disconfirming possibilities because they simply cannot imagine other hypotheses (Dunbar 1993).

Interpreting patterns in results from exploration of a science topic has confounded scientists and students over the history of science (Rutherford 1964). Students may either fail to discover regularities in data or fail to integrate anomalous data into their explanation. Chinn and Brewer (1993) documented that anomalous data can lead some students to substantially revise their initial claim and others to ignore the data, interpret the data as isolated or temporary, or assimilate evidence to fit prior beliefs. For example, when investigating what properties of a material determine if an object will sink or float, children tend to draw conclusions based on limited evidence and prior, incomplete ideas (e.g., mass alone determines sinking, Inhelder and Piaget 1958). In some cases, students' prior beliefs may even lead to errors in the recording or transcription of data (Klahr et al. 1993). Although these diverse inquiry strategies are often underdeveloped, a coherent instructional approach can guide students to learn and apply more productive strategies.

7.2.2 The Knowledge Integration Approach

The knowledge integration framework grew out of observations that students bring a repertoire of multiple, often conflicting, ideas to science class. While student ideas may not always be scientifically valid (i.e., "normative"), even naïve, "nonnormative" ideas can be leveraged and built upon – not replaced or discarded – to develop new, more scientifically accurate knowledge. Like content-specific ideas, exploration strategies exist as part of this repertoire (e.g. McElhaney and Linn 2011). Although these strategies may differ from scientific experts, many studies show that well-guided inquiry instruction, using tools that support meaningful experimentation and interpretation, can motivate effective exploration of complex science content (Hmelo-Silver et al. 2007; Krajcik et al. 2000). Research on authentic student inquiry suggests that successful activities engage students in critical analysis of multiple ideas and sources of evidence to develop coherent understanding (Linn 2006).

From the knowledge integration perspective, virtual models, by eliminating many physical limitations, offer unique opportunities for students to test their own ideas rather than those specified in a laboratory protocol. Likewise, automated guidance based on student behaviors can deliver personalized information that promotes further exploration (Vitale et al. 2015). Yet, while digital affordances can help guide students toward optimal strategies, underlying thought processes are fundamentally similar to those in a physical laboratory. Thus, research situated in a virtual laboratory may be creatively repurposed for physical laboratory settings.

7.2.3 *Density and Buoyancy*

The knowledge integration approach recognizes that intuitive ideas arise from real experiences and reflect legitimate observations and reasoning processes (Smith et al. 1993). Student ideas are diverse and reflect contextual circumstances (diSessa 2002). In the case of density and buoyancy, young children have difficulty differentiating between weight and size (Smith et al. 1985). In middle school relatively few students recognize that sinking and floating behavior is fundamentally determined by a trade-off between mass and volume (i.e., a proportion), but instead implicate a number of other factors, including mass (alone), shape, and presence of holes (Smith et al. 1992). Recognizing density as the central factor in buoyancy is conceptually challenging, it requires the learner to coordinate two independent features of an object, simultaneously. Colloquial language often implicates weight or mass in discussing buoyancy. For example, "rocks are heavy" and therefore sink, while "wood is light" and therefore floats.

Furthermore, many beliefs reflect real-world experiences in which objects of similar size are compared or simply the feeling of holding up something that is heavy (Hewson and Hewson 1984; Smith et al. 1992). Yet, while a mass-based understanding is nonnormative in general, it is correct in the special case where the volumes are equal. Therefore, this special case is one place to start in building coherent, accurate understanding.

7.2.4 This Study

This study applies design-based research methodology to evaluate and enhance the impact of a density-buoyancy curriculum. Design-based research is an emerging methodological perspective that facilitates initial theory building while concurrently addressing instructional goals (Barab and Squire 2004; Bell 2004). Although specific approaches vary, design-based research typically promotes attention to unexpected patterns and behaviors that are often minimized or overlooked in traditional research (Collins et al. 2004). Given the emergent and often surprising nature of classroom activities, multiple data sources are needed to illuminate the "mediating processes" that bridge material design and learning outcomes measured in formal assessments (Sandoval 2014). In the case of virtual experimentation, data logs can also reveal student thinking, based upon common patterns (Gobert et al. 2013).

Utilizing a design-based methodology, we present three design iterations of the density curriculum featuring a virtual laboratory. In order to evaluate the success of our curriculum, we apply the following performance criteria:

1. *Experimentation.* How well does the virtual laboratory enable students to gather evidence to explore a claim?
2. *Interpretation.* How well does the virtual laboratory support critical interpretation of evidence, thereby promoting dissatisfaction with nonnormative ideas?'
3. *Discovery.* How well does the virtual laboratory support the recognition and adoption of novel, normative ideas?

In an effort to conduct generalizable research, our iterative design process aims to improve the virtual laboratory interface without departing far from the materials and activities common to a physical laboratory. Additionally, we sought to improve the curriculum in order to provide better indicators in the areas of *experimentation*, *interpretation*, and *discovery*. Thus, building upon the initial design, the subsequent two redesigns include new features that provide better insight into student thinking.

7.3 Design Study Methods and Results

7.3.1 General Methodology

The three design iterations detailed below feature a set of common virtual materials, including an assessment given prior to and following the curriculum. While details of each curriculum design iteration vary (including the addition of new activities, as the design advances), the primary elements remain the same.

7.3.1.1 Materials

All computer materials, including pretest and posttest were constructed by the authors in the Web-based Inquiry Science Environment (WISE), which supports common assessment formats (multiple choice, open response, graph construction) and interactive learning tools, such as movies, diagrams, and simulations.

7.3.1.2 Pretest-Posttest

Although the pretest and posttest included several relevant items, we present one item that best addresses ideas represented in the target sequence of the curriculum.

7.3.1.2.1 Sink-Float Graph

To evaluate students' understanding of mass-volume ratio, we present the graph shown in Fig. 7.1. This graph displays the mass and volume of six points representing objects placed in water. Students are asked to choose from a set of multiple choice responses the set of three points that represents objects that would sink in water (correct answer: "A, B, C"). Incorrect choices reflect common nonnormative ideas, such as higher mass objects sink ("B, C, F") and higher volume objects sink ("C, E, F"). Students then explain their choices.

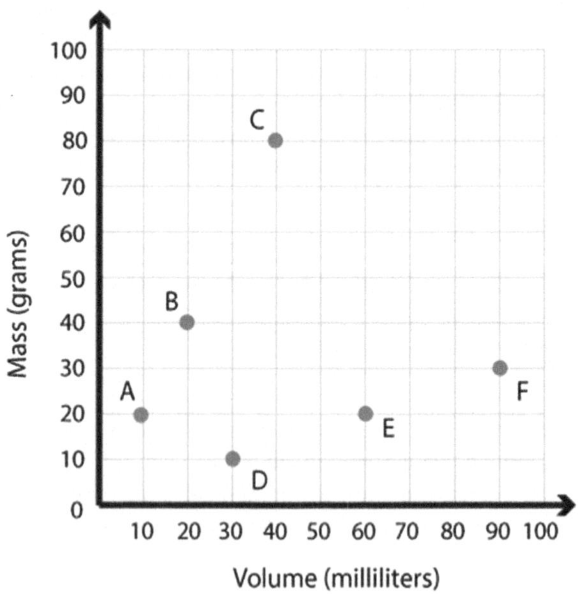

Fig. 7.1 Mass vs. volume graph displays six labeled points representing objects placed in water

7.3.1.3 Online Curriculum

The curriculum unit, "Sink or Float" was developed to facilitate exploration of density and buoyancy with a series of open investigations. Although the unit explores a number of advanced topics (e.g., how the density of a liquid can be measured), we focus on a target sequence of activities that center on developing a sense of how mass and volume proportionality impacts buoyancy.

7.3.1.3.1 Distinguishing ideas: Volume-Mass Debate

Prior to working with experimentation tools, students' ideas are elicited through the presentation of a debate between fictional student characters. A character named "Dan" states, "It's the total size of the object that matters. That's called volume! An object with a large volume will sink, an object with a small volume will float." Another character named "Aida" responds, "It's the mass that matters, not the volume. The more mass, the heavier an object feels. An object with a lot of mass will sink, an object with little mass will float." Additionally, another character, focusing on "flatness," was included in the initial design, but removed in later designs to allow for more time to focus on volume and mass. After reading this "debate," students were asked with whom they agreed. Students could select more than one or choose neither.

7.3.1.3.2 Virtual Laboratory

To further distinguish ideas, students have an opportunity to explore debate claims in the virtual laboratory (Fig. 7.2). For each debate character, students are presented with the nonnormative idea and asked to "conduct experiments to show [Dan, Aida] whether [he, she] is right or wrong."

Following the virtual laboratory, on a separate web page, students are presented with labeled images of all the objects they have constructed and a table displaying the results of their experiment with the headers: "ID" (to identify the object from the set of displayed images), "Materials" (all materials used in composition of the object), "Mass (g)", "Volume (ml)", and "Sinks in water?" In cases where students construct an object but do not submerge it in the liquid (and, thus, displacement cannot be observed), a "?" stands in place of the volume in the table. Likewise, in cases where students do not place their object on the scale, a "?" stands in place for mass. These features were included to emphasize realistic laboratory procedures.

Below the displayed images and table is an open response text box. Students are asked to "Use this data to explain to [Dan, Aida] how [he, she] is right or wrong. Write your response to [Dan, Aida] in the space below."

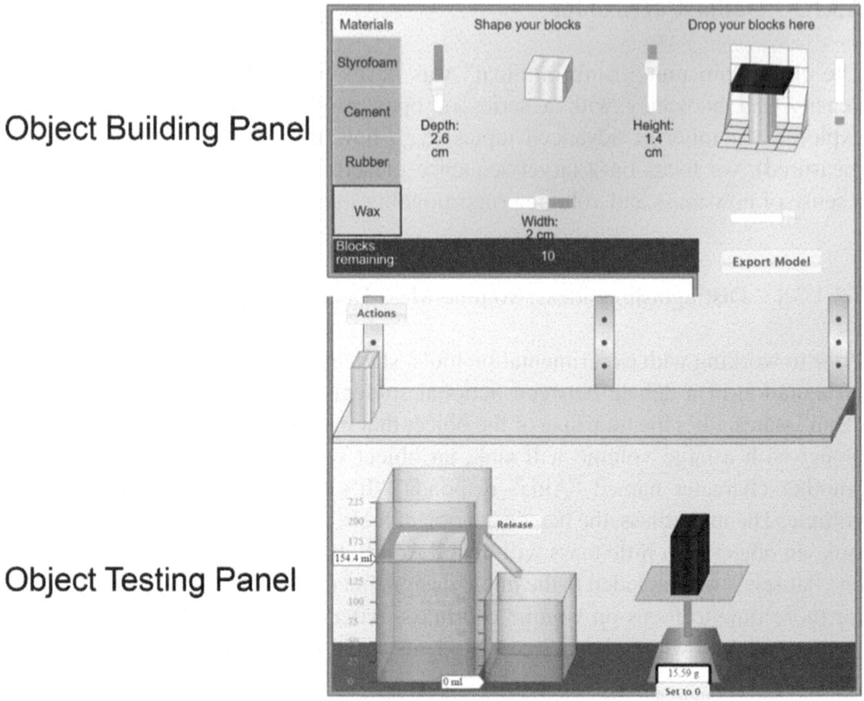

Fig. 7.2 Screenshot of virtual laboratory. The upper, "object building panel" is used to construct objects by selecting a material, shaping a block of the selected material by height, depth, and width, and then composing blocks into a single object within a grid. The lower, "object testing panel" is used to test objects by submerging them in a liquid, measuring displaced liquid, and finding mass with a scale

7.3.1.3.3 Reflecting on Ideas with Graphs: Construction and Critique

As an initial test of transfer of ideas explored in the virtual laboratory, students are asked to construct a graph similar to the one presented at pretest. Specifically, students are presented with an empty mass vs. volume grid (ranging from 0 to 100 ml on the x-axis and 0–100 g on the y-axis) and asked to plot four red points to represent objects that sink, four blue points to represent objects that float, and a green divider line to separate all objects that sink from all objects that float.

In this activity we utilized automated scoring (Vitale et al. 2015) to provide guidance based upon patterns predicted to indicate common ideas. For example, if a student places a horizontal line they are prompted to, "go back to the simulation and find out: do all heavy objects sink? Do all light objects float?" Thus, this item directs the students to address nonnormative ideas back in the virtual laboratory to gather more evidence.

Following graph construction (and a subsequent page asking students to describe the graph, which is beyond our scope), students are presented with an example of

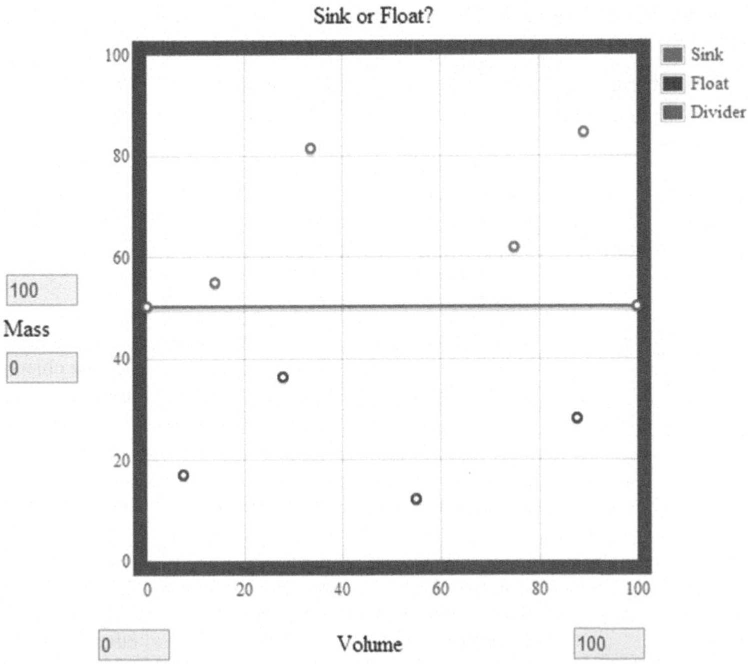

Fig. 7.3 "Aida's" graph of mass vs. volume displays four points representing sinking objects above 50 g, four points representing floating objects below 50 g, and a dividing line at $y = 50$

"Aida's" constructed graph that displays a mass-based understanding of buoyancy (Fig. 7.3). Students are asked to select whether they agree that Aida's graph is correct and explain her reasoning and their own interpretation.

7.3.1.4 Scoring and Analysis

To track student progress, we record how students' initial ideas (from pretest and debate) impact behaviors in the virtual laboratory and subsequently how behaviors in the virtual laboratory impact performance on later learning and assessment tasks. This data allows us to evaluate the efficacy of the virtual laboratory to promote effective *experimentation*, *interpretation*, and *discovery*.

To evaluate the virtual laboratory, we focus most closely on the idea that mass (or weight) solely determines buoyancy. At each step along the activity sequence, we can estimate whether a mass-based idea, a normative idea (e.g., density or material determines buoyancy), or other idea is dominant in students' thinking. Within a design iteration, changes in this distribution (in items of similar formats) may indicate changes in understanding due to the activity materials. Between design iterations, differences in distributions at any given step, or across a series of steps, suggests that changes to the curriculum impacted student performance. To evaluate

Table 7.1 Description of measures of conceptual categories for explanations, choices, and construction activities in pretest, posttest, and primary curriculum materials

Measure	Format	Normative	Mass-based	Other
Graph-pre	Explanation	Buoyancy determined by mass-volume ratio, density, material	Buoyancy determined by mass or weight	Buoyancy determined by volume or size, graph error, other, vague
Debate	Choice	Neither Dan (volume) nor Aida (mass)	Aida (mass) or both	Dan (volume)
Lab-mass materials	Construction	Maximum mass of floating objects tested is greater than or equal to the minimum mass of sinking objects	Maximum mass of floating objects tested is less than the minimum mass of sinking objects	Only sinking or floating objects tested
Lab-mass critique	Explanation	Aida is incorrect	Aida is correct	No clear indication
Lab-mass explain	Explanation	Buoyancy determined by mass-volume ratio or density	Buoyancy determined by mass or weight	Buoyancy determined by volume or size, other, vague
Graph-construct initial	Construction	Separates sinking and floating by line y = x	Separates sinking and floating points by horizontal line	Separates sinking and floating points by other line or unclear
Graph-construct final	Construction	Same as initial	Same as initial	Same as initial
Graph-critique	Explanation	Same as *lab-mass explain*	Same as *lab-mass explain*	Same as *lab-mass explain*
Graph-post	Explanation	Same as *graph-pre*	Same as *graph-pre*	Same as *graph-pre*

how students' ideas shift as they progress through the curriculum, we measure nine distinct measures of student thinking, categorized as *normative*, *mass-based*, and *other* (which may include incomplete or vague ideas) (Table 7.1).

Utilizing these measures, and others, we describe how we address each of our laboratory performance criteria.

7.3.1.5 Experimentation

What materials did students build and test in the virtual laboratory (i.e., *lab-mass materials*)? To what extent were the materials built determined by students' prior ideas (i.e., *graph-pre* and *debate*)? If materials constructed closely resemble prior distribution of ideas, then the virtual laboratory is mainly being used to confirm prior ideas. On the other hand, if materials conflict with prior ideas, then the design of the virtual laboratory has a clear impact on experimentation. However, if students construct a greater proportion of objects that indicate a mass-based understanding than prior ideas would suggest, this design impact can be viewed as negative.

7.3.1.6 Interpretation

To what extent do students utilize sufficient laboratory evidence to reject a mass-based understanding of buoyancy (i.e., *lab-mass critique*)? How does this differ in cases where students manufacture and test objects that support Aida's claim? If it is the case that students misinterpret evidence (either valid or invalid), then additional support for data interpretation may be necessary.

7.3.1.7 Discovery

To what extent do students adopt a novel, normative understanding of buoyancy either based on density, mass-volume ratio, or material properties while exploring the virtual laboratory (*lab-mass explain*)? If such a discovery does not occur during the virtual laboratory, does successful interpretation prepare students to discover normative ideas during the graph construction exercise (i.e., *graph-construct final*)?

7.3.2 Iteration 1

We present the initial design of the curriculum, with limited guidance and scaffolds, as a baseline for future revisions.

7.3.2.1 Methods

7.3.2.1.1 Participants and Procedure

This study was conducted with two 8th grade teachers and their 107 students (13- to 14-year-olds). Ms. S. (91 students) has over 20 years of science teaching experience. Ms. P. (16 students) was a second-year teacher. The school serves a diverse community (school demographics: 48% White, 4% Asian, 36% Hispanic, 5% Black; 37% reduced lunch) in a suburban area of the western United States. Students performed the pretest and posttest individually. For the curriculum, although students were expected to provide their own responses, they were advised to collaborate with an adjacent student. All study-related activities were conducted within the first month of the school year for a period of approximately six 50-min class periods. Students had not received any formal instruction on density or buoyancy concepts prior to this study.

In both classrooms students were arranged at tables of four and were typically expected to work with adjacent classmates. During the curriculum phase of the study, the teacher provided introductory discussion activities related to density. For example, in one opening activity, students were asked to describe whether a golf ball or ping-pong ball would sink in water. Students wrote a private response and then shared their ideas with the whole class. The teacher did not provide authoritative feedback in the discussion, but prompted her students to consider the problem as they engaged with the curriculum. While students worked autonomously, both the lead researcher and teacher circulated the classroom to offer guidance and address any technical issues that arose. Based upon prior discussion about the inquiry goals of the curriculum, the researcher and teacher agreed to focus one-on-one guidance on helping students interpret and implement written instructions. If students had difficulty answering a question, he or she would be referred back to an instructional tool to review, such as the virtual laboratory. In some cases, if a student did not have a sufficient set of objects in the virtual laboratory, he or she would be directed to make new, unique objects.

7.3.2.1.2 Materials

The main curriculum unit, "Sink or Float" and assessment items are described above. The virtual laboratory (Aida) can be previewed at this link: http://wise.berkeley.edu/previewproject.html?projectId=11723&step=2.10

7.3.2.2 Results

Observed values for the nine measured quantitates are displayed in the first three rows of Table 7.2. Percent values were calculated based upon the total number of students who responded to the particular assessment measure. Because some students did not complete all measures, the number of students varies across columns. Note that rows four through nine correspond to Iterations 2 and 3 and will be described in later sections of this chapter.

7.3.2.2.1 Experimentation

As Table 7.2 indicates, only 18 of 96 (19%) of students tested objects that could falsify Aida's claim in *lab-mass*, while 28 of 96 (29%) tested objects that directly supported her claim. The remainder did not test a sufficient number of objects. Of those who rejected Aida's mass-based and Dan's volume-based claims in *debate*, 3 of 15 (20%) tested objects that could falsify Aida's claim, while 2 of 15 (13%) tested objects that supported her claim. Of those who supported Aida's mass-based claim in *debate*, 10 of 59 (17%) tested objects that could falsify Aida's claim, while 18 of 59 (31%) tested objects that supported her claim. Comparing these distributions, there was no difference in materials produced by *debate* classification [$\chi^2(2) = 1.8$, $p > 0.2$].

Table 7.2 Summary results of each design reiteration

Iteration	Category	*Graph-pre*	*Debate*	*Lab-mass materials*	*Lab-mass critique*	*Lab-mass explain*	*Graph-construct initial*	*Graph-construct final*	*Graph-critique*	*Graph-post*
1	Norm	6 (6.1%)	15 (17%)	18 (18.8%)	18 (20.2%)	9 (10.1%)	–	25 (28.1%)	18 (19.8%)	17 (17%)
	Mass	65 (66.3%)	59 (67%)	28 (29.2%)	59 (66.3%)	50 (56.2%)	–	53 (59.6%)	20 (22%)	56 (56%)
	Other	27 (27.6%)	14 (15.9%)	50 (52.1%)	12 (13.5%)	30 (33.7%)	–	11 (12.4%)	53 (58.2%)	27 (27%)
2	Norm	22 (13.3%)	9 (14.5%)	20 (42.6%)	42 (46.7%)	14 (15.6%)	15 (16.7%)	46 (51.1%)	41 (45.6%)	73 (44%)
	Mass	100 (60.2%)	42 (67.7%)	2 (4.3%)	42 (46.7%)	35 (38.9%)	61 (67.8%)	35 (38.9%)	20 (22.2%)	71 (42.8%)
	Other	44 (26.5%)	11 (17.7%)	25 (53.2%)	6 (6.7%)	41 (45.6%)	14 (15.6%)	9 (10%)	29 (32.2%)	22 (13.3%)
3	Norm	22 (14.6%)	12 (17.6%)	37 (44%)	44 (52.4%)	15 (17.9%)	29 (34.1%)	57 (66.3%)	46 (54.8%)	67 (45%)
	Mass	106 (70.2%)	48 (70.6%)	24 (28.6%)	37 (44%)	35 (41.7%)	38 (44.7%)	24 (27.9%)	14 (16.7%)	64 (43%)
	Other	23 (15.2%)	8 (11.8%)	23 (27.4%)	3 (3.6%)	34 (40.5%)	18 (21.2%)	5 (5.8%)	24 (28.6%)	18 (12.1%)

Displays distribution of responses (frequency and percent) into three categories representing a normative (norm), mass-based, or other conceptualization of buoyancy for nine measures (columns starting at *Graph-pre*, ending at *Graph-post*) and three design iterations

Although this is in part due to the low number of students who actively rejected false claims at *debate* (15), it also suggests that it was difficult for students to construct objects that could falsify the mass claim, regardless of their prior belief. Indeed, of students who produced and tested at least two objects, 21 of 47 (45%) created objects of equal volume. Two-thirds of these 21 students (14) constructed all of their objects to be maximum size ($4 \times 4 \times 4$ cubes). It may be the case that students mistakenly believed that objects of comparable size were needed to test a hypothesis about mass. Maximizing the volume by shifting the html sliders to their furthest position was the simplest way to produce objects with equal volume.

7.3.2.2.2 Interpretation

As Table 7.2 indicates, 59 of 89 students (66%) validated Aida's mass-based claim based upon evidence collected in *lab-mass*. This is nearly identical to the proportion of students that agreed with Aida in *debate* (59 of 88, 67%), suggesting that *lab-mass* added little value, overall.

Looking at these distributions based upon the evidence students collected, for those who produced and tested objects that could falsify a mass based claim 8 of 16 (50%) did reject Aida. On the other hand, for those who produced mass supporting evidence, only 3 of 28 (11%) rejected Aida's incorrect claim. This significant difference between distributions [$\chi^2(2) = 9.7$, $p = 0.008$] suggests that students were able to interpret the data appropriately, regardless of whether the data was sufficient or misleading.

7.3.2.2.3 Discovery

As Table 7.2 indicates, only 9 of 89 (10%) of students gave a normative explanation of the evidence in *lab-mass explain*. Of these nine students, none had previously agreed with Aida in *debate*, indicating no clear cases of a conceptual shift. Although the first submission of the *graph-construct* item, prior to guidance, is another potential indicator of discovery, this data was not saved during the first design of the software. Yet, even in the case that performance on the final graph was due completely to discoveries in the virtual laboratory, and not to guidance while graphing, a similar proportion of students (53 of 89, 60%) created mass-based graphs as those who initially agreed with Aida in *debate* (59 of 88, 67%). Thus, from *debate* to final construction *graph-construct*, indicators of discovery are minimal.

7.3.3 Iteration 2

The results of Iteration 1 clearly indicate that students struggled to explore the virtual laboratory to investigate a mass-based claim. Over half did not test enough objects to even support the mass-based claim. Even when students actively attempted to test Aida's claim, interface features, such as the sliders, may have provided inconclusive evidence. In those cases, where objects are of equal volume, indeed the heaviest objects will sink.

Thus, for the initial design, interface features represented an obstacle for appropriate experimentation. To explore whether students could make valid interpretations and discoveries with this obstacle removed, we redesigned the virtual laboratory to simplify exploration. We provided a set of objects that, if tested in the virtual beaker and scale, would provide evidence that would challenge a mass-based claim.

7.3.3.1 Methods

7.3.3.1.1 Participants and Procedure

This study was conducted with two 8th grade teachers and their 171 students. Both teachers (Ms. G, 92 students; Ms. E, 79 students) have over 5 years of teaching experience and multiple prior experiences guiding online inquiry instruction. The school serves a diverse, middle-to-upper income community (school demographics: 63% White, 23% Asian, 9% Hispanic, 1% Black; 3% English learners; 11% reduced lunch) in a suburban area of the western United States. Students had not received any formal instruction on density or buoyancy concepts prior to this study.

Students performed the pretest and posttest individually. Following pretest, students were placed into dyads (workgroups) according to prior seating to work through the curriculum. Both teachers monitored the students to ensure that students remained on task. Both teachers and the lead researcher circulated the classroom to address logistical or conceptual difficulties. Students were encouraged to complete all activities and return to incomplete steps. All study-related activities were conducted over the course of six 50-min class periods.

7.3.3.2 Materials

The main curriculum unit, "Sink or Float," underwent the changes described below. The updated virtual laboratory can be previewed at this link: http://wise.berkeley.edu/previewproject.html?projectId=12981&step=2.2.9

7.3.3.2.1 Virtual Laboratory (Mass)

To improve exploration of Aida's mass-based claim, we provided students with a default set of objects. Table 7.3 displays the properties of these blocks.

These blocks were chosen such that students had multiple opportunities to compare objects with equal mass but different buoyancy. For example, the large Styrofoam block (row 2) and the small rubber block (row 6) have an equal mass

Table 7.3 Default blocks in *virtual laboratory (mass)*

Materials	Mass (g)	Volume (ml)	Sinks in water?
Styrofoam	1.6	8	Float
Styrofoam	9.6	48	Float
Wax	6.4	8	Float
Wax	9.6	12	Float
Rubber	28.8	24	Sink
Rubber	9.6	8	Sink
Cement	1.6	1	Sink
Cement	9.6	6	Sink

(9.6 g), but only the rubber block sinks. To view these numerical properties on the table following the laboratory, students were required to submerge the blocks in the water and place them on the scale.

7.3.3.2.2 Graph Construction

First, the graphing software was updated so that all intermediate submissions of student graphs for guidance were recorded. This provides a complete record of student progression from initial to final graph. Second, because in Iteration 1 few students appeared to return to the virtual laboratory to perform new trials, we presented a simulation directly adjacent to the graph.

Specifically, after submitting a graph with a sufficient number of points, a guidance message was displayed (like Iteration 1), and a simulation was positioned adjacent to the graph (Fig. 7.4). Instead of the material-based construction panel of the virtual laboratory, blocks in the simulation reflected the mass and volume of the point currently selected on the graph. If a point was dragged within the graph, the mass and volume of the linked graph changed accordingly. This "linked graph simulation" was designed to allow students to directly test hypotheses about the graph.

7.3.3.3 Results

Overall results can be viewed in Table 7.2, rows 4–6.

7.3.3.4 Experimentation

Although students were provided with sufficient materials to explore a mass-based claim, student workgroups still needed to interact with these objects to measure their volume and mass. In this case, 20 of 47 workgroups (42%) tested a sufficient set of blocks. Yet, only 2 of 47 workgroups (4%) tested only objects that supported a mass-based claim. While there still remained a large number of students who did not test

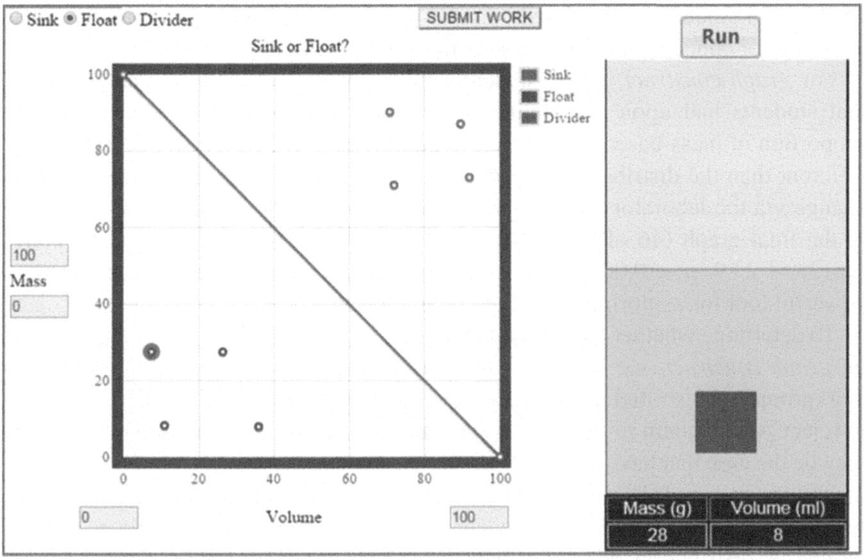

Fig. 7.4 Linked graph simulation. Students test buoyancy of objects with mass and volume specified by the graph. Selected point on the graph (e.g., blue point at $x = 8$ ml and $y = 28$ g) is displayed as a sinking block in the simulation

enough objects, the shift away from supporting a mass-based claim is a significant improvement from Iteration 1 [$\chi^2(2) = 16.1, p < 0.001$].

7.3.3.5 Interpretation

As Table 7.2 indicates, 42 of 90 students (46%) successfully rejected Aida's mass-based claim by exploring given blocks in *lab-mass*, which is a significant improvement from Iteration 1 [$\chi^2(2) = 14.5, p < 0.001$]. Focusing on workgroups who tested objects that could falsify a mass-based claim, 12 of 20 (60%) rejected Aida's claim, which is similar to the proportion of students in Iteration 1 that successfully interpreted evidence that contradicted Aida's claim (8 of 16, 50%; $\chi^2(1) = 0.1, p > 0.2$). Thus, an overall greater proportion of students who rejected Aida's claim in this iteration was primarily due to improved evidence, and not to better interpretation.

7.3.3.6 Discovery

As Table 7.2 indicates, only 14 of 90 (15.6%) of students gave a normative explanation, and 35 of 90 (39%) gave a mass-based explanation *lab-mass explain*, which is only a marginal improvement from Iteration 1 [$\chi^2(2) = 5.4, p = 0.07$].

Of these 14 students, 5 previously agreed with Aida in *debate*, providing minimal evidence for conceptual shifts due to actions in the virtual laboratory.

For *graph-construct*, initial student artifacts likely reflected the understanding that students had upon completing the virtual laboratory. In this case the high proportion of mass-based distribution of points (61 of 90, 68%) is not significantly different than the distribution at *debate* [$\chi^2(2) = 0.4$, $p > 0.2$], indicating no clear change via the laboratory. However, the higher proportion of normative responses in the final graph (46 of 90, 51%) represents a large shift from the initial graph [$\chi^2(2) = 23.9$, $p < 0.001$]. This suggests that the linked simulation provided a powerful tool for exploring new ideas about buoyancy.

To determine whether the virtual laboratory (mass) had any impact on these gains on *graph-construct*, we compared the performance on final graphs among those workgroups who shifted during the virtual laboratory (from their *debate* response) to reject Aida's claim to those who maintained their original nonnormative idea. It may be the case that having become dissatisfied with a mass-based claim, the virtual laboratory prepared them to discover mass-volume ratio in graph construction. On final graphs, those who changed views were significantly more likely to have a normative distribution of points than those who accepted Aida's claim [shifters (*normative*, 15; *mass*, 5; *other*, 0); maintainers (*normative*, 9; *mass*, 14; *other*, 5); $\chi^2(1) = 0.1$, $p > 0.2$]. This suggests that working with the virtual laboratory prepared students to learn in the graphing activity; however, it may be the case that students who were more likely to gather and interpret evidence correctly in the laboratory were more likely to make discoveries on the graph, without a causal link between the two.

7.3.4 Iteration 7.3

Iteration 2 demonstrated that narrowing student choices in an inquiry activity increases the chances that they will observe a specific phenomenon. In this case, students were more likely than in the first iteration to notice that objects of equal weight may have different buoyancy. Yet, simply observing this fact does not equate to learning the underlying, normative concept. Even students who correctly noticed that objects with equal weight could diverge in buoyancy were more likely than not to produce an initial graph that demonstrated a mass-based understanding.

Rather, the generative features of the linked graph simulation provided an opportunity for discovery by closely coupling the data representation in the graph with simulation feedback. This allowed students to directly test hypotheses. In particular, testing objects of equivalent masses was straightforward with the graph (e.g., by testing points on a vertical line), while it could only be performed indirectly using the slider controls in the first design.

For our final design of the virtual laboratory, our goal was to give students the opportunity to engage in evidence generation, but with a tool that would allow for more direct testing of hypotheses, thereby supporting discovery. To do this, we applied a graphing interface to the object construction panel.

7.3.4.1 Methods

7.3.4.1.1 Participants and Procedure

Iteration 3 was conducted within the same school as Iteration 2 in the following year with a new cohort of 153 8th grade students and the same two teachers (Ms. G, 94 students; Ms. E, 59 students). Students had not received any formal instruction on density or buoyancy concepts prior to this study. The procedures were the same as Iteration 2.

7.3.4.2 Materials

The main curriculum unit, "Sink or Float," underwent the changes described below. The updated virtual laboratory can be previewed at this link: http://wise.berkeley.edu/previewproject.html?projectId=15889&step=2.10.1

7.3.4.2.1 Virtual Laboratory

To take advantage of the success of graph construction in Iteration 2, we revised the construction interface for the virtual laboratory so that students could directly specify the mass and volume of their blocks. Instead of selecting a material, the student selects a point in the graph interface (Fig. 7.5). This point would immediately be represented as a block, which can be reshaped (maintaining mass

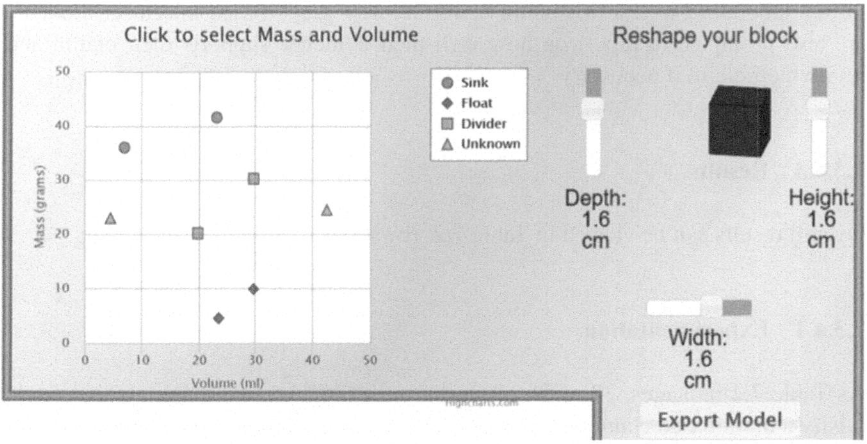

Fig. 7.5 Virtual laboratory with graphical interface. Students select mass and volume of material and are presented with a matching block, which can be reshaped (while maintaining mass and volume). Although the specified point is initially categorized as "unknown," after submerging in liquid, the point is recategorized as sink, float, or divider based upon its behavior in the liquid

and volume) and then placed into the testing area. Points initially placed in the graph are initially categorized as part of an "unknown" series, but after testing in the liquid, they are recategorized into the appropriate series (sink, float, divider). While the scale can be used for confirmation, the mass was determined by the user, and therefore unnecessary. Objects in the "divider" series neither sank nor floated upward toward the surface (at an appreciable speed).

Finally, because the evidence that was once displayed on a table in the subsequent page is now displayed as a graph on the current step, a text area was appended directly below the virtual laboratory to prompt data interpretation (to measure *mass-lab critique*, *mass-lab explain*).

7.3.4.2.2 Your Claim

To further distinguish between the affordances of the virtual laboratory to promote discovery and to prepare for future learning, we introduced a new open response item on the subsequent page following the mass investigation:

> By now you should have observed that BOTH Dan and Aida's claims ARE NOT QUITE RIGHT. If that's not what your found, please go back and take a closer look. Try making more objects with different masses and volumes. Now help Dan and Aida by creating your own claim... How can you predict if an object will sink or float based upon the object's mass and volume?

7.3.4.2.3 Virtual Laboratory: Your Own Idea (Henceforth Lab-Own)

Following the construction of their own claim, students are prompted to use the virtual laboratory to test this claim using the new graph-based interface. Students are also prompted to report on how well their evidence supports their claim, and revise their claim if necessary.

7.3.4.3 Results

Overall results can be viewed in Table 7.2, rows 7–9.

7.3.4.4 Experimentation

As Table 7.2 indicates, 37 of 84 of workgroups (44%) tested materials that could falsify a mass-based claim and 24 of 84 (29%) tested materials that support a mass-based claim in *lab-mass*. While this performance is not as strong as in Iteration 2, where the materials were given [$\chi^2(2) = 14.5$, $p < 0.001$], it exceeds performance in Iteration 1 [$\chi^2(2) = 16.1$, $p < 0.001$], where objects were constructed by setting block dimensions with sliders.

7.3.4.5 Interpretation

As Table 7.2 indicates, 44 of 84 workgroups (46%) successfully rejected Aida's mass claim based upon evidence collected in *lab-mass*, which is similar to Iteration 2 [$\chi^2(2) = 1.2$, $p > 0.2$]. Focusing on workgroups who tested objects that could falsify a mass-based claim, 26 of 37 (70%) rejected Aida's claim, which is similar to the proportion of students in Iterations 1 and 2 that successfully interpreted evidence that contradicted Aida's claim [Iteration 1, 8 of 16, 50%; $\chi^2(1) = 1.2$, $p > 0.2$; Iteration 2, 12 of 20, 60%; $\chi^2(1) = 0.1$, $p > 0.2$]. Thus, given a graph of data, students were as capable of interpreting the data correctly as when given a table in previous iterations.

7.3.4.6 Discovery

As Table 7.2 indicates, 15 of 84 workgroups (18%) gave a normative explanation and 35 of 84 (42%) gave a mass-based explanation *lab-mass explain*, which is similar to Iteration 2 [$\chi^2(2) = 0.5$, $p > 0.2$]. Of these 15 students, 5 previously agreed with Aida in *debate*, providing minimal evidence for discovery.

However, in a shift, students in Iteration 3 were more likely to construct normative initial graphs than nonnormative (mass or other) than in Iteration 2 [$\chi^2(2) = 10.2$, $p = 0.006$]. This indicates that exploration with the virtual laboratory facilitated discovery to a greater extent in Iteration 3 than Iteration 2.

Yet, the additional opportunity to explore the workgroup's "own idea" could be responsible for better initial graphs than the interface redesign. However, the distribution of responses when initially making their own claim (prior to testing) shows a plurality of normative claims [*normative*, 37; *mass-based*, 25; *other*, 21] and does not shift significantly after testing [*normative*, 44; *mass-based*, 24; *other*, 30; $\chi^2(2) = 1.0$, $p > 0.2$]. This suggests that their normative ideas were mostly established through activities in the virtual laboratory prior to testing their own ideas.

7.4 General Discussion

To evaluate the design of a virtual laboratory, we assessed students according to three dimensions of inquiry: *experimentation*, *interpretation*, and *discovery*. In our case, we found that *experimentation* (i.e., generating appropriate evidence to test a hypothesis) and *interpretation* (i.e., making valid causal inferences from data) were closely aligned. Generally, when students are expected to generate evidence, then their interpretations are dependent upon the quality of this evidence. Although alternative approaches could facilitate better interpretation, independent of experimental success (e.g., by using automated technologies to direct students to specific cases in their collected data), for our purpose, improvements in student *experimentation* were reflected in improvements in *interpretation*. On the other hand, improvements in *discovery* required a radical change in the interface. We discuss the improvements in the area of *experimentation*, *interpretation*, and *discovery*, in turn.

In the first iteration, students had difficulty testing the specific claim that mass alone determines buoyancy. This was, to a great degree, due to a flaw in the interface design; students could control size dimensions of blocks but not mass (directly). With this design, students often produced trials that controlled for volume, and made rational, but incomplete or invalid conclusions from this evidence (e.g., that heavy objects always float). In the second iteration, we revised the laboratory environment to help students test the mass hypothesis more directly. Specifically, by simplifying the range of artifacts to include objects of equivalent mass, but different density, we ensured that appropriate evidence would be generated in a table. As such there were clear improvements in the number of students who recognized that mass alone was not a complete explanation for why objects sink or float.

Yet, while this reduction of complexity between the first and second iterations improved *interpretation*, there was no clear impact on *discovery*. Very few students (approximately 15%) noticed that material or mass-volume ratio was the primary determinant of buoyancy. Without a better explanation for buoyancy, integrating mass and volume, students typically persisted with mass as the basis for sinking when plotting their graphs, initially. However, the linked graph simulation in Iteration 2 provided a novel opportunity for discovery. In particular, by testing different spaces of the graph, students often discovered how sinking and floating in water differed across a diagonal line in the graph. In these cases, the affordances of the graph simulation to rapidly test ideas led discovery.

Taking advantage of the linked graph simulation format, Iteration 3 reestablished meaningful exploration in the virtual laboratory. In turn, this enabled discovery of the relationship between mass and volume, as measured by the initial graph produced following virtual laboratory exercises. Yet, this came with a trade-off, as fewer students produced valid evidence in Iteration 3 than Iteration 2, where materials were given. Furthermore, students' valid conclusions in Iteration 3 did not necessarily derive from controlled experiments. For example, this workgroup's description demonstrates how they explored Aida's claim:

> Aida's claim that only mass determines whether an object will sink or float is incorrect. The mass to volume ratio, or density, determines if an object will sink or not. For example, an object with a mass of 28 grams and a volume of 7 milliliters sank, while an object with a mass of 60 grams and a volume of 64 milliliters floated. Even though it had a higher mass than the 28 gram block, it floated, while the lighter block sank. So, it is the mass to volume ratio (density) that determines whether or not a object floats, not just the mass."

In this case, while the students' evidence does contradict Aida's claim, the cases they use as evidence do not demonstrate a controlled variable strategy, which would have been possible by testing two points along a vertical line in the graph interface. Yet, it was still clear to the students that this data was sufficient. In this sense, using the graph as both an input and interpretation tool made a wider range of valid evidence available than is typically expected in controlled experiments.

Furthermore, in some cases simply having a novel interface for exploring ideas led to surprising discoveries. For example:

> Aida's claim is sort of correct. For Aida I did the same thing I did with Dan. I made a block with little mass but lots of volume and a block with lots of mass with little volume. The block with little mass and lots of volume floated, and the block with lots of mass and little volume sank. However, the block with average volume and average mass went into the middle and stopped and floated. I think that is why they are called 'dividers' on the graph, because they do not completely float and do not completely sink.

In this case, although the students initially constructed a test that would confirm Aida's claim (by choosing a high-mass, sinking object and a low-mass, floating object), the spontaneous discovery of objects with comparable mass and volume, which did not conform to expectations, provided an opportunity for learning. These results, which demonstrate the value of exploration without necessarily controlling variables, align with previous work that show that discovery can emerge from diverse student strategies (McElhaney and Linn 2011).

In addition to improving our curricular tools for the benefit of student learning, we also focused on developing better methods of tracking and assessing students. First, on the graph construction activity, we began saving intermediate submission of graphs. This allowed us to measure the immediate impact on the virtual laboratory on students' understanding of the role of mass and volume on buoyancy. Poor performance on initial graphs in Iteration 2 highlighted the need of further redesign of the virtual laboratory. Second, we added a new assessment item that allowed students to directly express their understanding following the virtual laboratory. While investigating Aida's mass-based claim, students rarely took the opportunity to express or test new ideas. Surprisingly, the difference in response distributions between *lab-mass explain* [*normative*, 15; *mass-based*, 35; *other*, 34] and their own claim [*normative*, 37; *mass-based*, 25; *other*, 21] was significantly different [$\chi^2(1) = 14.0$, $p < 0.001$]. Although we believe that the limitations in Iteration 2 were primarily driven by deficits in the interface (i.e., limited materials), it may be that they simply needed to be directed to use the virtual laboratory to explore their own idea.

In summary, across three design iterations we altered tools for exploration in the virtual laboratory. Although simplifying exploration by supplying materials enabled students to recognize, at least temporarily, that a mass-based claim was not supported by evidence, this activity had no clear impact on developing a normative understanding of buoyancy. On the other hand, by introducing an intuitive interface for exploring mass and volume meaningfully, students were able to take advantage of exploratory strategies to discover new ideas. This aligns with prior research that active, creative engagement with an activity fosters in-depth learning, while more linear approaches achieve, at best, superficial outcomes (Krajcik et al. 1998). Further research with alternative versions of exploratory tools is needed to fully understand how to best support exploration of density.

7.4.1 Limitations

The data presented here primarily compares the distribution of buoyancy concepts (normative, mass-based, and other) between different steps within the same project and across design iterations. These comparisons provide varying degrees of certainty depending on how different the steps being compared are or which two iterations are being compared. Regarding within-iteration comparisons, changes in distributions from initial to final graph constructions can be viewed as relatively conclusive evidence of gains because the data format and prompting remain constant. On the other hand, it may be the case that comparisons between distributions at *debate* and *lab-mass critique* reflect the specific characteristics of the task (e.g., *debate* presented two alternative conceptions simultaneously; *lab-mass* only refers to one). Thus, while a significantly larger proportion of students in Iterations 2 and 3 reject Aida's claim following *lab-mass* than in *debate*, this evidence should be viewed as supporting our argument, and not as a conclusive evidence of learning.

Likewise, comparisons between iterations are quasi-experimental in nature and may reflect differences between students and teachers. In particular, the students in Iteration 1 were drawn from a lower SES population than in Iterations 2 and 3. Therefore, it may be that the lack of appropriate experimentation and discovery results, relative to latter designs, reflected lower prior knowledge, engagement, or familiarity with inquiry practices. Yet, despite this possibility, evidence from students who were engaged and attempted to control variables, suggests that the initial design was a clear obstacle to learning. On the other hand, Iterations 2 and 3 were conducted in the same school, with the same teachers, in sequential years, and thus represent a more valid comparison.

Finally, our analysis does not capture the influence of spontaneous collaborations and mentoring that occurred throughout the study. Indeed, informal partnerships and assistance emerged routinely, and likely influenced learning behaviors. Likewise, the particular assistance given by the teachers varied according to the teachers' experience and personal relationships with the students. Further study, applying qualitative methods, is necessary to fully illustrate the emergent dynamics of the classroom and their impacts on learning.

7.4.2 Implications for Virtual and Physical Laboratories

The results described within this chapter, we believe, shed insights into both the value of virtual laboratories and the process of developing them to maximize their potential. First, the value of virtual laboratories is fundamentally determined by how it is coordinated with a coherent, meaningful sequence of activities. Robust online learning platforms such as WISE (Linn et al. 2003) or Go-Lab (Jong et al. 2014) facilitate design by offering students opportunities to make predictions, perform experiments, analyze data, and receive guidance in a manner that integrates each activity into a coherent whole. Furthermore, such platforms allow designers and researchers to employ well-established approaches to learning and inquiry. In our case, we applied the knowledge integration pattern (Linn and Eylon 2011) by preceding student experimentation with prediction and following experimentation with distinguishing and reflection activities. In this way, student engagement with the virtual laboratory is not isolated, but reflects an opportunity to address and challenge students' ideas.

Additionally, virtual laboratories offer an opportunity to test new interface designs and guidance approaches. Results from experimental tests of new designs may, in some cases, even provide insights about typical, physical laboratories. For example, we found that giving the students the ability to determine the size of objects they wish to test often led students to conclude, incorrectly, that heavier objects will always sink. Without guidance, students often assumed that they should be testing objects of equal size, leading to this conclusion. In a physical setting, students could perform a similar task (e.g., by cutting wax or other materials), but would likely encounter similar difficulties. In this case the teacher may want to either supply appropriate materials (as in Iteration 2) or guide students to construct materials that range in size.

On the other hand, we found that the mass-volume graph, as both an input for tests and output of results, was uniquely valuable in helping students discover the role of mass-volume ratio in buoyancy. This representational feature, particularly as an input device, has no clear counterpart in a physical laboratory (i.e., it required a theoretically infinite range of densities to specify an arbitrary mass and volume). In this case, the virtual laboratory may afford a conceptual activity that is limited or unavailable in physical space.

Beyond the specific affordances of our virtual laboratory, we believe that the design-based research process undertaken to develop features and guidance has clear implications for both design of virtual and physical learning activities. In particular, the use of logged process data is an important tool for discovering and verifying student strategies and unexpected obstacles. For example, in the 1st iteration, described in more detail in prior work (Vitale et al. 2016), we found that students were reluctant to return to the virtual laboratory when prompted in the graphing activity. We addressed this issue by incorporating simulation directly adjacent to the graph. In this case, overcoming a logistical obstacle inspired deeper inquiry with the materials. More generally, obtrusive demands of a laboratory can dampen enthusiasm and reduce students' willingness to undertake more productive struggles (e.g., testing diverse spaces of the graph). Indeed, virtual laboratories are an ideal tool for testing a broad range of critical pedagogical issues.

References

Anderson, J. R., Reder, L. M., Simon, H. A., Ericsson, K. A., & Glaser, R. (1998). Radical constructivism and cognitive psychology. *Education, 1*(1), 227–278.

Barab, S., & Squire, K. (2004). Design-based research: Putting a stake in the ground. *The Journal of the Learning Sciences, 13*(1), 1–14.

Bell, P. (2004). On the theoretical breadth of design-based research in education on the theoretical breadth of design-based research in education. *Educational Psychologist, 39*(4), 243–253.

Chinn, C. a., & Brewer, W. F. (1993). The role of anomalous data in knowledge acquisition: A theoretical framework and implications for science instruction. *Review of Educational Research, 63*(1), 1–49.

Collins, A., Joseph, D., & Bielaczyc, K. (2004). Design research: Theoretical and methodological issues. *The Journal of the Learning Sciences, 13*(1), 15–42.

De Jong, T., & Van Joolingen, W. R. (1998). Scientific discovery learning with computer simulations of conceptual domains. *Review of Educational Research, 68*(2), 179–201.

De Jong, T., Sotiriou, S., & Gillet, D. (2014). Innovations in STEM education: The go-lab federation of online labs. *Smart Learning Environments, 1*(3), 1–16.

diSessa, A. A. (2002). Why "conceptual ecology" is a good idea. In M. Limón & L. Mason (Eds.), *Reconsidering conceptual change issues in theory and practice* (pp. 29–60). Dordrecht: Kluwer. https://doi.org/10.1007/0-306-47637-1_2.

Dunbar, K. (1993). Concept discovery in a scientific domain. *Cognitive Science, 17*(3), 397–434.

Dunbar, K., & Klahr, D. (1989). Developmental differences in scientific discovery processes. In D. Klahr & K. Kotovsky (Eds.), *Complex information processing: The impact of Herbert A. Simon* (pp. 109–144). Hillsdale: Erlbaum.

Gobert, J. D., Sao Pedro, M., Raziuddin, J., & Baker, R. S. (2013). From log files to assessment metrics: Measuring students' science inquiry skills using educational data mining. *Journal of the Learning Sciences, 22*(4), 521–563.

Goldstone, R. L., & Wilensky, U. (2008). Promoting transfer by grounding complex systems principles. *The Journal of the Learning Sciences, 17*(4), 465–516.

Hewson, P. W., & Hewson, M. G. A. (1984). The role of conceptual conflict in conceptual change and the design of science instruction. *Instructional Science, 13*(1), 1–13.

Hmelo-Silver, C. E., Duncan, R. G., & Chinn, C. A. (2007). Scaffolding and achievement in problem-based and inquiry learning: A response to Kirschner, Sweller, and Clark (2006). *Educational Psychologist, 42*, 99–107.

Holbrook, J., & Kolodner, J. L. (2000). Scaffolding the development of an inquiry-based (science) classroom. *Fourth International Conference of the Learning Sciences*, pp. 221–227.

Inhelder, B., & Piaget, J. (1958). *The growth of logical thinking from childhood to adolescence.* New York: Basic Books.

Klahr, D., Fay, A. L., & Dunbar, K. (1993). Heuristics for scientific experimentation: A developmental study. *Cognitive Psychology, 25*, 111–146.

Koedinger, K. R., & Aleven, V. (2007). Exploring the assistance dilemma in experiments with cognitive tutors. *Educational Psychology Review, 19*(3), 239–264.

Kozma, R. B., Chin, E., Russell, J., & Marx, N. (2000). The roles of representations and tools in the chemistry laboratory and their implications for chemistry learning. *Journal of the Learning Sciences, 9*(2), 105–143.

Krajcik, J. S., Blumenfeld, P., Marx, R., Bass, K., Fredricks, J., & Soloway, E. (1998). Inquiry in project-based science classrooms: Initial attempts by middle school students. *Journal of the Learning Sciences, 7*(3), 313–350.

Krajcik, J. S., Blumenfeld, P. C., Marx, R., & Soloway, E. (2000). Instructional, curricular, and technological supports for inquiry in science classrooms. In J. Minstrell & E. H. Van Zee (Eds.), *Inquiring into inquiry learning and teaching in science* (pp. 283–315). Washington, DC: American Association for the Advancement of Science.

Linn, M. C. (2006). The knowledge integration perspective on learning and instruction. In R. Sawyer (Ed.), *The Cambridge handbook of the learning sciences* (pp. 243–264). Cambridge, MA: Cambridge University Press.
Linn, M. C., & Eylon, B.-S. (2011). *Science learning and instruction: Taking advantage of technology to promote knowledge integration*. New York: Routledge.
Linn, M. C., Clark, D., & Slotta, J. D. (2003). WISE design for knowledge integration. *Science Education, 87*(4), 517–538.
Mayer, R. E. (2004). Should there be a three-strikes rule against pure discovery learning? *American Psychologist, 59*(1), 14–19.
McElhaney, K. W., & Linn, M. C. (2011). Investigations of a complex, realistic task: Intentional, unsystematic, and exhaustive experimenters. *Journal of Research in Science Teaching, 48*(7), 745–770.
McNeill, K. L., Lizotte, D. J., Krajcik, J., & Marx, R. W. (2006). Supporting students' construction of scientific explanations by fading scaffolds in instructional materials. *Journal of the Learning Sciences, 15*, 153–191.
Perkins, D. N., & Grotzer, T. a. (2005). Dimensions of causal understanding: The role of complex causal models in students' understanding of science. *Studies in Science Education, 41*(1), 117–165.
Rutherford, J. (1964). The role of inquiry in science teaching. *Journal of Research in Science Teaching, 2*, 80–84. https://doi.org/10.1002/tea.3660020204.
Sandoval, W. (2014). Conjecture mapping: An approach to systematic educational design research. *Journal of the Learning Sciences, 23*(1), 18–36.
Schauble, L., Glaser, R., Raghavan, K., & Reiner, M. (1991). Causal models and experimentation strategies in scientific reasoning. *Journal of the Learning Sciences, 1*(2), 201–238.
Smith, C., Carey, S., & Wiser, M. (1985). On differentiation: A case study of the development of the concepts of size, weight, and density. *Cognition, 21*, 177–237.
Smith, C., Snir, J., & Grosslight, L. (1992). Models to facilitate using conceptual change: The case of weight-density differentiation. *Cognition and Instruction, 9*(3), 221–283.
Smith, J. P., DiSessa, A. A., & Roschelle, J. (1993). Misconceptions reconceived: A constructivist analysis of knowledge in transition. *The Journal of the Learning Sciences, 3*, 115–163.
States, N. L. (2013). *Next generation science standards: For states, by states*. Washington, DC: National Academy Press.
Tschirgi, J. E. (1980). Sensible reasoning: A hypothesis about hypotheses. *Child Development, 51*(1), 1–10.
Vitale, J. M., Lai, K., & Linn, M. C. (2015). Taking advantage of automated assessment of student-constructed graphs in science. *Journal of Research in Science Teaching, 52*(10), 1426–1450.
Vitale, J. M., Madhok, J., & Linn, M. C. (2016). Designing a data-centered approach to inquiry practices with virtual models of density. In C. Looi, J. Polman, U. Cress, & P. Reimann (Eds.), *Transforming learning, empowering learners: The International Conference of the Learning Sciences (ICLS) 2016* (pp. 591–598). Singapore: International Society of the Learning Sciences.
Wason, P. C. (1960). On the failure to eliminate hypotheses in a conceptual task. *Quarterly Journal of Experimental Psychology, 12*(3), 129–140.
Wilson, C. D., Taylor, J. A., Kowalski, S. M., & Carlson, J. (2010). The relative effects and equity of inquiry-based and commonplace science teaching on students' knowledge, reasoning, and argumentation. *Journal of Research in Science Teaching, 47*(3), 276–301.

Jonathan M. Vitale is a software developer and data analyst with the School District of Philadelphia and a former postdoc at the University of California, Berkeley.

Marcia C. Linn is Professor of Development and Cognition, specializing in science and technology in the Graduate School of Education, University of California, Berkeley.

Chapter 8
Real-Time Scaffolding of Students' Online Data Interpretation During Inquiry with Inq-ITS Using Educational Data Mining

Janice D. Gobert, Raha Moussavi, Haiying Li, Michael Sao Pedro, and Rachel Dickler

Abstract This chapter addresses students' data interpretation, a key NGSS inquiry practice, with which students have several different types of difficulties. In this work, we unpack the difficulties associated with data interpretation from those associated with warranting claims. We do this within the context of Inq-ITS (Inquiry Intelligent Tutoring System), a lightweight LMS, providing computer-based assessment and tutoring for science inquiry practices/skills. We conducted a systematic analysis of a subset of our data to address whether our scaffolding is supporting students in the acquisition and transfer of these inquiry skills. We also describe an additional study, which used Bayesian Knowledge Tracing (Corbett and Anderson. User Model User-Adapt Interact 4(4):253–278, 1995), a computational approach allowing for the analysis of the fine-grained sub-skills underlying our practices of data interpretation and warranting claims.

J. D. Gobert (✉)
Department of Educational Psychology, Rutgers Graduate School of Education, New Brunswick, NJ, USA

Apprendis, Berlin, MA, USA
e-mail: janice.gobert@gse.rutgers.edu; janice@apprendis.com

R. Moussavi
Teaching Systems Lab, Massachusetts Institute of Technology, Cambridge, MA, USA
e-mail: moussavi@mit.edu

H. Li · R. Dickler
Department of Educational Psychology, Rutgers Graduate School of Education, New Brunswick, NJ, USA
e-mail: haiying.li@gse.rutgers.edu; rachel.dickler@gse.rutgers.edu

M. Sao Pedro
Apprendis, Berlin, MA, USA
e-mail: mikesp@apprendis.com

M. E. Auer et al. (eds.), *Cyber-Physical Laboratories in Engineering and Science Education*, https://doi.org/10.1007/978-3-319-76935-6_8

Keywords Scaffold · Inquiry learning · Intelligent tutoring system · Microworld · Virtual lab

8.1 Introduction

Science educators and policy makers (NGSS Lead States 2013; OECD 2014) agree that richly integrating authentic inquiry with science content will promote well-honed learning strategies and allow students to apply and transfer their science knowledge in more flexible ways as is needed for tomorrow's jobs (Hilton and Honey 2011). As a result, as schools in the United States adopt the Next Generation Science Standards (NGSS), educators will need to (1) incorporate more inquiry experiences into instruction, (2) assess their students' inquiry practices/skills, and (3) ensure that each student demonstrates adequate progress on these.

Meeting these goals however poses significant challenges (Fadel et al. 2007). First, educators may not have adequate time, lab space, and/or physical materials for inquiry (Staer et al. 1998), particularly in schools with large class sizes (e.g., in Oregon there can be 50 students in a class). Second, grading inquiry is difficult, subjective, and time-intensive (Deters 2005). Third, teachers need *immediate* and *actionable data* to identify *which* of the many types of difficulties students are experiencing (Kuhn 2005) in order to foster students' growth (Shute 2008), but current assessments yield data too late for teachers to impact students' learning (Pellegrino et al. 2001). Fourth, developing authentic inquiry tasks and assessments is difficult due to its multifaceted, ill-defined nature (Williamson et al. 2006), and as a result, there are too few empirically tested resources to assess and support inquiry (Krajcik et al. 2000; Schneider et al. 2005). Lastly, since inquiry practices need to be honed over time, students need to engage in authentic inquiry multiple times across the school year, and without an automated solution, the burden on teachers to do grading is extremely onerous.

To add to these issues, the most recent student data on international comparisons of science performances show that American students continue to fall behind their peers. For example, in 2015, the United States ranked 25th worldwide on a key educational survey called the Program for International Student Assessment (PISA; Organization for Economic Cooperation and Development 2018). This is no doubt related, at least in part, to the many student difficulties that have been demonstrated for all of the inquiry skills identified by NGSS (2013). Specifically, students have trouble *forming testable hypotheses* (Chinn and Brewer 1993; Klahr and Dunbar 1988; Kuhn et al. 1995; Njoo and de Jong 1993; van Joolingen and de Jong 1997; Glaser et al. 1992) and difficulty *testing their hypotheses* (van Joolingen and de Jong 1991b, 1993; Kuhn et al. 1992; Schauble et al. 1991). They have difficulty *conducting experiments* (Glaser et al. 1992; Reimann 1991; Tsirgi 1980; Shute and Glaser 1990; Kuhn 2005; Schunn and Anderson 1998, 1999; Harrison and Schunn 2004; McElhaney and Linn 2008, 2010).

When interpreting data during inquiry, a key NGSS inquiry practice and the one addressed in this chapter, students have several different types of difficulties. They may draw conclusions based on confounded data (Klahr and Dunbar 1988;

Kuhn et al. 1992; Schauble et al. 1995), state conclusions that are inconsistent with their data (Kanari and Millar 2004), change ideas about causality (Kuhn et al. 1992), and/or have difficulty in making a valid inference and reconciling previous conceptions with their collected data, falling back on prior knowledge (Schauble 1990; Kanari and Millar 2004), thereby exhibiting confirmation bias during inquiry (Klayman and Ha 1987; Dunbar 1993; Quinn and Alessi 1994; Klahr and Dunbar 1988). They also fail to relate the outcomes of experiments to the theories being tested in the hypothesis (Schunn and Anderson 1999; Chinn and Brewer 1993; Klahr and Dunbar 1988).

When warranting their claims with evidence, one of the five essential features of classroom inquiry per NRC's (National Research Council 2011), they often provide little to no justification (McNeill and Krajcik 2011; Schunn and Anderson 1999) and create claims that do not answer the question posed (McNeill and Krajcik 2011). Students can also rely on theoretical arguments rather than on experimental evidence during warranting (Kuhn 1991; Schunn and Anderson 1999).

Lastly, they have difficulties developing rich explanations to explain *their findings* (Krajcik et al. 1998; McNeill and Krajcik 2007). When students provide reasoning for their claims, they often use inappropriate data by drawing on data that do not support their claim (McNeill and Krajcik 2011; Kuhn 1991; Schunn and Anderson 1999), make no mention of specific evidence (Chinn et al. 2008), or generally state that an entire data table is evidence (McNeill and Krajcik 2011; Chinn et al. 2008).

In this work, we sought to unpack the difficulties associated with data interpretation and warranting claims in particular.

8.2 Our Solution: Inq-ITS (Inquiry Intelligent Tutoring System; www.inqits.com)

In response to calls such as the Next Generation Science Standards, as well as teachers' assessment challenges and students' learning challenges, we have developed a solution that leverages schools' existing computing resources to help teachers with inquiry assessment by providing *automatic, formative data* and to help students learn these skills by providing *real-time, personalized scaffolds as they engage in inquiry*. Inq-ITS (Inquiry Intelligent Tutoring System) is a lightweight LMS, providing computer-based assessment and tutoring for science inquiry skills. It is a *no-install, entirely browser-based learning and assessment tool* created using evidence-centered design (Mislevy et al. 2012) in which middle school students conduct inquiry using science microworlds (Gobert 2015). Within Inq-ITS, which consists of different interactive simulations within microworlds, or virtual labs, for different domains in physical, life, and earth science, students "show what they know" by forming questions, collecting data, analyzing their data, warranting their claims, and explaining findings using a claim-evidence-reasoning framework, all key inquiry practices (NGSS Lead States 2013). As students work, the inquiry work products they create and processes they use are automatically assessed using our patented assessment algorithms (Gobert et al. 2016a, b). These assessment algorithms were built and validated using student data (Sao Pedro et al. 2010, 2012a, 2013b, c, 2014; Gobert et al. 2012, 2013, 2015; Moussavi et al. 2015, 2016a). They have been shown to be robust when tested across inquiry activities with diverse groups of students and match human coders with high precision (precision values ranging from 84% to 99%; Sao Pedro et al. 2012a, b, 2013a, b, 2014, 2015).

8.3 Others' Prior Research on Scaffolding Inquiry

Given student difficulties with inquiry as previously described, providing support to students for inquiry is critical if the Next Generation Science Standards (2013) or other policies emphasizing authentic science practices (e.g., OECD 2018) are to be realized. Scaffolds for inquiry can help students achieve success they could not on their own (Kang et al. 2014; McNeill and Krajcik 2011) and can lead to a better understanding of scientific concepts and the purpose of experimentation, as well as the inquiry skills used in experimentation (Kirschner et al. 2006). For example, providing scaffolding for a PhET simulation on circuit construction lead students to be more explicit in their testing (such as adding a voltmeter or connecting an ammeter in the circuit); this systematicity also transferred once scaffolding was removed (Roll et al. 2014). Additionally, the specific skill of collecting controlled trials, a lynchpin skill of inquiry, can be learned via strategy training and transfers to other topics (Klahr and Nigam 2004). Scaffolding can also be used to help students make connections between experimental data and real-world scenarios (Schauble et al. 1995). Lastly, scaffolding students' explanations during inquiry can yield positive effects on learning (Edelson et al. 1995; McNeill et al. 2006). Taken together, these results demonstrate the potential for deeper inquiry learning when students are provided with adequate support.

One drawback, however, to many of these studies is that the scaffolding is either provided by a teacher, is in the form of text-based worksheets, or in some other form that is either not scalable or fine-grained, i.e., operationalized at the sub-skill level. Additionally, these approaches typically require a student to know when they need help; however, students may not have the metacognitive skills needed to do so (Aleven and Koedinger 2000; Aleven et al. 2004).

In our system, by contrast, we use an automated approach that detects students' problems with inquiry and provides computer-based scaffolding in real time in order to support the acquisition and development of inquiry skills/practices (Gobert et al. 2013; Sao Pedro et al. 2013b, c, 2014; Gobert and Sao Pedro 2017). These scaffolds are designed to address specific aspects of scientific inquiry on a fine-grained level and can help students receive the help they need by targeting the exact sub-skill on which they are having difficulty. Our identification of each of the sub-skills underlying each of the science practices described by the NGSS (2013) is described elsewhere (Gobert and Sao Pedro 2017). This approach provides both scalable assessment of science inquiry practices as well scalable guidance so that students can get help while they are having difficulty. Scaffolding in real time has been shown to better support students' learning in general (Koedinger and Corbett 2006) and in inquiry learning in particular (Gobert et al. 2013; Gobert and Sao Pedro 2017). This approach has a great benefit over the others in that it is scalable so that NGSS practices, as described, can be learned.

8.4 Inq-ITS' Prior Work on Efficacy of Scaffolding

In our work, we have shown that our scaffolding can help students who did not know two skills related to planning and conducting experiments (NGSS Lead States 2013) – testing hypotheses and designing controlled experiments – acquire these skills and transfer them to a new science topic. These findings were robust both within the topic in which students were scaffolded *and* across topics for each domain studied (physical, life, and earth science), with scaffolded students maintaining and/or improving their skills in new topics when scaffolding was removed compared to those who did not receive scaffolding (Sao Pedro et al. 2013a, b, 2014).

With regard to the inquiry practices of interest in this chapter, namely, interpreting data and warranting claims, we recently conducted a systematic analysis of a subset of our data to address whether our scaffolding with Rex is supporting students in the acquisition and transfer of these inquiry skills. Later in the chapter, we provide an additional study, using Bayesian Knowledge Tracing (BKT) (Corbett and Anderson 1995), a computational approach allowing for the analysis of the fine-grained sub-skills underlying our practices of data interpretation and warranting claims.

Our data were drawn from 357 students in six middle school classes in the Northeast of the United States. Students completed two microworlds (Flower and Density) in either the Rex ($N = 156$) or No Rex ($N = 201$) condition. Mixed repeated measures ANOVAs on both interpretation skill and warranting skill were performed. An independent variable of time phase (repeated) was included in order to account for how participants consecutively completed two microworlds: Flower and Density. In the Flower virtual lab, none of the students received scaffolding from Rex, so performance in this virtual lab was used as the baseline. In the Density virtual lab, students were randomly assigned to either the Rex or No Rex condition. The Rex condition meant that Rex was available to assist students as they engaged in the microworld, whereas the No Rex condition meant that Rex was not available and could not be triggered. The results focused on the interactive effects of time × condition. Effect size was calculated using Cohen's d. All significance testing for the primary analyses was conducted with an alpha level of .05. Our main interest was the effect of Rex's scaffolding on learning.

Table 8.1 illustrates the estimated means of interpretation skill and warranting skill in the Rex and No Rex conditions as well as standard errors, lower and upper bound with 95% confidence interval, F values, and the effect size of Cohen's d in the pairwise analyses, respectively.

Table 8.1 Statistics for condition × time in the Flower and Density virtual labs

Skills	Time	Condition	Mean	SE	95% CI Lower	95% CI Upper	*F*	Cohen's *d*
Interpreting data	1	No Rex	0.68	0.02	0.65	0.71	0.48	0.074
		No Rex	0.66	0.02	0.63	0.70		
	2	No Rex	0.74	0.02	0.71	0.77	18.11***	0.454
		Rex	0.84	0.02	0.81	0.88		
Warranting claims	1	No Rex	0.37	0.02	0.32	0.41	3.63	0.203
		No Rex	0.30	0.03	0.25	0.35		
	2	No Rex	0.68	0.02	0.64	0.73	7.06**	0.283
		Rex	0.77	0.03	0.72	0.82		

$N = 714$; $df = 1, 710$. Time 1 is Flower and Time 2 is Density
SE standard error, *CI* confidence interval
***$p < .001$. **$p < .01$

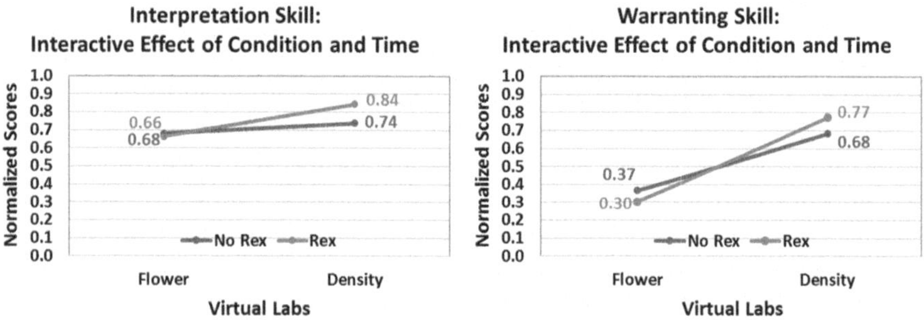

Fig. 8.1 Estimated means of *condition* × *time* in Flower and Density microworlds, respectively

8.4.1 Data Interpretation

There was a significant two-way interaction between condition × time for data interpretation skill, $F(2, 710) = 12.25$, $p < 0.001$ (see Table 8.1 and Fig. 8.1). The pairwise comparisons showed that students' interpretation substantially improved in both the No Rex (mean increased from .68 to .74, $p = .010$, $d = .26$) and Rex conditions (mean increased from .66 to .84, $p < .001$, $d = .79$). This implies that students' interpretation skills improved when they used the virtual lab *even without scaffolding from Rex*. In the second virtual lab, Density, students who received scaffolding from Rex achieved higher scores on interpreting data in the Rex condition than in the No Rex condition with a medium effect size. These findings confirm that students who received Rex's support experienced greater improvement on interpretation skills relative to students who did not receive support from Rex.

8.4.2 *Warranting Claims*

There was a significant two-way interaction between condition × time for warranting skill, $F(2, 710) = 10.40$, $p = 0.001$ (see Table 8.1 and Fig. 8.1). The pairwise comparisons showed that students' performance on warranting claims substantially improved in both the No Rex (mean increased from .37 to .68, $p < .001$, $d = 1.02$) and Rex conditions (mean increased from .30 to .77, $p < .001$, $d = 1.51$). This implies that students' skills at warranting claims improved when they used the virtual lab with or without scaffolding from Rex. Results also showed that there were no significant differences in students' skills at warranting claims when they conducted the first virtual lab, Flower, without Rex scaffolding. However, in the second virtual lab, Density, students who received scaffolding from Rex achieved higher scores on warranting claims in the Rex condition than in the No Rex condition with a small effect size. These findings further confirm that students who received Rex's support experienced greater improvement on warranting claims skills relative to students who did not receive support from Rex.

8.4.3 *Using Advanced Analytical Approaches to Study the Fine-Grained Effects of Scaffolding on Students' Data Interpretation and Warranting Claims*

In this study, we hypothesized that an automated scaffolding approach that provides personalized feedback would help students learn data interpretation skills and warranting claims skills. As such, we developed scaffolds within Inq-ITS that react when students have difficulty on these key skills and sub-skills (McNeill and Krajcik 2011; Gotwals and Songer 2009; Kang et al. 2014; Berland and Reiser 2009).

8.4.4 *Method*

8.4.4.1 Participants

Data were collected from 160 eighth grade students from the same school in the Northeast of the United States using Inq-ITS Density activities. All the students had previously used Inq-ITS, but not with this new scaffolding capacity.

8.4.4.2 Materials

Inq-ITS Density Virtual Lab Activities For each Inq-ITS virtual lab, there are typically three or four inquiry activities, consisting of driving questions that help guide students through the inquiry phases. Within each activity, students conduct inquiry by first articulating a testable hypothesis using a hypothesis widget with pulldown menus. They then experiment by collecting data with an interactive simulation through the manipulation of variables (Fig. 8.2a). Once they have collected all of their data, they interpret the results of their experiment by forming a claim in claim widget (similar to that used for hypothesizing) and selecting trials as evidence (Fig. 8.2b). Finally, students write a short open-response report that summarizes their findings from their inquiry using a claim-evidence-reasoning format (McNeill and Krajcik 2011).

In this study, three Density virtual lab activities were used. These activities aim to foster understanding about the density of different liquid substances (water, oil, and alcohol). In the first activity, the goal was to determine if the shape of the container affected the density of the liquid; the second was to determine if the amount of liquid affected the density; and the third was to determine if the type of liquid affected the density.

8.4.4.3 Procedure

Students worked on the Density activities in a computer lab at their school for the length of one science class (approximately 50 min). Each student worked independently on a computer at their own pace, meaning that not all students completed the entire set of activities by the end of the class period. Students were randomly assigned to one of two conditions: either the "Interpretation Scaffolding" ($n = 78$) or "No Interpretation Scaffolding" ($n = 82$) condition. For the first activity, none of the students, regardless of condition, received scaffolding. This allowed us to collect a baseline for each student on the targeted data interpretation sub-skills. For the next two activities, the students in the "Interpretation Scaffolding" condition received scaffolding during hypothesizing, data collection, and data interpretation. The students in the "No Interpretation Scaffolding" condition only received scaffolding during hypothesizing and data collection. The scaffolding during hypothesizing and data collection ensured that all students, regardless of scaffolding condition, had both a testable hypothesis and relevant, controlled data with which they could correctly undergo data interpretation (this design also allows us to isolate and systematically study the effects of scaffolding for data interpretation skills, as opposed to the two that proceed it in the inquiry process). Thus, students in both conditions worked in the same environment and on the same activities with access to hypothesizing and data collection scaffolding. The only difference was the presence of data interpretation scaffolding for one condition (Interpretation Scaffolding condition).

Evaluation of Inquiry Sub-skills For data interpretation and warranting claims, there are eight main sub-skills that are evaluated in the system using the work products students create. These work products are their claim (selecting the appropriate variables and relationship between them) and supporting evidence (selecting

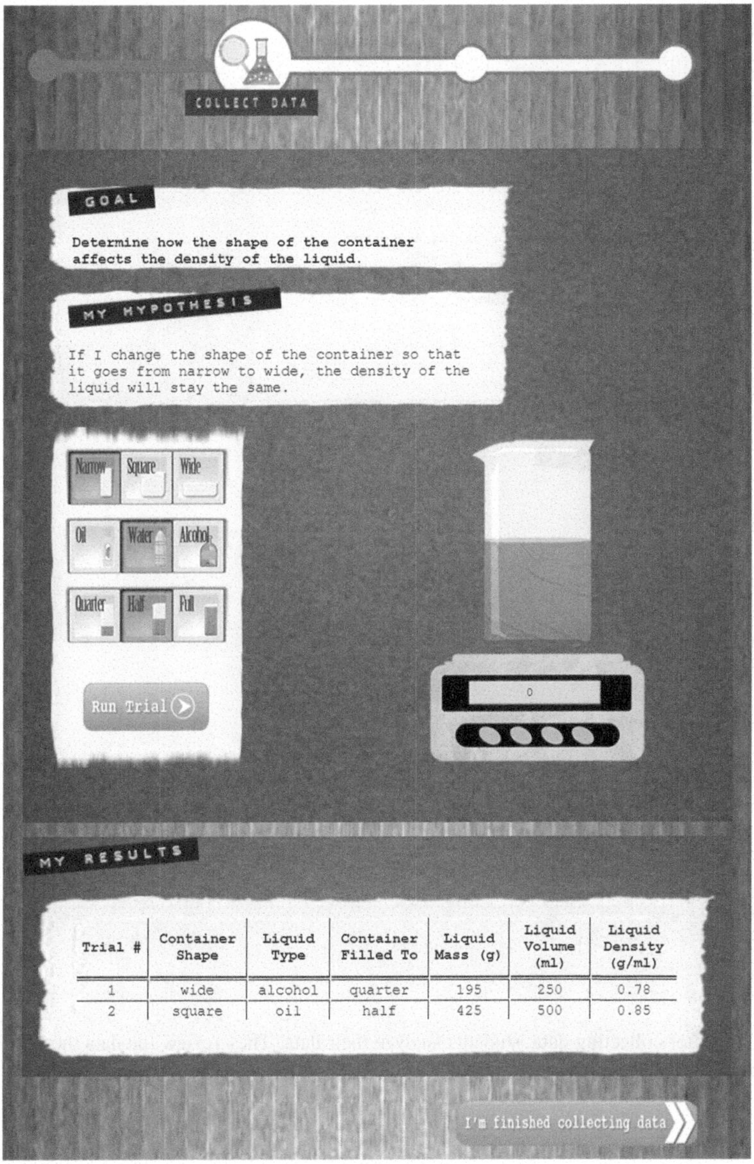

Fig. 8.2a In the "collect data" phase of the Inq-ITS Density virtual lab, students collect to test their hypothesis

relevant, controlled trials from their data table that reflect the relationship stated in their claim). These sub-skills and the specific criteria with which they are evaluated can be seen in Table 8.1. Since these sub-skills, defined in the context of this activity, are well-defined (Gobert and Sao Pedro 2017), they are evaluated using

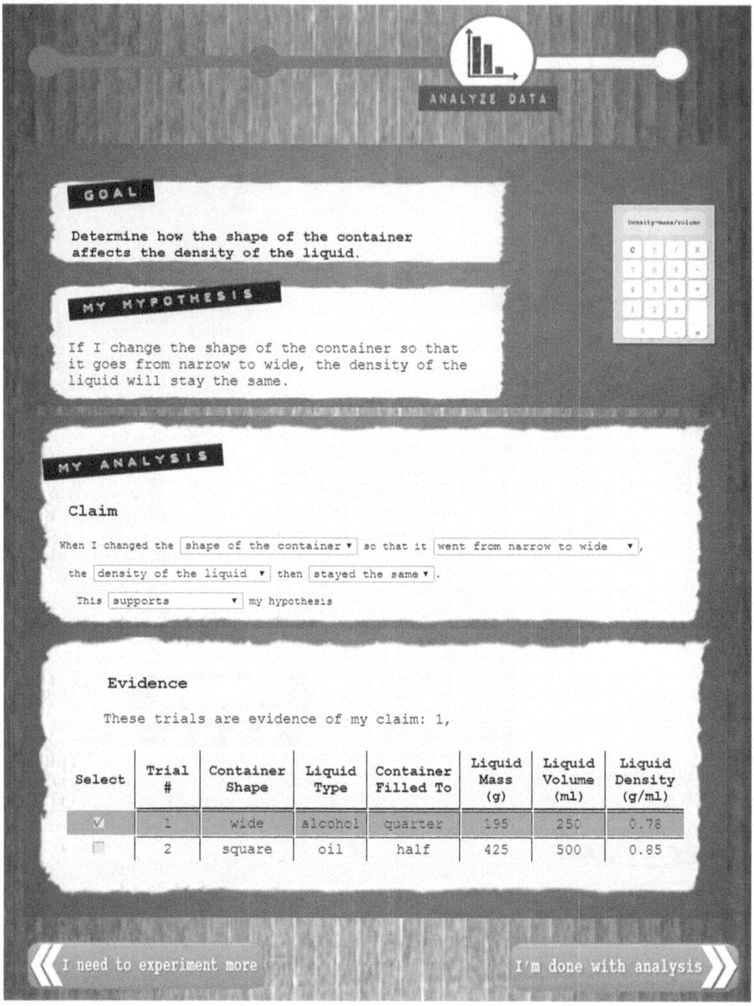

Fig. 8.2b After collecting data, students analyze their data. They review the data they collected, use pulldown menus to describe the trends found in their data, and select the evidence (trials) to support their claim

knowledge-engineered rules that specify if the sub-skill has been demonstrated. For example, for the sub-skill "Claim DV" shown in Table 8.2, the system evaluates whether or not the student has correctly chosen a variable that is measured, not changeable, within the simulation (a dependent variable) in the appropriate part of the claim. Within the context of the Density virtual lab, the appropriate dependent variable is "density of the liquid." So if the student states "density of the liquid" as the dependent variable, they would be marked as correctly demonstrating the DV sub-skill. However, if the student chooses another variable, such as one of

Table 8.2 Data interpretation sub-skills

Data interpretation sub-skills	Criteria
Interpreting the IV/DV relationship	Is the IV DV relationship correct?
Claim IV	Did the student correctly select an IV when making a claim?
Claim DV	Did the student correctly select a DV when making a claim?
Interpreting the hypothesis/claim relationship	Is the choice of whether the claim supports (or refutes) the hypothesis correct?
Controlled trials	Are all the selected trials controlled?
Warranting the IV/DV relationship	Do the selected trials support the stated IV/DV relationship?
Evidence	Did the student select more than one trial as evidence?
Warranting the hypothesis/claim relationship	Do the selected trials support the student's statement on whether their interpretation supports their hypothesis?

the independent variables like "type of liquid," as the dependent variable, then they would be scored as incorrectly demonstrating the DV sub-skill. As another example, for the sub-skill "interpreting the IV/DV relationship," a rule checks that the relationship between the independent and dependent variables specified in the claim is reflected in the data collected by the student. Elaborating further, if a student claims that "When I increased the size of the container the density of the liquid stayed the same" and their data reflects that relationship, that sub-skill would be scored as correct. If the data they collected did not reflect that relationship, the sub-skill would be scored as incorrect. The evaluation rules yield binary measure of correctness on each sub-skill (i.e., the results are presented as being correct or incorrect rather than having levels of correctness). This allows us to tease apart separate components (the sub-skills) within the broader skill of analyzing data.

Scaffolds in Inq-ITS Inq-ITS delivers scaffolds to students in text format via a pedagogical agent named Rex, a cartoon dinosaur (Fig. 8.3). Scaffolding is triggered automatically when a student completes their data analysis and at least one of the sub-skills is incorrectly demonstrated (evaluated by the knowledge-engineered rules discussed previously). This proactive scaffolding approach helps to support students in their inquiry processes (Schauble 1990; deJong 2006) by preventing students from engaging in ineffective behaviors (Buckley et al. 2006; Sao Pedro 2013). This proactive approach is also important because students may not be aware that they need help (Aleven and Koedinger 2000; Aleven et al. 2004). Once scaffolding is triggered, students may also ask Rex for additional clarification and support.

The scaffolds are designed to adapt to students' skill level by both providing multiple levels of automatic scaffolds and allowing students to request for further help or clarification (once support is auto-triggered), as needed. In this way, the

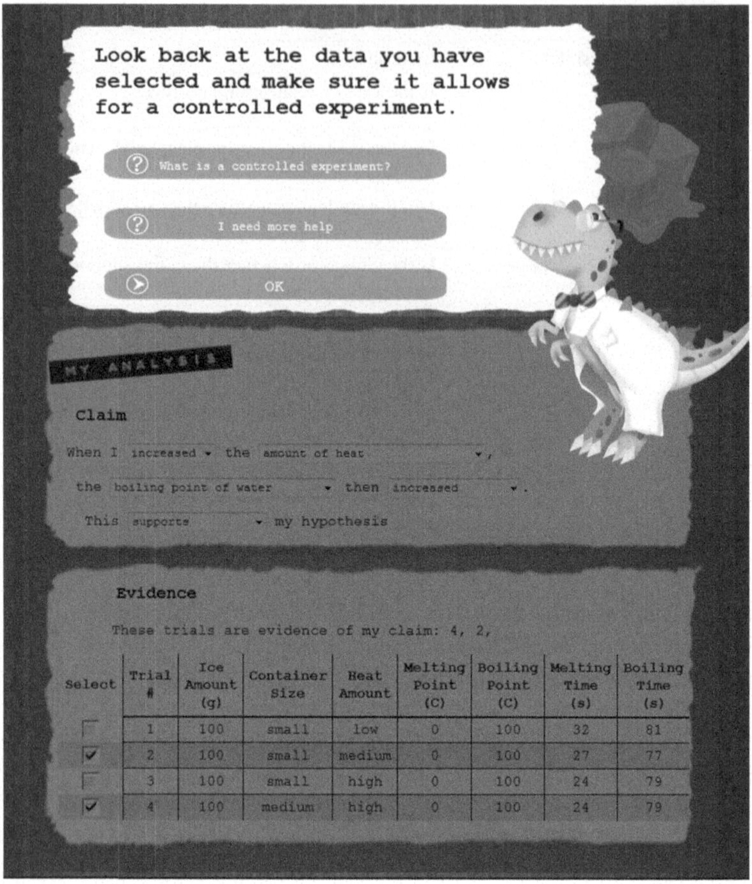

Fig. 8.3 Example scaffold delivered by Rex during data interpretation

scaffolds personalize each student's learning, recognizing that different students may need different amounts of help to successfully hone different sub-skills. The data interpretation scaffolds address four categories of procedurally-oriented difficulties that focus on the eight aforementioned sub-skills evaluated within data interpretation and warranting claims (Moussavi et al. 2015). These data interpretation and warranting claims scaffold categories are:

1. The Claim IV/DV does not match the hypothesis IV/DV.
2. The trials selected for warranting are not properly controlled or relevant to the claim.
3. The claim does not reflect the data selected.
4. The claim is incorrect as to whether it supports/does not support the hypothesis.

Since students may require scaffolding support for none, one, or many of these sub-skills, the scaffolds are designed to address these in the order listed above, so that each step of data interpretation is completed before moving onto the next. For example, it is impossible for students to correctly select relevant trials for warranting if they have not specified an appropriate IV and DV in their claim. Therefore, difficulty with creating a claim with the correct IV and DV (i.e., category 1) is scaffolded first until the sub-skill is demonstrated correctly before another difficulty is addressed. On the other hand, if a student also demonstrates difficulty with stating whether or not the claim supports the hypothesis, then the first three scaffolding categories are skipped and the student only receives the specific scaffolds that address category 4.

When students make multiple errors within the same category, we follow a sequence that increases the level of feedback given to the student. For the first error, a scaffold is provided to orient students to the current task. If the same error is repeated, they are then guided through the necessary procedural skills. Finally, the system provides a "bottom-out" hint telling students the procedure to follow. In this way, the student receives more and more targeted support, similar to cognitive tutors (e.g., Anderson et al. 1995; Corbett and Anderson 1995; Koedinger and Corbett 2006).

In sum, these scaffolds are designed to adapt to students' skill level by both providing multiple levels of automatic scaffolds and allowing students to request for further help or clarification (once support is auto-triggered), as needed. In this way, the scaffolds personalize each student's learning, recognizing that different students may need different amounts of help to successfully hone different sub-skills.

Data Analysis Approaches Due to the complexities and sub-skills inherent in the inquiry practices of data interpretation and warranting claims, an advanced analytical method using an extension of Bayesian Knowledge Tracing (Corbett and Anderson 1995) is better suited to address the effects of scaffolding on students' learning and transfer of sub-skills of inquiry under investigation here (Sao Pedro et al. 2013b). Bayesian Knowledge Tracing (BKT henceforth), a cognitive modeling approach to approximating the mastery of sub-skills in intelligent tutoring systems, is a powerful technique, and its prediction of student performance is as good as or better than similar algorithms that aggregate performance over time in order to infer student skill (e.g., Baker et al. 2011). Additionally, our group has shown that this approach is effective for modeling students' learning of inquiry, both with and without the presence of scaffolding (Sao Pedro et al. 2013b).

8.4.4.4 Bayesian Knowledge Tracing

Bayesian Knowledge Tracing (BKT) (Corbett and Anderson 1995) estimates the likelihood that a student knows a particular skill (or sub-skill) and disentangles between "knowing" and "demonstrating" that skill (or sub-skill) based on prior opportunities in which students attempt to demonstrate a particular skill. BKT assumes that knowledge of a skill is binary (either a student knows the skill or they do not) and that skill demonstration is also binary (either a student demonstrates a skill or they do not).

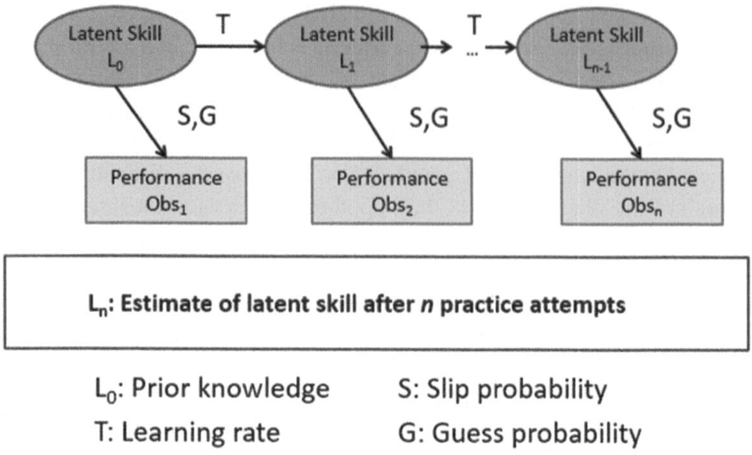

Fig. 8.4 Bayesian Knowledge Tracing model

Mathematically, four parameters are used to model whether a student knows a skill: L_0, T, G, and S (Corbett and Anderson 1995). L_0 is the probability of initial knowledge that the student is already in the "learned state," i.e., before they start the first problem. T is the probability of learning, i.e., the chance that the student goes from the "unlearned state" to the "learned state" over the course of doing all of the problems in the sequence. G is the probability of guessing, i.e., the chance that a student in the "unlearned state" answers the problem correctly. Lastly, S is the probability of slipping, i.e., the chance that a student in the "learned state" answers the problem incorrectly (Corbett and Anderson 1995). The parameters of G and S mediate the difference between "knowing" a skill and "showing" a skill. A student who shows the skill may not actually know it, contributing to G. Conversely, a student who knows the skill may not always show it, contributing to S. BKT, in this formulation, assumes that skills are not forgotten (Corbett and Anderson 1995); once a student is in the "learned state," they cannot forget and go back to the "unlearned state." Instead, if a student in the "learned state" does not "show" a skill at a specific practice opportunity, they are considered by the model to have "slipped," i.e., they were not able to show the skill at that time despite knowing it. This then affects the S parameter but does not change what state the student is considered to be in. See Fig. 8.4.

Prior work by Sao Pedro (2013) extended the traditional BKT model to account for the presence of a tutor intervention, similar to that of Beck et al. (2008). To incorporate scaffolding into the BKT framework, they introduced the dichotomous observable variable of Scaffolded = {True, False} and conditioned the learning rate (T) on that observable leading to two distinct learning rate parameters – $T_{\text{scaffolded}}$ and $T_{\text{unscaffolded}}$. This resulted in the following equations for computing $P(L_n)$, the likelihood of knowing a skill (Sao Pedro 2013):

$$P(L_n|\text{Scaffolded}_n = \text{True}) = P(L_{n-1}|\text{Prac}_n) + (1 - P(L_{n-1}|\text{Prac}_n))_* P(T_{\text{scaff}})$$

$$P(L_n|\text{Scaffolded}_n = \text{False}) = P(L_{n-1}|\text{Prac}_n) + (1 - P(L_{n-1}|\text{Prac}_n))_* P(T_{\text{unscaff}})$$

We follow this approach to determine whether data interpretation scaffolds are supporting students' learning.

One of the main assumptions of BKT is that skills are considered to be independent. This means that each skill that we want to track has to be modeled separately. Because of this, there were certain design considerations that we had to make when fitting our data to the BKT model, specifically with regard to how scaffolding condition was defined and practice opportunities were defined. These considerations are discussed in the following section.

8.4.4.5 Data Preparation Extensions to Leverage the BKT Framework

The data logged here differs from typical data logs due to how the data interpretation scaffolds were integrated into the system. In the system, all of the data interpretation sub-skills are designed to be evaluated at once. However, the data interpretation scaffolds are designed to only address one sub-skill at a time in order to give directed support, as described above. For example, if a student submits their analysis and is evaluated as both choosing an incorrect IV and an incorrect IV/DV relationship, even though they will have been evaluated on every data interpretation sub-skill, they will only receive the scaffold for one of their errors, in this case the error of the incorrect IV. Once the student revises their analysis and submits again, they are once more evaluated on all of the data interpretation sub-skills, regardless of what specific aspects of their analysis they changed.

Considering this and the fact that in BKT analysis every sub-skill is considered separately and has its own model, it became important to consider how the BKT framework defined the scaffolding condition and practice opportunity in order to create an accurate model. These design decisions for the BKT model are described in more detail below.

8.4.4.6 Determining Scaffolding Condition

Not all of the 78 students in the Interpretation Scaffolding condition needed the data interpretation scaffolds, and while some students only used one scaffold, others used multiple scaffolds targeting multiple sub-skills. Since BKT operates under the assumption of independence of skills, it would not be appropriate to label all of these students as having been scaffolded. Arguably, it is more important to model the scaffolds students received on a per skill basis, rather than simply considering them as scaffolded or not. Because of this, scaffolding was considered at the sub-skill level so that any scaffolds a student received for one specific sub-skill had no bearing on the student's scaffolding classification for the other sub-skills. This means that in the BKT model for the Claim DV sub-skill, for example, a student will only be considered to have been in the scaffolding condition if they ever received the specific scaffold directly addressing the Claim DV sub-skill, regardless of any other scaffold they may or may not have received. This makes it so that a student may only be in the scaffolding condition in the BKT model for one sub-skill or may be in the scaffolding condition in multiple BKT models on different sub-skills.

8.4.4.7 Determining Number of Practice Opportunities

In Inq-ITS, students click to submit their data interpretation after which the system records all of the actions as one practice opportunity and evaluates all of the sub-skills (Gobert et al. 2013). When scaffolding is being used, students who have been evaluated as "incorrectly demonstrating any sub-skill" receive scaffolding and are redirected to their data interpretation. Any subsequent actions students perform (up until submitting again) are considered part of a new practice opportunity for all sub-skills regardless of what specific sub-skill(s) were worked on, which can make it seem as though students require more practice opportunities to master a sub-skill than they actually do. For example, as shown in Table 8.3, based on the evaluations, it looks like after three practice opportunities, the student is still incorrectly demonstrating the "claim" and "support" sub-skills. However, if we look at the student's actions, we can see that the student was only focused on correctly demonstrating the "DV" sub-skill (due to the scaffolding received) and was not actually working on the other two sub-skills. Therefore, it would not be accurate to

Table 8.3 Example of practice opportunity succession

Student presses submit		
Sub-skills	**Evaluation**	**Practice opportunity**
IV	1	1
DV	0	1
Claim	0	1
Supports	0	1
Student receives scaffolding for DV, only changes DV (still incorrect), and submits		
Sub-skills	**Evaluation**	**Practice opportunity**
IV	1	2
DV	0	2
Claim	0	2
Supports	0	2
Student receives scaffolding for DV, only changes DV (correctly this time), and submits		
Sub-skills	**Evaluation**	**Practice opportunity**
IV	1	3
DV	1	3
Claim	0	3
Supports	0	3

Table 8.4 Example of collapsed evaluation

Sub-skills	Evaluation
IV	1
DV	0
Claim	0
Supports	0

say that the student had three practice opportunities for the "claim" and "support" sub-skills. This, then, needs to be accounted for in the BKT models in order to more accurately assess students' probability of learning.

The option considered here was to collapse student evaluations for each sub-skill within each activity into one practice opportunity. This acts as a "pre-smoothing" of data, and while it looks at the data in a slightly coarser way because of the rolling up of practice opportunities, it yields an easier model with fewer parameters. In collapsing students' evaluations, all of the evaluations for one sub-skill within an activity were examined, and a student would receive a correct evaluation for a particular sub-skill only if they always had correct evaluations for that sub-skill. This was done because if a student ever incorrectly demonstrated a sub-skill, it could be assumed that the student most likely did not know the sub-skill to begin with. This resulted in the student's evaluations in the above figure to be collapsed into one practice opportunity as shown in Table 8.4.

Therefore, the BKT analysis was performed for each of the assessed data interpretation and warranting claims sub-skills, using the scaffolding extension of the BKT framework developed by Sao Pedro (2013), as previously described.

8.4.4.8 Fitting BKT Model Parameters

To learn the parameters (L_0, T_{Scaff}, T_{Unscaff}, G, S) from student data for each of the BKT models (one model per targeted data interpretation sub-skill), we used a brute force grid search approach (Baker et al. 2010) to find the parameters that minimize the lowest sum of squared residuals (SSR) between the probability of demonstrating a skill and the actual data, as done in Sao Pedro et al. (2013b).

8.4.4.9 Determining Goodness of the BKT Models

Once the BKT parameters were determined, they were applied to the model, and then its predictive performance was tested against the same set of data used to construct the model. Although cross-validation helps to ensure that the models are accurate and can be applied to new students, it requires a held-out validation data set collected from a similar population. Since this work is exploratory in nature in that it is examining the first set of data collected with the data interpretation scaffolds, we did not have a held-out data set that could be used for this purpose. As such, the same set of data used for training was also used for validation, which can lead to over-fitting of the model. In ongoing work, we are addressing this limitation by using a held-out test set to test the models.

As in Sao Pedro et al. (2013b), performance was measured using A′ (Hanley and McNeil 1982), which is the probability that the detector will able to correctly label two examples of students' skill evaluation when in one the student is correctly demonstrating the skill and in the other the student is not. An A′ of 0.5 is indicative of chance performance, and an A′ of 1.0 is indicative of perfect performance.

8.5 Results

Our goal is to determine whether our automated scaffolding approach helps students acquire data interpretation sub-skills. We first look at a descriptive analysis of the frequency with which scaffolds were used across the activities. We also look at error rates for the sub-skills to get an initial look at students' progress with and without scaffolding. Then, as mentioned, we used our BKT extensions to approximate student learning of the data interpretation sub-skills and to make inferences about whether scaffolding was effective.

Descriptive Analysis Table 8.5 shows the number of students who received any data interpretation scaffold in an activity and the total number of scaffolds triggered in an activity. Not all the students were able to finish the third activity within the time frame of their science class, contributing to the lower number of students in Activity 3. Looking at these numbers, we can see that by the third activity, a fewer number of students received scaffolds, and that these students, overall, required less scaffolding support to successfully demonstrate the data interpretation sub-skills that we evaluate. This gives an initial indication that the scaffolding support, in its entirety, is helping students successfully interpret the data they collected and warrant their claims with data.

We next looked at the error rates for four of the data interpretation sub-skills most tightly related to the evaluations that trigger the scaffolds. Error rate is defined as the percentage of students who demonstrated that error in each activity. The graphs in Fig. 8.5 show the error rate of students in each of the two conditions (Interpretation Scaffolding condition and No Interpretation Scaffolding condition) as they worked through the three activities.

Table 8.5 Students using any data interpretation scaffold

	Activity 2	Activity 3
# of students in Interpretation Scaffolding condition who completed activity	76	64
# of students who used scaffolds	25	12
Total # of scaffolds triggered	207	32

Activity 1 is not presented, because scaffolding was not available in that activity

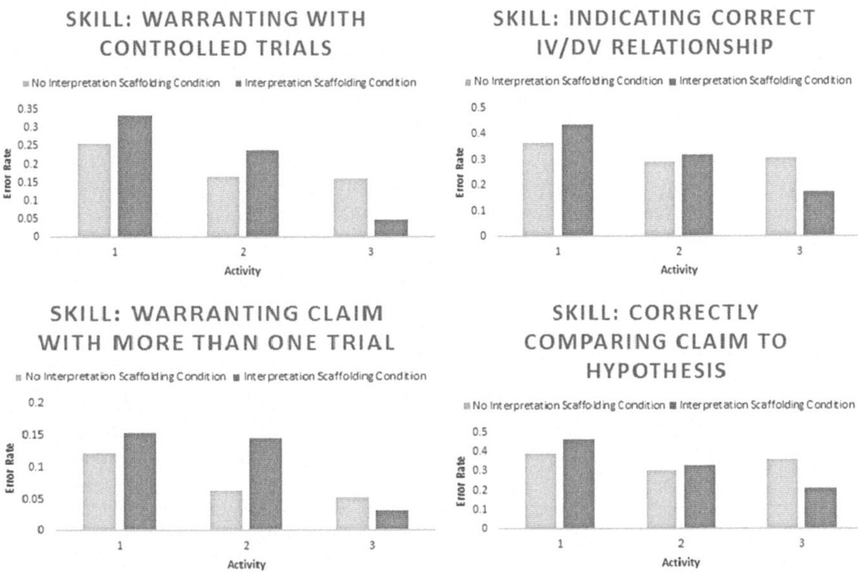

Fig. 8.5 Error rate analysis

As shown in these graphs (Fig. 8.5), student difficulty/error was present in each of these sub-skills, with the sub-skill "Interpreting correct IV/DV relationship" and "Interpreting hypothesis/claim relationship" having the highest initial error rates, regardless of condition. Furthermore, this analysis revealed that students in the "Interpretation Scaffolding" condition start with a higher error rate but end with a lower error rate. For example, for the sub-skill "Warranting with controlled trials," on their first opportunity, students in the Interpretation Scaffolding condition had an error rate of 0.33 compared to an error rate of 0.26 exhibited by the students in the No Interpretation Scaffolding condition. However, by their third opportunity, students in the Interpretation Scaffolding condition had a much lower error rate of 0.05, which was less than the error rate of 0.16 exhibited by the students in the No Interpretation Scaffolding condition. This indicates that students in the Interpretation Scaffolding condition are improving faster than the students in the No Interpretation Scaffolding condition.

The descriptive analyses suggest that scaffolding appears to be effective at helping students acquire these sub-skills. We next conduct a deeper inferential analysis using the BKT modeling framework described previously.

Inferential Analysis with Bayesian Knowledge Tracing As described previously, we fit BKT models using the student data collected and use A′ (Hanley and McNeil 1982) to measure the goodness of the models. Recall that an A′ of 0.5 is indicative of chance performance and an A′ of 1.0 is indicative of perfect performance. The A′ values for this analysis can be seen in Table 8.6. In this case, performance was measured to be relatively high for all of the sub-skills with A′ values between

Table 8.6 A′ values showing high performance of the BKT models

Sub-skill	A′
Interpreting the IV/DV relationship	0.73
Claim IV	0.70
Claim DV	0.69
Interpreting the hypothesis/claim relationship	0.72
Controlled trials	0.79
Warranting the IV/DV relationship	0.73
Evidence	0.81
Warranting the hypothesis/claim relationship	0.72

Table 8.7 BKT parameters for each sub-skill

				No Interpretation Scaffolding condition	Interpretation Scaffolding condition
Sub-skill	Probability of initial knowledge	Probability of guessing	Probability of slipping	Probability of learning	
Claim DV	0.72	0.30	0.04	0.69	0.71
Claim IV	0.94	0.21	0.01	0.83	0.36
Interpreting the IV/DV relationship	0.61	0.13	0.09	0.24	0.62
Interpreting the hypothesis/claim relationship	0.59	0.14	0.10	0.20	0.55
Controlled trials	0.71	0.12	0.04	0.27	0.79
Warranting the IV/DV relationship	0.62	0.10	0.10	0.22	0.64
Evidence	0.81	0.22	0.00	0.22	0.84
Warranting the hypothesis/claim relationship	0.59	0.13	0.10	0.20	0.53

0.69 and 0.81, allowing for parameter interpretation. However, again, since cross-validation was not done, it is possible that some of these models may be over-fitting to some student data (c.f. Sao Pedro et al. 2013).

The results from the BKT analysis indicate that the data interpretation scaffolds were effective in supporting the acquisition of data interpretation sub-skills. This can be seen through the values of the probability of learning. This value represents the chance that the student goes from the unlearned state to the learned state over the course of activities. As can be seen in the data in Table 8.7, the probability of learning for students receiving data interpretation scaffolding is higher for all but one of the evaluated sub-skills. This sub-skill, selecting an IV for the claim, also has a high probability of initial knowledge, which could indicate that the sub-skill is not being learned because so many students already know it (e.g., Sao Pedro et al. 2014). Also, compared to another sub-skill with a relatively high probability of initial knowledge – such as the Evidence sub-skill – the Claim IV sub-skill is noisier to assess, likely because it might be highly related to the content in each activity.

8.6 Discussion

The goal of this work was to test the efficacy of our data interpretation scaffolding on the sub-skills underlying the skill practices underlying data interpretation and warranting claims. We tested this in two ways, both using analysis of variance on the aggregate scores for each practice (data interpretation and warranting claims), as well as an innovative extension to Bayesian Knowledge Tracing (BKT) that considers the presence of scaffolding approximating mastery learning for each of the sub-skills of interest (Sao Pedro et al. 2013b). We also developed modifications to this framework, which allow it to be applied when condition and practice opportunity can be defined on different levels (i.e., activity level vs. skill level).

In developing our BKT extension, this work contributes a fine-grained method for unpacking the effect of scaffolding via logged, process data. Our extension to BKT was used as a modeling paradigm to track the sub-skills underlying data interpretation and warranting claims. This study was done within a complex domain of science inquiry whereby the student data, number of practice opportunity counts, and evaluated skills were not as clearly delineated as in previous studies in which BKT was used to evaluate educational interventions (Koedinger et al. 2010). This work provides a framework for how data in these complex environments can be treated before BKT can be used.

This work also explores modifying the BKT framework to represent and track students' learning of the targeted data interpretation sub-skills with and without scaffolding. Further analyses are needed to determine the efficacy of this model and its accuracy in comparison to other models. As the data used for this work was collected as an initial study of the data interpretation/warranting claims scaffolds, additional data will be used to cross-validate the predictive performance of the models used here and provide greater assurance in interpreting the parameters of the model. This method could then be used as students work through multiple domains with scaffolding to assess the efficacy of these scaffolds across a larger number of practice opportunities (e.g., Sao Pedro et al. 2014). This will also allow us to assess how scaffolding can impact the transfer of these skills from one science domain to another. Additionally, we will use this method on studies without scaffolding, which will give us data to better understand how this skill develops naturally.

Regarding inquiry, this work builds on prior research (Kang et al. 2014; McNeill and Krajcik 2011; Schauble 1990) on the nature of data interpretation and warranting claims skills, their assessment, and scaffolding. This work makes a contribution to the prior research on argumentation practices for inquiry by conceptualizing and framing the data interpretation and warranting claims practices as *necessary but not sufficient for* appropriate scientific argumentation.

When it comes to unpacking the broad components of explanation, Toulmin's (1958) model of argumentation is typically used (McNeill and Krajcik 2011; Gotwals and Songer 2009; Kang et al. 2014; Berland and Reiser 2009), breaking down argumentation into three main components: the use of claims, evidence, and reasoning. The interpretation of evidence and the creation of an evidence-based explanation or argument are both key practice in national science standards and essential for fostering students' science literacy (McNeill and Krajcik 2011; Kang et al. 2014).

We feel that unpacking the inquiry practices associated with data interpretation and warranting claims *separately* from students' data on claims, evidence, and reasoning, as expressed in open response format, is important because if students are having problems analyzing their data, they won't be able to successfully engage in explanation and argumentation. Our prior work has shown that a number of students are not able to articulate a correct explanation or argument despite knowing the data interpretation skills (Li et al. 2017). Moreover, there are large numbers of students who are being mis-assessed when their open responses are used as the only source of assessment: there are students who are skilled at science but cannot convey what they know in words (i.e., false negatives), as well as students who are skilled at parroting that they have read or heard but do not understand the science they are writing about (i.e., false positives; Gobert 2016). In short, using solely students' writing for assessment is only an accurate way of measuring what students know *if* they are good at articulating words.

To this end, we conceptualize/frame data interpretation and warranting claims practices as underlying the argumentation practices necessary for communicating science findings and thus find it necessary to study these skills separately from students' overall written explanations and arguments. Conceptualizing and supporting students on the components of the explanation framework – claim, evidence, and reasoning – in an automated and fine-grained way with appropriate sub-skills can help us unpack and target known difficulties documented by previous research (Gotwals and Songer 2009; McNeill and Krajcik 2011; Schunn and Anderson 1999). While we could make the assessment of these skills easier by designing activities that only target one skill at a time, this would be a much less authentic way of conducting inquiry. This work attempts to disentangle the effects of learning support delivered via automatic scaffolds that apply to individual sub-skills in an environment where multiple performance-based skills are being practiced and assessed at once. This gives us the nuance to examine these complex practices (as set forth by NGSS) and allows us to look at specifically what aspects students are having difficulty with and work to target those exact difficulties before moving on to students' claims, evidence, and reasoning.

Lastly, this work provides a scalable solution toward the assessment and scaffolding of these practices and in doing so represents a scalable solution to supporting teachers and students in NGSS practices.

References

Aleven, V., & Koedinger, K. (2000). Limitations of student control: Do students know when they need help? In G. Gauthier, C. Frasson, & K. VanLehn (Eds.), *Proceedings of the 5th international conference on intelligent tutoring systems* (pp. 292–303). Berlin: Springer.

Aleven, V., McLaren, B., Roll, I., & Koedinger, K. (2004). Toward tutoring help seeking: Applying cognitive modeling to meta-cognitive skills. In J. C. Lester, R. M. Vicario, & F. Paraguaçu (Eds.), *Proceedings of seventh international conference on intelligent tutoring systems* (pp. 227–239). Berlin: Springer.

Anderson, J. R., Corbett, A. T., Koedinger, K. R., & Pelletier, R. (1995). Cognitive tutors: Lessons learned. *The Journal of the Learning Sciences, 4*(2), 167–207.

Baker, R., Corbett, A., Gowda, S., Wagner, A., MacLaren, B., Kauffman, L., Mitchell, A., & Giguere, S. (2010). Contextual slip and prediction of student performance after use of an intelligent tutor. In *Proceedings of the 18th annual conference on user modeling, adaptation, and personalization* (pp. 52–63). Berlin: Springer.

Baker, R., Gowda, S., & Corbett, A. (2011) Automatically detecting a student's preparation for future learning: Help use is key. In *Proceedings of the 4th international conference on educational data mining* (pp. 179–188).

Beck, J., Chang, K. M., Mostow, J., & Corbett, A. (2008). Does help help? Introducing the bayesian evaluation and assessment methodology. In *Intelligent tutoring systems* (pp. 383–394). Berlin: Springer.

Berland, L. K., & Reiser, B. J. (2009). Making sense of argumentation and explanation. *Science Education, 93*(1), 26–55.

Buckley, B. C., Gobert, J. D., & Horwitz, P. (2006). *Using log files to track students' model-based inquiry.* Paper presented at the 7th international conference of the learning sciences, Bloomington, IN.

Chinn, C. A., & Brewer, W. F. (1993). The role of anomalous data in knowledge acquisition: A theoretical framework and implications for science instruction. *Review of Educational Research, 63*, 1–49.

Chinn, C. A., Duschl, R. A., Duncan, R. G., Buckland, L. A., & Pluta, W. J. (2008, June). A microgenetic classroom study of learning to reason scientifically through modeling and argumentation. In *Proceedings of the 8th international conference on International conference for the learning sciences* (Vol. *3*, pp. 14–15). International Society of the Learning Sciences.

Corbett, A., & Anderson, J. R. (1995). Knowledge tracing: Modeling the acquisition of procedural knowledge. *User Modeling and User-Adapted Interaction, 4*(4), 253–278.

de Jong, T. (2006). Computer simulations: Technological advances in inquiry learning. *Science, 312*, 532–533.

Deters, K. M. (2005). Student opinions regarding inquiry-based labs. *Journal of Chemical Education, 82*(8), 1178–1180.

Dunbar, K. (1993). Concept discovery in a scientific domain. *Cognitive Science: A Multidisciplinary Journal, 17*(3), 397–434.

Edelson, D. C., O'Neill, D. K., Gomez, L. M., & D'Amico, L. (1995). A design for effective support of inquiry and collaboration. In *The first international conference on computer support for collaborative learning* (pp. 107–111). Mahwah: Erlbaum.

Fadel, C., Honey, M., & Pasnick, S. (2007). Assessment in the age of innovation. *Education Week, 26*(38), 34–40.

Glaser, R., Schauble, L., Raghavan, K., & Zeitz, C. (1992). Scientific reasoning across different domains. In E. DeCorte, M. Linn, H. Mandl, & L. Verschaffel (Eds.), *Computer-based learning environments and problem-solving* (pp. 345–371). Heidelberg: Springer.

Gobert, J. (2015). Microworlds. In R. Gunstone (Ed.), *Encyclopedia of science education* (pp. 638–639). Netherlands: Springer.

Gobert, J. D. (2016). *Op-Ed: Educational data mining can be leveraged to improve assessment of science skills*. US News & World Report. http://www.usnews.com/news/articles/2016-05-13/op-ed-educational-data-mining-can-enhance-science-education.

Gobert, J. D., & Sao Pedro, M. A. (2017). Inq-ITS: Design decisions used for an inquiry intelligent system that both assesses and scaffolds students as they learn. Invited chapter in A. A. Rupp, & J. Leighton (Co-Eds), *Handbook of cognition and assessment*. New York: Wiley/Blackwell.

Gobert, J. D., Sao Pedro, M. A., Baker, R. S., Toto, E., & Montalvo, O. (2012). Leveraging educational data mining for real-time performance assessment of scientific inquiry skills within microworlds. *Journal of Educational Data Mining, 4*(1), 111–143.

Gobert, J. D., Sao Pedro, M., Raziuddin, J., & Baker, R. S. (2013). From log files to assessment metrics: Measuring students' science inquiry skills using educational data mining. *The Journal of the Learning Sciences, 22*(4), 521–563.

Gobert, J. D., Kim, Y. J., Sao Pedro, M. A., Kennedy, M., & Betts, C. G. (2015). Using educational data mining to assess students' skills at designing and conducting experiments within a complex systems microworld. *Thinking Skills and Creativity, 18*, 81–90.

Gobert, J. D., Baker, R. S., & Sao Pedro, M. A. (2016a). *U.S. patent no. 9,373,082*. Washington, DC: U.S. Patent and Trademark Office.

Gobert, J., Sao Pedro, M., Betts, C., & Baker, R. S. (2016b). *U.S. patent no. 9,564,057*. Washington, DC: U.S. Patent and Trademark Office.

Gotwals, A. W., & Songer, N. B. (2009). Reasoning up and down a food chain: Using an assessment framework to investigate students' middle knowledge. *Science Education, 94*(2), 259–281.

Hanley, J. A., & McNeil, B. J. (1982). The meaning and use of the area under a receiver operating characteristic (ROC) curve. *Radiology, 143*(1), 29–36.

Harrison, A. M., & Schunn, C. D. (2004). The transfer of logically general scientific reasoning skills. In K. Forbus, D. Gentner, & T. Regier (Eds.), *Proceedings of the 26th annual conference of the cognitive science society* (pp. 541–546). Mahwah: Erlbaum.

Hilton, M., & Honey, M. A. (Eds.). (2011). *Learning science through computer games and simulations*. Washington, DC: National Academies Press.

Kang, H., Thompson, J., & Windschitl, M. (2014). Creating opportunities for students to show what they know: The role of scaffolding in assessment tasks. *Science Education, 98*(4), 674–704.

Kanari, Z., & Millar, R. (2004). Reasoning from data: How students collect and interpret data in science investigations. *Journal of Research in Science Teaching, 41*(7), 748–769.

Kirschner, P. A., Sweller, J., & Clark, R. E. (2006). Why minimal guidance during instruction does not work: An analysis of the failure of constructivist, discovery, problem-based, experiential, and inquiry-based teaching. *Educational Psychologist, 41*(2), 75–86.

Klahr, D., & Dunbar, K. (1988). Dual search space during scientific reasoning. *Cognitive Science, 12*, 1–48.

Klahr, D., & Nigam, M. (2004). The equivalence of learning paths in early science instruction: Effects of direct instruction and discovery learning. *Psychological Science, 15*(10), 661–661.

Klayman, J., & Ha, Y. W. (1987). Confirmation, disconfirmation, and information in hypothesis testing. *Psychological Review, 94*(2), 211–228.

Koedinger, K., & Corbett, A. (2006). Cognitive tutors: Technology bringing learning sciences to the classroom. In R. Sawyer (Ed.), *The Cambridge handbook of the learning sciences* (pp. 61–77). New York: Cambridge University Press.

Koedinger, K., Pavlik Jr, P. I., Stamper, J., Nixon, T., & Ritter, S. (2010). Avoiding problem selection thrashing with conjunctive knowledge tracing. In *Educational data mining 2011*.

Krajcik, J., Blumenfeld, P., Marx, R., Bass, K., Fredricks, J., & Soloway, E. (1998). Inquiry in project-based science classrooms: Initial attempts by middle school students. *The Journal of the Learning Sciences, 7*, 313–350.

Krajcik, J., Marx, R., Blumenfeld, P., Soloway, E., & Fishman, B. (2000). *Inquiry based science supported by technology: Achievement among urban middle school students*. Paper presented at the Annual Meeting of the American Educational Research Association, New Orleans.

Kuhn, D. (1991). *The skills of argument*. Cambridge, MA: Cambridge Press.

Kuhn, D. (2005). *Education for thinking*. Cambridge, MA: Harvard University Press.

Kuhn, D., Schauble, L., & Garcia-Mila, M. (1992). Cross-domain development of scientific reasoning. *Cognition and Instruction, 9*(4), 285–327.

Kuhn, D., Garcia-Mila, M., Zohar, A., Andersen, C., White, S., Klahr, D., & Carver, S. (1995). Strategies of knowledge acquisition. *Monographs of the Society for Research in Child Development, 60*(4), 1–157.

Li, H., Gobert, J., & Dickler, R. (2017). Dusting off the messy middle: Assessing students' inquiry skills through doing and writing. In E. André, R. Baker, X. Hu, M. Rodrigo, & B. du Boulay (Eds.), *Lecture Notes in Computer Science* (Vol. 10331, pp. 175–187). Cham: Springer.

McElhaney, K., & Linn, M. (2008). Impacts of students' experimentation using a dynamic visualization on their understanding of motion. In *Proceedings of the 8th international conference of the learning sciences* (pp. 51–58). Netherlands: International Society of the Learning Sciences.

McElhaney, K., & Linn, M. (2010). Helping students make controlled experiments more informative. In K. Gomez, L. Lyons, & J. Radinsky (Eds.), *Learning in the disciplines: Proceedings of the 9th international conference of the learning sciences* (pp. 786–793). Chicago: International Society of the Learning Sciences.

McNeill, K. L., & Krajcik, J. (2007). Middle school students' use of appropriate and inappropriate evidence in writing scientific explanations. In M. Lovett & P. Shah (Eds.), *Thinking with data* (pp. 233–265). New York: Taylor & Francis Group, LLC.

McNeill, K. L., & Krajcik, J. S. (2011). *Supporting grade 5–8 students in constructing explanations in science: The claim, evidence, and reasoning framework for talk and writing*. Upper Saddle River: Pearson.

McNeill, K. L., Lizotte, D. J., Krajcik, J., & Marx, R. W. (2006). Supporting students' construction of scientific explanations by fading scaffolds in instructional materials. *The Journal of the Learning Sciences, 15*(2), 153–191.

Mislevy, R. J., Behrens, J. T., DiCerbo, K. E., & Levy, R. (2012). Design and discovery in educational assessment: Evidence centered design, psychometrics, and data mining. *Journal of Educational Data Mining, 4*(1), 11–48.

Moussavi, R., Kennedy, M., Sao Pedro, M. A., & Gobert, J. D. (2015). *Evaluating a scaffolding design to support students' data interpretation skills within a simulation-based inquiry environment*. Presented at the meeting of the American Education Research Association, Chicago.

Moussavi, R., Sao Pedro, M., & Gobert, J. D. (2016a). *Evaluating the efficacy of real-time scaffolding for data interpretation skills*. Paper presented at the meeting of the American Education Research Association, Washington, DC.

Moussavi, R., Sao Pedro, M., & Gobert, J. D. (2016b). *The effect of scaffolding on the immediate transfer of students' data interpretation skills within science topics*. Presented at the 12th International Conference of the Learning Sciences, Singapore.

National Research Council. (2011). *Successful K-12 STEM education: Identifying effective approaches in science, technology, engineering, and mathematics*. Washington, D.C.: National Academies Press.

Next Generation Science Standards Lead States. (2013). *Next generation science standards: For states, by states*. Washington, DC: The National Academies Press.

Njoo, M., & de Jong, T. (1993). Exploratory learning with a computer simulations for control theory: Learning processes and instructional support. *Journal of Research in Science Teaching, 30*, 821–844.

Organization for Economic Cooperation and Development. (2018). *PISA 2015 results in focus: What 15-year-olds know and what they can do with what they know*. Paris: Organization for Economic Cooperation and Development.

Pellegrino, J., Chudowsky, N., & Glaser, R. (2001). *Knowing what students know: The science and design of educational assessment*. Washington, DC: National Academy Press.

Quinn, J., & Alessi, S. (1994). The effects of simulation complexity and hypothesis-generation strategy on learning. *Journal of Research on Computing in Education, 27*(1), 75–91.

Reimann, P. (1991). Detecting functional relations in a computerized discovery environment. *Learning and Instruction, 1*(1), 45–65.

Roll, I., Yee, N., & Briseno, A. (2014). Students' adaptation and transfer of strategies across levels of scaffolding in an exploratory environment. In *Intelligent tutoring systems* (pp. 348–353). Switzerland: Springer International Publishing.

Sao Pedro, M. (2013). *Real-time assessment, prediction, and scaffolding of middle school students' data collection skills within physical science simulations* (Doctoral dissertation). Worcester: Worcester Polytechnic Institute.

Sao Pedro, M. A., Baker, R. S., Montalvo, O., Nakama, A., & Gobert, J. D. (2010). Using text replay tagging to produce detectors of systematic experimentation behavior patterns. In R. Baker, A. Merceron, & P. Pavlik (Eds.), *Proceedings of the 3rd international conference on educational data mining* (pp. 181–190).

Sao Pedro, M., Baker, R., & Gobert, J. (2012a). Improving construct validity yields better models of systematic inquiry, even with less information. In *Proceedings of the 20th conference on user modeling, adaptation, and personalization* (pp. 249–260). Berlin: Springer.

Sao Pedro, M., Gobert, J., & Baker, R. (2012b). *Assessing the learning and transfer of data collection inquiry skills using educational data mining on students' log files*. Paper presented at The Annual Meeting of the American Educational Research Association, Vancouver.

Sao Pedro, M., Baker, R., & Gobert, J. (2013a). Incorporating scaffolding and tutor context into bayesian knowledge tracing to predict inquiry skill acquisition. In S. K. D'Mello, R. A. Calvo, & A. Olney (Eds.), *Proceedings of the 6th international conference on educational data mining* (pp. 185–192).

Sao Pedro, M., Baker, R., & Gobert, J. (2013b). What different kinds of stratification can reveal about the generalizability of data-mined skill assessment models. In *Proceedings of the 3rd conference on learning analytics and knowledge.*

Sao Pedro, M., Baker, R., Gobert, J., Montalvo, O., & Nakama, A. (2013c). Leveraging machine-learned detectors of systematic inquiry behavior to estimate and predict transfer of inquiry skill. *User Modeling and User-Adapted Interaction, 23*(1), 1–39.

Sao Pedro, M. A., Gobert, J. D., & Betts, C. G. (2014). Towards scalable assessment of performance-based skills: Generalizing a detector of systematic science inquiry to a simulation with a complex structure. In *Intelligent tutoring systems* (pp. 591–600). Switzerland: Springer International Publishing.

Sao Pedro, M., Gobert, J., Toto, E., & Paquette, L. (2015). Assessing transfer of students' data analysis skills across physical science simulations. In I. Bejar (Chair), *The state of the art in automated scoring of science inquiry tasks.* Symposium conducted at the meeting of the American Education Research Association, Chicago.

Schauble, L. (1990). Belief revision in children: The role of prior knowledge and strategies for generating evidence. *Journal of Experimental Child Psychology, 49*, 31–57.

Schauble, L., Klopfer, L. E., & Raghavan, K. (1991). Students' transition from an engineering model to a science model of experimentation. *Journal of Research in Science Teaching, 28*(9), 859–882.

Schauble, L., Glaser, R., Duschl, R. A., Schulze, S., & John, J. (1995). Students' understanding of the objectives and procedures of experimentation in the science classroom. *The Journal of the Learning Sciences, 4*(2), 131–166.

Schneider, R., Krajcik, J., & Blumenfeld, P. (2005). Enacting reform-based science materials: The range of teacher enactments in reform classrooms. *Journal of Research in Science Teaching, 42*(3), 283–312.

Schunn, C. D., & Anderson, J. R. (1998). Scientific discovery. In J. R. Anderson (Ed.), *The atomic components of thought* (pp. 385–428). Mahwah: Lawrence Erlbaum Associates.

Schunn, C. D., & Anderson, J. R. (1999). The generality/specificity of expertise in scientific reasoning. *Cognitive Science, 23*(3), 337–370.

Shute, V. J. (2008). Focus on formative feedback. *Review of Educational Research, 78*(1), 153–189.

Shute, V., & Glaser, R. (1990). A large-scale evaluation of an intelligent discovery world: Smithtown. *Interactive Learning Environments, 1*, 55–71.

Staer, H., Goodrum, D., & Hackling, M. (1998). High school laboratory work in Western Australia: Openness to inquiry. *Research in Science Education, 28*(2), 219–228.

Toulmin, S. (1958). *The uses of argument*. New York: Cambridge University Press.

Tsirgi, J. E. (1980). Sensible reasoning: A hypothesis about hypotheses. *Child Development, 51*(1), 1–10.

van Joolingen, W. R., & de Jong, T. (1991a). Supporting hypothesis generation by learners exploring an interactive computer simulation. *Instructional Science, 20*(5), 389–404.

van Joolingen, W. R., & de Jong, T. (1991b). Characteristics of simulations for instructional settings. *Education and Computing, 6*(3-4), 241–262.

van Joolingen, W. R., & de Jong, T. (1993). Exploring a domain through a computer simulation: Traversing variable and relation space with the help of a hypothesis scratchpad. In D. Towne, T. de Jong, & H. Spada (Eds.), *Simulation-based experiential learning* (pp. 191–206). Berlin: Springer.

van Joolingen, W. R., & de Jong, T. (1997). An extended dual search space model of scientific discovery learning. *Instructional Science, 25*(5), 307–346.

Williamson, D., Mislevy, R., & Bejar, I. (2006). *Automated scoring of complex tasks in computer-based testing*. Mahwah: Lawrence Erlbaum Associates.

Janice D. Gobert, Ph.D., is a Professor of Learning Sciences and Educational Psychology at Rutgers Graduate School of Education. She is also a Co-founder and the CEO of Apprendis, whose main products are Inq-ITS and Inq-Blotter.

Raha Moussavi is a Ph.D. candidate in Learning Sciences & Technology at Worcester Polytechnic Institute. She is completing Ph.D. with Dr. Janice Gobert. She is presently employed in the Teaching Systems Lab, Massachusetts Institute of Technology.

Haiying Li, Ph.D., is a Postdoc with Dr. Janice Gobert in the Graduate School of Education at Rutgers University.

Michael Sao Pedro, Ph.D., is CTO and Co-founder of Apprendis and was a former graduate student of Dr. Janice Gobert at Worcester Polytechnic Institute. He is Co-inventor with Dr. Janice Gobert on the two patents for Inq-ITS and Inq-ITS.

Rachel Dickler is a Ph.D. student with Dr. Janice Gobert in the Graduate School of Education at Rutgers University.

Chapter 9
Pedagogical Support for Collaborative Development of Virtual and Remote Labs: Amrita VLCAP

Prema Nedungadi, Maneesha Vinodini Ramesh, Preeja Pradeep, and Raghu Raman

Abstract There is an explosive growth in e-Learning platforms, jointly developed by multiple institutions, which provide for virtual learning content. However, many are inadequate to support the complex requirements for collaborative development of distributed learning such as accommodation of wide-ranging technologies, servers, and remote equipment controlled by diverse software. Our solution is a multi-tier architecture that supports collaborative development, publishing in various online and print formats, security, audit, and access controls. Our design considerations include a highly scalable platform, use of open technologies, templates that provide pedagogical structure, multilingual functionality, and shared virtual availability of lab equipment from multiple geographic locations, along with secure access to remote equipment.

Our platform, VLCAP (Virtual Labs Collaboration and Accessibility Platform), provides structure yet flexibility to all users, includin developers, educators, and students. It offers extensive pedagogical support to lab developers in structuring the learning environment yet provides learners with a similar look and feel despite varying technologies used in the building of the labs. The learning environment consists of screens or tabs associated with various aspects of pedagogy such as conceptual background and theory, procedures, video demonstrations, animations, mathematically accurate simulations, and online access to remote equipment as

P. Nedungadi (✉)
Center for Research in Analytics & Technologies in Education (CREATE), Amrita Vishwa Vidyapeetham, Amrita University, Coimbatore, Tamil Nadu, India
e-mail: prema@amrita.edu

M. V. Ramesh · P. Pradeep
Center for Wireless Networks & Applications (AmritaWNA), Amritapuri, Amrita Vishwa Vidyapeetham, Amrita University, Coimbatore, Tamil Nadu, India
e-mail: maneesha@amrita.edu; preejapradeep@am.amrita.edu

R. Raman
School of Business, Amrita Vishwa Vidyapeetham, Amrita University, Coimbatore, Tamil Nadu, India
e-mail: raghu@amrita.edu

M. E. Auer et al. (eds.), *Cyber-Physical Laboratories in Engineering and Science Education*, https://doi.org/10.1007/978-3-319-76935-6_9

well as learning assessment tools. This allows learners to systematically master conceptual, procedural, experimental, and reporting skills. Lab developers may tailor both the order and the presentation style to suit various teaching styles as well as diverse student needs.

VLCAP is currently being used to drive two nationwide ICT projects under the Digital India initiative – Online Labs (OLabs) for school education and Virtual Labs for higher education. We include examples from OLabs and the remote triggered wireless sensor network lab (RT WSN Lab) to demonstrate the pedagogical structure available for lab developers to build effective environments for learners. Today VLCAP has over 350,000 registered learners and a few million unregistered users from around the world.

Keywords Open technologies · Remote lab · Physical lab · Simulation · Animation · Pedagogical tool · Authoring · Instructional design

9.1 Introduction

There is an explosive growth in e-Learning portals and MOOCs, many of which are developed by multiple institutions. Generally, in such environments, all developers are required to use the same technology to provide a consistent user experience. Though such models work for simple e-Learning content development, with primarily video lectures followed by online assessments, they are inadequate to support the more complex requirements for collaborative development of cross-institutional, distributed virtual labs that can accommodate a wide range of technology, servers, and remote equipment controlled by diverse software. Many existing learning content development platforms lack an easy-to-use interface to support the complex requirements for collaborative development and publishing of distributed learning content. An interface is required that allows a wide range of diverse technology, software, and remote equipment to be managed and hosted at geographically distributed institutions.

This chapter describes the Virtual Labs Collaboration and Accessibility Platform (VLCAP), which provides an authoring framework for developing and deploying virtual and remote labs. It details development process models that can be easily adapted to different institutional requirements and the pedagogical support to lab developers. [It details development process models that can be easily adapted to different institutional requirements and that provide pedagogical support to lab developers.]

Our design includes a highly scalable platform, the use of open technologies, and similar look and feel for users. This platform accommodates varying technologies, facilitates scheduling of lab equipment from multiple institutes, and allows multiple labs to share the same equipment. Secure access to remote equipment is also provided. In addition, the platform's multi-tier architecture supports collaborative development, publishing in various online and print formats, security, audit, and access controls. The data includes common format content in the central server while metadata from geographically distributed remote content can be stored at participating institutions.

VLCAP is currently used to develop and deploy two large-scale national impact Digital India projects: Online Labs (OLabs) for school education (Nedungadi et al. 2013) and Virtual Labs for higher education (Raman et al. 2014) that together have over 300,000 registered users. We describe various aspects of the system with examples from the science simulations in OLabs and remote triggered wireless sensor network lab (RT WSN Lab) (Pradeep et al. 2015). With an aim toward faster development and distribution of simulation and remote labs, VLCAP accommodates multiple types of stakeholders from educators with limited technical expertise to educational/curriculum developers, instructional designers, and software and rich media developers. The benefits of the platform include support for collaboration in developing and deploying new labs faster along with pedagogical support for lab developers.

9.2 Literature Review

Simulation and remote labs have enormous potential to promote pedagogical success for science and engineering students. Students can experience broader opportunities to experiment, explore, and understand the concepts with flexible hours. Research has shown that students benefit from the use of complex inquiry in virtual labs (McElhaney and Linn 2011; Nedungadi et al. 2015; Nedungadi and Raman 2016). Simulation labs may provide better outcomes due to repeatability and visual learning of concepts such as the flow of electricity that cannot be visualized with a physical or remote lab (Achuthan et al. 2014). A study by Zacharia and Constantinou (2008) indicated that virtual labs were equally or more effective than physical labs in terms of allowing learners to take control of their own learning.

Students indicated motivation, enjoyment, and the online experience of virtual labs as reasons for using it (Josephsen and Kristensen 2006). Research studies that compare physical and blended labs with both physical and simulation labs showed improved conceptual skills when compared to students who only used the physical lab (Kollöffel et al. 2011). Similarly, there was improved conceptual understanding by visualizing moving electrons in electric circuits (Finkelstein et al. 2006). Adaptive Learning Platforms have incorporated simulations and animations in Mathematics and Science to provide personalised learning environments to students (Nedungadi and Raman 2010)

For colleges with limited access to laboratories, remote engineering laboratories provide a flexible, efficient, and cost-effective solution (Rojko et al. 2009; Popescu and Odbert 2011; Freeman et al. 2012). Departments can provide access to higher-end equipment with minimal potential for damage of sensitive equipment (Gustavsson et al. 2009) and lower maintenance requirements. Additional advantages include sharing research equipment, improving classroom teaching, and enhancing the learning process. Not only do remote lab platforms increase lab availability to additional students, they allow for repetition of experiments from additional locations (Hutzel 2002).

Table 9.1 Virtual vs. remote labs

Type of lab	Merits	Demerits
Virtual	Ease of explaining concepts	Non-collaborative environment
	Interactive medium	No opportunity for real equipment interaction
	Cost efficiency	Absence of real data
	Anytime/anywhere availability	
	Ability to be used by many students simultaneously	
Remote	Realistic lab session	Virtual presence of the lab
	Interactive real equipment	Scheduled usage
	Collaborative opportunities	
	Debugging capability	
	Real-time data	
	Anytime/anywhere availability	
	Moderate cost	
	Controlled experimentation	
	Limited to one learner controlling equipment at a given time	

Certain aspects of distance learning and remote labs present challenges. Additional set up of cameras, controlling servers, controlled access, scheduling, and security to protect equipment is required. Also, the lack of access to physical equipment and activities may affect students' subjective experiences and perceptions (Cooper and Ferreira 2009). Experiments performed through remote labs may be considered "less effective" as the experimenters have to control some digital devices to obtain data. In addition, some research on remote labs reveals that, while students have positive responses in terms of usability of remote labs, their understanding and satisfaction is lower due to the feeling of distance from the real lab (Sousa et al. 2010). Other research shows that the learning outcomes of well-designed remote labs, supplemented by simulations and animations, are comparable to physical labs (Nedungadi and Raman 2010; Nair et al. 2012). Integrating videoconferencing into the remote labs using multi-conference sessions (Bochicchio and Longo 2009) can also improve the student experience that is otherwise affected by a limited view from the computer (Harward et al. 2008). The benefits and challenges of virtual and remote are described in Table 9.1 (Nedic et al. 2003; Ma and Nickerson 2006; Corter et al. 2004).

Many remote labs are still built without a shared interoperable approach (Canfora et al. 2004; Ferreira and Cardoso 2005). A few projects such as iLabs (Harward et al. 2008) and LiLa (Richter et al. 2011) aim to provide development and deployment that support multi-institutional remote labs. LiLa allows for experiments to be downloaded in the form of SCORM packages and then uploaded into an LMS. An additional constraint concerns the limited number and scope of authoring tools that content developers have at their disposal to create e-Learning content. Often a single tool may not be capable of performing all the functions needed for an application, and many developers are confined to using a set of tools for many different applications instead of a single, versatile tool.

Finally, many existing collaborative platforms lack an easy-to-use interface to support the complex requirements for collaborative development and publishing of distributed content. Ideally, such an interface would accommodate the wide range of diverse technology and software that lab developers have at their disposal, as well as remote equipment that can be managed and hosted at geographically distributed institutions.

9.3 VLCAP to Support Distributed Labs

Our solution is the platform called Virtual Labs Collaboration and Accessibility Platform (VLCAP), a multi-tier architecture that not only supports technological variations but also allows for collaborative development, publishing in various online and print formats, security, audit, and access controls. The system also provides the standard functions of a learning and content management system in ways that can accommodate effective development and use of virtual and remote labs.

VLCAP may reside in the cloud (Fig. 9.1), while remote equipment will be located at various geographically distributed labs behind institute firewalls (Raman et al. 2011). It maintains the technology-independent data of multi-institutional labs along with the metadata of parts of labs that are dependent on proprietary software technology or remote equipment in the cloud. Simulations that require proprietary software and remote equipment are maintained at local institutional servers. Thus, labs from different institutes may add existing simulations or remote labs to the system using customizable templates and protocols for remote labs. The platform also requires secure user access to remote equipment, thus ensuring the safety of expensive equipment. Lab owners may reuse components such as simulations, animations, videos, and assessments from other labs to create new components. The instructor can create groups, add students, assign learning modules, give assignments, monitor student usage, and evaluate student performance.

VLCAP offers a rich content management and collaborative authoring environment, with versioning of all changes along with automatic logging of data and related data analytics. Configurable templates offer a similar look and feel regardless of the source's technological constraints while allowing further customization by the developing institute. In addition, the system supports simultaneous deployment of multilingual labs and uses a modular format for text storage that lowers the cost of translation. Our platform supports content versioning and allows collaborative lab developers to revise, create, and manage versions while also allowing restoration to previous versions. The templates separate the content, the visuals, and the structure, allowing the final output to be defined based on the device type or publishing to media type such as web, app, or print.

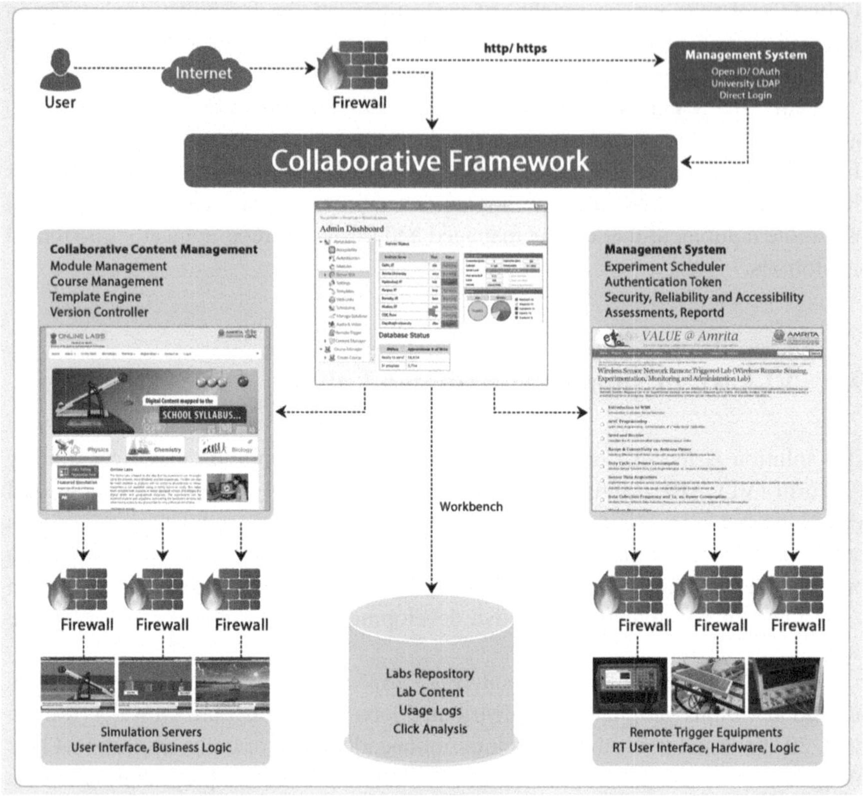

Fig. 9.1 Virtual Labs Collaborative Framework for Virtual Labs (VLCAP)

9.4 Pedagogical Support

VLCAP is designed to provide pedagogical support for both lab developers and learners. It supports collaborative development of labs for simulation, rich media, and content development. Those contributing to collaborative authoring include the lab owner, lead faculty, subject matter experts, instructional designers, simulation and animation developers, and video teams (Raman et al. 2014) (Fig. 9.2).

9.4.1 Customizable Templates

Designing effective learning labs requires both high quality learning content and a sound pedagogical approach. To facilitate optimal results, VLCAP offers customizable templates that support the workflow of the development process. The templates were designed based on discussions and analysis of learning methods of students in

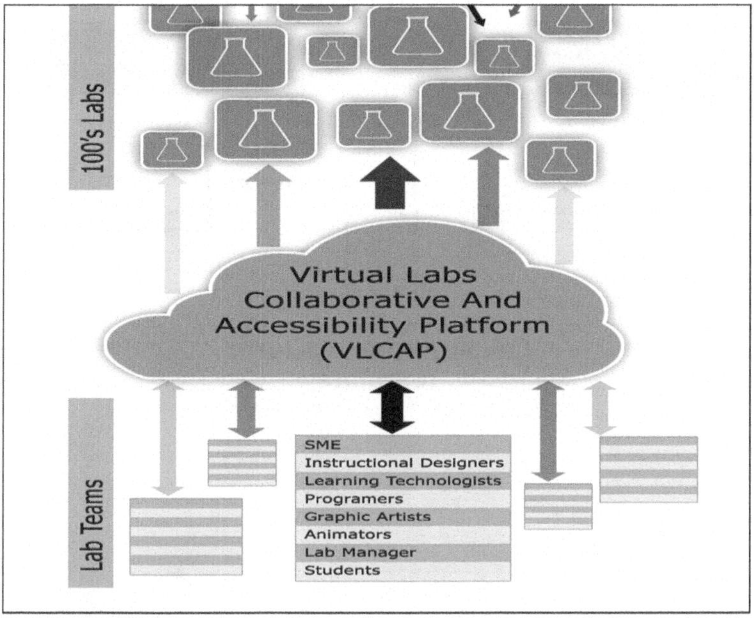

Fig. 9.2 VLCAP content development model

science labs and provide pedagogical support to lab developers to design a complete learning environment for each experiment. VLCAP provides a simple browser-based client environment that allows blending of virtual and physical aspects of an experiment so that students may learn from tutorials, perform simulations, or use remote equipment along with live-streaming video. Lab owners can add different types of content such as theory, simulations, videos, and assessments. They can easily add their equipment for secure access by online users and drag and drop their positions within the lab. Computer-naïve faculty can contribute to content and assessments, while media experts can add rich content.

VLCAP templates consist of pre-configured sets of screen layouts. The pedagogically structured templates guide educators to design their lessons based on a suggested order of material to be presented. Default screens are suggested in a typical order, but lab developers can reorder these. The template screens offer placeholders for text, images, videos, simulations, animations, and combinations of these. The pedagogical structure offered by the VLCAP templates may be designed using any combination of video tutorials, theory, procedure, animations, videos, and simulations and assessments. Although entirely new material may be created, existing simulations, animations, pictures, and other elements can be re-configured to build new labs. It is easy to modify a template by adding tabs, moving (drag to reposition), and deleting elements.

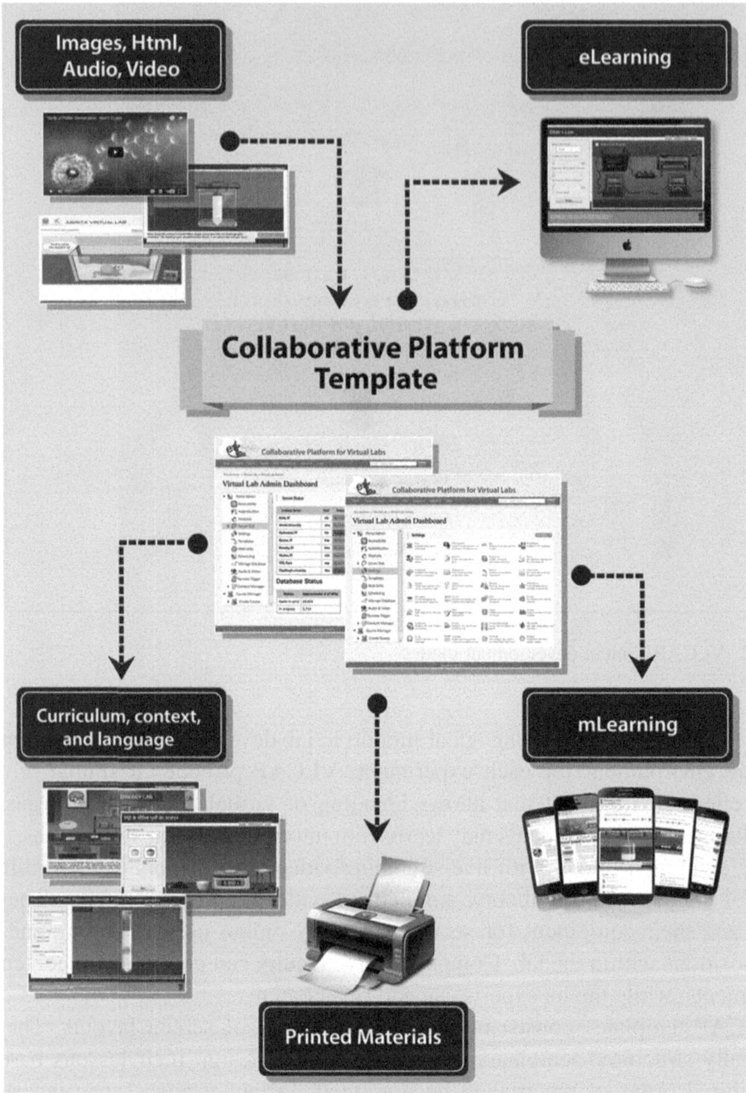

Fig. 9.3 Template-based pedagogical support

Because lab developers may type or drag and drop images, videos, or links to rich media, interactive learning environments can be produced without programming languages (Fig. 9.3).

Software developers, animation developers, and video editors may create and add rich media simulations, animations, and videos. Lab owners may also customize the workflow in a template to suit their preferred instructional methodology, change the icons of the tab, and include media from within the institute or from external

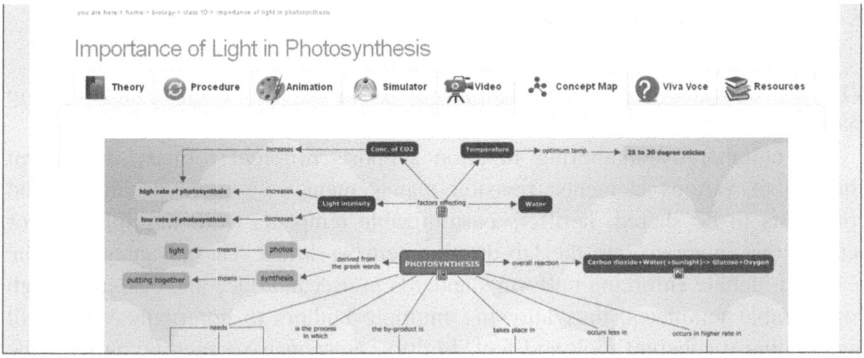

Fig. 9.4 OLabs concept maps

links (e.g., YouTube videos). Also available are customizable templates designed for reporting experiment results. The developer can add grids and worksheets for students to enter the observations of an experiment each time they perform it. They display the data in a scrollable and sortable table and save the data under the user ID that logged it. VLCAP provides for assessment-oriented templates that can accommodate multiple assessment types including multiple-choice, fill-in-the-blank, and drag and drop. Though the order of learning is recommended by the format and layout of the lab, learners can freely navigate various tabs and, for example, could start with the simulation before reviewing the procedure if that better accommodates their learning style. Furthermore, the templates allow teachers to customize the labs using various media content from both within the system and from external sources and assign specific labs to specific groups of students.

Pedagogical tools for educators such as concept map builders (Fig. 9.4) allow easy addition of various pedagogical tools for learning and assessment. Furthermore, they allow teachers to customize the labs using various media content from both within the system and from external sources and assign specific labs to specific groups of students.

The lab developers can create an entire learning environment. The templates reduce the technical overhead for content development using a WYSIWYG editor interface for easy addition of content without the need for programming. Even an untrained developer can easily start adding and modifying content. As noted, there are tools to include reports in tabular or other formats for capturing and reporting results. Different types of content, video, audio, image, animations, interactive simulations, and assessments can be added in the order determined by the lab owner.

9.4.2 *Collaborative Authoring*

VLCAP integrates collaborative authoring functions with content and learning management, thus allowing reuse of all types of learning elements (Fig. 9.5).

The collaborative authoring function supports different formats, concurrent changes of various elements, revision management, role-based privileges, and comments to be shared. It offers customizable templates with multiple tabs or screens that integrate into the lab development workflow. The advantages of the platform include enforcing uniform standards in user interface and design through configurable templates, thus allowing multiple authors to contribute while still maintaining a coherent look and feel. Multiple, responsive screen layouts for the simulation and remote labs are available for plug and play.

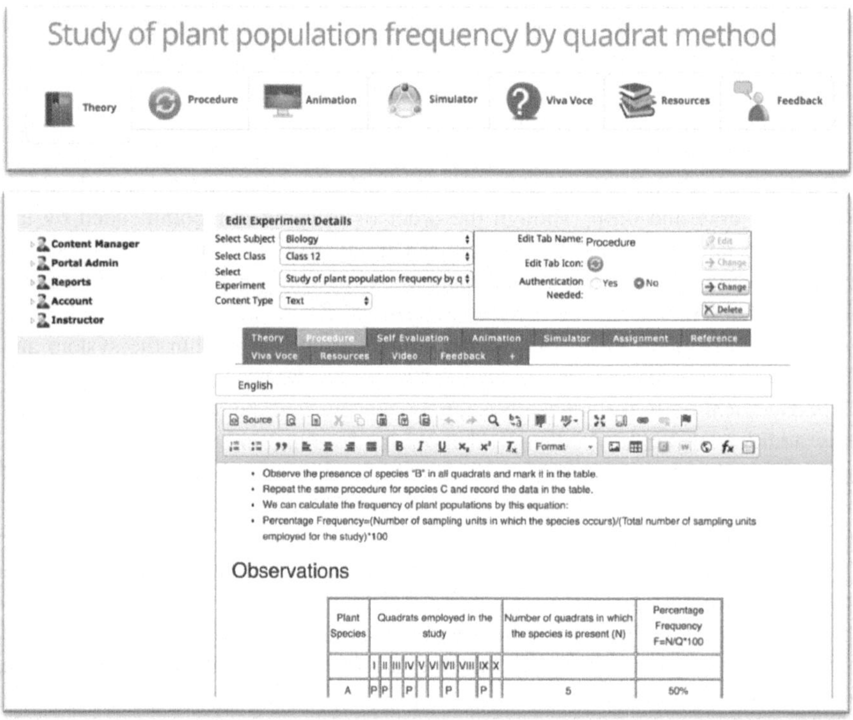

Fig. 9.5 Collaborative authoring lab developer interface

9.4.3 Editing Content and Flow

The interface for each experiment consists of different tabs (Fig. 9.6) such as theory, procedure, and simulation that allow learning via different formats. A lab developer can add, edit, delete, or manage these tabs by selecting the "Edit Experiment" on the navigation tree menu seen on the left side menu of the page and then selecting the particular tab to do the necessary editing.

Each experiment by default will have seven tabs. Selecting the desired content type from the drop-down list can change the content type of these tabs. The system supports multilingual content, and translations to a new language are easily supported in both development and deployment (Fig. 9.7).

9.4.4 Instructional Design Model

The design of the VLCAP labs (Fig. 9.8) is based on the ADDIE model that includes analysis, design, development, implementation, and evaluation phases (Lohr 1998). This was decided after comparing ADDIE with different instructional system models such as Dick and Carey (1990) and Kemp's instructional design model (Morrison et al. 2010). Kemp's model, though effective for traditional classroom learning, did not fit our need for a more systems-oriented approach. Dick and Carey's approach performs evaluations throughout each phase, but its "clear and measurable learning objectives" make it overly complex for our purposes

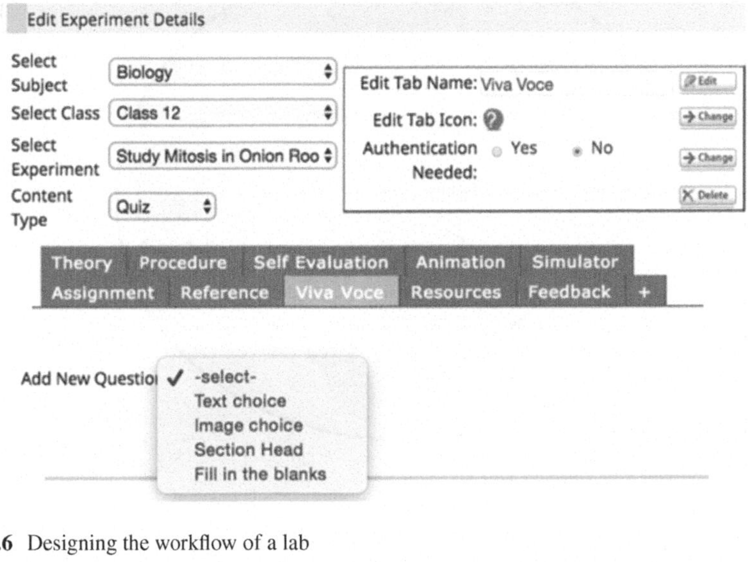

Fig. 9.6 Designing the workflow of a lab

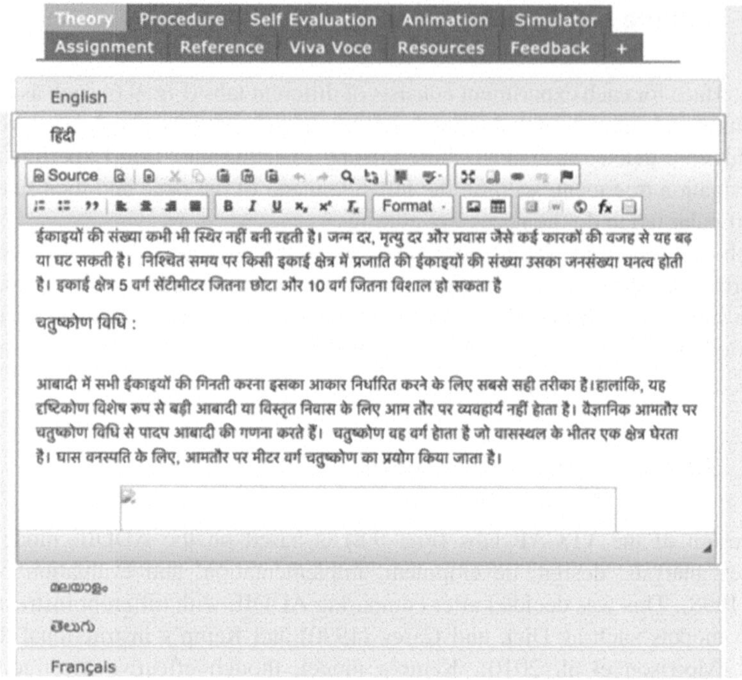

Fig. 9.7 Multilingual support

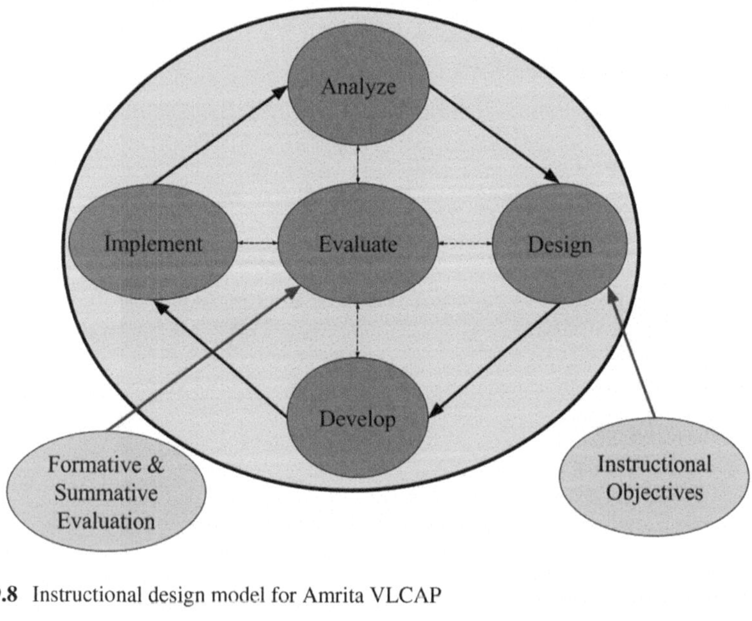

Fig. 9.8 Instructional design model for Amrita VLCAP

(Gustafson and Branch 2002). ADDIE is a flexible model with structured guidance for design and provides collaboration at every level (Allen 2006; Wang and Hsu 2009).

Identification of learning objectives, goals, audience, delivery options, and timeline of project was the main focus during the initial analysis phase. The second phase dealt with the design of the learning platform, experiment planning, contents to publish, media selection, arrangement of various formats of the contents, and prototyping by means of instructional objectives. Next, lab developers designed the experiments and performed continuous testing, validation, and debugging for the labs during the development phase. Ensuring the proper placement of contents, tools, and media as well as the procedure for training educators and students was carried out in the implementation phase. During the evaluation phase, both formative and summative evaluations of the lab were carried out. A formative evaluation was conducted for each individual stage of the process, and the resulting feedback information was immediately integrated into process revisions for that stage. The summative evaluation was performed by collecting feedback from users and incorporating their suggestions.

9.4.5 Pedagogical Support for Learners: A Learner-Centric Model

VLCAP incorporates a learner-centric pedagogical model that accommodates a variety of learning styles including visual aids (visual), learning by hearing in video demonstrations (audio), and learning through hands-on experience (kinesthetic).

Our learner-centric model includes the following characteristics:

- Learning by doing: VLCAP supports active engagement in lab activities via multiple rich media and text methods as an effective way to achieve learning objectives.
- Sequencing the learning: VLCAP allows for students to follow their preferred learning sequence. For example, they can skip the theory and video parts and directly attempt to perform the experiment and then later come back to theory topics, etc. The system will also maintain the learning sequence followed.
- Introducing concept maps: Implementation studies indicate that the students who used concept maps showed significant improvement on student achievement and positive feedback for the satisfaction survey, stating concept mapping would help them to learn new concepts in Biology (Nedungadi et al. 2015; Raman et al. 2015).

9.5 Case Study: Wireless Sensor Networks Remote Labs

VLCAP includes over 558 simulation and remote equipment-based labs developed by faculties in computer science, mechanical engineering, physical and chemical sciences, and biotechnology. In this example, the VLCAP platform was applied to development and deployment of the wireless sensor network (WSN) lab.

Many institutions lack facilities and infrastructure for WSN-based learning due to the higher cost of sensor nodes and other equipment needed for setting up a WSN laboratory or testbed. This lab provides a remote lab facility via an intuitive web-based platform (Sangeeth et al. 2015) where the learners perform experiments in the remote WSN testbed deployed in Amrita Vishwa Vidyapeetham campus via Internet.

Testbed-based remote learning was used as research comparing emulators, simulators, and testbeds for wireless sensor networks and results indicated that testbed-based remote learning is more efficient (Doddavenkatappa et al. 2011; Handziski et al. 2006; Hellbrück et al. 2011; Imran et al. 2010).

A web-based e-Learning platform with a WSN testbed (Fig. 9.9) including 11 sets of WSN experiments is provided to facilitate WSN learning and experimentation. Each WSN setup is comprised of sensors, sensor nodes, data acquisition boards, cameras, multimeters, computers, and servers for performing WSN experimentation.

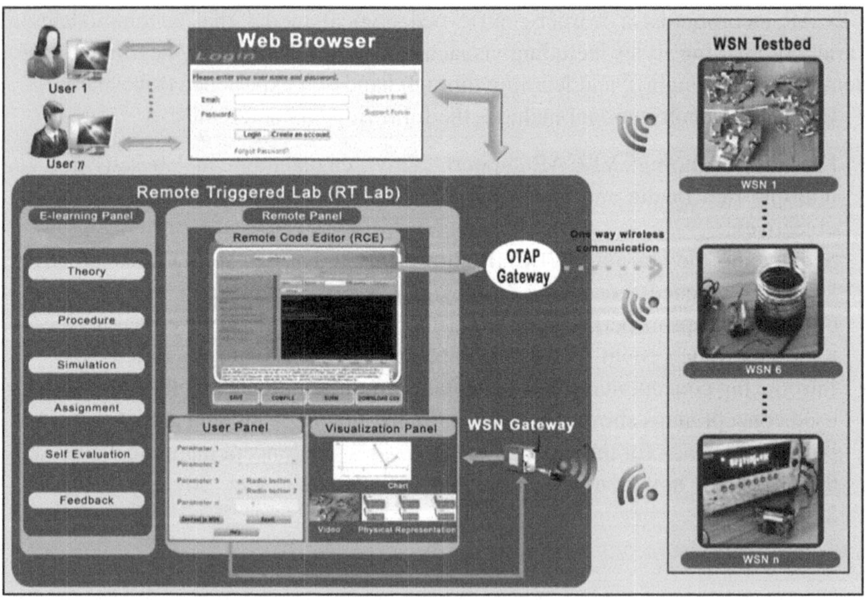

Fig. 9.9 Architecture of remote triggered wireless sensor network lab. (Reproduced from Sangeeth et al. 2015)

The lab offers an e-Learning panel, remote panel, and a visualization panel for the smooth conduction of experimentation. Users acquire both subject knowledge and a step-by-step procedure to perform the experiment by means of the e-Learning panel. The remote panel plays the most complex role in this lab and is used to view and perform the experiment by triggering the WSN testbed using the remote lab. It offers a user panel to run the experiment with the ability to change parameters and a remote code editor (RCE) for performing sensor node programming using an over-the-air programming (OTAP) mechanism. The visualization panel not only shows the experiment's results via charts and a physical representation but also includes a live-streaming video of the setup. The reprogramming facility using OTAP allows learners to observe the changes happening to WSN in real time by watching the live video.

VLCAP provides a user interface for reporting templates for benefits of remote labs. It provides remote access and practical experience in designing and deploying WSN. It further provides a web-based e-Learning platform for learning WSN concepts and a permanent testbed for development and testing of WSN concepts for those who don't have access to such an expensive lab infrastructure.

9.5.1 Course Delivery

The WSN course, based on pedagogical considerations, began with an introductory lecture on WSN to familiarize students with the topic and its applications in various domains. The students then reviewed the theory section and took an online self-evaluation to assess their knowledge level before performing the experiment. Next, they were introduced to the simulation and animation, followed by a question and answer session. This reinforced and improved understanding of concepts introduced in the theory session. Students then performed the experiment using the remote panel and subsequently were able to visualize and analyze their results. This process offered the added benefit of helping students to understand the debugging process in real time. Students were then given individualized assignments to deepen and extend their cognitive grasp of the concepts in each application along with assessments to evaluate their overall understanding of the subject knowledge.

To effectively utilize the RT Lab, each student who starts the experiment should follow the steps below:

(a) Become familiar with the aim, objective, and theory behind the experiment.
(b) Understand thoroughly the procedure, prerequisites, hardware details, and each step involved in conducting the experiment, along with the procedural details of how to effectively use the Remote Panel.
(c) Undertake self-evaluation to assess knowledge and understanding of theoretical concepts.
(d) View the animation to gain a procedural understanding of the experiment.

(e) Perform a simulation of the experiment to deepen understanding of the theory and its application.
(f) Perform the experiment in a real-world setting using the remote panel to achieve hands-on experience.
(g) Undertake assignments in various contexts to realize implications and broad applicability of the theory.
(h) Study the suggested references for additional information.

9.6 VLCAP Analytics

The platform provides rich visual and actionable analytics to lab developers, faculty members, students, and institution administrators on various aspects of usage of the experiments by learners. Analytics with both usage logs and page statistics are provided to educators and institutional administrators. It is possible to track a learner's progress through an experiment and share such data with both faculty and learner. By integrating with Google Analytics, VLCAP provides overall usage level including trends in number of users, page views, and most importantly the bounce rate (Figs. 9.10 and 9.11).

Faculty members are provided with analytics on which experiments are frequently accessed to plan their lab assignments (Table 9.2).

As a next step, a faculty member can get further details on a single experiment and review what aspects of the experiment the learners use. For example, in the case of Gram Stain Technique (Fig. 9.12), one can infer that the majority of students have viewed the theory and procedure tab.

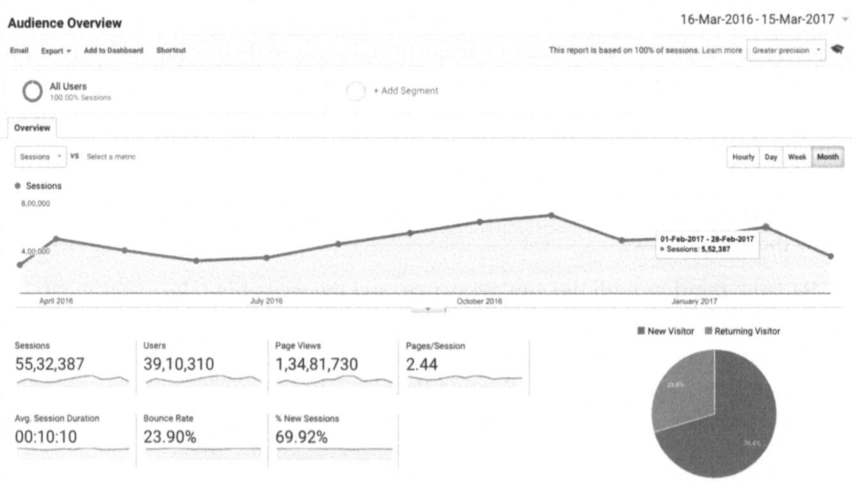

Fig. 9.10 VLCAP analytics for virtual labs

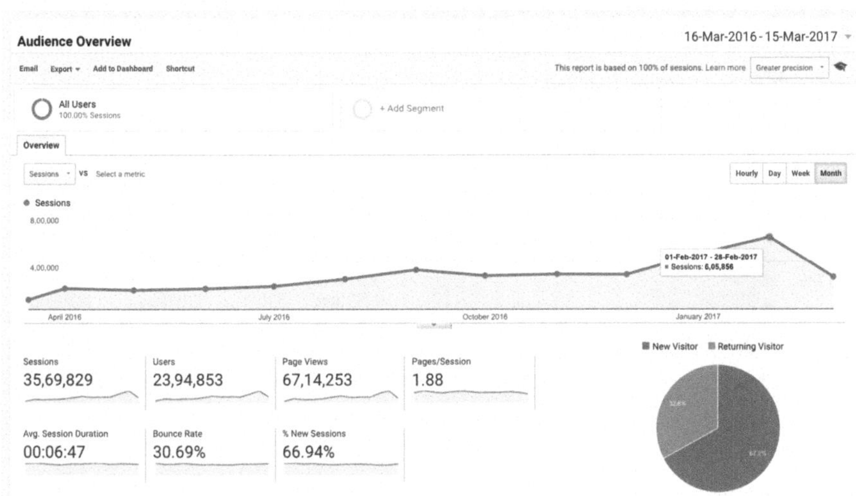

Fig. 9.11 VLCAP analytics for online labs

Table 9.2 VLCAP analytics: which experiments are most used?

Virtual labs	
Experiment name	Visits
Gram stain technique	252,862
Selective and differential media for identifying microorganisms	124,145
Bacterial growth curve	101,969
Zener diode as voltage regulator	87,198
Detection of functional groups	53,964

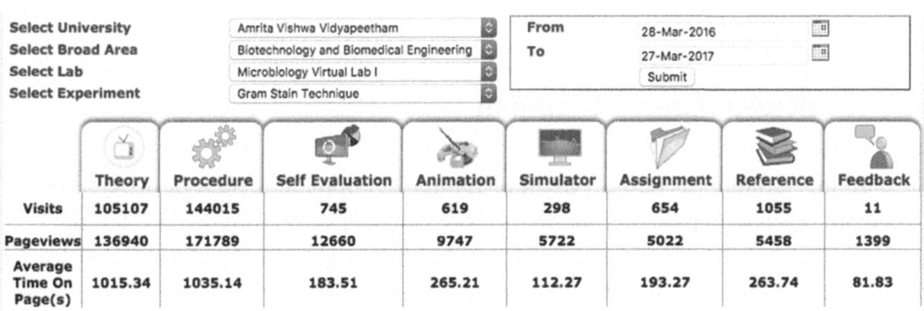

Fig. 9.12 VLCAP analytics: details for a single experiment

This lab has been operational for the past 3 years and actively used by various students, researchers, and faculty for learning WSN theory concepts and performing hands-on experience with sensor node programming. Figure 9.13 shows feedback from the lab users, with over 93% of users giving positive feedback (good, very good, or excellent) for all questions.

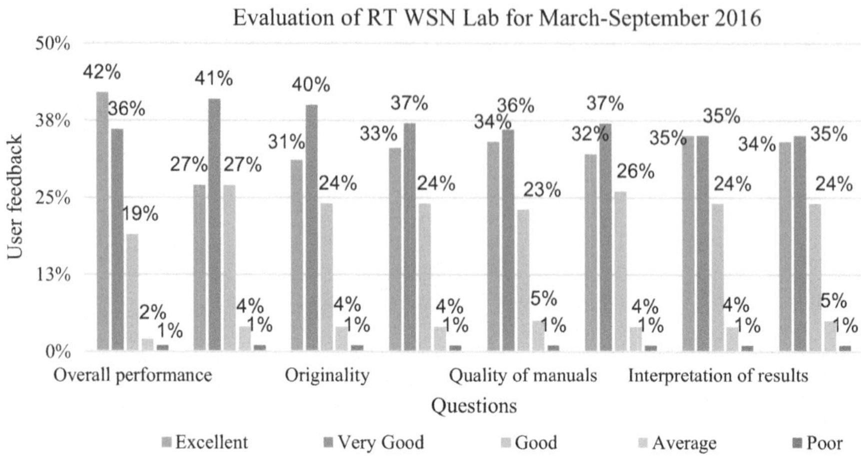

Fig. 9.13 Feedback statistics of RT WSN lab

In addition to direct access by online learners, RT WSN labs are actively used by nodal center partner institutions. Nodal centers do not have facilities to set up the WSN lab infrastructure and have expressed active interest in using them. The nodal center faculty members have been provided with training and instructions to incorporate RT WSN lab into their curriculum and teaching. As a result, each nodal center introduced this lab to their students, and they started learning WSN concepts and practicing hands-on experiments through our facility. Thus we started extending our reachability of WSN testbed initiative to other institutions in India. The statistics of nodal centers is shown in the figure (Fig. 9.14).

9.7 Conclusion and Discussion

Developing simulation-based labs and remote equipment labs require significant technical expertise and can come in the way of faster development and deployment of labs. The situation is further compounded when the lab developers and learners are geographically distributed. An easy-to-use collaborative development and deployment platform like VLCAP has encouraged faculty members with limited time and technical expertise to develop multiple experiments in a short time.

VLCAP allows remote lab developers and other educators to easily create sophisticated experiment-based lessons by providing plugins for secure access to their equipment, schedulers for reserving remote labs, and reusable templates that include a remote user interface as well as live-streaming video feedback to the lab interface. Learners can control the experiment and view changes on the testbed-based lab that she/he is controlling via live cameras. They can also obtain the data output from raw data that has been gathered by system and subsequently transmitted back and displayed with a graphical user interface.

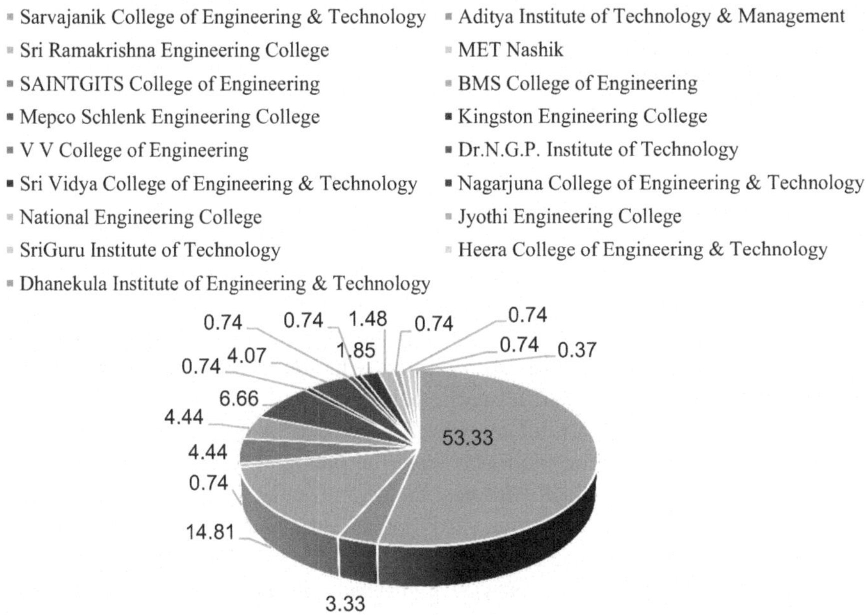

Fig. 9.14 RT WSN lab institution statistics

As of this writing, over 558 experiments across 9 disciplines have been developed and deployed using VLCAP. Today, between Online Labs and Virtual Labs, VLCAP has over 350,000 registered learners, plus several million unregistered users, from around the world. Our platform provides actionable learning analytics to educators, learners, and institutional administrators. In addition, 41 nodal centers in higher education are using VLCAP to include both simulation labs and remote labs in their curriculum and are regularly providing additional assessments for their students. VLCAP also allows for multilingual support, currently in four languages – English, Hindi, Malayalam, and Marathi – and is being scaled across 21 states of India.

In addition, VLCAP allows for an offline version that can be synchronized with the online version when the Internet is available. This feature allows the use of VLCAP even in remote areas with limited Internet access. Over 11,400 teachers from 3450 schools with approximately 1.55 million high school students have been trained in and provided access to the offline version of OLabs.

The VLCAP publishing system promotes collaboration and customizable processes at various stages and between educators, developers, and designers, allowing each author to create original content or to reuse content from the available resources. The pedagogical structure of the template guides lab developers in designing the labs. It allows reuse of rich media content, improves efficiency, and reduces cost while maintaining a similar look and feel for learners. It supports accountability as actions by all key stakeholders are logged, and there is an option to review and approve labs. Another benefit is that the content is easy to adapt for diverse users. Multilingual support follows a modular mapping of text, so as to reduce translation time and cost.

Simulation-based labs and remote equipment labs offer extensive benefits including the ability to significantly improve learning opportunities for school and higher education students regardless of geographic location, resource limitations, and challenges stemming from diverse languages. However, creating and maintaining such labs require substantial technical expertise, a factor which can potentially hinder their development and deployment. An easy-to-use collaborative platform like VLCAP directly addresses these issues. It encourages faculty members with limited time, resources, and technical expertise to develop multiple experiments in a short time, offers the expertise of experienced lab developers, makes expensive equipment broadly available to all participants, allows learners to optimize their participation, and provides useful analytics for assessment and improvement.

Thus VLCAP and similar platforms have the potential to extend the influence of excellent but technologically limited faculty, provide access to sophisticated but expensive equipment, and offer vastly expanded educational opportunities to students who would otherwise face very limited resources. Most importantly, these platforms have the potential to provide high-level resources and education to all students, regardless of geography, resource levels, or language.

Acknowledgments Our project is envisioned, inspired, and actively guided by Amrita University's Chancellor Sri Mata Amritanandamayi Devi. We acknowledge the support of the Ministry of Electronics and Information Technology (MeitY) and Ministry of Human Resources (MHRD), respectively, for the funding of Online Labs and Virtual Labs projects.

References

Achuthan, K., Bose, L. S., Francis, S., Sreelatha, K. S., Sreekala, C. O., Nedungadi, P., & Raman, R. (2014). Improving perception of invisible phenomena in undergraduate physics education using ICT. In *Information and Communication Technology (ICoICT)*, 2014 2nd International Conference on (pp. 226–231). IEEE.

Allen, W. C. (2006). Overview and evolution of the ADDIE training system. *Advances in Developing Human Resources, 8*(4), 430–441.

Bochicchio, M. A., & Longo, A. (2009). Hands-on remote labs: Collaborative web laboratories as a case study for it engineering classes. *IEEE Transactions on Learning Technologies, 2*(4), 320–330.

Canfora, G., Daponte, P., & Rapuano, S. (2004). Remotely accessible laboratory for electronic measurement teaching. *Computer Standards & Interfaces, 26*(6), 489–499.

Cooper, M., & Ferreira, J. M. (2009). Remote laboratories extending access to science and engineering curricular. *IEEE Transactions on Learning Technologies, 2*(4), 342–353.

Corter, J. E., Nickerson, J. V., Esche, S. K., & Chassapis, C. (2004). *Remote versus hands-on labs: A comparative study*. 34th ASEE/IEEE Frontiers in Education Conference.

Dick, W., & Cary, L. (1990). *The systematic design of instruction* (3rd ed.). New York: Harper Collins.

Doddavenkatappa, M., Chan, M. C., & Ananda, A. L. (2011). Indria: A low cost. 3d wireless sensor network testbed. In *International conference on testbeds and research infrastructures* (pp. 302–316).

Ferreira, J. M., & Cardoso, A. M. (2005). A Moodle extension to book online labs. In *Proceedings of the 2nd Remote Engineering and Virtual Instrumentation International Symposium (REV 05)*.

Freeman, J., Nagarajan, A., Parangan, M., Kumar, D., Diwakar, S., & Achuthan, K. (2012). Remote triggered photovoltaic solar cell lab: Effective implementation strategies for Virtual Labs. In *Technology Enhanced Education (ICTEE)*, 2012 IEEE International Conference on (pp. 1–7). IEEE.

Gustavsson, I., Nilsson, K., Zackrisson, J., Garcia-Zubia, J., Hernandez-Jayo, U., Nafalski, A., & Lago, T. (2009). On objectives of instructional laboratories, individual assessment, and use of collaborative remote laboratories. *IEEE Transactions on Learning Technologies, 2*(4), 263–274.

Handziski, V., Köpke, A., Willig, A., & Wolisz, A. (2006). TWIST: A scalable and reconfigurable testbed for wireless indoor experiments with sensor networks. In *Proceedings of the 2nd international workshop on Multi-hop ad hoc networks: From theory to reality* (pp. 63–70). ACM.

Harward, V. J., Del Alamo, J. A., Lerman, S. R., Bailey, P. H., Carpenter, J., DeLong, K., & Long, P. D. (2008). The iLab shared architecture: A web services infrastructure to build communities of internet accessible laboratories. *Proceedings of the IEEE, 96*(6), 931–950.

Hellbrück, H., Pagel, M., Köller, A., Bimschas, D., Pfisterer, D., & Fischer, S. (2011). Using and operating wireless sensor network testbeds with WISEBED. In *Ad Hoc Networking Workshop (Med-Hoc-Net)*, 2011. The 10th IFIP Annual Mediterranean (pp. 171–178). IEEE.

Hutzel, W. J. (2002). A remotely accessed HVAC laboratory for distance education. *International Journal of Engineering Education, 18*(6), 711–716.

Imran, M., Said, A. M., & Hasbullah, H. (2010). A survey of simulators, emulators and testbeds for wireless sensor networks. In *Information Technology (ITSim), 2010 International Symposium in* (Vol. 2, pp. 897–902). IEEE.

Josephsen, J., & Kristensen, A. K. (2006). Simulation of laboratory assignments to support students' learning of introductory inorganic chemistry. *Chemistry Education Research and Practice, 7*(4), 266–279.

Kollöffel, B., & de Jong, T. (2013). Conceptual understanding of electrical circuits in secondary vocational engineering education: Combining traditional instruction with inquiry learning in a virtual lab. *Journal of Engineering Education, 102*, 375–393. https://doi.org/10.1002/jee.20022.

Lohr, L. (1998). Using Addie to design a web-based training interface. In S. McNeil, J. Price, S. Boger-Mehall, B. Robin, & J. Willis (Eds.), *Proceedings of Society for Information Technology & Teacher Education International Conference 1998* (pp. 452–455). Chesapeake, VA: Association for the Advancement of Computing in Education (AACE).

Ma, J., & Nickerson J. V. (2006). Hands-on, simulated, and remote laboratories: A comparative literature review. *ACM Computing Surveys (CSUR), 38*(3), Article 7.

McElhaney, K. W., & Linn, M. C. (2011). Investigations of a complex, realistic task: Intentional, unsystematic, and exhaustive experimenters. *Journal of Research in Science Teaching, 48*(7), 745–770.

Morrison, G. R., Ross, S. M., Kemp, J. E., & Kalman, H. (2010). *Designing effective instruction*. New York: Wiley.

Nair, B., Krishnan, R., Nizar, N., Radhamani, R., Rajan, K., Yoosef, A., & Diwakar, S. (2012). Role of ICT-enabled visualization-oriented virtual laboratories in universities for enhancing biotechnology education–VALUE initiative: Case study and impacts. *FormaMente, 7*(1–2), 1–18.

Nedic, Z., Machotka, J., & Nafalski, A. (2003). Remote laboratories versus virtual and real laboratories. *33rd Annual Frontiers in Education* (Vol. 1, pp. T3E-T3E). IEEE.

Nedungadi, P., & Raman, R. (2010, August). Effectiveness of adaptive learning with interactive animations and simulations. In *Advanced Computer Theory and Engineering (ICACTE)*, 2010 3rd International Conference on (Vol. 6, pp. V6–40). IEEE.

Nedungadi, P., & Raman, R. (2016). The Medical Virtual Patient Simulator (MedVPS) platform. In *Intelligent systems technologies and applications* (pp. 59 –67). Cham: Springer.
Nedungadi, P., Haridas, M., & Raman, R. (2015). Blending concept maps with Online Labs (OLabs): Case study with biological science. *Proceedings of the Third International Symposium on Women in Computing and Informatics*. ACM.
Nedungadi, P., Malini, P., & Raman, R. (2015). Inquiry based learning pedagogy for chemistry practical experiments using OLabs. In *Advances in intelligent informatics* (pp. 633–642). Cham: Springer
Nedungadi, P., Raman, R., & McGregor, M. (2013). Enhanced STEM learning with Online Labs: Empirical study comparing physical labs, tablets and desktops. In *Frontiers in Education Conference, 2013 IEEE* (pp. 1585–1590). IEEE.
Popescu, D., & Odbert, B. (2011). *The advantages of remote labs in engineering education*. Educator's Corner-Agilent Technologies-application note, 11.
Pradeep, P., Divya, P., Devi, R. D. A., Rekha, P., Sangeeth, K., & Ramesh, M. V. (2015). *A remote triggered wireless sensor network testbed*. Wireless Telecommunications Symposium (WTS), 2015, 1–7. IEEE.
Raman, R., Nedungadi, P., Achuthan, K., & Diwakar, S. (2011). Integrating collaboration and accessibility for deploying virtual labs using vlcap. *International Transaction Journal of Engineering, Management, & Applied Sciences & Technologies, 2*(5), 547–560.
Raman, R., Achuthan, K., Nedungadi, P., Diwakar, S., & Bose, R. (2014). The VLAB OER experience: Modeling potential-adopter student acceptance. *IEEE Transactions on Education, 57*(4), 235–241.
Raman, R., Haridas, M., & Nedungadi, P. (2015). Blending concept maps with online labs for STEM learning. In *Advances in intelligent informatics* (pp. 133–141). Cham: Springer.
Richter, T., Boehringer, D., & Jeschke, S. (2011). Lila: A European project on networked experiments. In *Automation, Communication and Cybernetics in Science and Engineering 2009/2010* (pp. 307–317). Berlin: Springer.
Rojko, A., Debevc, M., & Hercog, D. (2009). Implementation, effectiveness and experience with remote laboratory in engineering education. *Organizacija, 42*(1), 23–33.
Sangeeth, K., Pradeep, P., Divya, P., Rekha, P., Sheeja, L., & Ramesh, M. V. (2015). A remote code editing framework for AMRITA remote triggered WSN laboratory. *5th International Conference on IT Convergence and Security (ICITCS)*, (pp. 1–5). IEEE.
Sousa, N., Alves, G. R., & Gericota, M. G. (2010). An integrated reusable remote laboratory to complement electronics teaching. *IEEE Transactions on Learning Technologies, 3*(3), 265–271.
Wang, S., & Hsu, H. (2009). Using the ADDIE model to design second life activities for online learners. *TechTrends, 53*(6), 76–81.
Zacharia, Z. C., & Constantinou, C. P. (2008). Comparing the influence of physical and virtual manipulatives in the context of the physics by inquiry curriculum: The case of undergraduate students' conceptual understanding of heat and temperature. *American Journal of Physics, 76*(4&5), 425–430.

Prema Nedungadi is the director of Amrita Center for Research in Analytics & Technology in Education (CREATE), Amrita Vishwa Vidyapeetham, India.

Maneesha Vinodini Ramesh is a professor and the director of Amrita Center for Wireless Networks and Applications, Amrita Vishwa Vidyapeetham, India.

Preeja Pradeep is a Ph.D. scholar at Amrita Center for Wireless Networks and Applications, Amrita Vishwa Vidyapeetham, India.

Raghu Raman is a professor and chair of Amrita School of Business at Amrita Vishwa Vidyapeetham, India.

Chapter 10
Model-Based Inquiry in Computer-Supported Learning Environments: The Case of Go-Lab

Tasos Hovardas, Margus Pedaste, Zacharias Zacharia, and Ton de Jong

Abstract This chapter focuses on model-based inquiry in computer-supported environments, especially through the use of the Go-Lab platform (www.golabz.eu). Go-Lab is an online learning platform that offers students the opportunity to engage in inquiry-based science learning, in a structured and supportive manner, by providing environments for learning (i.e., Inquiry Learning Spaces), where virtual or remote laboratories and software scaffolds (e.g., tools for generating hypotheses and designing experiments) that support inquiry learning processes have been integrated. The purpose of this chapter is to unravel how the Go-Lab platform, especially some of its virtual laboratories, can be used for model-based learning. In so doing, we discuss core requirements for model-based inquiry in expressing, testing, and revising models. Further, we present three examples of Go-Lab virtual laboratories, with modeling and simulation affordances, to explain how they could be used by educators as means for enacting model-based inquiry.

Keywords Affordance · Guidance · Model-based inquiry · Inquiry cycle · Modeling tool

T. Hovardas (✉) · Z. Zacharia
Department of Educational Sciences, Faculty of Social Sciences and Education, University of Cyprus, Nicosia, Cyprus
e-mail: hovardas@ucy.ac.cy; zach@ucy.ac.cy

M. Pedaste
Centre for Educational Technology, Institute of Education, Faculty of Social Sciences, University of Tartu, Tartu, Estonia
e-mail: margus.pedaste@ut.ee

T. de Jong
Department of Instructional Technology, Faculty of Behavioral, Management and Social Sciences, University of Twente, Enschede, The Netherlands
e-mail: a.j.m.dejong@utwente.nl

M. E. Auer et al. (eds.), *Cyber-Physical Laboratories in Engineering and Science Education*, https://doi.org/10.1007/978-3-319-76935-6_10

10.1 Introduction

In this chapter we attempt to address a series of instructional and design challenges of enacting model-based inquiry with virtual laboratories of the Go-Lab platform (http://www.golabz.eu/labs). Go-Lab learning materials have been designed and instantiated in an inquiry-based context while using virtual and remote laboratories as means of exploration and experimentation (e.g., see the "learning spaces" at www.golabz.eu). For the purposes of this chapter, we attempt to show that the Go-Lab platform could move beyond the general inquiry-based approach (e.g., Pedaste et al. 2015) and support other inquiry-oriented learning approaches. In particular, we focus on the enactment of model-based inquiry, which has been reported in the literature as a rather challenging approach, but with a lot to offer learning-wise to the students (Windschitl et al. 2008a). Computer-supported learning environments, such as Go-Lab, can provide the means for a model-based inquiry enactment. Besides the virtual laboratories with modeling affordances, they can also offer guidance which can provide support to the students throughout a model-based enactment. To fulfill this purpose, we organized the chapter in the following sections: First, we define model-based inquiry and associate it with recent research of the domain. Next, we introduce our inquiry framework and explain how this framework fits the model-based inquiry approach. Then, we discuss the Go-Lab guidance tools available for supporting students when enacting model-based inquiry. We also report on the experience of the Go-Lab project to outline specific recommendations for fine-tuning guidance offered to students during their inquiry. In the next section, we present three examples which instantiate model-based inquiry in the context of Go-Lab. Finally, we draw some conclusions coming out of the three examples and discuss how these examples could inform practice.

10.2 Model-Based Inquiry in Computer-Supported Learning Environments

Models and model-based inquiry have been a primary teaching and research focus in science education during the last three decades (Clement 2000; Gobert and Buckley 2000; Louca and Zacharia, 2008, 2012, 2015; Hovardas 2016). Models are understood as scientific representations of systems or phenomena, which allow for tracing and monitoring the interrelations and interactions among the structural components that compose the system or the phenomenon at hand (e.g., McComas 2002; Matthews 2005). In science education, the term "model" might refer to mental models (e.g., Clement 2000, pp. 1042–1043; Gobert and Buckley 2000, p. 892) or external/concrete models (e.g., Louca and Zacharia 2012). A mental model reflects the initial ideas of students for a phenomenon under study. Such a mental model might be depicted by students via several means, for instance, as a paper-and-pencil drawing or by a modeling tool. Indeed, science instruction often engages students in expressing their mental models. The idea is to construct models that align with scientific accounts of the targeted systems or phenomena (i.e., "scientific models" or "expert consensus models"). However, models employed in science education as desired outcomes of instruction might differ from fully fledged scientific models, and they might be simplified to suit learning goals, without losing their epistemological rigor (i.e., "teaching models" or "target models"). The desired transition from initial mental models of students toward target models might involve a series of "intermediate models" (Clement 2000, p. 1042). At the end of an educational intervention, student competence might be readily evaluated by the convergence of the updated mental models of students with scientific models. Student knowledge and skills would be assessed through a direct comparison of the primary aspects of the models constructed by the students with the corresponding aspects of the target models at task.

Testing and revision of models has been a prominent avenue for model-based inquiry (Campbell et al. 2013; Clement 2000; Grünkorn et al. 2014; Halloun 2007; van Joolingen et al. 2005; Windschitl et al. 2008a)[1]. A basic distinction noted in this direction has been between testing and retesting models constructed by students, on the one hand, and using ready-made models, on the other (see also Mellar and Bliss 1994). In both cases, a considerable difficulty has been to bridge models depicting student ideas, on the one hand, with scientific explanations of the systems or phenomena under focus, on the other (Soulios and Psillos 2016). A first challenge for educators has been to align target models in accordance with students' capabilities and knowledge and, at the same time, configure target models so that they retain core aspects and functionalities of scientific models. A further challenge for educators has been to plan an effective learning activity sequence (or "learning progression" for longer or larger teaching units), which would support the transition from initial models to target models. All options that have been proposed, in that direction, have involved a series of intermediate steps in modeling pedagogies (Oh and Oh 2011), in an attempt to foster reflection on alternative or gradually advancing models of the same system or phenomenon and to elaborate on their strengths and weaknesses. This has also included the utilization of empirical data to validate a model (van Joolingen et al. 2005). Overall, a trajectory would be traced from students' initial mental models, through testing and revision of intermediate models, up to the target models, namely, the scientific version of models employed for educational purposes (Campbell et al. 2013).

A recent review has revealed that the most technological support to modeling pedagogies in computer-supported learning environments has been offered for

[1]Broadly approached, terminology on modeling would separate among different modeling pedagogies (van Joolingen et al. 2005; Campbell et al. 2013), i.e., "expressive" modeling has been largely related to elicitation of students' initial ideas, namely, students' initial mental models, "experimental" modeling would necessitate empirical data to validate a model, "evaluative" modeling would involve screening among rival versions of a model, "exploratory" modeling would be operationalized by means of a ready-made model (i.e., a model which was not created by students themselves), and "cyclic" modeling would include model revision.

"expressive" modeling (elicitation of students' mental models) and "exploratory" modeling (operationalized by means of a ready-made model) (Campbell et al. 2015). Once again, the idea here is to bridge the apparent instructional and technological interface between student initial ideas and the target model of instruction that is aligned to core scientific assumptions of the modeled system or phenomenon. The concern for educators and designers to better operationalize and support the transition from student first mental models to sound target models has been echoed in the model-based inquiry perspective proposed by Windschitl et al. (2008b). This latter perspective has been quite critical to school practice that does not give credit to images of the world that precede observations. Student representations of phenomena prior to observations correspond to student mental models that will first need to be expressed and made explicit, in order to guide exploration or experimentation later on. This view is in line with an epistemological position, according to which, the formulation of hypotheses can be taken as interrelation of variables. Since hypotheses link dependent variables to independent ones, multiple hypotheses might be processed to study multiple dimensions of a phenomenon under study, as these dimensions are described by the variables tested. A scientific model of the phenomenon would provide a coherent whole structure by these variables, and it would comprise a solid reference base for variable identification and hypothesis generation. In this regard, hypotheses would incorporate and interrelate structural components (i.e., variables) of a model (e.g., Giere 1991; Nersessian 2002, 2005). Model-based inquiry is compatible to nature-of-science approaches that interpret scientific theories as constellations of models, especially in facilitating the epistemological rigor of theories by elaborating on model attributes (Ariza et al. 2016; Develaki 2007; Lefkaditou et al. 2014). Such an approach would challenge a stand-alone view of exploration or experimentation with ready-made models, and it would direct educators and designers toward embedding the sequence of learning activities needed to plan and execute an exploration or an experiment (i.e., formulation of hypothesis, designing an experiment, executing the experiment) within the broader frame of model building and testing (see, for instance, Windschitl et al. 2008b, p. 311)[2].

Within computer-supported learning environments, certain virtual laboratories (i.e., open-ended virtual labs that allow the preparation/building of an experiment

[2]Close-ended simulations do not offer students the option of expressing their mental models, because the model is already there. In this case, possible relations between variables would have to be assumed/discovered. It is an issue whether this variable-by-variable approach would allow the student to grasp a complete picture of the whole phenomenon under study, as if one would have expected based on a modeling procedure, during which the whole phenomenon would be modeled and remodeled. After all, the design rationale behind any modeling tool has been to first give students the opportunity to create a model and then simulate it. It could be that we might isolate a limited number of variables to study a phenomenon. However, nonlinear thinking and system dynamics with feedback mechanisms and delay cannot be easily addressed with matching variables in pairs of two, where we mostly presuppose linear relationships between two variables at a time. Here we come across epistemological issues linking model-based inquiry to systems thinking, where the latter cannot be facilitated without the former.

setup) are resources that could facilitate model-based inquiry. For instance, many virtual laboratories offer affordances that allow for outlining the basic components of a system or a phenomenon, enacting modeling tasks, and using scientific models (i.e., simulations) for exploration or experimentation purposes (de Jong et al. 2013; Zacharia and de Jong 2014). Further, virtual laboratories may enable speeding up or slowing down phenomena running at varying speed. Another important aspect of using virtual laboratories in computer-supported environments is that they carry affordances that make non-visible components of systems or phenomena visible (e.g., de Jong et al. 2013; Olympiou et al. 2013; Zangori and Forbes 2015). For instance, virtual laboratories may allow for zooming in or out in small-scale or large-scale systems, respectively. Identifying and distinguishing between readily observable (i.e., visible) as well as hidden (i.e., non-visible) elements is crucial for being able to use a model as an explanatory device and for following underlying causes and effects that relate to that model (Hmelo-Silver and Azevedo 2006; Jacobson and Wilensky 2006; Olympiou et al. 2013; Zangori et al. 2015). At the same time, however, some types of virtual laboratories might not be suitable for enacting model-based inquiry. For instance, when virtual laboratories do not offer modeling options or when the modeling options they provide are minimal (i.e., close-ended simulations that do not allow the users to build their own experiment setup; see Sect. 10.6 for examples of open-ended virtual labs and close-ended simulations), then model revision cannot be effectively operated, or it may be even heavily impaired. Further, when basic modeling assumptions are not readily traceable for the user of the virtual laboratory and when simulation is the only option, then the user might perform multiple simulation tasks, but he/she would still fail to acknowledge the core underlying principles of the model, on which all simulations depend.

In what follows, we will attempt to address a series of instructional and design challenges of model-based inquiry with virtual laboratories by presenting the work undertaken within the Go-Lab project (http://www.go-lab-project.eu/). As modeling would go along with quite demanding learning tasks, students would need to be substantially supported in their learning trajectories while constructing and revising models. Computer-supported learning environments can offer valuable guidance to students during their learning routes throughout an inquiry procedure based on modeling. Initial mental models would largely overlap with prior knowledge of students. Virtual laboratories that offer modeling affordances would allow for an exploration of the basic structural compartments involved in a model. This would enable students to identify the variables, which would be needed later on for the formulation of research questions or hypotheses. Virtual laboratories would allow students to simulate models and generate data based on these models. Students would then continue their inquiry as long as they would be able to use simulation data to accept or reject their hypotheses. At the latter stages of the inquiry procedure, students will need to reach conclusions, report their work to peers and the teacher, as well as reflect on the whole learning activity sequence. For all these tasks, the Go-Lab platform can offer a series of virtual laboratories and software scaffolds to design and enact model-based inquiry (see Sect. 10.4 and Table 10.1).

Table 10.1 Scaffolds available in the Go-Lab platform for all inquiry phases and their main affordances

Phase (sub-phase) of the inquiry cycle	Software scaffold/application	Main affordances of the software scaffold/application
Orientation	Concept Mapper (https://www.golabz.eu/app/concept-mapper)	Predefined terms provided to students to construct a concept map
Conceptualization; Questioning (sub-phase)	Question Scratchpad (https://www.golabz.eu/app/question-scratchpad)	Predefined variables provided to students to formulate their questions
Conceptualization; Hypothesis generation (sub-phase)	Hypothesis Scratchpad (https://www.golabz.eu/app/hypothesis-scratchpad)	Predefined variables provided to students to formulate their hypotheses
Investigation; Exploration (sub-phase)	Observation Tool (https://www.golabz.eu/app/observation-tool)	Students can record and arrange observations during an experiment
Investigation; Experimentation (sub-phase)	Experiment Design Tool (https://www.golabz.eu/app/experiment-design-tool)	Predefined variables provided to students to design their experiment
Investigation; Data interpretation (sub-phase)	Data Viewer (https://www.golabz.eu/app/data-viewer)	Variables and data sets provided to students to construct graphs
Conclusion	Conclusion Tool (https://www.golabz.eu/app/conclusion-tool)	Learning products of prior activities provided to students to draw their conclusions
Discussion; Reflection (sub-phase)	Reflection Tool (https://www.golabz.eu/app/reflection-tool)	Feedback provided to students about their use of an Inquiry Learning Space
Discussion; Communication (sub-phase)	Report Tool (https://www.golabz.eu/app/report-tool)	Learning products of prior activities provided to students to report on their inquiry

10.3 Inquiry Phases and Learning Trajectories in Model-Based Inquiry with Virtual Laboratories

In their review of inquiry-based learning, Pedaste et al. (2015) identified five phases that define an inquiry cycle (Fig. 10.1). These phases include fundamental tasks of scientific inquiry and streamline learning activities so as to achieve optimal learning gains. The first phase has been called "Orientation," and it involves learning activities aimed at arousing student interest toward the domain. In this phase, the research topic and the driving questions about a system or phenomenon should be also clarified. The next phase is "Conceptualization," which includes tasks related to the identification of variables about the system or the phenomenon at hand and which will be handled by students. "Conceptualization" might take two forms, depending on students' prior knowledge about the domain or experience in inquiry learning. Novice learners, who would have their first encounter with the topic, would

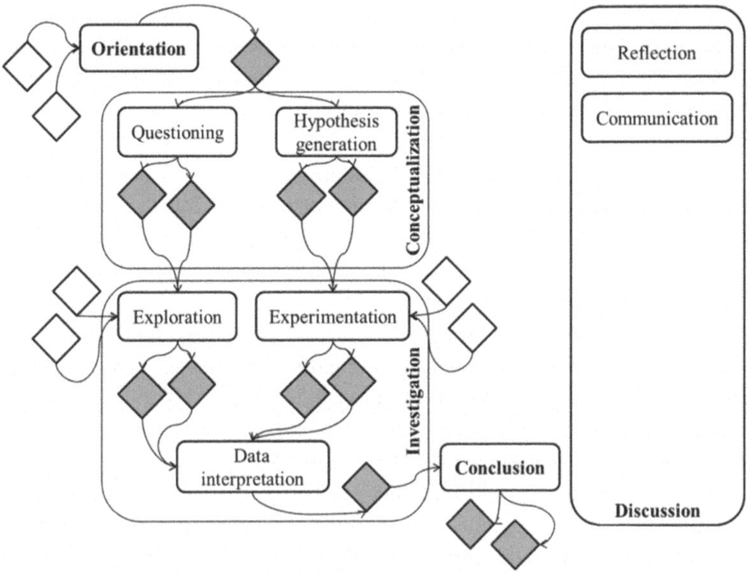

Fig. 10.1 Phases and sub-phases of the inquiry-based learning framework (Adapted from Pedaste et al. 2015). Rectangles depict sub-phases or phases (the latter in bold), dark rhombuses learning products, and white rhombuses any reference material offered to students by the teacher or the learning environment. Arrows show the sequence of phases. The flow presented is only indicative (for a complete report of relations between phases and sub-phases, see Pedaste et al. 2015)

pose questions with main variables outlined ("Questioning" sub-phase), while more experienced learners, who would be familiar with the topic, would be able to formulate hypotheses ("Hypothesis generation" sub-phase). This duality would be continued in the "Investigation" phase, where novice learners would proceed to an exploration of the topic ("Exploration" sub-phase), while experienced learners would execute an experiment ("Experimentation" sub-phase). Expressing, testing, and revising a model would be integrated in the "Investigation" phase in either sub-phase, namely, either as exploration, to detect indications of a relation between variables identified, or experimentation, to verify a hypothesized relation between variables and address a research hypothesis (de Jong 2015)[3]. After modeling and data generation, students would go on to the third sub-phase of "Investigation," where they would have to interpret their data ("Data interpretation" sub-phase). The main challenge in this latter sub-phase would be to arrive at meaningful results out of

[3]With regard to the inquiry cycle, "exploratory" modeling (i.e., students working with ready-made models) might not always equate to the exploration trajectory in the inquiry cycle as defined by Pedaste et al. (2015). For instance, the exploration trajectory is distinguished from the experimentation trajectory in the inquiry cycle in that the first incorporates research questions, while the latter presupposes hypotheses. However, "exploratory modeling" might accommodate both questions and hypotheses.

the data students had collected and analyzed. "Conclusion" is the phase that follows, with students drawing main conclusions out of their exploration or experimentation. In this phase, students also need to align their conclusions with research questions or hypotheses formulated earlier in their inquiry. The fifth phase of the inquiry cycle is termed "Discussion" and includes the sub-phases of "Communication" and "Reflection." In "Communication," students interact with peers or teachers to share outcomes and experiences and to receive or offer feedback on their inquiry. In "Reflection," each student reflects on his or her learning tasks and the learning route taken. These sub-phases might be activated within or between other phases, as well as at the end of an entire inquiry cycle.

The inquiry cycle could be completed via two alternative pathways, which are split in the "Conceptualization" and "Investigation" phases (Fig. 10.1; "Questioning" and "Exploration" sub-phases, for novice learners, "Hypothesis generation" and "Experimentation" sub-phases, for more experienced learners). These two alternative trajectories would involve the construction of different learning products[4] by students, as they would undertake learning activities. For instance, in the "Questioning" sub-phase, students will produce questions, and these questions will be used later on as part of the input students will dispose of in the "Exploration" sub-phase to construct or revise a model. This model will be another example of a learning product. Alternatively, students would need to formulate a hypothesis (i.e., learning product in "Hypothesis generation" sub-phase), before proceeding to an experimentation with a model in a virtual laboratory ("Experimentation" sub-phase), where data generated and organized in tables or figures would be the next learning products of students. All input necessary for processing learning activities has been given in Fig. 10.1 either as dark rhombuses, which denote learning products, or as white rhombuses, which denote any other reference material offered by the teacher or the learning environment.

With regard to model-based inquiry, virtual laboratories with modeling and simulation functionalities might be used by educators and designers for structuring the whole inquiry cycle. The heuristic value of models has been frequently underlined, especially in terms of generating predictions, hypotheses, and explanations (Coll and Lajium 2011; Forbes et al. 2015; Justi and Gilbert 2003; Hovardas and Korfiatis 2011; Lefkaditou et al. 2014; Petridou et al. 2013; Schwarz and White 2005; Schwarz et al. 2009; Verhoeff et al. 2008; Windschitl et al. 2008a). The model of a phenomenon under study can provide an insightful reference base for examining various dimensions of the phenomenon, as they can be operationalized by the variables included in the model. In this direction, the multifarious compatibilities

[4]Learning products that are created by students themselves as they go through a learning activity sequence have been characterized as "emerging learning objects (ELOs)" in the frame of the Science Created by You (SCY) project (see de Jong et al. 2010, 2012). These can include concept maps, models, questions, hypotheses, experimental designs, tables or figures with simulation data, and any other artifact that is the product of student work and can be stored and recalled upon demand for educational purposes. Learning products provide a core alignment of computer-supported learning environments with the theoretical and operational framework of constructivism.

of modeling and inquiry-based learning have been frequently highlighted to single out testing, revising, and retesting models (e.g., Lehrer and Schauble 2006). Models constructed by students themselves as learning products would constitute expressed models at the initial steps of their inquiry. If these models can be simulated to generate data, then student inquiry would build on elaboration of research questions and hypotheses via model simulation. In that direction, model construction and revision might be seen as a strategy of configuring the whole inquiry cycle in model-based inquiry, where models and modeling would comprise an indispensable device for promoting student knowledge and skills as well as their epistemological understanding[5]. This design would ultimately lead to a possible way of resolving the challenge in facilitating intermediate steps in model-based inquiry and supporting the transition from initial models of students to target models. A first task for educators, where they might need considerable assistance, is to select or configure target models suited for model-based inquiry (Windschitl et al. 2008b). Then, students might take the trajectory delimited for novice learners and explore the system or phenomenon under study in their first modeling tasks ("Exploration" sub-phase). To begin with, students would need an adequate backing in the "Orientation phase," so that they would be guided to mark out one or two core variables, with which they will also encounter when using the virtual laboratory. Such an assistance would foster an acknowledgment of variables that would be shared between initial models of students and target models. Moreover, this option would provide the necessary bridge between the initiation of model-based inquiry and the desired learning outcome.

If the first trajectory in our design was exploration of a system or phenomenon, the next trajectory involves experimentation, which might need the articulation of a new inquiry cycle. Learners in that cycle would have had a familiarization encounter with model, modeling, and the virtual laboratory. Such an experience might allow them to formulate hypotheses. In turn, generating simulation data would prove crucial for any model revision, namely, for being able to validate the model constructed by students on the basis of the data it can generate. A manifest assumption in that approach of ours is that educators would need to schedule at least two subsequent inquiry cycles (i.e., one cycle involving exploration and another one involving experimentation), which largely overlap with the two alternative learning trajectories depicted in Fig. 10.1. This option might reflect the well-documented fact that experimentation has been for long a primary focus of science education and it has therefore attracted the attention of educators and designers (van Joolingen and Zacharia 2009). However, if we conceive of hypotheses as statements that interrelate variables identified in models (see, for instance, Windschitl et al. 2008a), then the ability to formulate a hypothesis content-wise would depend on the ability to employ a basic model of the system or phenomenon under study. Offering the option of simulation (i.e., trajectory involving experimentation) without

[5]In that regard, our approach presents a marked resemblance with learning by design; see Kolodner et al. (2003), de Jong and van Joolingen (2007), and Weinberger et al. (2009).

delineating a basic model of the phenomenon under study first (i.e., trajectory involving exploration) might eventuate in trial-and-error attempts of students that would hardly be informed by a comprehensive ability to reflect on models and modeling and on testing and retesting their models, accordingly. Furthermore, the precedence of exploring before experimenting would provide the opportunity to students to familiarize themselves with the virtual laboratory they would use, the main modeling skills, and the main variables to begin with.

10.4 Laboratories and Applications in the Go-Lab Platform

For either learning trajectory, the production, storage, retrieval, and reprocessing of learning products are supported by the Go-Lab platform with tools (software scaffolds/applications), which can be embedded in all phases and sub-phases of the inquiry cycle, in order to provide necessary guidance and scaffolding to students (see Table 10.1 for an indicative list of software scaffolds across phases and sub-phases of the inquiry cycle[6]). For instance, students can use the Question Scratchpad to formulate research questions and the Hypothesis Scratchpad to formulate hypotheses (Figs. 10.2 and 10.3, respectively). The entire arrangement with a selected laboratory[7], support in the form of software scaffolds, and all other instructional guidance in the form of reference material offered to students, comprises an Inquiry Learning Space (ILS; http://www.golabz.eu/spaces). An ILS is a learning environment structured along the phases and sub-phases of the inquiry cycle and serviced with the support needed so that students will be able to choose a learning activity sequence and have an optimal inquiry route[8].

The integration of virtual laboratories in Inquiry Learning Spaces might allow for much more flexibility in student inquiry than when using virtual laboratories in a stand-alone fashion[9]. Implementation studies in the frame of the Go-Lab project have revealed that there seems to be a minimum amount of time that should be spent on a task, while working with a virtual laboratory or software scaffolds, so that students would effectively execute a series of learning activities (Hovardas et al. 2017). When less time than this threshold is spent, then students might have quite

[6]All software scaffolds available at the Go-Lab platform can be found at http://www.golabz.eu/apps. For a comprehensive review of guidance provided to students in computer-supported learning environments with virtual and remote laboratories, see Zacharia et al. (2015).

[7]The Go-Lab platform offers online an entire array of laboratories for supporting inquiry-based learning, including virtual laboratories and remotely operated educational laboratories (http://www.golabz.eu/labs). In this contribution, we have focused on virtual laboratories.

[8]Inquiry Learning Spaces available in the Go-Lab platform can be found at http://www.golabz.eu/spaces

[9]Educators can use the Go-Lab authoring tool to select virtual laboratories and software scaffolds/applications and embed them in phases and sub-phases of the inquiry cycle in order to create an Inquiry Learning Space (de Jong et al. 2014).

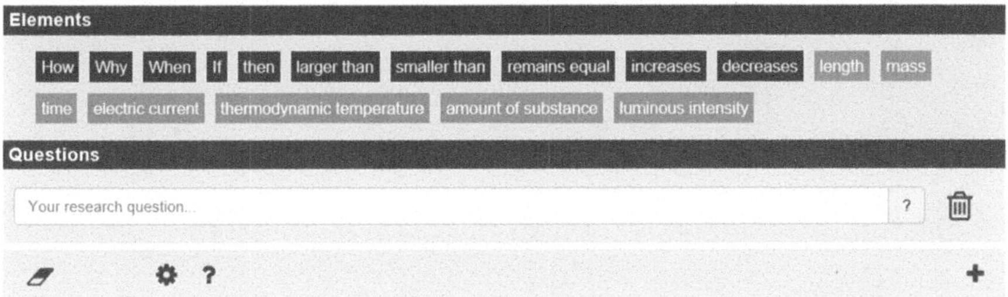

Fig. 10.2 The Question Scratchpad (https://www.golabz.eu/app/question-scratchpad)

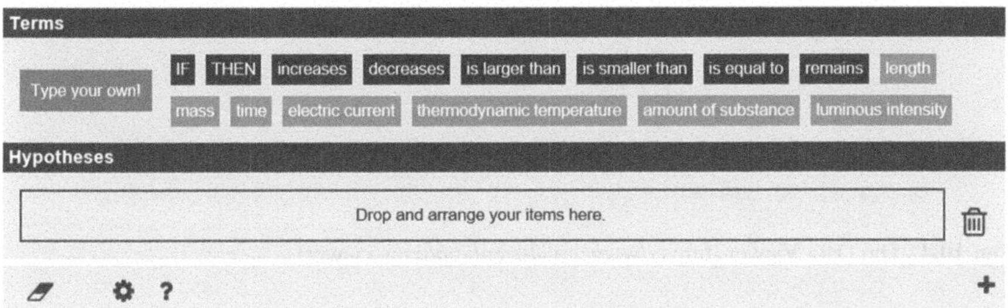

Fig. 10.3 The Hypothesis Scratchpad (https://www.golabz.eu/app/hypothesis-scratchpad)

low contextual or task and process awareness that leads to insufficient learning gains (Pedaste and Sarapuu 2006a, b). In this case, students should revisit former steps in their trajectories and rework their learning products to account for the remainder. For instance, if students had not identified all variables needed to undertake an exploration or an experimentation, then they would need to move backward in the activity sequence and devote additional time to working with the virtual laboratory and software scaffolds. This retrospective action might compensate for the time required to complete basic assignments. There can be multiple designs, which might foster such retrospective action and which might build on synergies between virtual laboratories and software scaffolds. For instance, when students would be ready to construct a graph in the Data Viewer (https://www.golabz.eu/app/data-viewer) (Fig. 10.4), the tool could offer students only one variable (e.g., the dependent variable) to construct their graph, and in this case students would need to identify the independent variable to plot. This option could be operationalized by linking the Data Viewer to a virtual laboratory (e.g., the Electrical Circuit Lab; http://www.golabz.eu/lab/electrical-circuit-lab) (Fig. 10.5; see Sect. 10.6.1 and Table 10.2 for a detailed account of model-based inquiry for electrical circuits) with a data set container. In an alternative linkage, students might be offered more than two variables to construct their graph, and in this case they would need to screen among variables and select the dependent and independent variable to accomplish the

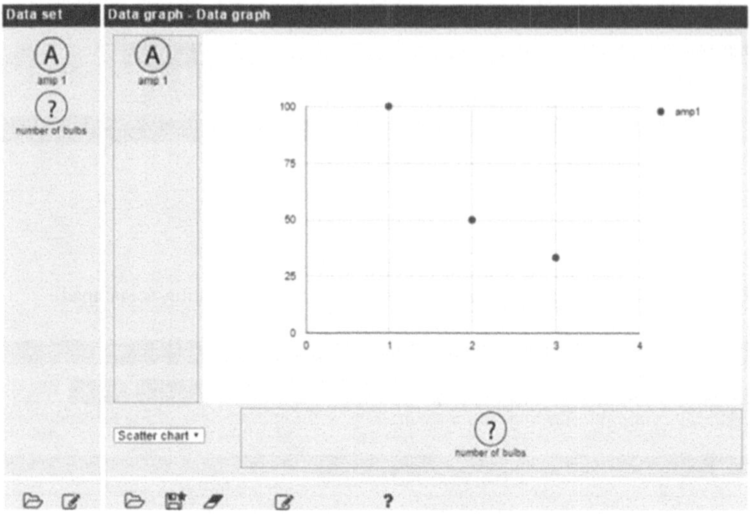

Fig. 10.4 The Data Viewer (https://www.golabz.eu/app/data-viewer)

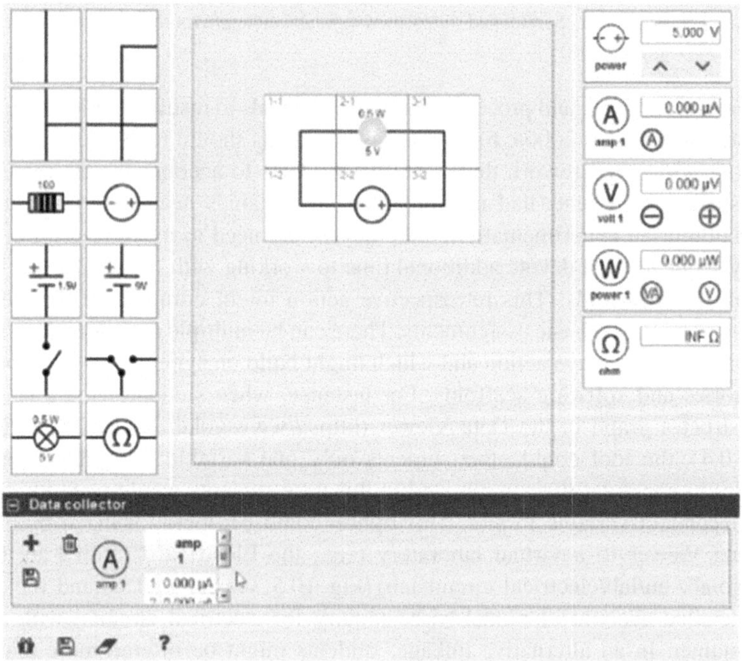

Fig. 10.5 Electrical Circuit Lab (http://www.golabz.eu/lab/electrical-circuit-lab)

Table 10.2 Subsequent cycles of model-based inquiry for electrical circuits

Inquiry cycle	Main modeling rationale	Operationalization of the Investigation phase of the inquiry cycle	Main exploration/experimentation rationale
First	Model a simple electrical circuit	Hands-on exploration	Interrelate basic structural components (e.g., power source, wire, bulb) and monitor the simplest function of a simple electrical circuit (e.g., the bulb lights up)
Second	Model a simple electrical circuit	Exploration with online lab (e.g., Electrical Circuit Lab; http://www.golabz.eu/lab/electrical-circuit-lab)	Interrelate basic structural components (e.g., power source, wire, bulb) and monitor the simplest function of a simple electrical circuit (e.g., the bulb lights up)
Third	Model electrical circuits in series and in parallel	Experimentation with online lab (e.g., Electrical Circuit Lab; https://www.golabz.eu/lab/simple-pendulum-1)	Interrelate basic structural components (e.g., power source, wire, multiple bulbs) and monitor the brightness of bulbs in the two types of circuits (e.g., in the circuit in series, brightness decreases when number of bulbs increases; in the circuit in parallel, brightness remains constant when number of bulbs increases)
Fourth	Model electrical circuits in series and in parallel	Experimentation with online lab (e.g., Electrical Circuit Lab; http://www.golabz.eu/lab/electrical-circuit-lab)	Monitor number of bulbs and total electric current in the two types of circuits (e.g., in the circuit in series, total electric current decreases when number of bulbs increases; in the circuit in parallel, total electric current increases when number of bulbs increases)
Fifth	Model electrical circuits in series and in parallel	Experimentation with online lab (e.g., Electrical Circuit Lab; http://www.golabz.eu/lab/electrical-circuit-lab)	Monitor voltage and electric current in the two types of circuits (e.g., in both types of circuits, electric current increases with voltage)

The sequence of inquiry cycles presented is only indicative; multiple other sequences might be possible, depending on main modeling and exploration/experimentation rationales

graphing task. This option might be operationalized through a linkage of the Data Viewer with the Experiment Design Tool (https://www.golabz.eu/app/experiment-design-tool) (Fig. 10.6). Both designs would trigger retrospective action, which is easier to enact in computer-supported learning environments and might open novel avenues in inquiry-based learning.

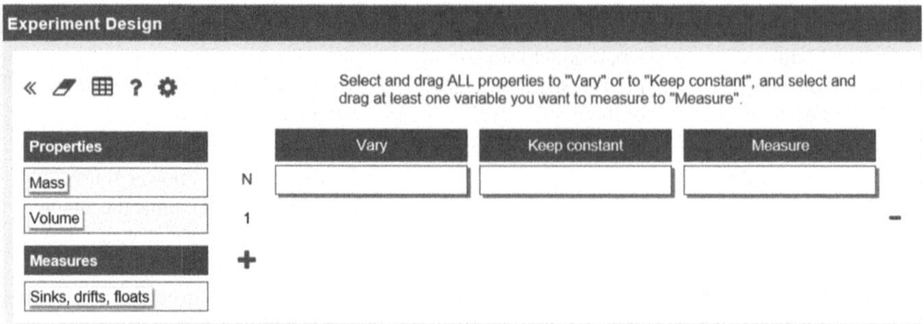

Fig. 10.6 The Experiment Design Tool (https://www.golabz.eu/app/experiment-design-tool)

10.5 Support Offered to Students in Model-Based Inquiry with Virtual Laboratories

Within the frame of the Go-Lab project, we have arrived at specific recommendations for fine-tuning guidance across the phases of the inquiry cycle, which will be presented in this section. The overall aim behind these recommendations is to achieve an optimum effect on student performance through the use of all the resources offered via the Go-Lab platform (Tasiopoulou and de Jong 2016). Guidance is provided through a number of tools and throughout the whole inquiry process (see Table 10.1). Specifically, guidance tools have been developed to support students in each inquiry phase (for details see Zacharia et al. 2015). On top of these guidance tools, we have noticed through previous studies (Tasiopoulou and de Jong 2016) that special support should be provided for any inquiry-oriented enactment, including model-based inquiry, and enhance peer and teacher feedback. Moreover, we have noticed that alternative configurations of certain guidance tools could further optimize the support provided. Below we discuss all these aspects in detail.

In terms of providing teacher feedback and enacting on-the-fly formative assessment, teachers might focus on one or two crucial learning products along the learning activity sequence. For instance, hypotheses formulated by students or their experimental designs would give an overview of their progression. This can involve the variables which students would have identified, how they would have categorized these variables (e.g., dependent variables, variables remaining constant, independent variables), and how many experimental trials they would have planned. The learning products, which would be depicted by the teacher for such a procedure, would reveal student performance, and they would denote student progression up to a certain point in the learning activity sequence. These learning products would also play a crucial role in the forthcoming activities. For instance, if a student had not identified the variables involved in an experimentation, then tasks undertaken while building or simulating a model in a virtual laboratory would carry along that weakness. The teacher would diagnose student progression by concentrating on these learning products, and he/she would be ready to provide timely feedback, when this would be required. Although a substantial number of formative assessment formats have been using a wide array of instruments to diagnose student performance, such as multiple-choice items, data collection by means of these instruments would necessitate allocation of additional time for data analysis, and this would endanger the proper timing of teacher feedback. Using learning products for the purpose of enacting formative assessment would shorten considerably the time frame from diagnosis of student performance to provision of teacher feedback (for more details, in this direction, see Hovardas 2016). Future research might shed more light on how much and what kind of feedback provision might be undertaken by computer-supported learning environments without the direct involvement of the teacher. Additionally, there is a need to examine options for configuring upcoming cycles of model-based inquiry based on student performance in former cycles, so that support would be as much as learner-tailored as possible. Across all these options, target models would prove crucial for outlining the optimal form of all learning products expected along learning trajectories.

Subsequent rounds of model-based inquiry would necessitate adequate and effective configuration of guidance tools, such as scaffolds. There might be different versions of the same tool, which would correspond to varying degrees of guidance. A challenge for designing computer-supported learning environments has always been to find a balance between structuring student work (De Boer et al. 2014; Zacharia et al. 2015), for instance, partitioning tasks and letting them be processed serially (Clarke et al. 2005; Kalyuga 2007; Pollock et al. 2002; van Joolingen et al. 2011) and problematizing student inquiry, namely, directing student attention to aspects (e.g., mistakes made by the students during their inquiry enactment) that would remain unaccounted for if students would not have been alerted (Reiser 2004; Sweller et al. 1998). For students with less prior knowledge, scaffolds need to be configured so as to provide increased support and guidance. For instance, in a tool such as the Question Scratchpad, all words need to be provided for students with relatively less prior knowledge so that they can formulate their research questions. As student knowledge advances, this support might be gradually removed (see Pea 2004; McNeill et al. 2006 for a detailed account on "fading" scaffolds). Accordingly, lesser words in the Hypothesis Scratchpad would be enough for more experienced students to formulate their hypotheses. If students succeeded in formulating their hypotheses with lesser words, then this would be an indication that they had progressed in the corresponding inquiry skills. All scaffolds, together with their introduction and fading, need to refer to target models and to fuel the desired transition from initial models of students to target models.

10.6 Working Examples of Subsequent Cycles of Model-Based Inquiry with Virtual Laboratories

In this section we will provide three working examples of subsequent cycles of model-based inquiry, which center on working with virtual laboratories. We will need to underline, first, that our level of analysis will not be an inquiry cycle itself but it will refer to a higher grain size, namely, the movement from one cycle to the next so as to foster an analogous transition from initial models built by students, through intermediate model versions, to target models. Second, we should highlight that we will take advantage of the two learning trajectories we have already identified when presenting phases and sub-phases of the inquiry cycle, that is, the path through questioning and exploration, on the one hand, and the alternative path leading through hypothesis generation and experimentation, on the other. It can be that some virtual laboratories might support student inquiry along both creating a model and simulating it. Other virtual laboratories, however, might allow only for executing simulations. A laboratory might be informed by a ready-made model, where students might not be able to intervene and change model compartments or their interrelations. These types of laboratories let students only change parameters of variables and monitor model behavior through these alterations, but they do not offer a remodeling option. To enable model building and exploration, educators would need to plan a preceding inquiry cycle with another laboratory which would enable model building. Laboratories that enable model building as well as model simulation and data generation would be eligible for both inquiry cycles, namely, the first cycle, where the model has to be constructed, and the second one, where the model will be simulated. Another note that we need to make here is that there would be multiple options of planning subsequent cycles of model-based inquiry, where model building and exploration or experimentation with a virtual laboratory would alternate with hands-on activities or outdoor activities to facilitate optimal learning gains. The working examples, which will follow, will illustrate this perspective, too. One among our main points will be to exemplify model-based inquiry aiming at unraveling hidden assumptions in virtual laboratories.

10.6.1 Electrical Circuits

The Electrical Circuit Lab (Fig. 10.5; http://www.golabz.eu/lab/electrical-circuit-lab) can be used by students to build and simulate simple or more complex electrical circuits. Building a simple electrical circuit is already a modeling task, while more complex electrical circuits in series or in parallel might increase the complexity of the modeling exercise. In the same vein, when a student adds structural compartments available in the Electrical Circuit Lab to advance a circuit, which had been constructed previously, then this might be considered as model revision. The Electrical Circuit Lab provides simulation and data generation capabilities, which would guide model testing, revision, and retesting. Students might begin their inquiry in electrical circuits with a hands-on (physical lab) exploration followed by a subsequent exploration in the virtual laboratory. In either case, the main exploration rationale would be to interrelate basic structural components of an electrical circuit (e.g., power source, wire, bulb) and monitor the simplest function of a simple electrical circuit (e.g., the bulb lights up). The transition from hands-on exploration to an upcoming exploration of a simple electrical circuit within a virtual laboratory could serve as a task for aligning basic structural components of models between the two modeling contexts. It can also include a discussion of basic assumptions behind the functionalities offered by the virtual laboratory. Such a contradistinction would be scheduled so as to unravel assumptions in the virtual laboratory which might remain hidden and unaccounted for. More inquiry cycles can be enacted with the Electrical Circuit Lab by having the students experimenting with in series and parallel circuits, while examining the differences between these two types of more complex circuits along a series of variables (e.g., number of bulbs, brightness of bulbs, total electric current, and voltage). Overall, the sequence of cycles of model-based inquiry presented in Table 10.2 has been planned to present an increasing complexity in modeling tasks and inquiry skills. After that sequence, student inquiry might go on by adding further inquiry cycles, which might again alternate between the virtual laboratory, hands-on exploration and experimentation, or exploration and experimentation outside the classroom (e.g., school experiment or home experiment).

10.6.2 Bicycle Gearing

The GearSketch is another virtual laboratory included in the Go-Lab platform (Fig. 10.7; http://www.golabz.eu/lab/gearsketch). It can be used to model the motion of the gearing mechanism of a bicycle (Table 10.3). Namely, students can insert the basic structural components of the gearing mechanism (e.g., front and back gear, chain, back wheel) and monitor its simplest function. The basic exploration rationale here is to follow how pedaling effort is setting the front gear in motion and how that motion is transmitted through the chain to the back gear and then to the back

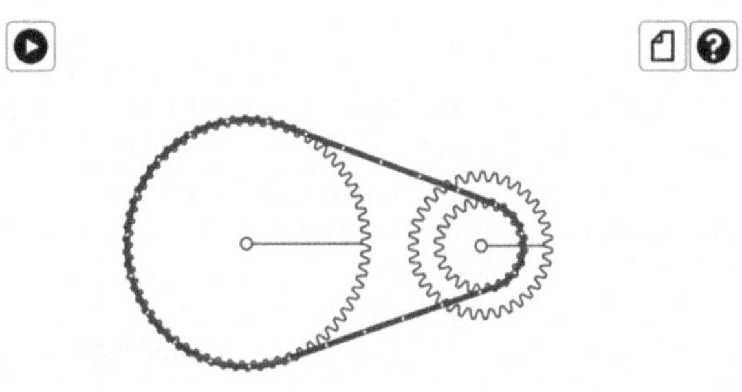

Fig. 10.7 GearSketch (http://www.golabz.eu/lab/gearsketch)

Table 10.3 Subsequent cycles of model-based inquiry for bicycle gearing

Inquiry cycle	Main modeling rationale	Operationalization of the Investigation phase of the inquiry cycle	Main exploration/experimentation rationale
First	Model the motion of the gearing mechanism of a bicycle	Exploration with online lab (e.g., GearSketch; http://go-lab.gw.utwente.nl/production/gearsketch/gearsketch.html)	Interrelate basic structural components (e.g., front and back gear, chain, back wheel) and monitor the simplest function of the gearing mechanism of a bicycle (e.g., motion from the front gear is transmitted through the chain to the back gear)
Second	Model bicycle gearing for a bicycle with a single gear	Outdoor exploration (turn bicycle upside-down and perform hand-powered pedaling; friction between ground and bicycle wheels removed)	Interrelate basic structural components (e.g., front and back gear, chain, back wheel) and monitor the transmission of motion from the gearing mechanism of the bicycle to the back wheel
Third	Model bicycle gearing for a bicycle with multiple gears	Outdoor experimentation (turn bicycle upside-down and perform hand-powered pedaling; friction between ground and bicycle wheels removed)	Interrelate basic structural components (e.g., one front and multiple back gears, chain) and monitor the speed of the back wheel of the bicycle for higher vs. lower gears (e.g., the higher the gear, the higher the speed of the back wheel for the same pedaling force)
Fourth	Model bicycle gearing for a bicycle with multiple gears	Outdoor experimentation (use bicycle and perform foot-powered pedaling; friction between ground and bicycle wheels added to the system)	Monitor rider effort for higher vs. lower gears (e.g., the higher the gear, the higher the pedaling force needed due to static friction)
Fifth	Model bicycle gearing for a bicycle with multiple gears	Experimentation with online lab (e.g., GearSketch; http://go-lab.gw.utwente.nl/production/gearsketch/gearsketch.html)	Interrelate basic structural components (e.g., one front and multiple back gears, chain) and monitor routes of chains for gears of varying radiuses (e.g., the higher the gear, the longer the route)

The sequence of inquiry cycles presented is only indicative; multiple other sequences might be possible, depending on main modeling and exploration/experimentation rationales

wheel of the bicycle. A first point to note is that the GearSketch provides modeling functionalities that are much closer to the initial representations of learners, in contrast to the Electrical Circuit Lab, which enables the construction of more abstract models. This is why it can be readily used as a virtual laboratory before any other inquiry cycle preceding it. Of course, that would not exclude outdoor exploration or experimentation, which can follow. Indeed, students might employ a real bicycle and turn it upside-down to perform hand-powered pedaling (Fig. 10.8). In that configuration of the bicycle, friction between ground and bicycle wheels

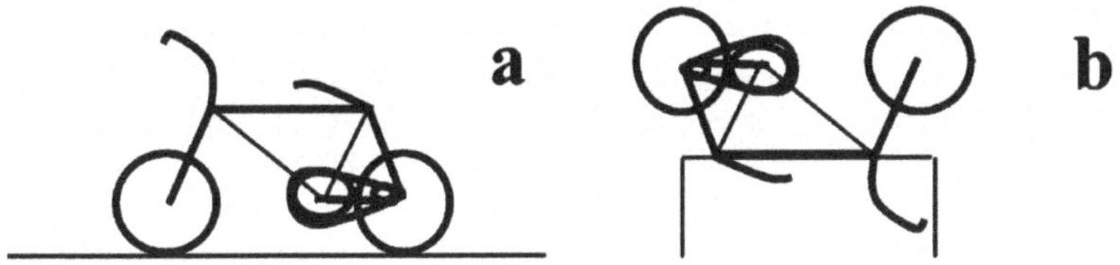

Fig. 10.8 Bicycle standing in the upright position (**a**) and bicycle turned upside-down (**b**)

would have been removed. Students can identify the basic structural components of the gearing mechanism in the real bicycle (e.g., front and back gear, chain, back wheel) and monitor the transmission of motion from the gearing mechanism of the bicycle to the back wheel. In an upcoming experimentation, students will be able to continue using the real bicycle turned upside-down and monitor the speed of the back wheel of the bicycle for higher vs. lower gears. It is expected that the higher the gear, the higher the speed of the back wheel for the same pedaling force. In a next inquiry cycle, students might use the bicycle and perform foot-powered pedaling. In this case, the friction between ground and bicycle wheels would have been added to the system. The students would be able to monitor rider effort for higher vs. lower gears. It is expected that the higher the gear, the higher the pedaling force needed due to static friction. A last inquiry cycle would return the students back to the GearSketch to model bicycle gearing for a bicycle with multiple gears. Students would need to interrelate basic structural components of the new system (e.g., one front and multiple back gears, chain) and monitor the routes of chains for gears of varying radiuses. It is expected that the higher the gear, the longer the route. Student inquiry can go on further by modeling a tandem bicycle for two riders.

10.6.3 Simple and Inverted Pendulums

Our third example concerns simple and inverted pendulums (Table 10.4). Students can first use modeling software like Algodoo to interrelate basic structural compartments of the simple pendulum (e.g., pivot and weight) and prepare a first draft of their model (Fig. 10.9). Students will be able to monitor the simplest function of a simple pendulum, where the weight performs oscillations of standard width after displacement. The next inquiry cycle might involve an experimentation with the Simple Pendulum (Fig. 10.10; https://www.golabz.eu/lab/simple-pendulum-1). This is a virtual laboratory, where students can study the motion of a simple pendulum motion with damping and follow the motion of the weight back to rest position after displacement (e.g., after the weight has performed oscillations of decreasing width after displacement). Further inquiry into pendulums might involve an outdoor exploration with a child swing (Fig. 10.11a). If a person swings, then he or she might not move his/her legs, and in this case there is a damping effect. If the person moves his or her legs, however, then this resupplies the system with energy

Table 10.4 Subsequent cycles of model-based inquiry for simple and inverted pendulums

Inquiry cycle	Main modeling rationale	Operationalization of the Investigation phase of the inquiry cycle	Main exploration/experimentation rationale
First	Model the motion of a simple pendulum	Exploration with modeling and simulation software (e.g., Algodoo; http://www.algodoo.com/)	Interrelate basic structural components (e.g., pivot and weight) and monitor the simplest function of a simple pendulum (e.g., weight performs oscillations of standard width after displacement)
Second	Model the motion of a simple pendulum motion with damping	Experimentation with online lab (e.g., Simple Pendulum; https://www.golabz.eu/lab/simple-pendulum-1)	Interrelate basic structural components (e.g., pivot and weight) and monitor the motion of weight back to rest position after displacement (e.g., weight performs oscillations of decreasing width after displacement)
Third	Model the motion of a child swing	Outdoor exploration (person swings first without moving his/her feet and then with his/her feet moving)	Interrelate basic structural components (e.g., pivot and weight) and monitor energy transformations in a pendulum (e.g., movement of the child's legs resupplies the system with energy lost due to damping)
Fourth	Model the motion of a Segway (inverted pendulum)	Exploration with online lab (e.g., Segway Control Simulation; http://www.golabz.eu/lab/segway-control-simulation)	Interrelate basic structural components (e.g., center of mass above the pivot point) and monitor the simplest function of an inverted pendulum (e.g., vehicle starts when driver shifts body slightly forward or backward; upright position retained through calibration provided by a digital control system including gyroscopic sensors and accelerometer-based leveling sensors, which drive the wheels of the Segway forward or backward, respectively)
Fifth	Model the motion of the human body when walking (inverted pendulum)	Outdoor exploration (lean forward up to the point that one's foot needs to move also forward in order not to fall)	Interrelate basic structural components (e.g., upper part of the human body behaves as an inverted pendulum with weight center of the body as its pivot) and monitor the simplest simulation of an inverted pendulum (upright position retained through calibration provided by semicircular canals in the inner ear)

The sequence of inquiry cycles presented is only indicative; multiple other sequences might be possible, depending on main modeling and exploration/experimentation rationales

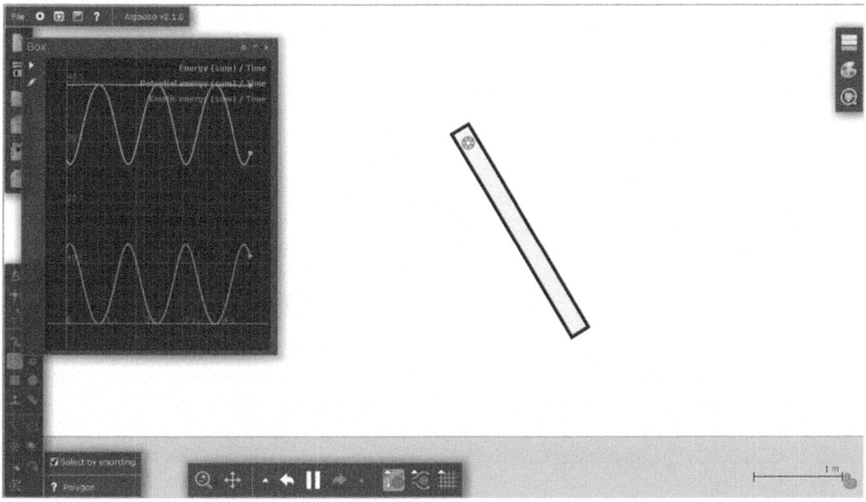

Fig. 10.9 Simple pendulum modeled in Algodoo

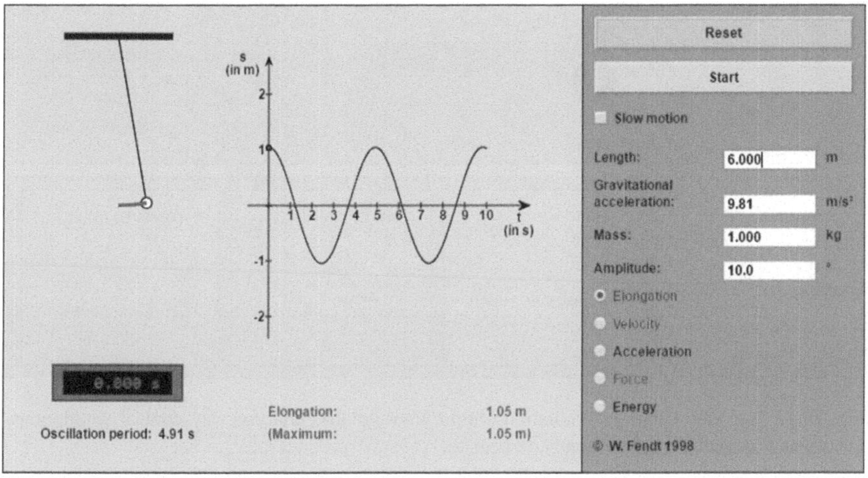

Fig. 10.10 Simple Pendulum (https://www.golabz.eu/lab/simple-pendulum-1). Lab owner: Walter Fendt; http://www.walter-fendt.de/html5/phen/pendulum_en.htm

lost due to damping effects. In a next inquiry cycle, the Segway Control Simulation (Fig. 10.12; http://www.golabz.eu/lab/segway-control-simulation) can be used to interrelate basic structural components of the inverted pendulum (e.g., center of mass above the pivot point) and monitor the simplest function of an inverted pendulum. The vehicle starts moving, when the driver shifts his or her body slightly forward (Fig. 10.11b). Upright position is retained through calibration provided by a digital control system that drives the wheels of the Segway forward. In contrast

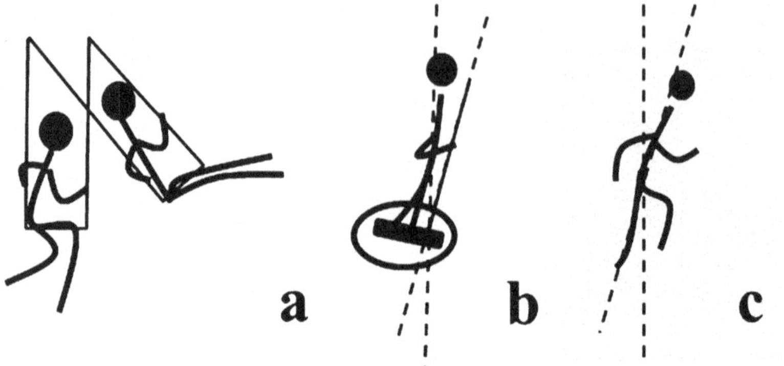

Fig. 10.11 Simple and inverted pendulums: swing (a), Segway (b), walking (c)

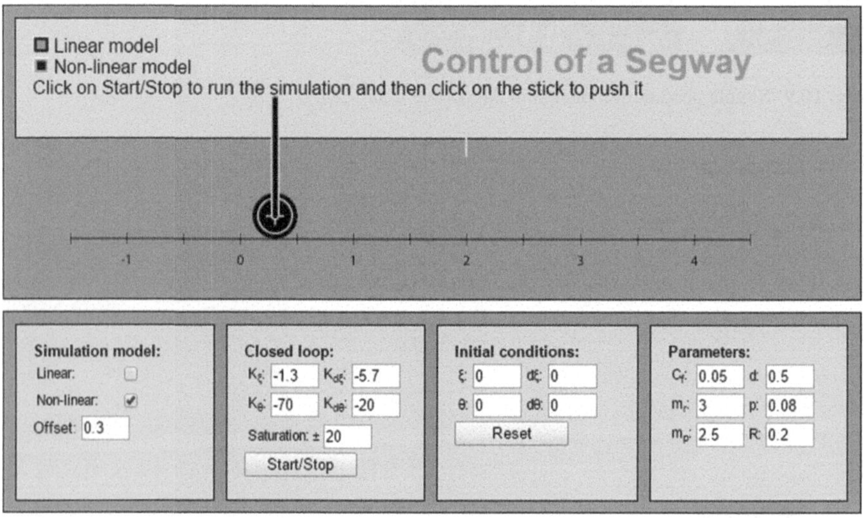

Fig. 10.12 Segway Control Simulation (http://www.golabz.eu/lab/segway-control-simulation). Lab owner: Benjamin Fankhauser

to simple pendulums, inverted pendulums involve a mechanism of correcting for any divergence from the upright position. These mechanisms are responsible for initiating movement, on the one hand, but also for returning the weight to the upright position, when needed. The case of the human body, when walking, is another exemplification of the inverted pendulum. An outdoor exploration can let students lean forward up to the point that one of their feet needs to move also forward in order not to fall (Fig. 10.11c). The upper part of the human body behaves as an inverted pendulum with weight center of the body as its pivot. Upright position is retained through calibration provided by semicircular canals in the inner ear.

10.7 Conclusion and Implications for Practice

In the three examples presented above, we tried to showcase how Go-Lab virtual laboratories could be used for enacting model-based inquiry. We have attempted to highlight "virtues" of virtual laboratories in terms of their modeling affordances and how instructional arrangements could instantiate them. In so doing, teachers need to employ at least two inquiry cycles in their instruction in order to address both model building as well as using models for exploration and experimentation of the system or phenomenon under study.

Some laboratories offer the option of constructing a structure (e.g., an electrical circuit in the Electrical Circuit Lab; a gear mechanism in the GearSketch) and, then, simulate that structure to derive a simulation outcome. Indeed, the simulation would not be possible unless the first step would be completed. This would align with the most basic modeling requirement of any modeling tool, namely, a two-step process of first constructing a model and, then, simulating that model. In that direction, the Electrical Circuit Lab and GearSketch could be seen as laboratories that enable model-based inquiry, meaning that models of electrical circuits or gear mechanisms could be constructed, tested, and revised to progress gradually to more complex models. Other laboratories (e.g., Simple Pendulum; Segway) do not allow for this two-step process. Students can only change parameters and observe the simulation outcome, but they are not able to construct a model and simulate their model or revise it. Students cannot even add new variables and thus test these new variables. For this second category of labs (i.e., close-ended simulations), in order to incorporate them in any model-based inquiry paradigm, we would need to accompany them with software that would allow modeling the phenomenon included in the close-ended simulation. This relates to the third example we have included in the paper, i.e., the case of the simple pendulum, where we used the Algodoo software to allow students to model the simple pendulum before using our close-ended, ready-made simulation.

All sequences took into account the modular nature of model-based inquiry (i.e., building an initial model and, then, testing and revising this model to arrive at the target model of instruction), which might be quite adaptable to curricula and school practice. However, teachers would need substantial support to screen among resources available, arrange them along phases and sub-phases of inquiry, and plan their instruction accordingly. The Go-Lab platform offers user manuals and online courses, tutorials, and a community forum for teachers to interact (http://www.golabz.eu/support). To further build on teacher input, design-based research might provide valuable insight for model-based inquiry in computer-supported learning environments through evidence-based learning progressions (e.g., Cobb et al. 2003; Duschl et al. 2011; Shea and Duncan 2013; Lehrer and Schauble 2015). The iterative nature of design-based research might be perfectly compatible with successive inquiry cycles in model-based inquiry, and it might give considerable opportunities for refining learning trajectories. Designing virtual laboratories and embedding them in adequately configured learning environments must incorporate evolving student and teacher needs and desires so that student performance might be optimized.

Acknowledgments This book chapter reports on work undertaken within the frame of the Go-Lab project (Project title: Global Online Science Labs for Inquiry Learning at School; Call identifier: FP7-ICT-2011-8; Project number: 317601) funded by the European Union. The present document does not represent the opinion of the European Union, and the European Union should not be held responsible for any use that might be made of its content.

References

Ariza, Y., Lorenzano, P., & Adúriz-Bravo, A. (2016). Meta-theoretical contributions to the constitution of a model-based didactics of science. *Science & Education*. https://doi.org/10.1007/s11191-016-9845-3.

Campbell, T., Oh, P. S., & Neilson, D. (2013). Reification of five types of modeling pedagogies with model-based inquiry (MBI) modules for high school science classrooms. In M. S. Khine & I. M. Saleh (Eds.), *Approaches and strategies in next generation science learning* (pp. 106–126). Hershey: IGI Global.

Campbell, T., Oh, P. S., Maughn, M., Kiriazis, N., & Zuwallack, R. (2015). A review of modeling pedagogies: Pedagogical functions, discursive acts, and technology in modeling instruction. *Eurasia Journal of Mathematics, Science & Technology Education, 11*, 159–176.

Clarke, T., Ayres, P., & Sweller, J. (2005). The impact of sequencing and prior knowledge on learning mathematics through spreadsheet applications. *Educational Technology Research & Development, 53*, 15–24.

Clement, J. (2000). Model based learning as a key research area for science education. *International Journal of Science Education, 22*, 1041–1053.

Cobb, P., Confrey, J., diSessa, A., Lehrer, R., & Schauble, L. (2003). Design experiment in educational research. *Educational Researcher, 32*, 9–13.

Coll, R. K., & Lajium, D. (2011). Modeling and the future of science learning. In M. S. Khine & I. M. Saleh (Eds.), *Models and modeling cognitive tools for scientific enquiry* (pp. 3–21). New York: Springer.

De Boer, G. E., Quellmalz, E. S., Davenport, J. L., Timms, M. J., Herrmann-Abell, C. F., Buckley, B. C., Jordan, K. A., Huang, C.-W., & Flanagan, J. C. (2014). Comparing three online testing modalities: Using static, active, and interactive online testing modalities to access middle school students' understanding of fundamental ideas and use of inquiry skills related to ecosystems. *Journal of Research in Science Teaching, 51*, 523–554.

de Jong, T. (2015). Go-Lab classroom scenarios handbook. Deliverable D1.4. The Go-Lab Project: Global Online Science Labs for Inquiry Learning at School.

de Jong, T., & van Joolingen, W. R. (2007). Model-facilitated learning. In J. M. Spector, M. D. Merrill, J. J. G. van Merriënboer, & M. P. Driscoll (Eds.), *Handbook of research on educational communication and technology* (3rd ed., pp. 457–468). New York: Lawrence Erlbaum.

de Jong, T., van Joolingen, W. R., Giemza, A., Girault, I., Hoppe, U., Kindermann, J., Kluge, A., Lazonder, A. W., Vold, V., Weinberger, A., Weinbrenner, S., Wichmann, A., Anjewierden, A., Bodin, M., Bollen, L., D'Ham, C., Dolonen, J., Engler, J., Geraedts, C., Grosskreutz, H., Hovardas, T., Julien, R., Lechner, J., Ludvigsen, S., Matteman, Y., Meistadt, Ø., Næss, B., Ney, M., Pedaste, M., Perritano, A., Rinket, M., von Schlanbusch, H., Sarapuu, T., Schulz, F., Sikken, J., Slotta, J., Toussaint, J., Verkade, A., Wajeman, C., Wasson, B., Zacharia, Z. C., & van der Zanden, M. (2010). Learning by creating and exchanging objects: The SCY experience. *British Journal of Educational Technology, 41*, 909–921.

de Jong, T., Weinberger, A., van Joolingen, W. R., Ludvigsen, S., Ney, M., Girault, I., Kluge, A., Lazonder, A., Pedaste, M., Wasson, B., Wichmann, A., Geraedts, C., Hovardas, T., Julien, R., Lejeune, A., Manoli, C., Matteman, Y., Sarapuu, T., Vold, V., Wanders, B., & Zacharia, Z. (2012). Using scenarios to design complex technology-enhanced learning environments. *Educational Technology Research & Development, 60*, 883–901.

de Jong, T., Linn, M. C., & Zacharia, Z. C. (2013). Physical and virtual laboratories in science and engineering education. *Science, 340*, 305–308.

de Jong, T., Sotiriou, S., & Gillet, D. (2014). Innovations in STEM education: The Go-Lab federation of online labs. *Smart Learning Environments, 1*, 1–16.

Develaki, M. (2007). The model-based view of scientific theories and the structuring of school science programmes. *Science & Education, 16*, 725–749.

Duschl, R., Maeng, S., & Sezen, A. (2011). Learning progressions and teaching sequences: A review and analysis. *Studies in Science Education, 47*, 123–182.

Forbes, C. T., Zangori, L., & Schwarz, C. V. (2015). Empirical validation of integrated learning performances for hydrologic phenomena: 3rd-grade students' model-driven explanation-construction. *Journal of Research in Science Teaching, 52*, 895–921.

Giere, R. N. (1991). *Understanding scientific reasoning* (3rd ed.). New York: Harcourt Brace Jovanovich College Publishers.

Gobert, G. D., & Buckley, B. C. (2000). Introduction to model-based teaching and learning in science education. *International Journal of Science Education, 22*, 891–894.

Grünkorn, J., Upmeier zu Belzen, A., & Krüger, D. (2014). Assessing students' understandings of biological models and their use in science to evaluate a theoretical framework. *International Journal of Science Education, 36*, 1651–1684.

Halloun, I. (2007). Mediated modeling in science education. *Science & Education, 16*, 653–697.

Hmelo-Silver, C. E., & Azevedo, R. (2006). Understanding complex systems: Some core challenges. *The Journal of the Learning Sciences, 15*, 53–61.

Hovardas, T. (2016). A learning progression should address regression: Insights from developing non-linear reasoning in ecology. *Journal of Research in Science Teaching, 53*, 1447–1470.

Hovardas, T., & Korfiatis, K. J. (2011). Towards a critical re-appraisal of ecology education: Scheduling an educational intervention to revisit the 'Balance of Nature' metaphor. *Science & Education, 20*, 1039–1053.

Hovardas, T., Xenofontos, N., & Zacharia, Z. (2017). Using virtual labs in an inquiry context: The effect of a hypothesis formulation tool and an experiment design tool on students' learning. In Levin, I., & Tsybulsky, D. (Eds.), *Digital tools and solutions for inquiry based STEM learning* (pp. 58–83). Hershey: IGI Global.

Jacobson, M. J., & Wilensky, U. (2006). Complex systems in education: Scientific and educational importance and implications for the learning sciences. *The Journal of the Learning Sciences, 15*, 11–34.

Justi, R. S., & Gilbert, J. K. (2003). Teachers' views on the nature of models. *International Journal of Science Education, 25*, 1369–1386.

Kalyuga, S. (2007). Expertise reversal effect and its implications for learner-tailored instruction. *Educational Psychology Review, 19*, 509–539.

Kolodner, J. L., Camp, P. J., Crismond, D., Fasse, B., Gray, J., Holbrook, J., Puntambekar, S., & Ryan, N. (2003). Problem-based learning meets case-based reasoning in the middle-school science classroom: Putting learning by design (TM) into practice. *The Journal of the Learning Sciences, 12*, 495–547.

Lefkaditou, A., Korfiatis, K., & Hovardas, T. (2014). Contextualizing the teaching and learning of ecology: Historical and philosophical considerations. In M. Matthews (Ed.), *International handbook of research in history, philosophy and science teaching* (pp. 523–550). Dordrecht: Springer.

Lehrer, R., & Schauble, L. (2006). Scientific thinking and scientific literacy: Supporting development in learning in context. In W. Damon, R. M. Lerner, K. A. Renninger, & I. E. Sigel (Eds.), *Handbook of child psychology* (Vol. 4, 6th ed., pp. 153–196). Hoboken: Wiley.

Lehrer, R., & Schauble, L. (2015). Learning progressions: The whole world is NOT a stage. *Science Education, 99*, 432–437.

Louca, L., & Zacharia, Z. (2008). The use of computer-based programming environments as computer modeling tools in early science education: The cases of textual and graphical program languages. *International Journal of Science Education, 30*, 287–323.

Louca, T. L., & Zacharia, C. Z. (2012). Modeling-based learning in science education: A review. *Educational Review, 64*, 471–492.

Louca, L. T., & Zacharia, Z. C. (2015). Examining learning through modeling in K-6 science education. *Journal of Science Education and Technology, 24*(2–3), 192–215.

Matthews, M. R. (2005). Idealization and Galileo's pendulum discoveries: Historical, philosophical and pedagogical considerations. In M. R. Matthews, C. F. Gauld, & A. Stinner (Eds.), *The pendulum: Scientific, historical, philosophical & educational perspectives* (pp. 209–235). Dordrecht: Springer.

McComas, W. F. (2002). *The nature of science in science education*. Dordrecht: Kluwer Academic Publishers.

McNeill, K. L., Lizotte, D. J., Krajcik, J., & Marx, R. W. (2006). Supporting students' construction of scientific explanations by fading scaffolds in instructional materials. *The Journal of the Learning Sciences, 15*, 153–191.

Mellar, H., & Bliss, J. (1994). Introduction: Modeling and education. In H. Mellar, J. Bliss, R. Boohan, J. Ogborn, & C. Tompsett (Eds.), *Learning with artificial worlds: Computer based modeling in the curriculum* (pp. 1–7). London: The Falmer Press.

Nersessian, N. (2002). The cognitive basis of model-based reasoning in science. In P. Carruthers, S. Stich, & M. Siegal (Eds.), *The cognitive basis of science* (pp. 17–34). Cambridge: Cambridge University Press.

Nersessian, N. (2005). Interpreting scientific and engineering practices: Integrating the cognitive, social, and cultural dimensions. In M. Gorman, R. D. Tweney, D. Gooding, & A. Kincannon (Eds.), *Scientific and technological thinking*. Hillsdale: Lawrence Erlbaum Associates.

Oh, P. S., & Oh, S. J. (2011). What teachers of science need to know about models: An overview. *International Journal of Science Education, 33*, 1109–1130.

Olympiou, G., Zacharia, Z. C., & de Jong, T. (2013). Making the invisible visible: Enhancing students' conceptual understanding by introducing representations of abstract objects in a simulation. *Instructional Science, 41*, 575–587.

Pea, R. D. (2004). The social and technological dimensions of scaffolding and related theoretical concepts for learning, education, and human activity. *The Journal of the Learning Sciences, 13*, 423–451.

Pedaste, M., & Sarapuu, T. (2006a). Developing an effective support system for inquiry learning in a Web-based environment. *Journal of Computer Assisted Learning, 22*, 47–62.

Pedaste, M., & Sarapuu, T. (2006b). The factors influencing the outcome of solving story-problems in a Web-based learning environment. *Interactive Learning Environments, 14*, 153–176.

Pedaste, M., Mäeots, M., Siiman, L. A., de Jong, T., van Riesen, S. A., Kamp, E. T., Manoli, C. C., Zacharia, Z. C., & Tsourlidaki, E. (2015). Phases of inquiry-based learning: Definitions and the inquiry cycle. *Educational Research Review, 14*, 47–61.

Petridou, E., Psillos, D., Hatzikraniotis, E., & Kallery, M. (2013). A study on the exploratory use of microscopic models as investigative tools: The case of electrostatic polarization. In G. Tsaparlis & H. Sevian (Eds.), *Concepts of matter in science education* (pp. 199–212). Dordrecht: Springer.

Pollock, E., Chandler, P., & Sweller, J. (2002). Assimilating complex information. *Learning and Instruction, 12*, 61–86.

Reiser, B. J. (2004). Scaffolding complex learning: The mechanisms of structuring and problematizing student work. *The Journal of the Learning Sciences, 13*, 273–304.

Schwarz, C. V., & White, B. Y. (2005). Metamodeling knowledge: Developing students' understanding of scientific modeling. *Cognition and Instruction, 23*, 165–205.

Schwarz, C. V., Reiser, B. J., Davis, E. A., Kenyon, L., Archer, A., Fortus, D., & Krajcik, J. (2009). Developing a learning progression for scientific modeling: Making scientific modeling accessible and meaningful for learners. *Journal of Research in Science Teaching, 46*, 632–654.

Shea, N. A., & Duncan, R. G. (2013). From theory to data: The process of refining learning progressions. *The Journal of the Learning Sciences, 22*, 7–32.

Soulios, I., & Psillos, D. (2016). Enhancing student teachers' epistemological beliefs about models and conceptual understanding through a model-based inquiry process. *International Journal of Science Education, 38*, 1212–1233.

Sweller, J., van Merriënboer, J. J. G., & Paas, F. G. W. C. (1998). Cognitive architecture and instructional design. *Educational Psychology Review, 10*, 251–296.

Tasiopoulou, E., & de Jong, T. (2016). Integrated validation and evaluation report and recommendations. Deliverable D8.4 (Complete version). The Go-Lab Project: Global Online Science Labs for Inquiry Learning at School.

van Joolingen, W. R., & Zacharia, Z. C. (2009). Developments in inquiry learning. In N. Balacheff, S. Ludvigsen, T. de Jong, & S. Barnes (Eds.), *Technology-enhanced learning* (pp. 21–37). Dordrecht: Springer.

van Joolingen, W. R., de Jong, T., Lazonder, A. W., Savelsberg, E. R., & Manlove, S. (2005). Co-Lab: Research and development of an online learning environment for collaborative scientific discovery learning. *Computers in Human Behavior, 21*, 671–688.

van Joolingen, W. R., Giemza, A., Bollen, L., Bodin, M., Manske, S., Engler, J., & Halik, K. (2011). *SCY cognitive scaffolds and tools (DII.2)*. Twente: SCY Consortium.

Verhoeff, R. P., Waarlo, A. J., & Boersma, K. T. (2008). Systems modelling and the development of coherent understanding of cell biology. *International Journal of Science Education, 30*, 543–568.

Weinberger, A., Dolonen, J., Hovardas, A., Pedaste, M., Kluge, A., Ludvigsen, S., et al. (2009). *SCY scenario handbook and pedagogical plans, version 1*. Enschede: University of Twente.

Windschitl, M., Thompson, J., & Braaten, M. (2008a). Beyond the scientific method: Model-based inquiry as a new paradigm of preference for school science investigations. *Science Education, 92*, 941–967.

Windschitl, M., Thompson, J., & Braaten, M. (2008b). How novice science teachers appropriate epistemic discourses around model-based inquiry for use in classrooms. *Cognition and Instruction, 26*, 310–378.

Zacharia, Z. C., & de Jong, T. (2014). The effects on students' conceptual understanding of electric circuits of introducing virtual manipulatives within a physical manipulatives-oriented curriculum. *Cognition and Instruction, 32*(2), 101–158.

Zacharia, Z. C., Manoli, C., Xenofontos, N., de Jong, T., Pedaste, M., van Riesen, S. A. N., Kamp, E. T., Mäeots, M., Siiman, L., & Tsourlidaki, E. (2015). Identifying potential types of guidance for supporting student inquiry when using virtual and remote labs in science: A literature review. *Educational Technology Research and Development, 63*, 257–302.

Zangori, L., & Forbes, C. T. (2015). Exploring third-grade student model-based explanations about plant relationships within an ecosystem. *International Journal of Science Education, 37*, 2942–2964.

Zangori, L., Forbes, C. T., & Schwarz, C. V. (2015). Exploring the effect of embedded scaffolding within curricular tasks on third-grade students' model-based explanations about hydrologic cycling. *Science & Education, 24*, 957–981.

Tasos Hovardas is a Senior Research Associate in the Research in Science and Technology Education Group at the University of Cyprus, and he is an Editorial Board Member of the Journal of Research in Science Teaching (2017–2020).

Margus Pedaste is Professor of Educational Technology and head of Pedagogicum in the Institute of Education at the University of Tartu, Estonia.

Zacharias Zacharia is Director of the Research in Science and Technology Education Group at the University of Cyprus, and an Associate Professor of Science Education at the Department of Educational Sciences of the University of Cyprus.

Ton de Jong is Full Professor of Instructional Technology, head of the departments of Instructional Technology and Educational Science and Director of the Master program Educational Science and Technology at the University of Twente, the Netherlands.

Part III
Cyber-Physical Laboratories: Best Practices and Case Studies

Section Introduction

This section highlights a number of remote laboratory case studies covering a range of application areas that can be considered as representative best practices. There is a total of six chapters highlighting remote laboratories for life science experiments, automation engineering, hardware-in-the-loop systems, integration of augmented reality and haptic devices, heat transfer experiments, and additive manufacturing. The contributions provide an insight from a different perspective, and each discussion leads the reader to understand the rationale behind the approaches taken and obtain further information of interest.

Almost all reported remote laboratory developments are related to engineering, technology, and physics topics; our first chapter for this section is introducing a remote laboratory development in biological science. This chapter is titled as "Life-Science Experiments Online: Technological Frameworks and Educational Use Cases" by Zahid Hossain and Ingmar Riedel-Kruse. The chapter describes and compares four biology cloud laboratories demonstrating different user interaction modes, underlying hardware and software architecture, biological content, and scalability issue. In addition to the educational use, the chapter describes research applications. The authors illustrate the general design rules for biology cloud experimentation laboratory along with open questions regarding future technology and opportunities for scalability and wide deployment.

The second chapter titled "A CPS Integration Platform as a Framework for Generic Remote Labs in Automation Engineering" by Reinhard Langmann describes the development of a generic or customizable remote laboratory utilizing a web-oriented automation system (WOAS). This uses the new paradigms from cyber-physical systems and service-based automation. The platform allows one to develop a remote laboratory with given requirements using web-based tools. The author describes the use of the WOAS portal as a framework for creating user-specific remote laboratory for automation technology training and demonstrates its effectiveness through three applications.

The third chapter of this section presents the development and utilization of an additive manufacturing (AM) laboratory environment. The chapter is titled as "The Development and Implementation of Instruction and Remote-Access Components of Additive Manufacturing" by Ismail Fidan and his co-authors. The chapter starts with highlighting the historical funding support the team has received for developing this remote laboratory. This is followed by a discussion on AM technologies and how this system has been accessed over the network for remote communication. The authors then illustrate the details of their developed remote laboratory facility as well as the instruction materials used for course delivery. Finally the chapter concludes with the presentation of student feedback while utilizing the facility for educational delivery.

The fourth chapter "Design and Implementation of a Remote Laboratory for Heat Transfer Experiments" by Ridha Ennetta and his co-authors describes the design and the implementation of a remote laboratory for heat transfer learning purposes. It summarizes the work carried out to adapt and redesign a heat exchanger bench to be remotely accessed and controlled. This laboratory introduced many fundamental aspects of heat transfer, both theoretically and practically. An evaluation procedure was also carried out for this development, while focusing on technical and pedagogical aspects. The evaluation results demonstrated that the expected learning outcomes of this remote laboratory seem to be very interesting compared to conventional laboratories.

The fifth chapter "Collaborative Virtual Laboratory Environments with Hardware in the Loop" by Zhou Zhang and his co-authors highlights a virtual laboratory system with experimental hardware in the loop. The chapter discusses the concept, history, and current status of virtual laboratories as well as techniques used to create those. This is followed by presenting its shortcomings and promising approaches for overcoming those. The authors closed the chapter with a pilot implementation of two laboratory experiments along with evaluation studies of participating students. The results indicated that the developed virtual laboratory environments were well received by the students.

The last chapter of this section which is titled as "Mobile Cyber-Physical Labs: On the Integration of Mobile Devices with Laboratory Test-Beds to Teach Dynamic Systems and Control Concepts" by Jared Frank and his co-authors proposes the use of mobile cyber-physical laboratories in which the hardware and software of mobile devices are leveraged in the measurement, control, monitoring, and interaction with physical test-beds in the laboratory. Two separate approaches for developing cost-effective and portable educational test-beds are proposed. These utilize the sensing, storage, computation, and communication capabilities of mobile devices to facilitate inquiry-based educational experiences.

Chapter 11
Life-Science Experiments Online: Technological Frameworks and Educational Use Cases

Zahid Hossain and Ingmar H. Riedel-Kruse

Abstract We review remote (or "cloud") lab technologies for life-science experimentation. Compared to other remote labs such as for physics, a particular challenge arises from the variability and stability of biological materials. We describe and compare four biology cloud labs that demonstrate different user interaction modes, i.e., real-time and turn-based interactive, programmed, and augmented batch, respectively, and furthermore regard their underlying hard and software architecture, biological content ("bio-ware") (i.e., microswimmer phototaxis, slime mold chemotaxis, bacterial growth under antibiotics, RNA folding), and various other features such as the time required for one experiment or scalability to large user numbers. While we generally focus on educational use cases, research applications are included as well. General design rules for biology cloud experimentation labs are derived; open questions regarding future technology and opportunities for wide deployment are discussed. We hope that this review enables stakeholders from the life sciences, engineering, and education to join this relevant and exciting field.

Keywords Biology · Life sciences · Remote experimentation · Online experimentation · Cloud lab · Education · Biotic processing unit (BPU)

11.1 Introduction

Being able to perform versatile biology experiments online has many applications for research and education. Many access barriers to life-science experimentation exist for academic and commercial research, mainly due to professional training needs, cost of equipment purchase and operation, and safety considerations (Sia and

Z. Hossain
Departments of Bioengineering and Computer Science, Stanford University, Stanford, CA, USA

I. H. Riedel-Kruse (✉)
Department of Bioengineering, Stanford University, Stanford, CA, USA
e-mail: ingmar@stanford.edu

M. E. Auer et al. (eds.), *Cyber-Physical Laboratories in Engineering and Science Education*, https://doi.org/10.1007/978-3-319-76935-6_11

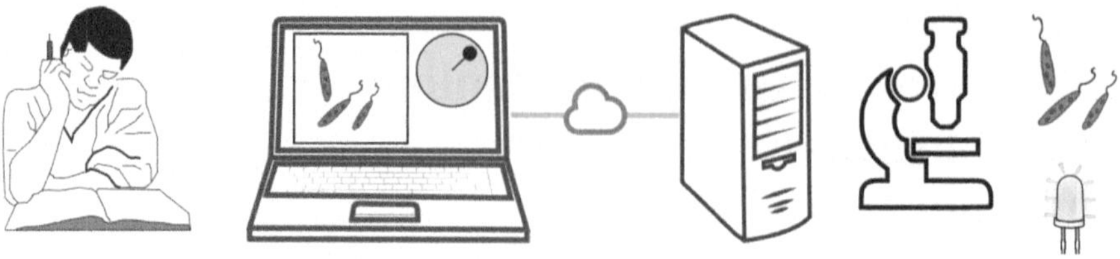

Fig. 11.1 Biology cloud experimentation labs enable remote users (scientists and students) to conveniently carry out life-science experiments online

Owens 2015). Remote operation of biology experiments in the cloud (Fig. 11.1) has been suggested to help lower these barriers (Hossain et al. 2015, 2016). Since biological investigations are diverse—unlike general-purpose computing, there is no clear foundation (e.g., binary 1s and 0s) for executing all types of experiments—different types of back-end instruments and online architectures are needed to address the duration of an experiment, the response time of the biological material, and the frequency of user interactions.

Multiple approaches to implement biology cloud labs have been developed but only rather recently (i.e., over the past ~4 years): We previously developed two interactive biology cloud lab architectures that were real-time (Hossain et al. 2016) and turn-based (Hossain et al. 2015); commercial and academic entities developed noninteractive biology cloud labs where experiments can be programmed (Riedel-Kruse 2017; Transcriptics 2015; Klavins 2017), and online citizen science games have been deployed that provide the user with experimental feedback (EteRNA) (Lee et al. 2014). All these labs have been used in educational contexts to various extends.

These four approaches can be categorized based on their directness and flexibility of the user interactions, which is enabled and constrained by the underlying architecture: (1) "Real-time interaction" enables direct experimentation and adaptive user input on the sub-second time scale, while the experiment is running (Hossain et al. 2016). This is suited for biological phenomena with response times on the scale of seconds. Experiment duration is typically short (minutes), and a user obtains sole and direct control of a single instrument for a time period on the scale of minutes (although both requirements could be relaxed, in principle). (2) "Turn-based interaction" also enables direct experimentation and adaptive user input, while the experiment is running, but now on more discrete time scale, e.g., every few minutes (Hossain et al. 2015). The biological response time of interest is significantly longer than 1 s, and no real-time interaction is required. Experiment duration might be multiple hours, and experiments of multiple users can be multiplexed and parallelized on a single machine or on multiple machines (again, these requirements can be relaxed). (3) "Programmed batch" enables code-based instruction of one or multiple instruments to execute a more complex series of experiments. Here, all instructions are completely predefined before the experiment starts (Riedel-Kruse 2017; Transcriptics 2015; Klavins 2017), and no interaction or adaptions during the experiment are possible. This approach is particularly geared toward academic and industrial research, where robots shuttle biological samples between fully automated pieces of equipment, thereby enabling highly complex experiments on the scale of hours. (4) "Augmented batch" enables the user to focus on higher level experimental design tasks while abstracting away the particularities of controlling an instrument. This is particularly useful for citizen science games (Lee et al. 2014) that provide experimental feedback to online players. (Note that these four examples provided here do not map exclusively onto these four categories, e.g., interactive labs can be used for batch processing (Hossain et al. 2016), or pre-programmable labs could be converted into turn-based ones (Riedel-Kruse 2017) depending on the exact hardware setup. Furthermore, these approaches can be categorized along other dimensions, and we will discuss throughout the paper.)

The goal of this paper is to provide an overview of these existing biology cloud labs with a particular focus on educational uses, although we also consider professional and citizen science. We highlight their architectures, practical implementation, and user testing of these approaches; detailed descriptions of these studies can be found in the original publications (Hossain et al. 2015, 2016; Riedel-Kruse 2017; Lee et al. 2014). We also briefly mention purely virtual approaches, i.e., simulations of biology experiments (de Jong et al. 2013; Heradio et al. 2016). We provide a systematic comparison between these four approaches (Table 11.1), and we discuss open questions for future larger-scale deployment and for increasing the availability of distinct experimentation types.

11.2 Background and Motivation

Cloud labs are poised to help solve significant educational challenges. Familiarity with advanced scientific practices and "authentic inquiry" (Chinn and Malhotra 2002; Pedaste et al. 2015; States 2013) are imperative for K-12 and college education (Next Generation Science Standards, NGSS; States 2013; Bybee 2013) but are difficult to achieve in real-world classrooms given logistics and cost (Chinn and Malhotra 2002; Wellington 2007). In addition to traditional physical hands-on labs, virtual and remote labs have been successfully deployed recently, particularly in engineering and physics (de Jong et al. 2013; Heradio et al. 2016). User studies have shown that hands-on, remote, and virtual modalities each have distinct advantages given educational goals and situational contexts, but ultimately, the question is how to best use these approaches synergistically (de Jong et al. 2013; Heradio et al. 2016; Wieman et al. 2008; Bonde et al. 2014; Sauter et al. 2013). Remote experiments in the life sciences have been lacking compared to these other disciplines, in particular due to the added challenges and necessary logistics for keeping biological materials healthy and readily available for extended periods of time.

Modern biotechnology and life sciences are poised to provide solutions to these challenges. Of particular importance are liquid-handling robotics (Kong et al. 2012)

Table 11.1 Comparison of four biology cloud labs

User instruction mode	Real-time interaction	Turn-based interaction	Programmed batch	Augmented batch
Biological substrate	*Euglena gracilis*	*Physarum polycephalum*	*Escherichia coli*	RNA
User controlled variable (stimulus)	Light	Food solution	Antibiotics	Nucleotide sequence
Raw output data	Image sequence of *Euglena* in microfluidic chip	Image sequence of Petri dish with *Physarum*	Optical density of bacterial population	Single-nucleotide-resolution chemical reactivity measurements
Processed data output	Cell tracks	Binarized image	Growth curves	Graphical display of secondary RNA structure
Interactive experimentation?	Yes (real-time)	Yes (turn-based)	No	No
# Experiments per run per BPU	1	6	96	10,000
# BPUs in cluster	6	3	1	1 (incl. manual labor)
Duration of one experimental run	~1 min	~48 h	~24 h	~1 month
# Exp. in 24 h	~5000	~10	~100	~0.1
Cost per experiment	~US $0.01	~US $10	~US $1	~US $0.2
Maximum frequency of updated user input	600/run	~250/run	1/run	1/run
	(10/s)	(6/h)	(1/day)	(1/month)
Actual # of updates users made per run	~5/run	~3/run	1/run	1/run
# perceived available choices per update	~16	~400	~10	$\sim 4^{100}$
# choices per experiment	>1000	>100	~10	$\sim 4^{100}$
Dimensionality of experimental design space	~100	~5	~1	$\sim 4^{100}$
Extendability to other experiments	Medium	Low	Very high	Low

and integrated microfluidic devices (Balagaddé et al. 2005; Melin and Quake, 2007) that incorporate sensing and actuation devices, achieving very complex liquid handling (often at high throughput) to fully automate sophisticated life-science experiments (Fig. 11.2). These technologies are increasingly impacting our society through their academic and industrial use, will potentially also soon lead to devices

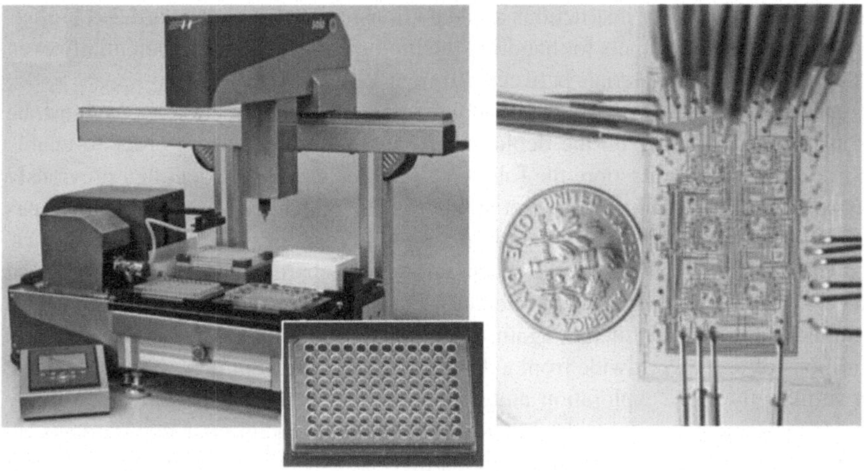

Fig. 11.2 Automation and cost reduction in life-science experiments via (left) liquid-handling robotics and (right) microfluidics. (Images adapted from Kong et al. (2012) and Balagaddé et al. (2005))

of personal use, and may ultimately transform our daily lives as radically as modern computing technology has done previously (Riedel-Kruse et al. 2011; Gerber et al. 2016). Hence, the life sciences and associated technologies should also be put at the forefront of formal and informal education in order enable modern citizens to navigate these new realities.

These new technologies and new educational needs both enable and motivate the field of interactive biology (Riedel-Kruse et al. 2011; Gerber et al. 2016), in which human users interact with microscopic organisms and processes in real time. In addition to cloud labs (Hossain et al. 2015, 2016), these interactive technologies have been implemented as biotic games (Riedel-Kruse et al. 2011, self-builder smartphone kits (Kim et al. 2016), and interactive museum exhibits (Lee et al. 2015). College-level device classes have been deployed around such interactive biology and game project themes (Cira et al. 2015), and we expect future synergy as students build interactive biology devices and put them online as remote labs (Hossain et al. 2016). User studies associated with these previous projects often identified standout features of a real biological system compared to pure simulation (Hossain et al. 2015, 2016), although ultimately we believe that both real and simulations should be combined synergistically for better educational outcomes. Advantages of real biology labs include the chance of genuine discovery and also illustrating biological noise and variability (Hossain et al. 2015, 2016).

To aid the design of instruments suitable for biological cloud labs (and interactive biology in general), we previously introduced the conceptual abstraction of biotic processing units (BPUs) (Hossain et al. 2015; Riedel-Kruse et al. 2011; Hossain and Riedel-Kruse 2017; Lam et al. 2017). BPUs are instruments that have both sensors and actuators that interface with the biological material, with standardized digital input/out channels for instructions and data transfer as well as standardized biological input/output channels for handling the biological material (and potentially even moving biological materials between different BPUs).

When setting up a biology cloud lab, several design specifications must be considered depending on the deployment needs. In particular, in order to enable K-12 and college education, the following features have been identified previously as particularly valuable (Hossain et al. 2016): The system must (1) enable the types of inquiry mandated (which would be very different for professional science vs. educational K-12 purposes); (2) have a low entry barrier and be usable even at the K-12 level; (3) be real-time interactive; (4) have a fast turnaround time (within minutes); (5) be fault tolerant against biological variability and failure; (6) scale to millions of users worldwide from a design as well as economic viewpoint; (7) have a sufficiently large exploration and discovery space; and (8) generalize to many other experiment types easily. For research purposes, additional requirements do apply, such as high fidelity and reproducibility of the results, furthermore significant versatility of instruments, and biological materials that can be processed.

11.3 System 1: Real-Time Interaction (*Euglena* Phototaxis, Light)

This system was developed with the goal to allow direct, real-time interactivity with microbiological systems—at cost and scale (Hossain et al. 2016) (Fig. 11.3). This goal required a short overall experimental duration (at the scale of minutes) and full automation to enable 24/7 access without much manual labor at the back end.

11.3.1 Architecture

On this platform, a single user becomes—for a limited amount of time—the sole actuator of a remotely placed piece of equipment (BPU). The user management system was implemented as a real-time queue. The primary new affordance of this platform is a direct and closed interactive feedback loop between the user and the biological system, but submitting fully preprogrammed batch experiments that are executed serially at a later time is also possible.

The BPU for this implementation consisted of a simple microfluidic chip (Whitesides 2006) housing the phototactic single-celled organism *Euglena gracilis* (Fig. 11.3a, b) (Barsanti et al. 2012). The chamber on this chip is a square (approximately 1 mm long, 1 mm wide, and 150 μm high) and has an inlet and outlet for fluid and organism exchange. These organisms are imaged from above via a webcam microscope. On each of the four sides of the chip, an LED shines light of varying intensity onto the chip and where this intensity can be controlled by the user.

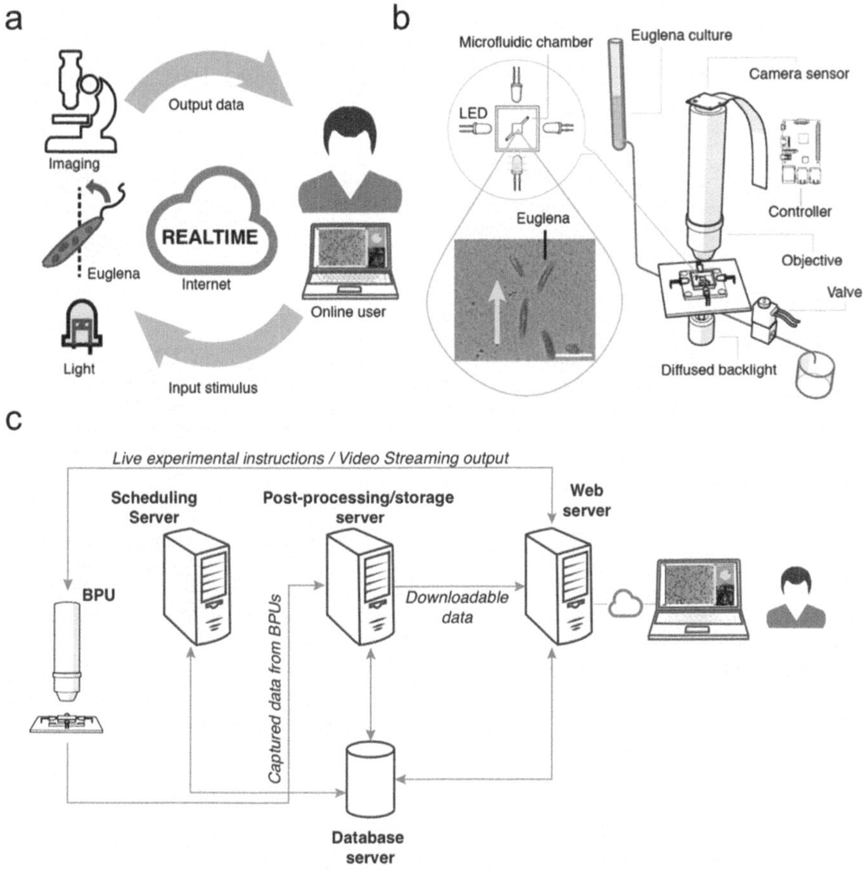

Fig. 11.3 Real-time biology online lab architecture for light-based interaction with photoresponsive microorganisms. (**a**) Online users send light stimuli to *Euglena* and observe the response in real time. (**b**) Back-end hardware. *Euglena* are replenished automatically from an upstream reservoir. Scale bar, 50 μm. (**c**) System architecture. (Images adapted from Hossain et al. (2016))

Euglena responded to these stimuli by swimming away from high light intensities (Barsanti et al. 2012). Many more subtle responses to light are detectable in this system, such as cells spinning around their own axes. *Euglena* cells respond to a change in light conditions on the time scale of seconds, making them particularly attractive for interactive experiments for students and even children.

A cluster of six such BPUs was set up, each of which was controlled by its own microcomputer to control the LEDs, to stream live video, to post-process data, and to communicate with the central server. The task scheduling concepts of high-performance computing. The work of Etsion and Tsafrir (2005) was adopted to design the central server. This server assigns BPUs and remote users according to a non-exclusive group allocation policy, handles distinct BPU types, routes

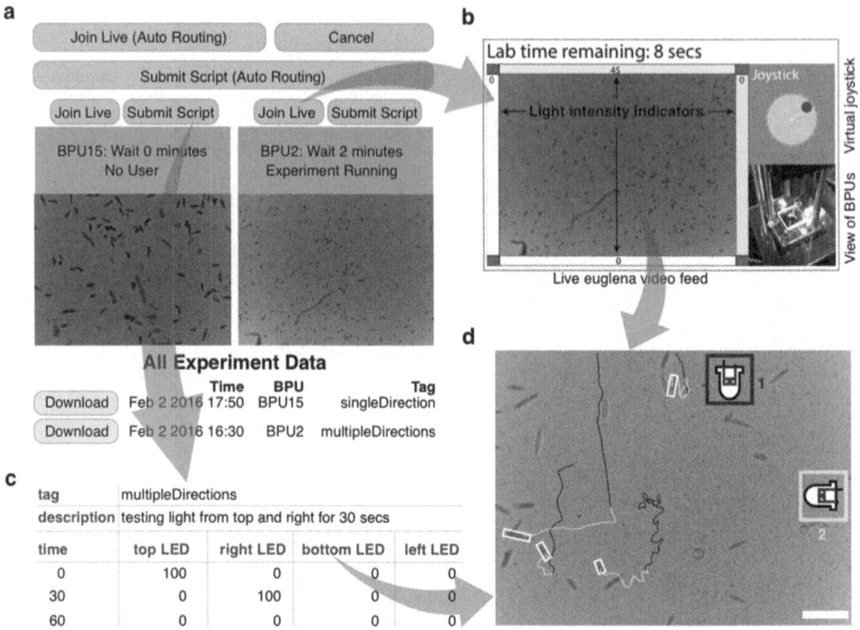

Fig. 11.4 The *Euglena* cloud lab. (**a**) Landing webpage. (**b**) Live mode, with a virtual joystick to control the intensities of the four LEDs. (**c**) Example of preprogrammed instructions for batch mode. (**d**) Example of the cellular response to a light stimulus sequence from top to right (blue, yellow). Scale bar, 100 μm. (Images adapted from Hossain et al. (2016))

experiments to the best-suited BPU, and optimizes wait time through load balancing. A webserver including databases then connects to the user on the client side.

Users perform real-time exploratory as well as preprogrammed experiments that are executed at a later time, and users can download the data for analysis (Fig. 11.4). The user controls the intensity and direction of the two-dimensional light stimulus via a simple online joystick.

A particular affordance of this BPU and organism is the opportunity to implement a low-cost, fully automated cloud lab. *Euglena* cultures are typically stable over long periods (multiple weeks) without much care given appropriate growth medium and light for photosynthesis. The microfluidic chip is connected to an external *Euglena* culture, and hence fresh *Euglena* can be automatically exchanged into the culture via an automated valve whenever needed, typically every few days, yielding a fully automated platform that requires <15 min maintenance once each week per BPU. Another important feature is an automonitoring framework in which each BPU runs an experiment automatically every hour, thereby determining the density of cells as well as their velocity and responsiveness to light. If these parameters are outside the desired regime, then the system attempts to correct itself by autoflushing fresh organisms into the chip. If the system still is not appropriate, then lab personnel are notified to service the BPU. Given that there are multiple BPUs in the cluster, remote users have a very high chance (>99%) of finding at least one functional BPU available at any time; the webserver then also routes users to a "good" BPU. Such automonitoring and self-correcting schemes are essential for delivering cloud labs containing variable, fragile biological materials at low cost and high scale.

11.3.2 Deployment in K-12 Education and Assessment

This platform has been used and tested in multiple middle schools (Hossain et al. 2016). During one study, the cloud lab was projected to the front of a class (27 students, seventh and eighth grade; Fig. 11.5 left), so that all students could do the experiments together. Students then analyzed their data in pairs on their own computer and finally engaged with a virtual modeling environment (see also details in Sect. 11.7, Fig. 11.16) to fit parameters. In another study, 34 students (eighth grade; Fig. 11.5 right) working individually or in pairs used the iLab (Harward et al. 2008) batch interface to submit instructions for light stimuli. The system ran experiments for these students, and the students received movies for analysis. Students chose a diverse set of designs: some explored light intensity, some tuned the light direction, and other students were less systematic.

In both middle-school deployments, it became clear that students liked the activities overall, that the students felt empowered, and that there was a positive educational outcome. While it is possible to introduce the system in one or two class sessions, there should be sufficient time for each student to understand the system and to run multiple experiments. Due to restrictions on class time, firewall restrictions, and the number of available setups, it was not always possible to let each student run as many experiments as desired. In general, it appeared that five to ten experiments lasting 1 min each would be ideal for each pair of students.

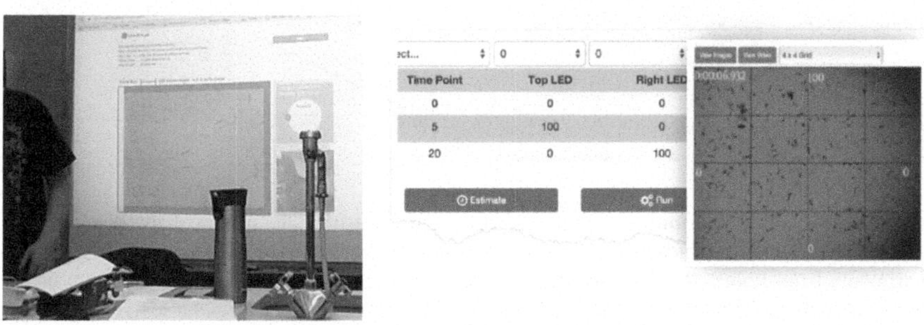

Fig. 11.5 Middle-school deployment of the *Euglena* cloud lab. Left, projection of the setup to the front of the class. Right, *Euglena* cloud lab use through the iLab platform via batch mode. (Images adapted from Hossain et al. (2016))

11.3.3 Deployment in College Education and Assessment

It was also tested whether university students taking a professor-led theory-based biophysics class could successfully carry out experiments and sophisticated quantitative data analysis from home in a self-paced manner on this platform (Fig. 11.6) (Hossain et al. 2016). Over 14 days, ten students, working individually, completed a homework project focusing on concepts regarding microswimmers, diffusion, and low Reynolds number hydrodynamics (Purcell 1997). Using the live mode (Fig. 11.4b), students explored *Euglena* light response behavior and made cells swim along geometric paths (Fig. 11.6a). Students were able to self-discover semiquantitative relationships, e.g., reporting that the "fraction of *Euglena* participating in the directed motion seems to increase as you hold the joystick longer, and depending on the intensity of the light." They performed back-of-the-envelope analyses of *Euglena* size (~50 μm), speed (~50 μm/s), and drag and propulsion forces (~10 pN) (Purcell 1997), experimentally confirming lecture content. Students then analyzed self-generated large-scale batch data (Fig. 11.6b) in MATLAB to test two hypotheses: (1) Do *Euglena* behave like passive Brownian particles? (2) Does the population-averaged velocity differ between dark and light conditions? These results demonstrate that even 1 min experiments provide students with rich experimental data including hundreds of auto-traced cells, supporting sophisticated statistical analysis. The logged data also revealed that students accessed the system at their own convenience at day and at night and that they engaged in different modes of experimentation.

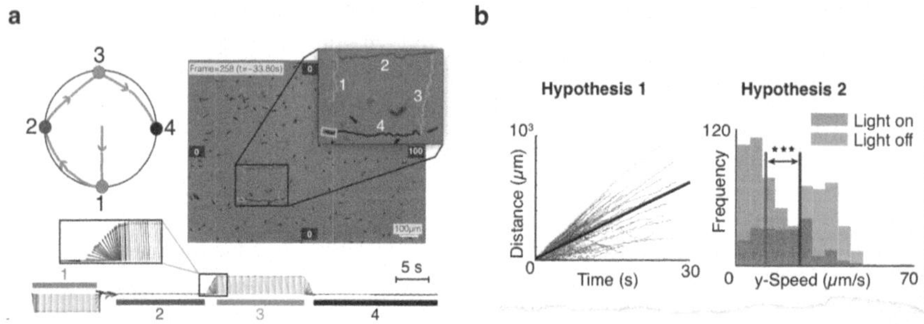

Fig. 11.6 User studies in middle school and college demonstrate the utility of the platform for face-to-face and online education. (**a**) University students performed exploratory joystick-based experiments from home. (**b**) Automatically generated large-scale data (hundreds of cells) using batch mode allowed students to investigate two hypotheses. Left: Are *Euglena* active or passive particles? Right: Does the population-averaged swimming speed depend on light conditions? (Images adapted from Hossain et al. (2016))

11.3.4 Deployment in a MOOC Setting and Assessment

An open online course was developed around this *Euglena* online lab and deployed via the Open edX platform (Hossain et al. 2017). This online course with a remote biology lab engaged >300 remote learners worldwide (Fig. 11.7 left) in the scientific practices of experimentation, modeling, and data analysis to investigate phototaxis of a microorganism. Participants typically took 2–6 h to complete the course during a 1-week period. The course was reoffered weekly, which allowed to respond to user feedback and to iterate on the course content. Overall, >2300 experiments were run by these participants.

In contrast to the deployments on this platform described earlier, here students were completely autonomous in their actions, although the course itself was significantly scaffolded. In addition to the previously offered activities, this online course incorporated data handling via Google Sheets (Fig. 11.7 right), which was more amenable than MATLAB, especially since even middle schools are increasingly using Google Sheets. Online users were asked to execute a final open research project (a voluntary option in order to not overburden the students within a 1-week period). Twenty-one students engaged in their own research projects, for example, exploring how *Euglena*'s response depends on light intensity or duration of the applied light. These students made discoveries that appear in the literature (e.g., how *Euglena* sometimes "freeze" for ~1 s if the light intensity increases very suddenly (Ozasa et al. 2014)). Thus, users on such a platform can engage in realistic scientific inquiry and make genuine discoveries.

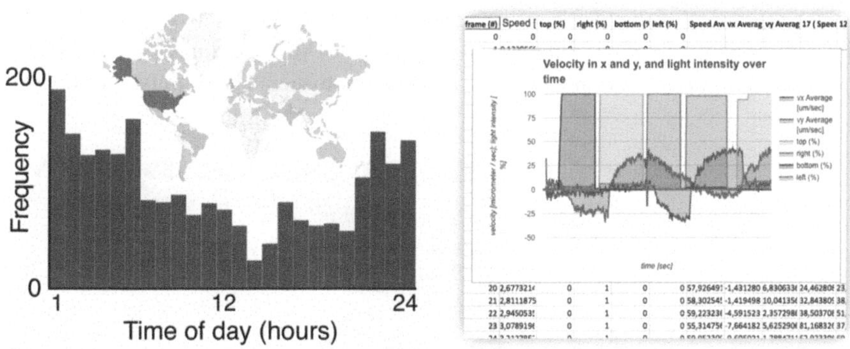

Fig. 11.7 MOOC-scale deployment of the *Euglena* cloud lab. Left: System access pattern. Inset, density of traffic sources by location. Right: Students exported data into Google Sheets, where relationships could be plotted easily. (Images adapted from Hossain et al. (2017))

11.3.5 Reflections and Next Steps

These deployments and user studies have shown that this *Euglena*-based platform enjoys high educational affordances by enabling students to go through the major components of the scientific inquiry paradigm, that the challenge level can be adapted to specific educational needs (middle school to advanced college), that the experimentation and discovery space is sufficiently rich, that the students and teachers overall like these activities, and that the experiment duration and associated costs are such that large-scale deployment seems feasible. Students performed scientific practices and engaged in inquiry-based learning within a short time span without logistical effort, which was impossible before. Our findings also suggest that classrooms could be flipped in the future, with the students operating the lab as homework (Fig. 11.7).

The experimental throughput and cost of such a *Euglena*-based platform scales to massive user numbers and diverse curricular demands, from middle school to college to MOOCs. There are >15 million high-school students in the USA alone, and hundreds of millions of users in developing countries and remote locations could access such platforms via increasingly ubiquitous smartphones (Ozcan 2014). It was estimated that implementing lesson plans in which $\sim$1 million students each run five to ten experiments per year could be achieved with $\sim$250 BPUs, a modest back-end footprint of $\sim 10\ m^2$, and standard 1 Gb/s internet connectivity. Importantly, each experiment would cost less than 1 US cent; hence, cloud lab access for all students in a class (34 students, 10 experiments each) would be less than one live *Euglena* sample ($\sim$US \$7 plus shipping).

Given the generality of the BPU paradigm, other biological specimens, stimuli, and experimental frameworks are amenable to this cloud lab framework. The platform already supports complex investigations of microswimmers and microecologies that are of current interest to the biophysics community (Romensky

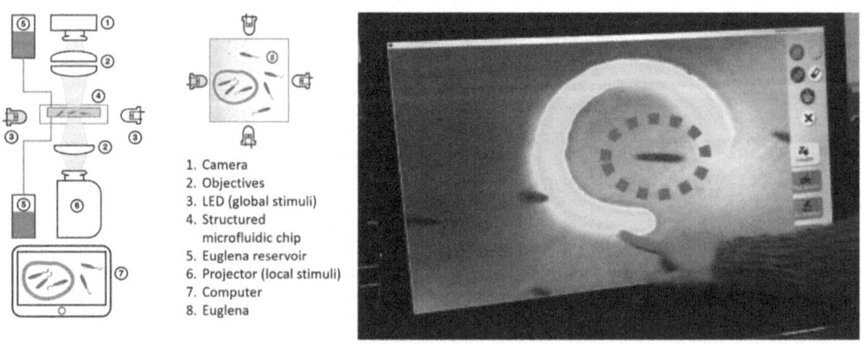

Fig. 11.8 Expanding the *Euglena* cloud lab. Left: Setup to projects light patterns onto a microfluidic chamber housing light-responsive *Euglena* cells. Right: Patterns drawn by user onto a touchscreen are projected onto phototactic *Euglena* that accumulate in colored regions. (Images adapted from Lee et al. (2015))

et al. 2015; Goldstein 2015). Image data are information-rich (e.g., this platform unexpectedly captured cell-division events); combined with a rich stimulus space, many phenomena can be identified and systematically studied. Projector-based setups for *Euglena* (Lee et al. 2015) enable a much richer set of spatiotemporal stimuli, including the use of colors and more complex "mazes" for *Euglena* (Lam et al. 2017). The communication and data protocols are not domain-specific; hence, this platform is expandable beyond *Euglena* and light stimuli to a general class of increasingly automated and low-cost/high-throughput experiments, such as those involving valve switching in microfluidic devices (Balagaddé et al. 2005) and cloud chemistry (Skilton et al. 2015).

The obvious next step is to deploy the current *Euglena*-based platform in more classrooms, particularly in a teacher-autonomous fashion in which the teacher creates the desired lesson plans, and where all students have enough time and opportunity to operate the platform by themselves. The first studies along these lines are currently under way. In order to achieve this goal, the platform must also be scaled up from the current 6 to 20 online microscopes to enable all student pairs in a typical classroom to work concurrently.

It would also be important to synergistically complement these online activities with local hands-on activities, e.g., observing *Euglena* directly through a hands-on microscope. Further, the modeling and simulation aspects should be extended, such as demonstrated previously with the programming language Scratch (Resnick et al. 2009; Kim et al. 2016). Having students build their own interactive microscopes (Cira et al. 2015; Kim et al. 2016), which could even be put online in the long run, and empowering students to self-publish their experiments are other future objectives.

Notably, since these experiments are controlled with a Raspberry Pi, a camera, and a simple electronic board, other experiments outside biology, such as a physics pendulum, could be amenable to investigation. Conversely, given that the back-end experiments are kept sufficiently modular, integration into other cloud lab frameworks is possible (Heradio et al. 2016).

11.4 System 2: Turn-Based Interaction (Slime Mold Chemotaxis, Food)

This biology cloud lab architecture was motivated by the idea of enabling real-time interaction between a remote user and a biological organism in a turn-based manner. This interaction was intended to be visually intuitive, with the back-end hardware being so simple that it could potentially be reproduced by students as a mini-cloud.

11.4.1 Architecture

The architecture of this cloud lab is optimized to allow multiple users to share multiple instruments (BPUs), each of which carries out multiple biology experiments in parallel (Fig. 11.9) (Hossain et al. 2015). In order to enable turn-based interactivity, an underlying batch processing framework was developed. Batch processing is increasingly common in the life sciences, including usage of high-throughput hardware in which each machine typically handles only a specific type of experiment with a specific set of instructions—many experiments can be executed in parallel. Each BPU has its own controller and operates synchronously on its own clock while querying the central database for updated instructions and for sending the biological measurements back to the database. Multiple users access their experiments remotely in an asynchronous manner, sending instructions and checking for experimental updates at arbitrary times. This architecture enables collaborative experimentation and optimal user distribution among BPUs. Users are assigned their experiment slot prior to the experiment run, and they can change the experimental instructions multiple times throughout the run. Hence, this architecture coordinates asynchronous user actions with synchronous equipment cycles to optimally utilize parallelized equipment.

As a specific demonstration, an experimental paradigm was developed for studying the spatiotemporal chemotactic response of the slime mold *Physarum polycephalum* to an oatmeal solution food trail (Fig. 11.9) (Hossain et al. 2015). *Physarum* is a single-celled, multi-nuclei, cytoplasmic organism that forms active and dynamic tube networks to search for food (Alim et al. 2013; Tero et al. 2010; Adamatzky 2010). Food trails of liquid oatmeal that are pipetted onto the agar surface stimulate the growth and behavior of the organism, offering a scientifically interesting as well as educational relevant experimental paradigm with high-dimensional input and output spaces.

For this implementation, liquid handling-imaging robots (BPUs) were developed from Lego Mindstorms (Fig. 11.9) (Hossain et al. 2015; Gerber et al. 2017); each of three such robots could run six experiments in parallel. The organism was housed in an open Petri dish, which was imaged from below and chemically stimulated from above via dispensing droplets of nutrient solution. The BPUs communicated with a Python-based webserver. The front-end user interface (UI) (Fig. 11.10) enabled remote users to select a specific experiment (either one that is currently running or one that had already finished and was archived). The experimental interaction consisted of users graphically determining where and when liquid food stimuli would be administered by the robot onto the Petri dish (Fig. 11.10b, c). Before the experiment, a lab technician prepared fresh Petri dishes with *Physarum*. The BPU then administered new stimuli (as determined by the remote user) and obtained images every 10 min over an experiment that typically lasted 24 or 48 h. At the end of the experiment, all data were archived, and the dishes and *Physarum* were discarded.

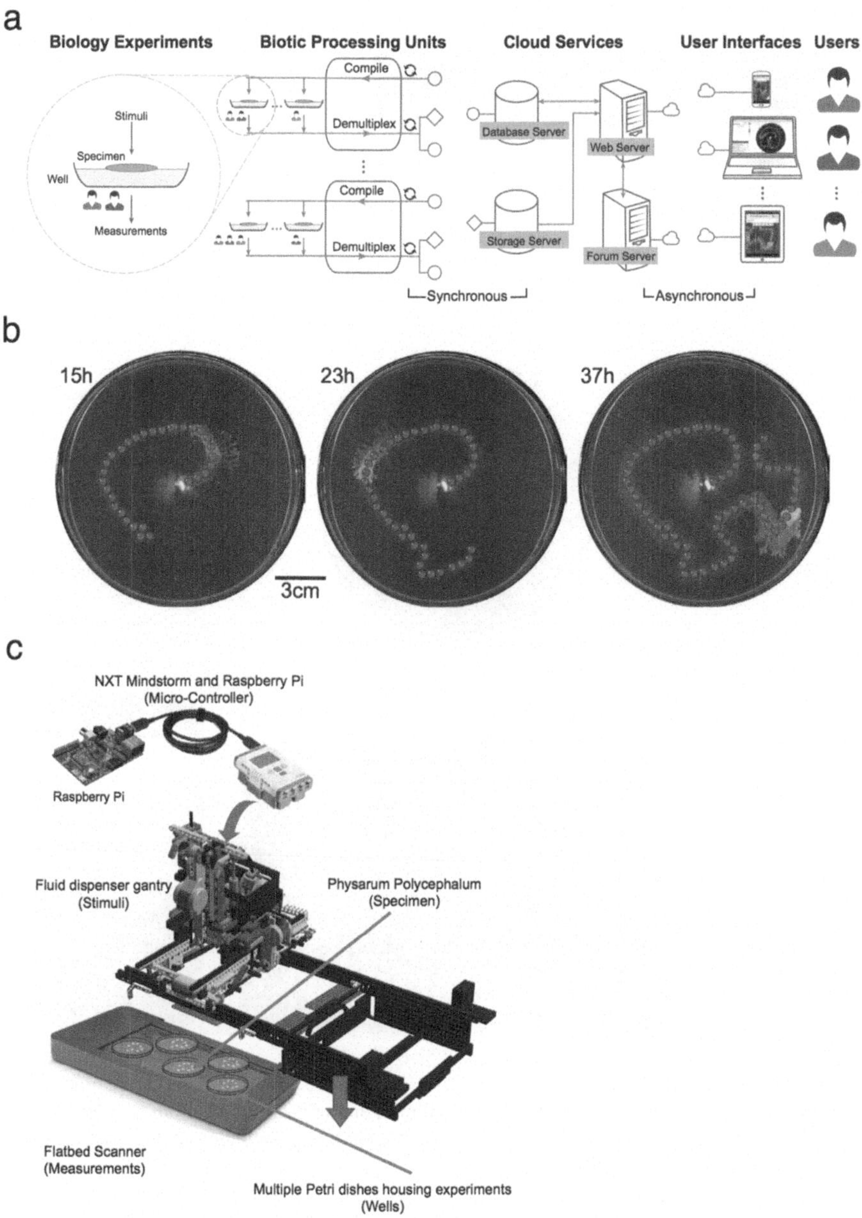

Fig. 11.9 Experiments and hardware for interaction with a slime mold. (**a**) A turn-based cloud lab allows multiple asynchronous users to share equipment for synchronous experimentation. (**b**) The spatiotemporal chemotactic growth response of *Physarum* (yellow) to an oatmeal solution trail (red). (**c**) BPU consisting of a Lego pipetting robot and a flatbed scanner. (Images adapted from Hossain et al. (2015))

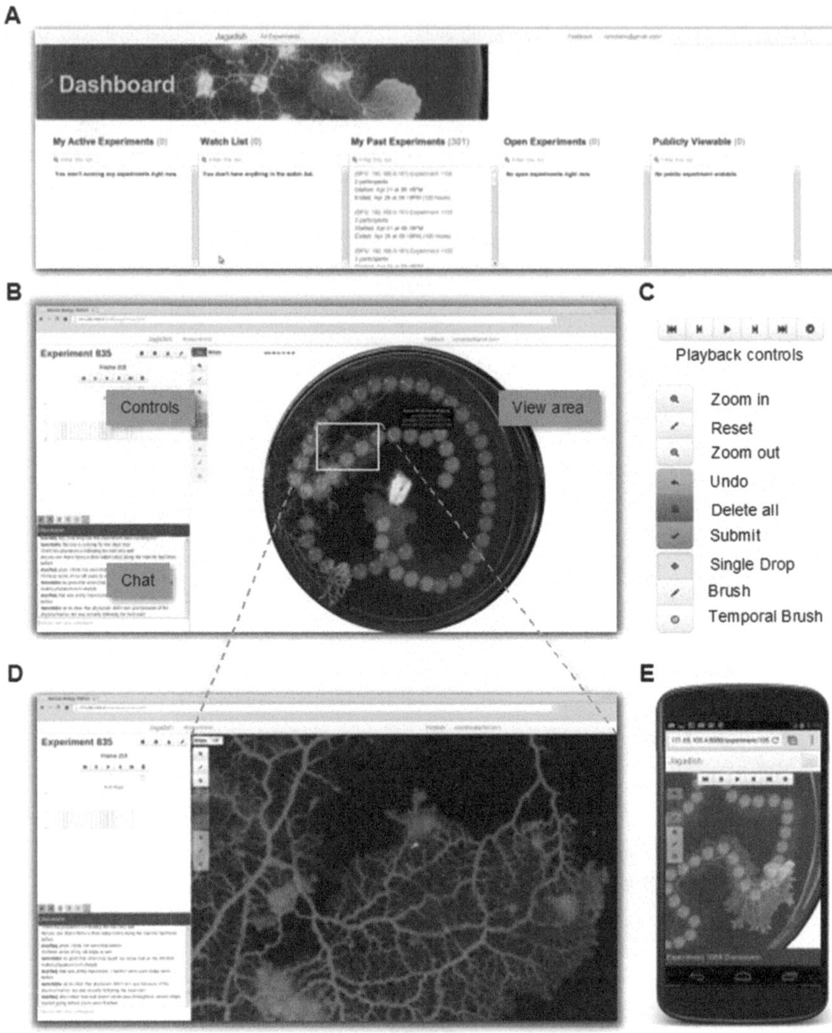

Fig. 11.10 UIs for the *Physarum* cloud lab. Users (**a**) choose a current or past experiments, (**b, c**) dictate the position and timing of chemical stimuli, and (**d**) scroll and zoom through existing image data. (**e**) Experiments are operable from multiple platforms, including smartphones. (Images adapted from Hossain et al. (2015))

11.4.2 Deployment in College Education and Assessment

This *Physarum*-based system was tested in a 10-week lecture-based graduate-level biophysics class at a university. Four students had access to the cloud experimentation platform throughout the course and performed ~20 experiments each. The students typically logged in two to three times during the run of each experiment. Throughout the course, the students progressed from guided work to free exploration to self-motivated experiments that led to a final course project. Students reported interesting observations in their experimental data and then developed a biophysical model (which was the learning objective of the course) to explain various aspects of their experimental data.

This user study revealed that the system was fully stable for the 10-week period. Students self-reported that they liked the online experimentation system and that it was a valuable addition to the otherwise theory-based class. Students also expressed that using real biology experiments (rather than simulations) significantly increased their motivation to explore these biological specimens. The analysis of all user actions revealed differences in student behavior, for example, how much of the previous experimental data was analyzed before conducting the next experiment. Thus, this study highlights the potential of biology cloud labs for educational use as well as for learning analytics (Romero and Ventura 2010).

As a final class project, the students were tasked to engage in the relevant parts of genuine scientific practice: exploration, making observations, formulating hypotheses, designing experiments, and developing a biophysical model. During this project, two students made interesting observations on how the network structure of *Physarum* depends on the overall size of the organism, as well as how the shape of the organisms (number of branches and length distribution of branches) dynamically changes over time (Fig. 11.11). One (nonbiology) student was particularly struck by his observation that organisms with smaller masses had fewer branches (Fig. 11.11i), which seemingly went against the notion of "self-similarity across scales" in fractals that had been discussed earlier in the course. The corresponding phenomena had not been described in the literature. The students then collected more data and iteratively developed and improved a biophysical model capturing these phenomena (Fig. 11.11ii–iv). These students are currently in the process of submitting a full research paper detailing their biophysical model (Cira and Riedel-Kruse 2017). Thus, biology cloud labs also show potential to be used for genuine research, enabling students to perform deep inquiry over the internet.

11.4.3 Reflections, Lessons Learned, and Next Steps

A major challenge of this particular implementation was the back-end logistics supporting these experiments. For example, approximately 30–60 min was always required for a lab technician to prepare all fresh biological material before starting the next round of experiments. Further, the overall footprint of the platform (a server rack filled with three BPUs executing 18 experiments in parallel over 48 h) does not easily scale to very large numbers of remote users in multiple institutions. Nonetheless, this platform would be beneficial as a local cloud lab within a school, for example, and where the chosen Lego Mindstorms implementation would allow students to build and modify their own instruments (Danahy et al. 2014; Gerber et al. 2017). Swapping out the hardware (BPUs) for more professional, higher-throughput

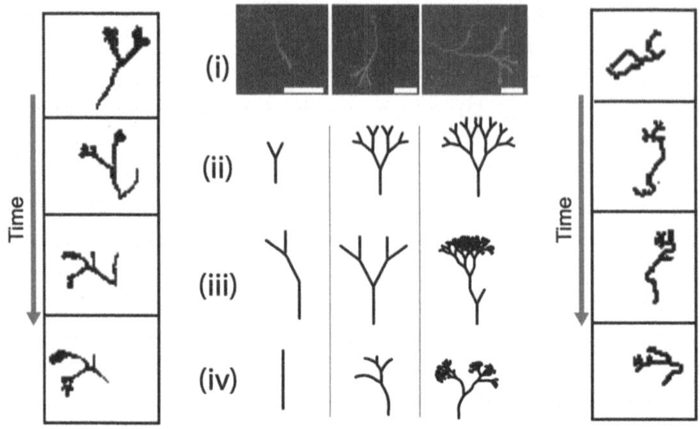

Fig. 11.11 Experimentation (left) and modeling (right) by graduate students using this cloud lab. Middle: Iterative modeling motivated by the cloud lab. (*i*) Image data reveal size-dependent network structures. Scale bars, 3 mm. (*ii*) Static symmetric bifurcation model. (*iii*) Static random bifurcation model. (*iv*) Dynamic growth-retraction model. Right: Time sequence of the model in (*iv*), compare to the sequence at left (Images adapted from Hossain et al. 2015)

instruments would allow the execution of different types of experiments at much larger scale and lower cost.

The cost and back-end logistics per experiment were significant and could be estimated as follows (assuming this type of system would be deployed at much larger scale). A BPU costs approximately US $500 in parts, and each BPU houses six experiments, with three runs per week for 1 year, leading to 50*3*6 = ~1000 experiments. Additional costs include lab personnel to maintain *Physarum* colonies, prepare the agar plates, and prepare each experimental run, which is estimated at 2 h per week or ~US $100 in labor cost/week. Lab space would cost ~US $10/experiment. This estimate does not include the initial development of the platform.

Overall, this system successfully supported students in their learning activities, enabled the introduction of an experimental component into a theory-based class, and empowered nonbiologist students to carry out biology experiments in depth, effectively lowering access barriers.

11.5 System 3: Programmed Batch (Bacterial Growth, Antibiotics)

The computational cloud and time-sharing paradigms (Fox 2011) have recently inspired the development and deployment of biology cloud experimentation labs for research, such as commercial platforms that can execute experiments semiautomatically (Transcriptic, Emerald Cloud Lab) (Sia and Owens 2015; Riedel-Kruse 2017; Transcriptics 2015; Hayden 2004). These commercial platforms provide a large suite of instruments and reagents, with the ultimate vision of enabling the automated execution of any academic or industrial experiment, in particular when it comes to molecular and cell biology.

Here we also point to "Aquarium" (Klavins 2017), an academic "Laboratory Operating System" where online users can choose from prespecified laboratory protocols and experimental workflows via an online web interface; these experiments are then executed (in large part by manual labor, i.e., undergraduate technicians that can be easily trained), enabling students and researchers to build, e.g., transgenic strains online.

These platforms are different from the ones described earlier in this article as they are not interactive during the experiment. Instead, all experimental instructions must be provided before the start of the experiment. The experiments have turnaround times on the scale of days or more. None of these labs had been used for education previously; hence, a collaboration with the company Transcriptic was initiated to test one of these platforms with students. These investigations are described in detail in Riedel-Kruse (2017).

11.5.1 Architecture

Transcriptic has been developing a "Workcell" platform in which a robot shuttles biological specimens, for example, contained in 96-well plates, between experimental instruments such as liquid-handling robots, imaging devices, and incubators (Fig. 11.12). Experiments can be fully programmed in Python. This overall framework is under constant development; for example, some experimental steps are still executed by hand but will eventually be automated. Hence, the general vision and roadmap to full and flexible cloud experiment automation is clear.

11.5.2 Deployment in College Education and Assessment

To test the platform's educational potential, bacterial growth under the influence of antibiotics was chosen given its relevance for college level classes the relative ease of implementation on the existing platform (Riedel-Kruse 2017). Initially, bacteria were loaded into 96-well plates. Each student could claim 6 wells on that plate, allowing 15 students to work at once, leaving a few wells as controls. Prior to the start of the experiment, students defined the concentration of antibiotics in each well, leading to different growth rates over $\sim$8 h (Fig. 11.13). Every 20 min, the amount of bacteria in a dish was measured via spectrophotometry. This cloud lab did not allow for interactive experimentation, i.e., users were not able to add antibiotics throughout the experiment, but this could be added to this framework in the future.

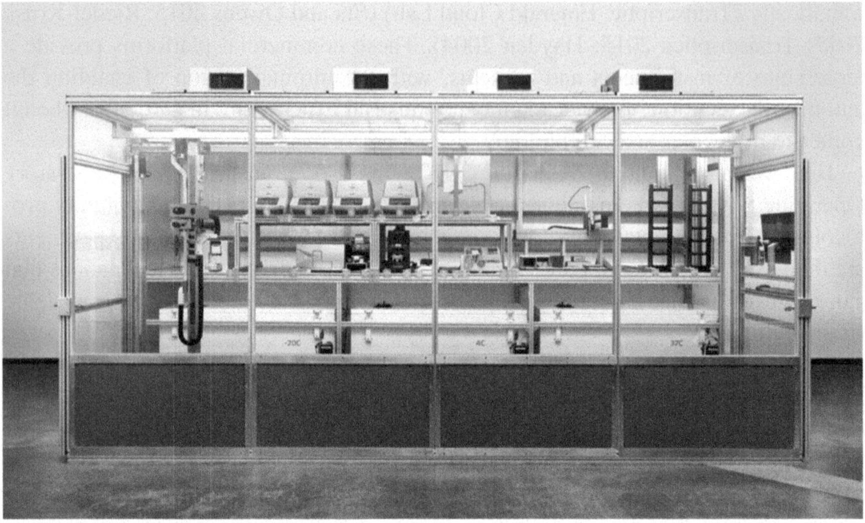

Fig. 11.12 Transcriptic Workcell, a custom robotic cellular and molecular biology laboratory. (Image adapted from https://www.transcriptic.com/)

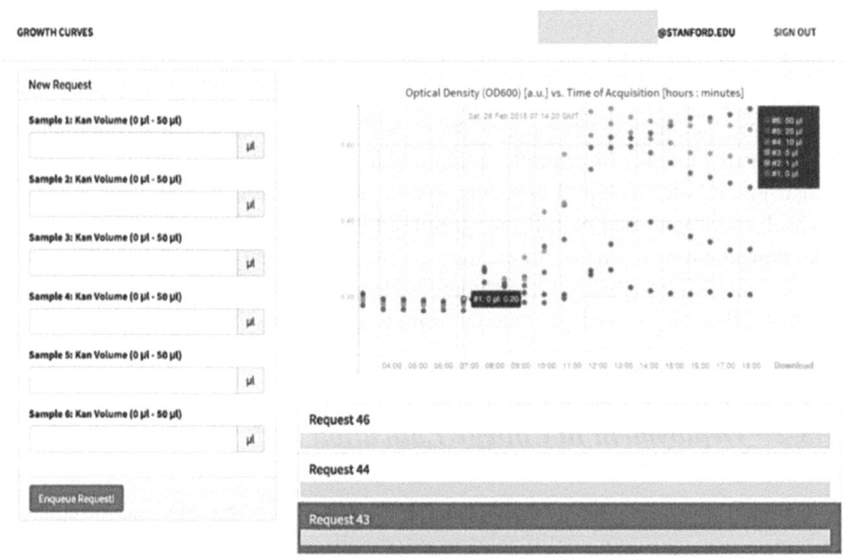

Fig. 11.13 Customized Transcriptic UI for educational deployment. Left: Six antibiotic amounts can be submitted. Right: Batch of data at the end of the experiment (time and optical density appear on the x and y axes, respectively). (Images adapted from Riedel-Kruse (2017))

A user study was run where 13 students could run 6 wells over 6 successive rounds of experimentation (36 experiments in total). It was found that one to two

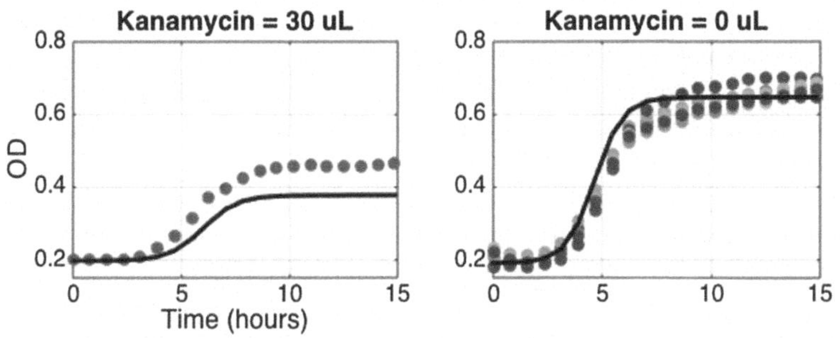

Fig. 11.14 Student-generated data (dots) and fit to models (solid lines) of two antibiotic concentrations. (Images adapted from Riedel-Kruse (2017))

rounds were needed to familiarize the students with the platform. Students then chose one of many growth models that had been discussed previously in class and fit their data to that model in MATLAB. Unfortunately, mutations arose in some of the bacteria over successive experiments, and there were some other technical issues, leading to not fully consistent results between experiments that made it more challenging for the students to interpret their data. These technical challenges were due to the early stage of platform development at the time but were later rectified by the company with updated equipment and protocols.

11.5.3 Reflections and Next Steps

Overall, the activities were successful and allowed students to design and run their own experiments, collect their data, and post-process the data. The exploration space available to the students was rather low dimensional compared to the cloud labs discussed earlier. The student's only option was to choose one of six antibiotic concentrations (they explored only a one-dimensional space). The concentration was determined before the experiment started—there was no interactivity during the experimental run. Further, the data consisted of zero-dimensional measurements points, which is much less information rich than the image data in the other cloud labs, rendering the experience more abstract than a classic experiment.

Based on Transcriptic's business model, the cost of these of experiments was ~US $70 per 96-well plate. This cost depends on the experiment type and is likely to decrease in the future given advancements in the technology. This platform would also offer a much higher variety of experimental types due to the diverse set of instruments in the Workcell. The 96-well experiment suggests that high-throughput experiments could be virtually partitioned between many users. The challenges encountered due to early-phase technology also suggested opportunities for students to confront the real messiness of biological experiments in an educational context, at the same time point to the importance of stability and robustness of cloud labs, which is a particular challenge for biological systems.

In conclusion, commercial cloud labs are on the rise and afford interesting opportunities to run very complex life-science experiments in the cloud. The cost per student, in the range of US $30 (6 wells, 6 experiments), is not cheap, but in a reasonable range for lab classes. Given existing technologies (robotics, range of instruments, underlying scripting language), future opportunities should open up to pool the experiments of many more students (e.g., using 1536-well plates), enabling higher-dimensional experimentation (e.g., choosing from multiple antibiotics) as well as interactivity (allowing users to change experimental parameters such as antibiotic concentration throughout the run based on current experimental results). To achieve these goals, corresponding UIs must be developed that also account for educational requirements. From the company's perspective, enough students must use such a system to support the initial investment. Alternatively, the educational value of this platform could begin with graduate-level research, and as these platforms become less expensive and user friendly, their usage could expand even into K-12 education.

11.6 System 4: Augmented Batch (RNA Folding, Nucleotide Sequence)

A fourth set of biology cloud labs relates to citizen science games such as Foldit (Cooper et al. 2010) and EteRNA (Lee et al. 2014). Both games enabled tens of thousands of online players to participate in research by solving puzzles regarding protein and RNA folding, respectively. EteRNA is special in that it additionally provided experimental feedback for a smaller subset of (more expert) players. Foldit was primarily virtual but has also been used in projects where player suggestions were experimentally tested (Eiben et al. 2012). It should be noted that for these projects, the experimental work at the back-end was not fully automated. Instead, there was significant hands-on work by lab scientists—which does not matter much from the remote user's perspective.

11.6.1 Architecture

The EteRNA platform revolves around the scientific question of how a particular RNA folds into its secondary structure based on its primary RNA structure (its nucleotide sequence). Here, the online user is provided with a gamified graphical UI displaying an RNA strand with the four nucleotides marked by letter (CGAU) and color (Fig. 11.15). The user can change individual nucleotides and then instruct the computer to calculate the currently predicted folding structure due to the base

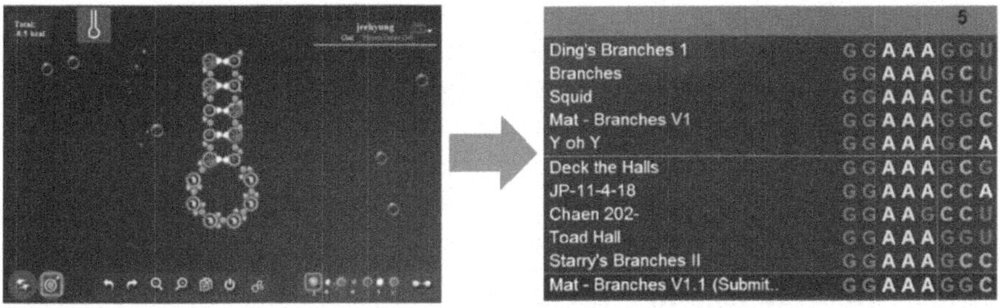

Fig. 11.15 EteRNA lets players explore the relationship between RNA sequence and secondary structure. (Image adapted from Lee et al. (2014))

pairing based on lowest energy considerations. Users are guided through a number of puzzles of increasing difficulty. After users have gained sufficient understanding of the platform and the RNA folding features by solving ~120 puzzles, they are allowed to participate in the lab.

Lab participation means that users are asked to come up with nucleotide sequence that will fold into a desired target shape and which will then be tested experimentally. For any given lab puzzle, each user makes her suggestion, and then based on certain criteria, the most promising designs are chosen to be tested experimentally. At the time of the first major deployment (Lee et al. 2014), the experimental throughput was only eight designs per week and carried out in significant part by manual labor; throughput has improved since then to ~10,000 designs per month through parallelized microfluidic chip technology (Bida and Das 2012; Seetin et al. 2014) but still operated in part manually. The particular RNA sequences are synthesized, and the nucleotide base pairings are assessed via single-nucleotide-resolution chemical reactivity measurements (SHAPE) (Lee et al. 2014). The experimental results (secondary structure and base pairing) are conveyed back to the user with single-nucleotide resolution through an in-game visualization that is similar to the original design interface (Fig. 11.15).

11.6.2 Citizen Science (and Educational) Deployment

During the first major deployment, >37,000 players experimented with this platform (Lee et al. 2014). During each weekly round, players submitted their proposals for designs to be tested, of which eight were chosen to be synthesized and tested experimentally. Over successive iterations, the designs suggested by the best players eventually consistently outperformed current RNA prediction algorithms, enabling the development of better prediction algorithms that took into account the new rules that players had identified. This development demonstrates the power of citizen science, in particular when coupled with experimental feedback.

So far, EteRNA has not been formally used nor assessed for formal education, to our knowledge. However, Nova Labs (http://www.pbs.org/wgbh/nova/) created a version of the simulation to support students learning about RNA in middle and high schools, and we are aware of many K-12 and college instructors who use EteRNA with their students.

11.6.3 Reflections and Next Steps

The costs for any experiment are due to labor and reagents, which for EteRNA were estimated to be ~US $2.000 per month or ~US $0.2 per design. The experimental design space of the platform is arguably very large since each position of the RNA strand of given length N can be any of four nucleotides (4^N, where N is already given for a given lab, but could modified). The virtual part of the platform has been deployed in various educational settings (unpublished results and personal communication by Prof. Das).

It is interesting to note that "designing an experiment" through a highly augmented user interface (including game elements) rather than operating or instructing a scientific instrument directly. These citizen science projects (EteRNA and Foldit) clearly demonstrate a very different avenue by which non-experts can be empowered to do experiments and participate in research. The success of these projects certainly motivates more fully automated and versatile cloud lab designs in the future.

11.7 Virtual Biology Cloud Labs and Interactive Simulations/Models

Although it is not the primary goal of this article to extensively address virtual biology labs, we would like to mention a few approaches (Fig. 11.16). (1) For the *Euglena* online lab discussed in Sect. 11.3, a modeling environment had been co-deployed (Hossain et al. 2016) that primarily allows students to perform parameter fitting. (2) Modeling environments like Scratch (Resnick et al. 2009) have been explored to enable students to program simple models of cellular behavior (Kim et al. 2016). (3) Other groups have developed gamified laboratories (such as Labster) that fully animate all lab components (Bonde et al. 2014). A number of other life-science simulations exist, for example, as part of the PhET project (Wieman et al. 2008). We note that both real and virtual labs have their distinct advantages and limitations, e.g., less cost at scale, and "running every experiment within seconds" in virtual labs versus the potential for novel discoveries or changes in student motivation in a real labs. Ideally, both approaches would be deployed synergistically (de Jong et al. 2013).

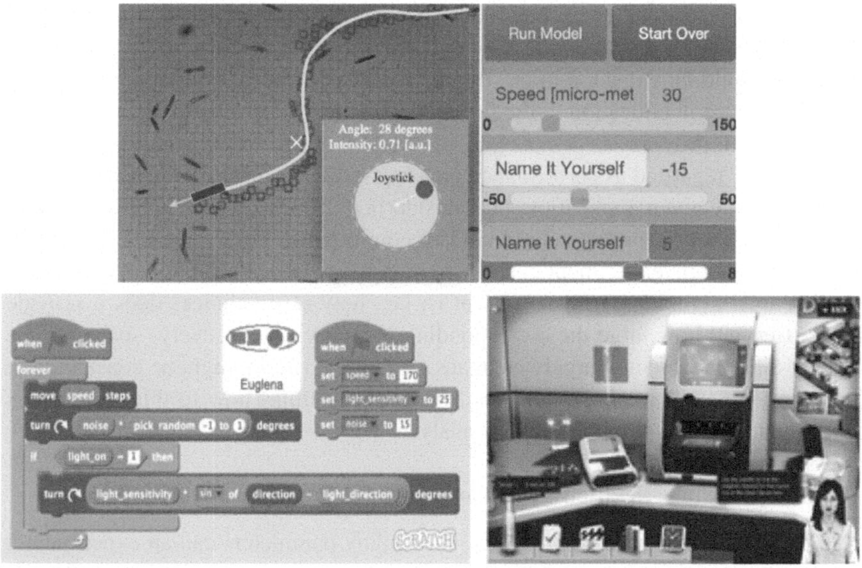

Fig. 11.16 Examples of virtual biology labs and simulations. Top: Middle-school students modeling *Euglena* phototaxis. Bottom left: Modeling *Euglena* behavior in Scratch. Bottom right: Gamified laboratory (Labster). (Images adapted from Bonde et al. (2014))

11.8 Lessons: Performance Metrics for "Interesting" Cloud Labs

Given that now a small but distinct and versatile number of biology cloud labs exist, we are in the fortunate situation to be able to compare these labs (Table 11.1) and to extract overarching themes and generalizable rules. A significant portion of these insights would also apply to cloud labs outside the life sciences.

The four cloud-lab architectures we presented are all rather different from a conceptual point of view. (1) The *Euglena* lab allows a single user to "own" an instrument for a short period of time. The experiment is real-time interactive, and biological responses are apparent within seconds. The low-cost and short experiment duration make this approach scalable. Parallelization is achieved by deploying multiple BPUs. (2) The *Physarum* lab shows how multiple experiments that belong to different users are executed in parallel on a single instrument. These experiments are interactive—the user makes changes while the experiment is running, but there is a delay of a few minutes. The individual user does not have direct control of this instrument. (3) The Transcriptic experiment is parallelized but not interactive during the run at all. Given that the Workcell moves samples between instruments automatically, it allows for essentially infinitely complex experiments (all other platforms described here are confined to a specific experiment type). (4) EteRNA is also parallelized, noninteractive, and provides feedback on the scale of weeks; the UI abstracts the process of experimental design into a game, although it eventually becomes scientific research for the dedicated user. Each of these platforms could be extended in the future. We expect that all four approaches will have their place in education in the future, depending on particular applications.

Table 11.1 provides a comparison of the four labs, many features of which could also be regarded as performance metrics. Which of these features are relevant depends on the given application, but considering all of them in the planning phase of developing a new biology cloud lab is recommended. For example, one can ask how much a single experiment costs, how many experiments can be run per unit of time, or how complex each experiment is, i.e., how many choices does it provide to the user and how large the corresponding response (or discovery) space is. The numbers in this table are largely estimate, and other criteria might be considered in the future—overall we hope that this overview illustrates how to think about this performance issue. A more detailed analysis will also be published in the future (Hossain et al. 2017).

In the following, we discuss these and other considerations in more detail:

1. *The size of the exploration space.* How many parameters can an experimenter change and how many distinct experiments can be run? For example, we found that the *Euglena* experiments allowed for changes in light intensity and direction on a 10-ms time scale. For simplicity, assuming a 1-min experiment with 0.1 s resolution, four LEDs with ten intensity settings would generate an astronomical amount of measurements $(10^4)^{600}$ distinct experimental sequences. In contrast, the Transcriptic experiments allowed users to choose among ~10 antibiotic concentrations before the start of the experiment.
2. *The size of the discovery space.* Despite a large exploration space, many experimental designs can have equivalent outcomes. Hence, we need to ask how many experimental outcomes ("discoveries") are possible. For example, if *Euglena* essentially reorients to the light stimulus on a 10-s time scale, once the user has discovered that behavior, he is done. In reality, image data may capture a much richer range of responses to different light intensities, yielding a larger and more complex discovery space. For example, *Euglena* displays many subtle changes in behavior due to light stimuli; it even changes its shape due to strong light. In general, it will be challenging to quantify the discovery space completely, as exhaustive exploration is usually not practical. Choices should be made whether to provide the user with the information-rich, raw data (e.g., raw movies of *Euglena* behavior) versus processed and information-reduced data (e.g., a table with positional information for the cells).
3. *Combined exploration/discovery space.* Combining both the input and output possibilities for an experiment would quantify how "interesting" a platform is. For example, in the MOOC deployment (Hossain et al. 2017), online learners were asked to propose their own investigations. Approximately, 10 dependent and 10 independent variables were identified, implying ~100 experimental investigations that could be executed, which for an educational setting is certainly very interesting. We also refer to the paradigm of low floor, high

ceiling, wide walls (Resnick and Silverman 2005), which describes how easy it is to engage in a particular platform, but also how diverse and complex an investigation can become. For example, in order to enable "authentic inquiry" in the classroom (Chinn and Malhotra 2002), this amount of freedom is desired.

4. *Biological variability as a challenge.* For all four architectures, biological variability requires significant consideration. On the one hand, keeping the user experience and experimental outcomes consistent (within defined bounds) is important, and not always easy. Significant layers of automonitoring, self-correction, and controls can be deployed, as, for example, in the *Euglena* lab (and which could still be improved). We therefore also recommend that each instrument provides the user with quality measures for their experiments (such controls are good practice for experimentation in general). Even when a system has been stable for months, biology may still hold surprises, such as mutations.
5. *Biological variability as an opportunity.* On the other hand, this variability provides interesting phenomena that are absent from more deterministic physics labs, potentially making the experiments more interesting and "lifelike." Variability and noise in biological systems are active areas of research (Elowitz et al. 2002). Students must be prepared to encounter variability, which can be exploited to great educational effect. In either case, this variability needs to be delivered within the proper educational context.
6. *The benefits of "living" labs.* Why not just simulate? Unlike pure simulations, live biological organisms are highly complex systems with emergent, unpredictable properties, providing educational opportunities for novel discoveries. Student feedback captured this aspect, for example, with "It was fun to play around with real organisms ... " (Hossain et al. 2016). Implementations should also aim to harness this unpredictability and to convey it to the user. We note that simulations and experimentation should be used in synergy. Cloud labs should also utilize and feature the "realisms," e.g., information-rich image data (as in the *Physarum* lab) may be more enticing and interesting than a processed graph of single-point measurements (as in the bacterial growth lab). The entire instrumentation architecture should be conveyed so that the user can understand it and feel agency. Real labs also provide students to be confronted with experimental noise, anomalous data, and even failed experiments. Interacting with living matter can also provoke ethical discourse that does not arise from simulation alone, which again could be put to good use in an educational context (Cira et al. 2015; Harvey et al. 2014).
7. *Potential safety and ethical issues.* The safety aspect should be considered. Although remote experimentation can generally be considered much safer than hands-on experimentation, remote users could potentially cause harm, e.g., by hacking the system or generating dangerous biological material. Compared to other science disciplines, biological experiments are special given that particular biological organisms or types of experiment may fall under ethical

regulations, e.g., animal rights. Additionally, users and bystanders may voice their own concerns about what kinds of experiments with a given organism are in good taste. Ethical analysis of biotic games (Harvey et al. 2014) has provided some general guidelines and insights, even though the value of an "educational experiment" is likely considered of higher priority than "game play."

8. *Time for executing one experiment (and time of one user interaction).* A lower time limit exists for any given biological process based on how fast the experiment can be executed. For example, the effect of antibiotics on bacterial populations can only be detected after hours, while *Euglena* responses due to light are apparent within seconds. Note that these time limits can be pushed to some extent by using instruments with higher spatial or temporal resolution, e.g., the effect of antibiotics on bacterial cells can be observed within <1 h when imaging individual cells directly (Kong et al. 2012).
9. *Time required for experiment reset.* The biological and instrument downtime between experiments needs to be considered. In the case of the *Euglena* lab, after the light stimuli have been turned off, the *Euglena* go back to their prior state on the scale of 15–60 s. In the case of the *Physarum* lab, all biological material must be replenished for each new experimental run. One should also discriminate between the time it takes for the biological material to reset and some other downtime of the instrument, such as processing the last rounds of image data. Additional downtime results from instrument and biology maintenance.
10. *Experimental throughput.* Many of these issues ultimately point to how many experiments can be run in a given time. Experimental throughput can be increased by shortening the duration of a given experiment (including the necessary downtime between experiments), by parallelizing the number of experiments on a given instrument (BPU), by increasing the number of instruments in a cluster, and by replicating these BPU clusters at different sites.
11. *Number of experiments and time required for user familiarization with the platform.* When deploying the experiment, students generally should do five to ten experiments on a platform to allow for familiarization with the experiment, to explore, and to collect controlled data. Even if the platform allows many experiments in parallel, the student should have the opportunity for iterative, successive operations. Hence, it should be determined how many experiments are minimally required to promote a meaningful experience on the platform. If the experiments are expensive, then training experiences (as in EteRNA) could lower the load on the physical cloud lab.
12. *Logistics and automonitoring.* A major challenge compared to other online platforms (such as remote operation of physics experiments) is the maintenance required for biological material. Accordingly, choices must be made at the start of the project to account for these logistics and—if possible—to make use of specimens and hardware that minimize these challenges. The implementation of automation and automonitoring is crucial and has been significantly achieved with the *Euglena* cloud lab. Working with biological material and protocols that show consistent behavior is important. Back-up instruments should also be

considered. The increasing advances and cost reductions in biotechnological automation (including high-throughput machines) will enable increasingly more robust platforms, including commercial ones, in the future.

13. *Cost per experiment (and the business model).* The total cost of any individual experiment (or a set of experiments that would provide a coherent investigation) should be considered. These costs are driven by consumables, maintenance, and service, as well as by the initial development efforts. The numbers from Transcriptic may be the most reliable information currently available, as they have an underlying business model. These numbers can be in flux, and as technology improves and the concept becomes more common, costs will certainly go down. Generally, a benchmark for comparison is the cost of a similar experiment in a conventional, hands-on setting. As a relevant comparison, shipment of living organisms from a school supply company starts at ~US $20 for ~20 students; consumables for more sophisticated biology experiments can easily go well above US $100.
14. *Complexity and investment for initial setup, flexibility for future adaptions, and ease of replication by others.* Significant effort is required to initially set up a platform. In the simplest case, remote screen sharing is a very fast and easy way to enable remote biology experimentation and to prototype a platform. How easily this platform can be operated and modified for other experimental types is another important consideration. In that sense, the Workcell approach is inherently much more flexible. Open source code and building instructions could foster incentives for others to replicate and innovate. We also expect that general operation and data handling standards for cloud labs will emerge.

Conclusions on Specifications: The importance of each of these properties depends on the application. Providing a fast and simple biology experiment to millions of high-school students (e.g., to enable students to experience *Euglena* phototaxis) has a very different requirement than providing a community of hundreds of scientists with a platform to execute complex, versatile, and highly precise experiments (as a company like Transcriptic may seek to achieve).

11.9 Next Steps and Open Research Questions

The educational effectiveness of the presented platforms has been demonstrated to varying extents, but undoubtedly all platforms deserve more assessment through wider student and teacher participation as well as controlled studies. The individual studies for these cloud labs indicate learning gains, especially as self-reported by students, but more systematic pre- and posttests are warranted. The *Euglena* and *Physarum* cloud labs enabled students to perform biology experiments at a level of sophistication that is absent from presential and online education. Empowering students to perform inquiry-based practices in which they construct knowledge like professional scientists is a major achievement of these biology cloud labs.

We see several important avenues for future research and development on these biology cloud labs.

1. Refining and testing course content for specific learner groups on the existing platforms, such as middle- and high-school biology students, ultimately paving the way for usage by several thousands or millions of students
2. Including other relevant scientific practices, such as collaborative teamwork and model building
3. Having participants implement more complex projects all the way to geographically distributed team projects
4. Utilizing these platforms for deeper analysis using learning analytics to aid instructors and educational researchers
5. Extending these platforms to other experiment types (other stimuli, other organisms, and distinct types of microbiology experiments)
6. Updating BPU performance protocols, for example, to achieve automatic LED brightness adjustment for optimal negative phototaxis and feedback to users on "current instrument quality"
7. Exploring optimal UIs and scripting languages for online experiments and data handling
8. Open standards that enable easier setup and modifications of biological cloud labs
9. Ultimately bringing experts from different areas closer together, especially bioengineers, software engineers, researchers into human-computer interactions, and educators

11.10 Conclusions

We have presented four distinct user interaction modes and architectures for biology cloud labs and discussed the importance of biological variability, automonitoring, and domain-specific BPUs. These best practices could also be implemented for cloud labs in other engineering disciplines (Heradio et al. 2016) in which labs are currently mostly oriented toward single users and single devices. We primarily focused on educational use cases, but emerging high-end research cloud labs were included in our discussion. We conclude that the requirements and approaches for such goals are very different but will be complementary and synergistic in the long run.

Biological cloud (or remote) labs are particularly challenging, as the long-term robustness of the biological matter requires additional manual work or automation to provide a consistent experience. On the other hand, complex biological phenomena—especially when utilizing information-rich image data—constitute very rich discovery spaces. Enabling students to perform inquiry-based practices in which students construct knowledge like professional scientists is another major achievement of these biology cloud labs. Given that at least four biology cloud labs have been successfully tested with hundreds of students (tens of thousands for EteRNA), we are confident that biology cloud labs are feasible and useful.

Deploying biology cloud labs in education could help solve significant educational challenges and simultaneously provide economies of scale to help these technologies to mature. With the more than 15 million high-school students in the USA alone as well as the rise of MOOCs, education will be an important driver of the development of biology cloud labs. Curricula are usually offered repeatedly, allowing technologies to be developed iteratively and tested with many users. These cloud labs provide a cost-effective and practical means to implement inquiry-based learning and ultimately to accomplish the visions of NGSS (Bybee 2013) and the National Research Council (2012).

Critically, the data-logging capabilities of any cloud system constitute a unique opportunity to delve into how learners explore biological experiments that typically have a great deal of natural variability. Learning outcomes can be thoroughly investigated, e.g., in the context of bifocal modeling (Blikstein et al. 2012), when real experiments are juxtaposed with modeling. Several studies have indicated that combining reality (with variability and noise) and modeling (typically clean data) is more beneficial for learning content than either strategy in isolation (Heradio et al. 2016; Blikstein et al. 2012). Moreover, there are indications that students typically explore experiments in novel ways when data are shared with other students. These affordances could be further investigated in a quantifiable manner by implementing data-sharing capabilities in the application layer of our cloud lab.

Biological cloud labs open many interesting avenues for human-computer interactions but require carefully designed UIs. Some experimentation styles benefit from visual programming, while others may benefit from textual descriptions. Biotic games (Riedel-Kruse et al. 2011; Lee et al. 2015; Kim et al. 2016) are another interesting application of BPUs that may foster interest in biology in a playful manner through gamification. Excitingly, games could be implemented in the top UI layer of biology cloud labs. Phone-based internet-of-things instrumentation and diagnostics provide another paradigm for distributed instrumentation (Ozcan 2014).

In summary, we foresee that the iterative development and deployment of biology cloud labs in educational contexts will greatly benefit education and facilitate the development of individual BPU clusters (one experiment type at a time). Certainly, not all experiments can be carried out this way, but with cloud labs, a significant portion of standard biological experiments can likely be implemented much more cost-effectively and without complex logistics. Hence, an investigator (student or professional scientist) can concentrate on experimental design and data analysis, rather than on logistics and the hands-on skills required of a successful experimenter. We expect that there will be synergy between educational and scientific research performed in centralized facilities. We look forward to a future that fosters interdisciplinary participation and democratization of biology experimentation through cloud labs.

Acknowledgment This work was supported by Stanford BioX IIP, Stanford VPOL, Stanford MediaX, and NSF Cyberlearning (NSF 1324753). We would like to thank from R. Das, B. Keep, and R. Waters. Note: This review article summarizes information from various sources; in case where this information was from our own lab's previous publications we often edited from the original text, and while we cited those original sources, we did not put those text pieces in quotes.

References

Adamatzky, A. (2010). Routing Physarum with repellents. *The European Physical Journal E, Soft Matter, 31*(4), 403–410.

Alim, K., Amselem, G., Peaudecerf, F., Brenner, M. P., & Pringle, A. (2013). Random network peristalsis in Physarum polycephalum organizes fluid flows across an individual. *Proceedings of the National Academy of Sciences, 110*(30), 13306–13311.

Balagaddé, F. K., You, L., Hansen, C. L., Arnold, F. H., & Quake, S. R. (2005). Long-term monitoring of bacteria undergoing programmed population control in a microchemostat. *Science, 309*(57310), 137–140.

Barsanti, L., Evangelista, V., Passarelli, V., Frassanito, A. M., & Gualtieri, P. (2012). Fundamental questions and concepts about photoreception and the case of Euglena gracilis. *Integrative Biology, 4*(1), 22–36.

Bida, J. P., & Das, R. (2012). Squaring theory with practice in RNA design. *Current Opinion in Structural Biology, 22*(4), 457–466.

Blikstein, P., Fuhrmann, T., & Greene, D. (2012). Bifocal modeling: Mixing real and virtual labs for advanced science learning. In *Proceedings of the 11th International Conference on Interaction Design and Children,* Germany, June 12–15.

Bonde, M. T., Makransky, G., Wandall, J., Larsen, M. V., Morsing, M., Jarmer, H., & Sommer, M. O. A. (2014). Improving biotech education through gamified laboratory simulations. *Nature Biotechnology, 32*(7), 694–697.

Bybee, R. W. (2013). The next generation science standards and the life sciences. *Science & Children, 50*(6), 7–14.

Chinn, C. A., & Malhotra, B. A. (2002). Epistemologically authentic inquiry in schools: A theoretical framework for evaluating inquiry tasks. *Science Education, 86*(2), 175–218.

Cira, N. J., Chung, A. M., Denisin, A. K., Rensi, S., Sanchez, G. N., Quake, S. R., & Riedel-Kruse, I. H. (2015). A biotic game design project for integrated life science and engineering education. *PLoS Biology, 13*(3), e1002110–e1002118.

Cira, N., Ma, E., & Riedel-Kruse, I. H. (2017). Network dynamics of slime molds, *in preparation*.

Cooper, S., Khatib, F., Treuille, A., Barbero, J., Lee, J., Beenen, M., Leaver-Fay, A., Baker, D., Popović, Z., & Players, F. (2010). Predicting protein structures with a multiplayer online game. *Nature, 466*(7307), 756–760.

Danahy, E., Wang, E., Brockman, J., Carberry, A., Shapiro, B., & Rogers, C. B. (2014). LEGO-based robotics in higher education: 15 years of student creativity. *International Journal of Advanced Robotic Systems, 11*(27), 1–15.

de Jong, T., Linn, M. C., & Zacharia, Z. C. (2013). Physical and virtual laboratories in science and engineering education. *Science, 340*(6130), 305–308.

Eiben, C. B., Siegel, J. B., Bale, J. B., Cooper, S., Khatib, F., Shen, B. W., Players, F., Stoddard, B. L., Popović, Z., & Baker, D. (2012). Increased Diels-Alderase activity through backbone remodeling guided by Foldit players. *Nature Biotechnology, 30*(2), 190–192.

Elowitz, M. B., Levine, A. J., Siggia, E. D., & Swain, P. S. (2002). Stochastic gene expression in a single cell. *Science, 297*(5584), 1183–1186.

Etsion, Y., & Tsafrir, D. (2005). A short survey of commercial cluster batch schedulers. *The Hebrew University of Jerusalem, 13*, 44221.

Fox, A. (2011). Cloud computing – What's in it for me as a scientist? *Science, 331*(6016), 406–407.

Gerber, L. C., Kim, H., & Riedel-Kruse, I. H. (2016). Interactive biotechnology: Design rules for integrating biological matter into digital games. In *Proceedings of the First International Joint Conference of DiGRA and FDG*. Dundee, Scotland, UK.

Gerber, L. C., Calasanz-Kaiser, A., Hyman, L., Voitiuk, K., Patil, U., & Riedel-Kruse, I. H. (2017). Liquid-handling Lego robots and experiments for STEM education and research. *PLoS Biology, 15*(3), e2001413–e2001419.

Goldstein, R. E. (2015). Green algae as model organisms for biological fluid dynamics. *Annual Review of Fluid Mechanics, 47*(1), 343–375.

Harvey, H., Havard, M., Magnus, D., Cho, M. K., & Riedel-Kruse, I. H. (2014). Innocent fun or 'Microslavery'? *Hastings Center Report, 44*(6), 38–46.

Harward, V. J., del Alamo, J. A., Lerman, S. R., Bailey, P. H., Carpenter, J., DeLong, K., Felknor, C., Hardison, J., Harrison, B., Jabbour, I., Long, P. D., Mao, T., Naamani, L., Northridge, J., Schulz, M., Talavera, Varadharajan, C., Wang, S., Yehia, K., Zbib, R., & Zych, D. (2008). The iLab shared architecture: A web services infrastructure to build communities of internet accessible laboratories. *Proceedings of the IEEE, 96*(6), 931–950.

Hayden, E. C. (2004). The automated lab. *Nature, 516*(7529), 131–132.

Heradio, R., de la Torre, L., Galan, D., Cabrerizo, F. J., Herrera-Viedma, E., & Dormido, S. (2016). Virtual and remote labs in education: A bibliometric analysis. *Computers & Education, 98*(C), 14–38.

Hossain, Z., & Riedel-Kruse, I. H. (2017). Concept and characterization of biotic processing units (BPUs), *in preparation*.

Hossain, Z., Blikstein, P., Riedel-Kruse, I. H., Jin, X., Bumbacher, E., Chung, A. M., Koo, S., Shapiro, J. D., Truong, C. Y., Choi, S., & Orloff, N. D. (2015). Interactive cloud experimentation for biology. In *Presented at the the 33rd Annual ACM Conference* (pp. 3681–3690). New York

Hossain, Z., Bumbacher, E., Chung, A. M., Kim, H., Litton, C., Pradhan, S., Walter, A., Jona, K., Blikstein, P., & Riedel-Kruse, I. H. (2016). A real-time interactive, scalable biology cloud experimentation platform. *Nature Biotechnology, 34*(12), 1293–1298.

Hossain, Z., Bumbacher, E., Blikstein, P., & Riedel-Kruse, I. H. (2017). Authentic science inquiry learning at scale enabled by an interactive biology cloud experimentation lab. In *Proceedings of the first ACM conference on learning scale conference*. ACM.

Kim, H., Gerber, L. C., Chiu, D., Lee, S. A., Cira, N. J., Xia, S. Y., & Riedel-Kruse, I. H. (2016). LudusScope: Accessible interactive smartphone microscopy for life-science education. *PLoS One, 11*(10), e0162602–e0162616.

Klavins, E. (2017). The aquarium project. Retrieved from http://klavinslab.org/aquarium.html

Kong, F., Yuan, L., Zheng, Y. F., & Chen, W. (2012). Automatic liquid handling for life science: A critical review of the current state of the art. *Journal Laboratory Automation, 17*(3), 169–185.

Lam, A. T., Samuel-Gama, K. G., Griffin, J., Loeun, M., Gerber, L. C., Hossain, Z., Cira, N. J., Lee, S. A., & Riedel-Kruse, I. H. (2017). Device and programming abstractions for spatiotemporal control of active micro-particle swarms. *Lab on a Chip, 15*(10), 351.

Lee, J., Kladwang, W., Lee, M., Cantu, D., Azizyan, M., Kim, H., Limpaecher, A., Yoon, S., Treuille, A., Das, R., & EteRNA Participants. (2014). RNA design rules from a massive open laboratory. *Proceedings of the National Academy of Sciences, 111*(6), 2122–2127.

Lee, S. A., Bumbacher, E., Chung, A. Cira, N. J., Walker, B., Park, J. Y., Starr, B., Blikstein, P., & Riedel-Kruse, I. H. (2015). Trap it!. Presented at the *33rd Annual ACM Conference* (pp. 2593–2602). New York.

Melin, J., & Quake, S. R. (2007). Microfluidic large-scale integration: The evolution of design rules for biological automation. *Annual Review of Biophysics and Biomolecular Structure, 36*, 213–231.

National Academies Press. (2012). *Committee on a Conceptual Framework for New K-12 Science Education Standards, Board on Science Education, National Research Council, A Framework for K-12 Science Education: Practices, crosscutting concepts, and core ideas*. Washington, DC: National Academies Press.

Ozasa, K., Lee, J., Song, S., & Maeda, M. (2014). Transient freezing behavior in photophobic responses of Euglena gracilis investigated in a microfluidic device. *Plant and Cell Physiology, 55*(10), 1704–1712.

Ozcan, A. (2014). Mobile phones democratize and cultivate next-generation imaging, diagnostics and measurement tools. *Lab on a Chip, 14*(17), 3187–3188.

Pedaste, M., Mäeots, M., Siiman, L. A., de Jong, T., van Riesen, S. A. N., Kamp, E. T., Manoli, C. C., Zacharia, Z. C., & Tsourlidaki, E. (2015). Phases of inquiry-based learning: Definitions and the inquiry cycle. *Educational Research Review, 14*(C), 47–61.

Purcell, E. (1997). Life at low Reynolds number. *American Journal of Physics, 45*(3), 11.

Resnick, M., & Silverman, B. (2005). Some reflections on designing construction kits for kids. In *Wearable computers, the IEEE international symposium on* (pp. 1–6), March.

Resnick, M., Silverman, B., Kafai, Y., Maloney, J., Monroy-Hernández, A., Rusk, N., Eastmond, E., Brennan, K., Millner, A., Rosenbaum, E., & Silver, J. (2009). Scratch: Programming for all. *Communications of the ACM, 52*(11), 60.

Riedel-Kruse, I. H. (2017). Incorporating a commercial biology cloud lab into online education. In *Proceedings of the International Conference on Remote Engineering and Virtual Instrumentation*. New York.

Riedel-Kruse, I. H., Chung, A. M., Dura, B., Hamilton, A. L., & Lee, B. C. (2011). Design, engineering and utility of biotic games. *Lab on a Chip, 11*(1), 4–22.

Romensky, M., Scholz, D., & Lobaskin, V. (2015). Hysteretic dynamics of active particles in a periodic orienting field. *Journal of the Royal Society Interface, 12*(108), 20150015–20150015.

Romero, C., & Ventura, S. (2010). Educational data mining: A review of the state of the art. *IEEE Transactions Systems, Man, and Cybernetics Part C, 40*(6), 601–618.

Sauter, M., Uttal, D. H., Rapp, D. N., Downing, M., & Jona, K. (2013). Getting real: The authenticity of remote labs and simulations for science learning. *Distance Education, 34*(1), 37–47.

Seetin, M. G., Kladwang, W., Bida, J. P., & Das, R. (2014). Massively parallel RNA chemical mapping with a reduced bias MAP-seq protocol. *Methods in Molecular Biology, 1086*(6), 95–117.

Sia, S. K., & Owens, M. P. (2015). Share and share alike. *Nature Biotechnology, 33*(12), 1224–1228.

Skilton, R. A., Bourne, R. A., Amara, Z., Horvath, R., Jin, J., Scully, M. J., Streng, E., Tang, S. L. Y., Summers, P. A., Wang, J., Pérez, E., Asfaw, N., Aydos, G. L. P., Dupont, J., Comak, G., George, M. W., & Poliakoff, M. (2015). Remote-controlled experiments with cloud chemistry. *Nature Chemistry, 7*(1), 1–5.

States, N. L. (2013). *Next generation science standards: For states, by states*. Washington, DC: The National Academies Press.

Tero, A., Takagi, S., Saigusa, T., Ito, K., Bebber, D. P., Fricker, M. D., Yumiki, K., Kobayashi, R., & Nakagaki, T. (2010). Rules for biologically inspired adaptive network design. *Science, 327*(5964), 439–442.

Transcriptics. (2015). *Discovery biology on demand*. Retrieved from https://www.transcriptic.com

Wellington, J. (2007). America's lab report: Investigations in high school science. *Science Education, 91*(3), 514–515.

Whitesides, G. M. (2006). The origins and the future of microfluidics. *Nature, 442*(7101), 368–373.

Wieman, C. E., Adams, W. K., & Perkins, K. K. (2008). PHYSICS: PhET: Simulations that enhance learning. *Science, 322*(5902), 682–683.

Zahid Hossain was a PhD student in Computer Science and Bioengineering at Stanford (2011–2017) researching Realtime Interactive Biology Cloud Labs and Computer Graphics; he now works at Meta Co on Augmented Reality.

Ingmar H. Riedel-Kruse is an Assistant Professor of Bioengineering at Stanford University, USA.

Chapter 12
A CPS Integration Platform as a Framework for Generic Remote Labs in Automation Engineering

Reinhard Langmann

Abstract The paper describes generic or convertible remote labs, which use the new paradigms from cyber-physical systems (CPS) and service-based automation for their creation. As a result, any remote lab can be created, operated and utilised simply and quickly via web-based engineering. This was realised through a freely accessible CPS integration platform using already available industrial process interfaces and specially developed remote lab services.

Keywords Remote lab · Automation engineering · Generic remote labs · CPS integration platform · Remote lab services · Service-based automation

12.1 Introduction

Remote labs for technical systems are in use throughout the world in the field of education and are also utilised for engineering and technical training. However, studies reveal that remote labs are usually complex proprietary systems requiring a high level of development expense, which practically cannot be recovered via commercial use (Seiler 2013). The reasons for this include:

- The lack of reproducibility and interoperability.
- Restricted usability by outside parties.
- The absence of standards for interfaces and components.

The almost exclusive concentration of the development work for remote labs at universities and technical universities without noteworthy corporate investment leads to makeshift setup of very different remote lab structures, which always merely serve the specific interests of the developing technical university. Without

R. Langmann (✉)
Duesseldorf University of Applied Sciences, Düsseldorf, Germany
e-mail: langmann@ccad.eu

M. E. Auer et al. (eds.), *Cyber-Physical Laboratories in Engineering and Science Education*, https://doi.org/10.1007/978-3-319-76935-6_12

a sufficiently high level of corporate participation, sustainable progress cannot be reasonably expected for the standardisation of remote labs.

A new approach aims to improve at least the reproducibility and usability by outside parties by using new methods from the world of web technologies, e.g. by introducing web services for laboratory functionalities (Caminero et al. 2014) or by adding techniques like mashups, gadgets and OpenSocial widgets (Tawfik et al. 2014). But again, in this case practically, no industrial standards are considered. The results again are very specific and complex web-based solutions that can hardly be reproduced by outside parties. The solutions remain in the domain of the academic developers. A standardisation for the new, specific LaaS (Lab as a Service) interfaces is also being ignored, owing to the limited distribution and lack of corporate/industrial support.

If we look at the problem in relation to the field of automation technology, the following situation results in practical laboratory tasks in automation technology that are mostly carried out on normal industrial systems, i.e. the systems that are used for industry-specific production automation. These are used as laboratory devices in training directly or as components, sometimes also in modified form. This also affects the corresponding control and operating software, which also have to be compatible (e.g. WinCC, LabView, WebFactory). In principle, these systems, in particular process control (SCADA) and operating (HMI) systems, could also be used for calling up remote labs. At present, most of these systems permit remote access via the Internet and contain all elements (visualisation, webcam, access to process data, text fields, etc.) for calling up a remote lab. However, the following factors currently impede their use for educational remote labs:

- High licencing costs and limited number of remote clients.
- Access and administration functions for anonymous users (learners) are not available.
- Unlimited web capability is so far only provided for a few systems.

Thus, specific and proprietary remote labs have been developed – in automation technology too – at considerable expense, and they only have a limited application range, as discussed above. Examples of this can be found in Langmann (2011) and Coquard et al. (2008).

12.2 Motivation

The introduction of the new paradigms in production automation since 2012, such as Industry 4.0 (National Academy of Science and Engineering 2013), Industrial Internet (Evans and Annunziata 2012) and Cyber-Physical Production Systems (CPPS) (VDI/VDE-GMA 2013), however, gives automation technology increasingly better opportunities to utilise the new web-based technology and networked industrial technologies directly and without extensive changes for educational purposes in remote labs. An essential prerequisite for this is represented by the future structure of a CPS-based automation system as shown in Fig. 12.1.

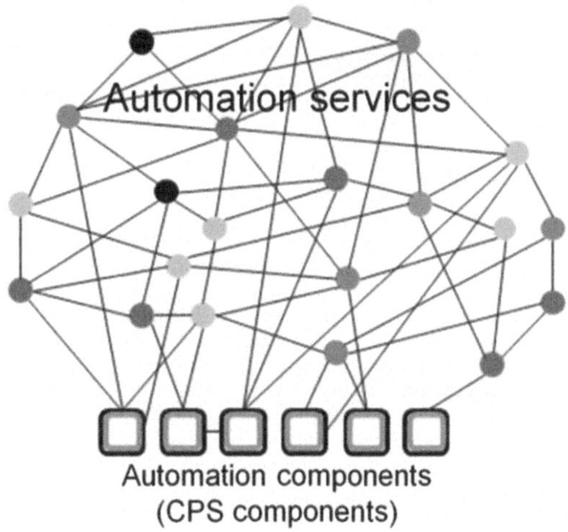

Fig. 12.1 CPS-based automation [8]

According to VDI/VDE-GMA (2013), a CPS is understood as a system which is characterised by the linking of real (physical) objects and processes with information-processing (virtual) objects as well as processes via open, partially global information networks, which are connected to one another at all times. Automation devices, which are provided with local intelligence and are also being networked globally via the IP network, for example, can be considered CPS components. In a further variant, these CPS components can be structured as Industry 4.0 (I40) components, taking into account the Reference Architectural Model Industry 4.0 (RAMI) (Hankel 2016). An example of PLC control as an I40 component is given in Langmann and Rojas-Pena (2016).

If we view an automation-related remote lab from the perspective of Fig. 12.1, and if such a remote lab also simultaneously structures remote lab services and remote lab device components that are built as CPS, we could also use CPS-based automation systems, as shown in Fig. 12.1, for the setup of remote labs. Currently, however, there is still a problem: The new automation structure shown in Fig. 12.1 is a vision for the future. At present there are no application-ready systems on the market which implement CPS-based and service-oriented automation. The situation shown in Fig. 12.2 is based on the use of industry standard application systems for the construction of remote labs.

Until now, there was a big gap between the application system's characteristics in automation technology and the requirements of remote labs for global networking (Internet networking). With the new CPS-based application systems, however, this gap is closing and the industry standard application systems can increasingly be used for the construction of didactic remote labs. Currently, the first application-ready systems, which implement CPS-based and service-oriented automation, are being offered or prepared for the market. Such systems are currently platforms and portals, which are known as IoT (Internet of Things) platforms or IIoT (Industrial

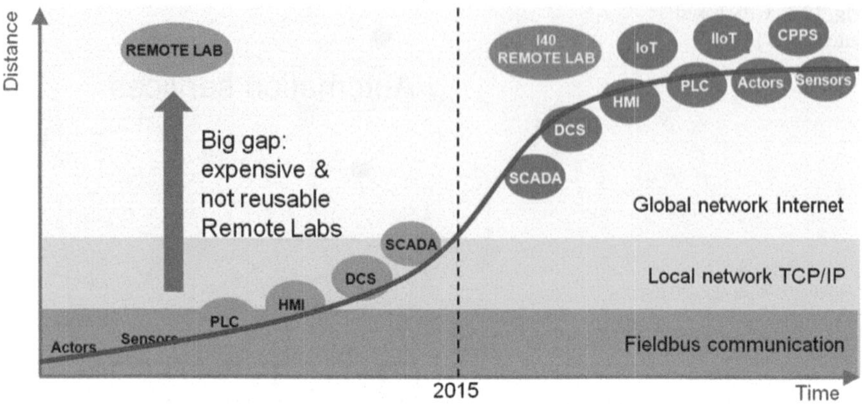

Fig. 12.2 Remote labs in the light of Industry 4.0 (I40)

Internet of Things) platforms. These platforms already have essential properties that are required to build remote labs:

- Consistent implementation with web technologies and seamless integration into the web world.
- The development of functional systems is carried out in the web browser without additional software or plug-ins.
- In part, functions can already be used as services.

Using these systems to build remote labs, the new remote labs can also be referred to as I40 remote labs.

12.3 State of Technology

In the introduction to this article, we have already looked at the general state of technology regarding remote labs. In particular, available solutions from the IoT world need to be examined for their applicability in constructing automation technology remote labs. Table 12.1 lists some of the currently known IoT systems with which remote labs could technically be created.

The IoT platforms listed in Table 12.1 can be regarded scientifically as a CPS integration platform. The systems do not represent an application system, but provide only a framework that can be used to construct and operate functional systems in a web browser for various industries. In this way, these platforms basically realise the structure shown in Fig. 12.1 (at least in large parts) and could also be used as a framework for the projection of remote labs. However, there are a number of problems regarding automation technology, especially for remote labs:

Table 12.1 Features of selected known IoT platforms

IoT platform	Configuration and operability	Device connection/integration	Expandability through third parties
ThingWorx	Elaborate, training required	Possible, but elaborate	Yes, through ThingWorx marketplace
Bolt IoT	Complex, intended for developers	Only via special Bolt Card	Yes, through HTML and JavaScript
Carriots	Elaborate, training required	Possible, but elaborate via REST API	Yes, via special APIs
Ayla	Elaborate, training required	Only via special "design kit"	No
Everything IoT	Elaborate, programming required	Possible, via Raspberry Pi	Yes, via SDKs
Grove Streams	Elaborate, training required	Possible, via Raspberry Pi	Yes, via special APIs
Bosch IIoT	So far only product announcement (as of: October 2016)		
Zatar	Elaborate, training required	Possible, via special tool kit	Yes, via REST APIs

- The integration of devices with automation interfaces (e.g. OPC, Modbus) is very elaborate or not possible at all.
- The available services (functions) are not sufficient for automation purposes. An expansion with own services is complex or not possible.
- All systems are fee-based according to the cloud computing model, i.e. monthly charges and/or a consumption-dependent service charge are payable. This makes it difficult and expensive for public education institutions.
- The platforms themselves are only operated by the providers themselves. The operation of a user's own platform is not intended.

The tested systems, as seen in Table 12.1, are also not intended for training purposes, but for industrial applications. Up to now it is not described in the state-of-the-art the use of such IoT platforms for the establishment of remote labs for education.

One exception is the IoT Platform IoTool (2016). The system, also referred to as *smartphone as an IoT gateway*, is offered for academic training and teaching and can be used for the design and operation of remote experiments for various sensor applications. However, this is a proprietary system with non-open device interfaces, in which the utilisation costs for the complete system with the configuration of their own functional systems are relatively high. IoTool can also not be expanded with a user's own services or functionalities.

Between 2011 and 2014, however, the prototype of a web-based platform was created, which can integrate worldwide distributed services also with worldwide distributed CPS components (device components) into a functional system. This was within the framework of the R&D project "Architecture and Interfaces of a

Web-Oriented Automation System (WOAS)" and involved the participation of ten corporations (Langmann and Meyer 2013). In doing so, it basically does not matter whether this functional system involves a system for the automation of a technical process or a remote lab. The developed platform is called *IIoT platform WOAS* (short, WOAS portal) and works in principle as a CPS integration platform and thus as a framework for the generation of new user-specific functional systems (Langmann and Jacques 2016). The WOAS portal is available and free of charge for educational purposes. Upon request, educational institutions can receive access for administrators (can configure new remote labs) and to users (can use new remote labs). In addition, it is possible for an educational institution to operate its own WOAS portal. The portal is easily portable within a software container.

This paper will now describe the use of the WOAS portal as a framework for creating user-specific remote labs for automation technology training and will demonstrate this with the help of three examples.

12.4 Concept

For CPS-based automation, the automation devices (controls, sensors, actuators, etc.) can be considered CPS components after adding corresponding interfaces to the IP network. Accordingly (Langmann 2014a, b), this results in the CPS structures as shown in Fig. 12.3 for different automation devices (AD). A virtual device (VD), which maps the real device in the virtual world, functions as an interface from the CPS components to services distributed in the network.

The virtual device is used to map the process data of an automation device via event-based channels onto web- or Internet-suitable objects in a uniform manner and makes this available in a web browser. Via integrated protocol or device gateways, any industrial interfaces (OPC, Modbus TCP, etc.) can therefore be made available in the IP network. For the data transfer between a VD and an automation device, a

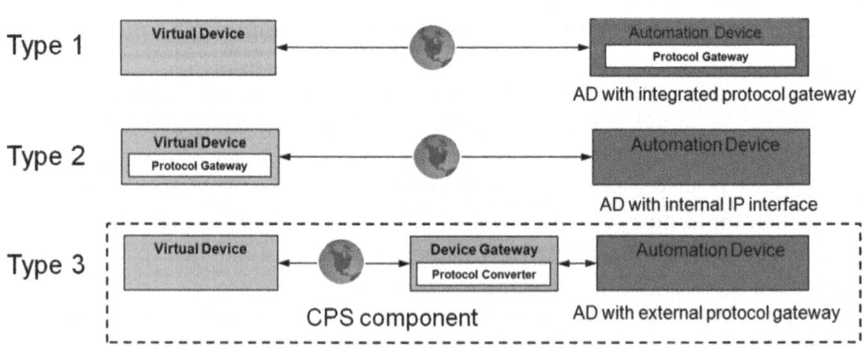

Fig. 12.3 Automation devices or lab devices as CPS components

simple and pragmatic *WOAS device protocol* was developed, which utilises JSON (JavaScript Object Notation) as a data format for the process data transfer via HTTP.

The protocol or device gateways can be characterised as follows:

- The *WOAS Protocol Gateway* (WPG) realises the implementation of the WOAS device protocol into the required industrial protocol for communication with the real automation device. The WPG can be a direct component of the VD directly as a JavaScript protocol converter or it is implemented on an external hardware (Industrial PC, embedded server).
- The *WOAS Device Gateway* (WDG) is the external hardware that allows the implementation of one or more WPGs. If an Industrial PC is used, other required device/protocol drivers (e.g. OPC servers) can also be installed on the WDG.

Three types of virtual devices are distinguished for adapting to different device conditions and industrial protocols, which differ in particular in the manner of their integration into the WOAS portal:

- *VD type 1*: These virtual devices are already implemented as a plug-in in the WOAS server (internal VD) and directly convert kernel calls to the required device protocol information. This is advantageous if the automation device already has a suitable IP interface (e.g. OPC UA) or an internal protocol gateway.
- *VD type 2*: The virtual device includes a separate protocol gateway in JavaScript, which translates the WOAS protocol commands into the required industry protocol/device interface.
- *VD type 3*: The virtual device communicates with an external device gateway, which generally is connected to the VD via WebSockets and through which WOAS device protocol messages are exchanged. Other protocols can also be used for communication.

Figure 12.3 illustrates the different types of connection of automation devices via the various VD types in the WOAS portal. The VD and protocol gateway or device gateway together form a *web connector* for the automation/lab device as a CPS interface. As the laboratory devices are also suitable for an automation technology remote lab, the CPS components can also be used identically for remote labs, as seen in Fig. 12.3.

Various automation functions, such as HMI elements, real-time trend and cloud-distributed webcam services, have already been developed in the WOAS project as automation services (Fig. 12.1). These services can also be used directly for remote labs. Moreover, if specific remote lab services are required (e.g. remote desktop service), in that case such services can be integrated without any problem into the IIoT platform WOAS considering the WOAS development guidelines. Competence Centre Automation Dusseldorf (2014a, b) describes the WOAS guidelines for the development of a new virtual device, and new services are described in detail. The creation of a new remote lab therefore comprises the following steps:

- Installation of a suitable web connector in the laboratory device to make it CPS compatible.

Table 12.2 Types of automation services

Service type	Visibility in the web browser	Process data input/output	Functionality	Example
Type 1	No	Input OR (input and output)	Implementation of an algorithm	Measurement value processing, data archiving, sequence control
Type 2	Yes	Input	Visualisation of process data	Real-time trend,
Type 3	Yes	Output OR (output and input)	Operation of process data	Slider, switch, keypad
Type 4	Yes	Input and output	Operation and visualisation of process data	HMI, alarm handling
Type 5	Yes	–	Evaluation of historical data	Trend analysis, alarm analysis

- Check whether the services available in the WOAS portal are sufficient for the remote lab. If supplementary services are required, these must be newly developed using web technology, saved on any server and integrated in the service directory of the platform.
- Configuration of the remote lab in the WOAS portal taking educational principles into account. The configuration here is completely in the web browser.

No additional specific software development is needed for simple remote experiments and remote labs. It is sufficient to configure the new remote lab in the browser.

A development kit with examples and corresponding documentation is available for the development of new services, for example, for user-specific remote labs. The integration of new services into the system is easily carried out by copying the new services into the service directory of the WOAS portal (similar to the expansion of the Moodle learning management system with new modules).

To use the automation functions as services, they are divided into five service object types in the WOAS system depending on the respective process data required in each case (Table 12.2). In contrast to classic web services, the automation services in the WOAS portal can be used on the server side, on the client side or in a mixed runtime version. Client technologies (e.g. mashups) can be combined with server technologies (web services).

The uniform and pragmatic interfaces form the basis for the simple expandability and integration of third-party devices and services in the WOAS portal, which are freely available for both the devices and the services, and can thus be used by third parties. Both interfaces (device, service) are very similar and are based on an object-oriented JavaScript call interface, event-based data processing and a channel concept for the transmission of process data (Langmann and Meyer 2013).

12.5 Implementation

The realisation of remote labs using the IIoT platform WOAS will be described in more detail below. The capabilities of the currently available platform are revealed and demonstrated based on three examples.

12.5.1 IIoT Platform WOAS

The IIoT platform WOAS (WOAS portal) is a client-enabled and role-based web portal, which can be used to connect services and devices to and between one another (Langmann and Meyer 2013). The implementation of the platform is realised using HTML5, Java, JavaScript (JS), PHP and MySQL database. The data transfer between the core components occurs via WebSocket (WS) and JSON. The freely available jWebsocket server with additionally developed plug-ins is used for this purpose. A remote lab is created in a configuration process using a WOAS creator (= WOAS portal in EDIT mode). Figure 12.4 shows the WOAS creator in the web browser.

The left area in Fig. 12.4 contains the navigator for the user-specific workspaces/views; the middle area shows the work area of the creator; and the right part contains the parameterisation menu for the services. The lower section contains additional tabs for services and virtual devices. A remote lab is represented in the platform as a WORKSPACE with a variable number of VIEWs (websites).

The ergonomic principle WYSIWYG (what you see is what you get) has been implemented to operate the creator. There are no extensive menu bars with difficult to understand icons, but only a few control elements depending on the visible context in the configuration window.

The WOAS portal is publicly available and can be used under http://woas.ccad.eu. According to the Competence Centre Automation Dusseldorf (2014b), a public admin access (guest admin) for the configuration of, for example, a new remote lab can be found in the user manual for the platform. At present, any remote labs can be configured without any extra effort. These use the systems (assembly system for model cars, processing station with rotary table) provided by CCAD (http://www.ccad.de) as laboratory equipment (Langmann and Jacques 2016). In the WOAS portal, users act in different roles:

- *Company*: Administrators and users can be defined in the WOAS portal for a company (customer). An administrator (e.g. a university) has the rights to

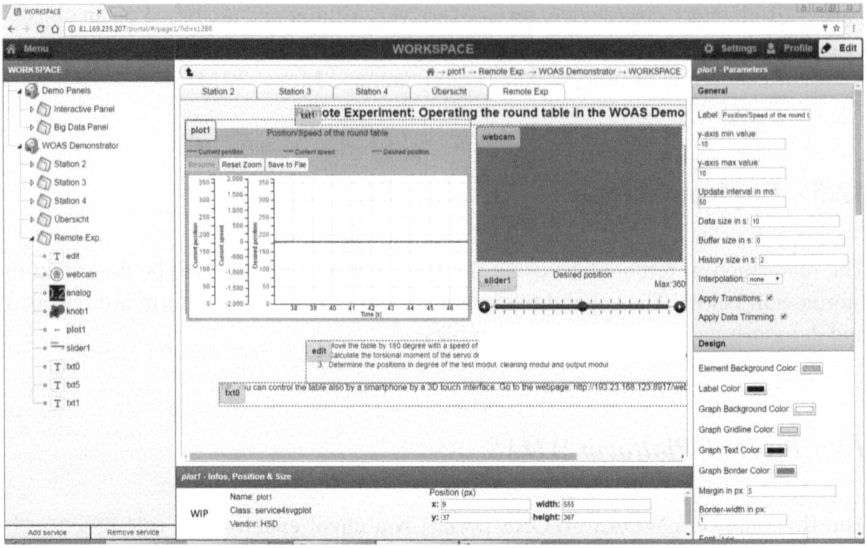

Fig. 12.4 Projecting of a remote lab in the WOAS portal (EDIT mode)

configure customer-specific functional systems (e.g. remote labs) via the creator. Company users (e.g. students) can access and work with these remote labs in their web browser.

- *Publisher*: In the WOAS portal, the publisher has the rights to integrate a new automation service into the service directory.
- *Manufacturer*: The manufacturer has the right to integrate new device classes into the portal and to describe the general device part in the creator.

The roles in the WOAS portal can be fine-tuned. For example, specific services can be exclusively assigned to corresponding companies. It is thus possible that services and devices developed by third parties can only be made available to special user groups. Accounts for companies can be requested free of charge by public education institutions through the homepage of the IIoT platform WOAS (http://woas.ccad.eu).

Generally, it is also possible for educational institutions to host a WOAS portal for training purposes themselves. The consistent alignment of the implementation of the WOAS portal on HTML5 and WebSockets as well as a "responsive" design makes the entire system independent of the client computer. The web browsers of mobile devices (smartphone, tablet) can also function as a runtime/edit environment for a remote lab in the WOAS portal. However, since a WOAS remote lab functions as a "rich" client in the specific case, a corresponding CPU performance level of the client is required, which is not always the case for mobile devices.

Table 12.3 Available web connectors for connecting laboratory devices with industrial interfaces to the WOAS portal (FBD – Function Block Diagram)

Industrial interface	Implementation	Communication	Type (see Fig. 12.3)	Response time (measured reference value)
OPC DA	JS/Java	WS	3	80 ms
Modbus TCP	JS/PHP	WS	3	200 ms
IEC 61131	JS/FBD	WS	2	40 ms
Vathauer frequency converter (proprietary)	JS/Java	WS	1	250 ms

12.5.2 Lab Devices as CPS Components

To connect laboratory devices with industrial interfaces to the web as CPS components, different web connectors are currently available (Table 12.3). Third parties can integrate their laboratory devices in the WOAS portal without any problem via these web connectors. Information on the installation of the web connectors can be found in the corresponding documentations. Detailed information on the measured time response of the device accesses per Table 12.3 is published in Langmann (2014a, b).

Most automation devices that are used as laboratory devices already have an OPC and/or a Modbus TCP interface. An integration in the WOAS portal and the use of these devices for a remote lab are therefore especially easy. It only requires two steps:

- Installation of the web connector software on a local PC, where the relevant OPC server is installed. An installation in an embedded device (e.g. with Linux) is possible for Modbus TCP.
- Specification of a public IP address to access the web connector.

All further work for creating the remote lab involves configuration in the WOAS portal. In order to integrate new industrial interfaces or proprietary communication protocols of third parties, access to the WOAS portal has been created via two special virtual device classes. Table 12.4 shows the main characteristics of these two VD classes.

According to Node-RED (2016), all nodes of the Node-RED can be integrated into the WOAS portal via a corresponding flow. This makes it possible to quickly create new device protocols via graphically configurable Node-RED flows as a protocol gateway. To date, Modbus TCP flows and flows for the device connection of I/O devices (with digital and analog inputs) have been tested using proprietary device protocols.

Table 12.4 Available virtual device classes for the integration of any device protocols into the WOAS portal

VD class	Web protocol	Communication	Type (s. Fig. 12.3)	Description
VD for Node-RED	WOAS device protocol	WS	3	Uses Node-RED as a universal protocol gateway
VD for MQTT	MQTT	WS	3	Enables access from the WOAS portal to any MQTT broker

The *VD for MQTT* allows the use of any data from an MQTT broker for WOAS applications. Corresponding devices (laboratory equipment), however, must publish their process data in an MQTT broker. This is currently still not standard, but with the increasing spread of the Internet of Things, suitable gateways are also offered to publish device data in an MQTT broker. First I/O modules (e.g. WISE-5231) also already have an original MQTT access.

For the use of smartphone sensors in remote experiments, an app can be downloaded from the homepage of the WOAS portal (section: Resources), which can publish the measured values of selected sensors in real time on any MQTT broker. Thus, remote experiments can be implemented easily and quickly with smartphone sensors for technical and physical training (Langmann and Ferfers 2014). In principle, devices with any protocols can be used to set up remote labs.

12.6 Remote Lab Services

The IIoT platform WOAS utilises the service paradigm for implementing the required functionality (Langmann and Jacques 2016). The required service interface is well structured and clear via a data model and a call-up interface, so that a third-party provider can provide further services. From the viewpoint of someone configuring the system, a service always comprises two description parts:

- *General*: This part is defined by the service provider (publisher). This includes, e.g. service name, version, IP address and description.
- *Specific*: A user defines this part during the configuration process of a remote lab. This includes, e.g. visualisation parameters and process data assignments.

At present, the WOAS portal provides 50 services for the configuration of a remote lab. Table 12.5 illustrates some examples for this.

Competence Centre Automation Dusseldorf (2014b) provides a complete overview and description of the current services. The service catalogue is continuously updated and extended. Any functionality, which can be compiled via client- and/or server-end web programming and which implements the WOAS

Table 12.5 Examples of simple services from the WOAS portal

Service name	Description
16-segment display	Display element for showing numerical values as a 7-segment indicator
Pie chart	Display of max. 5 process data as a pie chart
LED	Digital display element using parameterisable images for visualisation of the 1/0
Angular gauge	Measuring instrument for displaying analog process data
Analog display	Dynamic text display of process data with alarm message
Bar graph	Bar graph display of an analog process value
Table display	Dynamic text display of process data in table form with interactive input for writable process data
Switch	Switch with feedback using parameterisable images for visualisation of the switch positions
Slider	Slide control for the input of analog process values
Knob	Rotary knob for the input of analog process values
Rotary switch	Toggle switch for a maximum of 8 digital and/or analog process data
HTML editor	HTML editor for integrating HTML elements (text, images, tables, etc.) into a WOAS view

service interface, can be realised as a service as part of the WOAS concept. Some of the more complex services important for remote labs will be described in more detail below.

12.6.1 Webcam

The webcam service implements the transfer of a video stream from a video server. HTML5 video streaming is used. A special video player is not required on the client side. Any HTML5 video streaming server can be used as the video server, e.g. implemented in a Rasperry Pi (JSMPG 2016). By using HTML5 streaming, the video image in a WOAS view can easily be overlaid by other dynamic services (e.g. different displays), so it is very easy to configure *augmented reality* applications with the WOAS portal. An example is shown in Fig. 12.5.

12.6.2 Real-Time Plotter

The plotter service (Fig. 12.6) allows the display of up to three process data simultaneously in real time. The service can be parametrised in many different ways and can thus be optimally adapted to the respective measuring signals. The real-time signal display can be stopped in order to examine and print certain signal ranges more precisely.

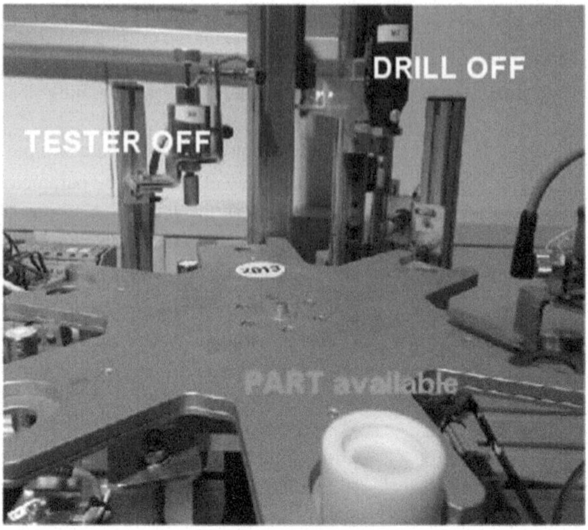

Fig. 12.5 Overlay of a webcam video stream with dynamic process information

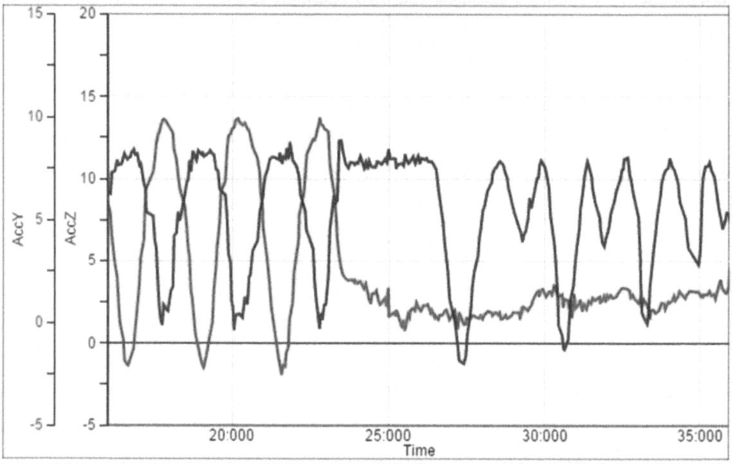

Fig. 12.6 Real-time plot for Y and Z acceleration with a smartphone sensor (Langmann and Ferfers 2014)

12.6.3 Remote Desktop

In a remote lab, complete remote access to a PC's programming system is often required at the laboratory site. The remote desktop service can be used for this purpose. This is a VNC service, which similarly to a webcam uses HTML streaming over WebSockets and therefore also doesn't require a special player. This service uses noVNC (2016) which is an HTML5-based remote desktop web client which can communicate with a remote VNC server via WebSockets.

12.6.4 SVG Animation

The SVG animation service can be used to animate any SVG graphics with dynamic process data. Eight channels are currently available, which can be used to change the visibility, translation, scaling and colour of process data elements (nodes) of the SVG graphics. In addition, the SVG graphics nodes can also be used as interactive input elements, e.g. buttons.

12.6.5 Big Data Store

The Big Data store service (Fig. 12.7) allows to store process data from a WOAS application in Google Cloud. The user needs a corresponding project account with Google. Thereby, the process data can be stored virtually to any extent and over long periods of time. The service uses a block channel, through which all process data of any VD instance can be stored in real time. Currently, the web connector for OPC DA can be used for that purpose (see Table 12.3).

12.6.6 Big Data Analysis

The data that has been stored with the Big Data store service can be evaluated by the Big Data analysis service. This service uses Google Charts to display the analysis results. In addition to eight predefined analysis algorithms, the user can also freely configure his own analysis queries. Figure 12.8 shows a sample evaluation of the main usage times of a remote lab during the winter semester 2015/2016 by 75 students. All process data changes on the system were recorded throughout the semester.

Fig. 12.7 Representation of a Big Data store instance on a WOAS application

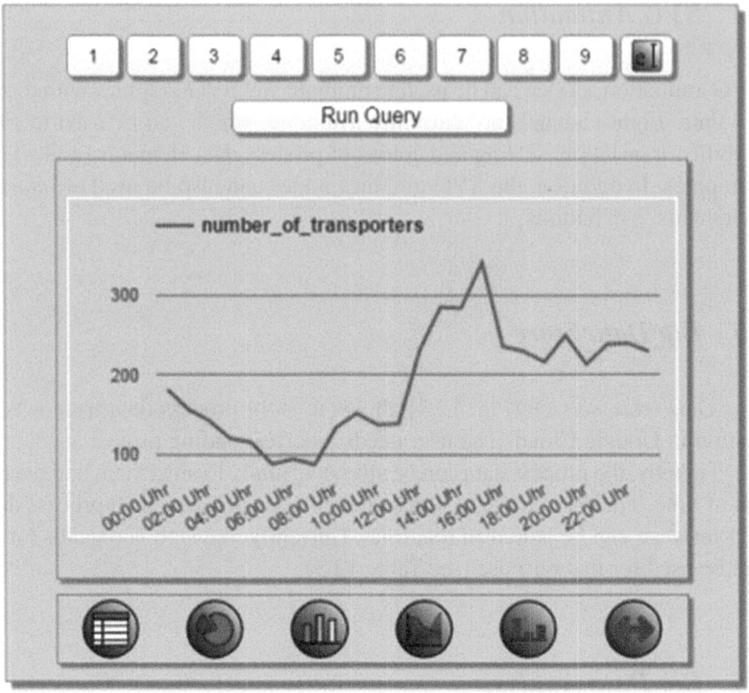

Fig. 12.8 Representation of a Big Data analysis instance on a WOAS application

12.6.7 Chat

With the chat service, all logged-in users can communicate with each other in the WOAS portal. This can be particularly useful for setting up remote labs with collaborative learning tasks. An example of remote collaborative learning with the WOAS portal within an automation system is shown in Fig. 12.9 (see also Example III below).

12.7 Application Examples

In connection with the use of the WOAS portal for the establishment and operation of remote labs and remote experiments, the two types, open remote lab and fixed remote lab, can be distinguished as follows:

Open Remote Lab: The WOAS portal is made available to a student as an open tool environment for the configuration of functional systems. As a learning environment, this is comparable to a freely accessible classical laboratory

Fig. 12.9 According to Example for Education 4.0 (2015), four students from different locations work remotely on an assembly station for model cars

workplace. The student has all possibilities to create a required functional system. The learning task is described in corresponding didactic learning documents. A tutor chaperones the student during the remote work in the system, either through a remote collaboration tool or via face-to-face consultations. For a learning task with an open remote lab, the student needs access as an "Administrator" to the WOAS portal as well as suitable real systems that are connected to the web world as CPS components via a gateway. Typical learning tasks for automation technology are, for example, development and testing of user interfaces (Human Machine Interface).

Fixed Remote Lab: The remote lab in the WOAS portal presents itself as a predefined didactic learning environment, in which the student has to solve predetermined tasks remotely on a real system. For this scenario, the student only logs in to the WOAS portal as a "user".

Usually, the setup of a fixed remote lab environment requires high development efforts and corresponding know-how, which means that a teacher will not create and modify such a remote lab by himself. A teacher can also configure remote labs himself in a relatively short time and adapt and modify these at short notice, when using the WOAS portal as a framework for the creation of remote labs. For this purpose, he needs to be logged in as an "Administrator" in the system. Same as with an open remote lab, the required real technical equipment and devices are connected to the Internet via gateways as CPS components.

The three following application examples from the Duesseldorf University of Applied Sciences are intended to illustrate the use of the IIoT platform WOAS for the design and operation of open and fixed remote labs.

12.7.1 Example I: Fixed Remote Experiment with a Rotary Table

The example uses a processing and test station with a rotary table (Fig. 12.10) (Langmann and Jacques 2016). In the experiment, a student is asked to operate the position-controlled rotary table and record and evaluate the angular speed and acceleration as well as determine the positions of processing modules in the station.

The process data access to the system per Fig. 12.10 occurs via a web connector for Modbus TCP as CPS interface. Figure 12.11 shows the remote experiment in the web browser.

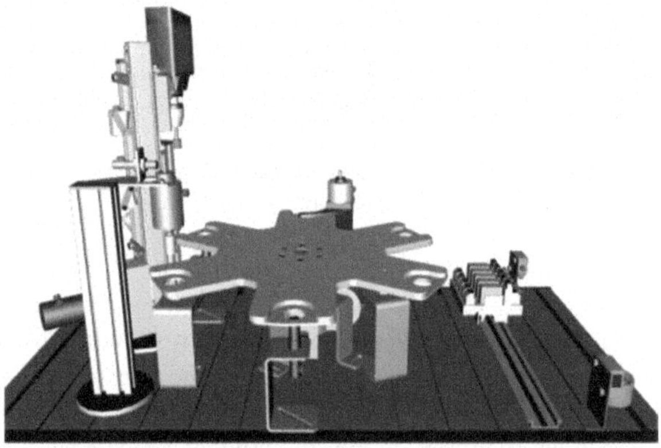

Fig. 12.10 Processing and test station with position-controlled rotary table

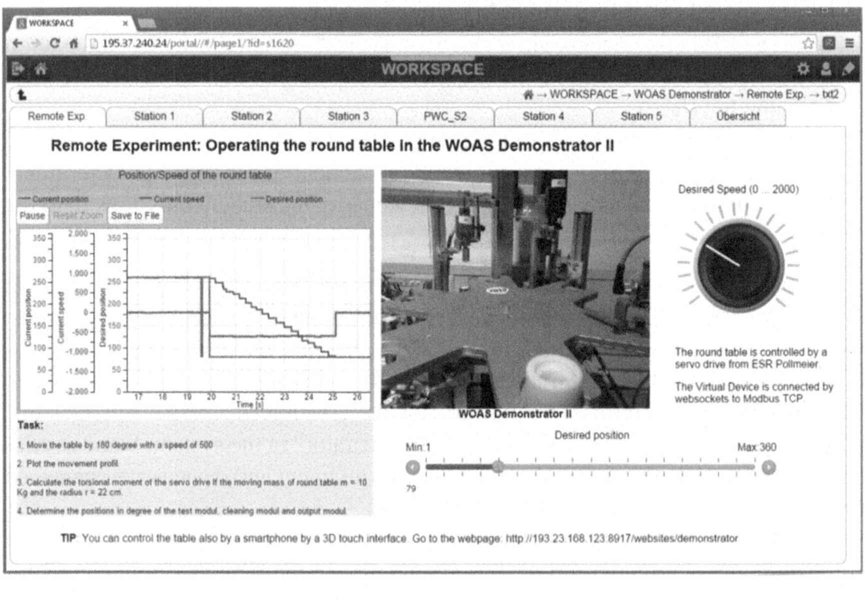

Fig. 12.11 Fixed remote experiment for a position-controlled rotary table

As remote lab services, the following are available to the student on the website:

- Video image of the rotary table.
- Two-channel real-time plotter. The plotter enables the saving of displayed values on the client PC.
- Rotary knob for specifying the angular speed.
- Slide control for specifying the position target value.

The example is available publicly via the guest access for the WOAS portal and can be used in, for example, a course on "drive technology".

12.7.2 *Example II: Fixed Remote Experiment with a Gripper Arm*

In a laboratory test for pneumatics, students are supposed to measure the movement time of a gripper arm, controlled by a single-acting pneumatic valve, and then develop proposals for optimising the adjustment of the valve. Figure 12.11 shows the gripping arm, for which the measurements are supposed to be carried out (Fig. 12.12).

A teacher would now like to integrate a fixed remote experiment. The following prerequisites are presumed:

- The assembly station is connected to the Internet via an OPC gateway as a CPS component, and all sensors and actuators are accessible.
- A webcam is installed.
- The teacher has access to the WOAS portal as an "Administrator".

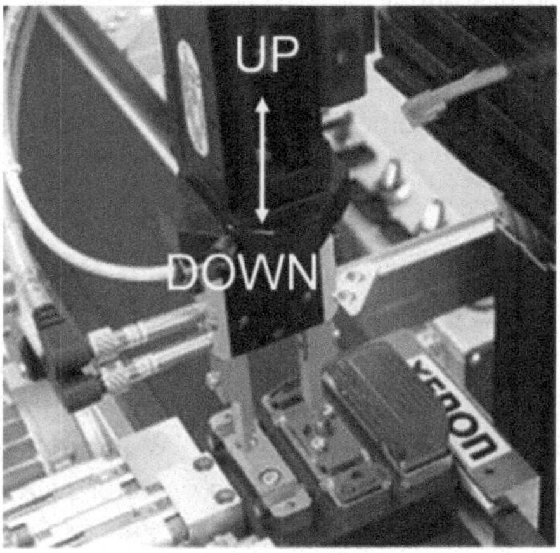

Fig. 12.12 Pneumatically operated gripper arm at an assembly station for model cars

Fig. 12.13 Fixed remote experiment for measurement of the movement time of a pneumatic valve

With these preconditions, the teacher can configure the remote experiment in a short time in the WOAS portal. An evaluation showed that a teacher with sufficient knowledge of using the WOAS portal can create this remote experiment in approximately 12 min. Figure 12.13 shows the view with the remote lab in the WOAS portal.

This makes it clear that, when the WOAS portal is used as a framework, remote labs can be set up and used quickly and easily. Even existing remote labs can be quickly modified by the teacher and adapted to changed learning situations.

12.7.3 Example III: Open Remote Lab for an Assembly Station

The goal of this example is the use of the WOAS portal as an open remote lab in the course "Human-Machine Communication" (bachelor, fifth semester, specialisation in Automation Technology). Within the framework of a project task, the student must configure and test an operator panel for the stations of an assembly line for model cars. Figure 12.14 shows the assembly line. Each learning group implements an operator panel for testing and program operation of stations 2–5.

Fig. 12.14 Assembly line for model cars with five stations (Station 1: test station for the assembly parts, Station 2: assembly of the axis modules, Station 3: assembly of the body, Station 4: dismantling the body, Station 5: dismantling of the axis modules)

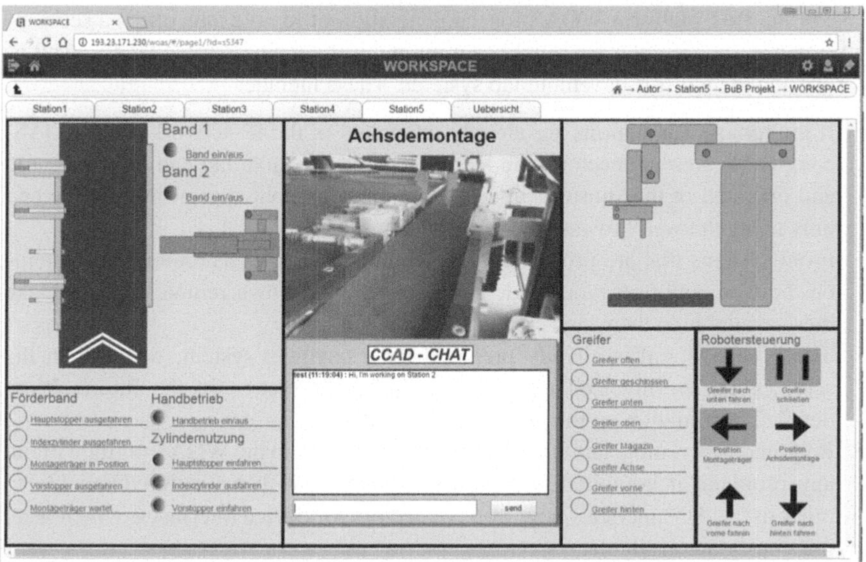

Fig. 12.15 Operator panel for Station 5 in the manual mode

The work on the real system is exclusively carried out remotely via corresponding accesses to the WOAS portal. Each project group consists of four students, who collectively have access to one "Administrator" login. In the WOAS portal itself, it is stipulated that a maximum of four administrators may work simultaneously (but at different stations) using this access login.

The operation of the complete assembly process can only be solved by a closed cooperation between all the students in the project group during remote work. To accomplish this, the chat service described above is used. Every winter semester, the learning task is carried out successfully by approximately 75 students. As an example, Fig. 12.15 shows one result of a project group from the winter semester 2015/2016.

Up to now there is no systematic pedagogical evaluation for the described learning task. But, because the content of the examination in the above mentioned course "Human-Machine Communication" is the implementation and documentation of the project work in the remote lab, the marks of the examination represent also a kind of pragmatic evaluation. On average the marks of the students are very high (86 scores of 96 max achievable), and it is also well known from different discussions with the students that they are very likely to work in this remote lab.

12.8 Benefits and Problems

Use of the IIoT platform WOAS for the generation of remote labs offers a series of advantages, in particular for quickly setting up user-specific remote labs as well as for worldwide multi-user remote lab systems. These include:

- If the remote lab stations are already integrated in the IP network with WOAS-compatible web connectors, the learning-specific remote lab can be configured and operated in the shortest time (a few hours) without special knowledge and only using the web browser.
- From systems that are provided by educational institutes as remote lab systems on the web, any user worldwide can configure their own remote lab and make these available to their own students.
- The WOAS portal is already prepared for a payment system, with which the services can be invoiced using a micropayment system in the future. More detailed information on this in (Langmann 2014b).
- Required new remote lab functions can be integrated in the WOAS portal without any problem, if compiled as a WOAS service, and hence offered quickly to other users. This includes new web connectors for device interfaces, which were previously not available.
- The WOAS portal is compiled in HTML5 and CSS 3 in responsive design and can be executed in all browsers, operating systems and display sizes. If only services are used, which do not require any special plug-ins in the browser (e.g. Java applets), remote labs for mobile clients (smartphone, tablet, etc.) could also be configured very easily.

The previously publicly available WOAS portal is a prototype and hence not yet optimised in its functionality. There are still various bugs, although these are gradually being eliminated.

The absence of a typical remote lab user management represents a greater problem. Up to now there was no reservation system for the remote labs, and a user-specific assessment of the results of a student's work in a remote lab is so far also not possible. Since the WOAS portal was developed as a general CPS integration platform without considering specific remote lab requirements, the corresponding functionalities are also not provided structurally in the platform.

Table 12.6 CPU load for operation of the workspace "Demo Panel" in the WOAS portal

Client computer	CPU load [%]
Intel Core i7-3770 CPU @ 3,4 GHz, 64-Bit- Windows 7, 8 GB RAM, Windows performance index 5,6, (PC)	15...20
Intel Core i5-4300U CPU @ 1,9 GHz, 64-Bit-Windows 8, 4 GB RAM, (surface tablet)	35...40
Athlon 3 GHz, 32-bit-Windows 7, 4 GB RAM, Windows performance index 4,1, (PC)	55...65

A remote lab in the WOAS portal is a "rich" client application and requires, irrespective of the amount of dynamic process data and the complexity of the services that are to be used, a corresponding processing power on the client computer. Consequently, newer client PCs and tablets have hardly any problems, but it might not be possible to use older PCs due to excessively slow execution of the JavaScript programs.

Table 12.6 illustrates the processing power required on the client side (browser = Google Chrome) for the RUN of the workspace "Demo Panel" in the guest access of the WOAS portal, in which 18 process data sets are processed with an update rate of approximately 50 ms (graphic-dynamic visualisations).

12.9 Summary and Future Works

As a CPS integration platform, the IIoT platform WOAS enables fully browser-based configuration and operation of functional systems, consisting of technical devices and systems as CPS components and associated services. Originally developed for use in automation technology, this platform can also be used to configure and operate applications as remote experiments and remote labs, which access technical equipment and systems over the Internet. The type of technical device generally does not matter. The only requirement is that the device is connected and accessible to the Internet as a CPS component.

As a multi-user-enabled and roll-based platform, the WOAS portal allows a virtually unlimited number of different users to design and operate various remote labs. Once a device pool is available on the Internet, it can be used by different remote labs and different students.

With the WOAS portal, both fixed remote labs with a predefined didactic structure and open remote labs can be built as a counterpart for flexibly usable classical laboratory workstations.

Up to now the IIoT platform WOAS is used only for creation of remote labs in automation engineering. But in general the system is open to use it as a framework for remote labs for all technical subjects. Maybe in this case depending on the specific application, it is required to extend the system by some new services and new virtual devices.

The IIoT platform WOAS can be updated and extended further using the following steps, depending on the available R&D resources:

- Revision of the entire platform for rectifying bugs and completing the absent functionality (e.g. automatic refresh when creating new device instances).
- Extension of the platform by implementing the prepared clearing system for fine-tuned termination of service usage.
- Completion of the planned functions for the user-friendly integration of new services by third-parties (at present new services still must be uploaded via FTP onto the WOAS server).
- Compilation of new services, e.g. runtime machine for IEC 61131 control programs, alarm management, process data on Google Maps, Twitter publishing service for process data (these services are already being partially tested).
- Development of additional web connectors for industrial interfaces, in particular for OPC UA.
- Increased use of Node-RED as a universal gateway and development of additional flows for use in remote labs.

Depending on the available resources in the CCAD as well as possible partners and interested parties, an instance of the IIoT platform WOAS is intended exclusively for learning purposes. The services and VD classes, available in this platform, are meant to be published on the Internet as part of an open source project. Third parties can then modify these modules/components and also develop new services and device accesses for the WOAS learning portal.

In addition, a list is supposed to be drawn up to show which technical facilities/devices are provided by educational facilities in the network, which can be used via a WOAS application (creation of a WOAS device pool).

References

Caminero, A. C., Robles-Gomez, A., Ros, S., Tobarra, L., Hernandez, R., Pastor, R., & Castro, M. (2014). On the creation of customizable laboratory experiments: Deconstruction of remote laboratories to create Laboratories as a Service (LaaS). *International Journal of Online Engineering, 10*(6), 35–42.

Competence Centre Automation Dusseldorf. (2014a). *Virtual devices – description*. Retrieved from http://woas.ccad.eu

Competence Centre Automation Dusseldorf. (2014b). *WOAD – interface description*. Retrieved from http://woas.ccad.eu

Coquard, P., Guillemot, M., Leleve, A., Noterman, D., & Benmohamed, H. (2008). AIP-Primeca RAO remote laboratories in automation. *International Journal of Online Engineering, 4*(1), 12–18.

Evans, P. C., & Annunziata, M. (2012). *Industrial internet: Pushing the boundaries of mind and machines (Whitepaper GE)*. Retrieved from http://www.ge.com/docs/chapters/Industrial_Internet.pdf

Example for Education 4.0. (2015). Retrieved from https://www.youtube.com/watch?v=Og-ICNiDwig

Hankel, M. (2016). *Reference architecture model industry 4.0 (RAMI 4.0). Industrie 4.0*. Retrieved from https://www.plattform-i40.de/I40/Redaktion/EN/Downloads/Publikation/rami40-an-introduction.pdf?__blob=publicationFile&v=4

IoTool. (2016). *Make your own IoT solutions*. Retrieved from https://iotool.io/

JSMPG. (2016). MPEG1 video & MP2 audio decoder in JavaScript. Retrieved from https://github.com/phoboslab/jsmpeg

Langmann, R. (2011). E-Learning & doing in training for automation engineers. In J. G. Zubia & R. G. Alves (Eds.), *Using remote labs in education* (pp. 271–289). Spain: University of Deusto.

Langmann, R. (2014a). *An interface for CPS-based automation devices*. In *Proceedings Konferenz für Angewandte Automatisierungstechnik in Lehre und Entwicklung (AALE 2014)* (pp. 133–142). München: DIV Deutscher Industrieverlag. 2014.

Langmann, R. (2014b). Rent an automation service – Smarte Geschäftsmodelle für Industrie 4.0. In *Proceedings GMA-Kongress AUTOMATION 2014*, [CD] VDI.

Langmann, R., & Ferfers, T. (2014). *Smartphone sensors in the Internet of Things (IoT)*. In *Proceedings Konferenz für Angewandte Automatisierungstechnik in Lehre und Entwicklung (AALE 2017)* (pp. 115–123). Berlin und Offenbach: VDE Verlag GmbH. 2017.

Langmann, R., & Jacques, H. (2016). *Generic remote labs in automation engineering*. Paper presented at EDUCON 2016, Abu Dhabi, UAE.

Langmann, R., & Meyer, L. (2013). *Architecture of a web-oriented automation system*. Paper presented at ETFA IEEE International Conference on Emerging Technology & Factory Automation, Cagliari, Italy.

Langmann, R., & Rojas-Peña, L. (2016). *A PLC as an Industry 4.0 component*. Paper presented at International Conference on Remote Engineering and Virtual Instrumentation, Madrid, Spain.

National Academy of Science and Engineering. (2013). *Recommendations for implementing the strategic initiative INDUSTRIE 4.0* (Final report). Frankfurt: Kagermann, H., Wahlster, W., & Helbig.

Node-RED. (2016). *Flow-based programming for the Internet of Things*. Retrieved from http://nodered.org

NoVNC. (2016). *How to access VNC remote desktop in web browser*. Retrieved from http://xmodulo.com/access-vnc-remote-desktop-web-browser.html

Seiler, S. (2013). Current trends in remote and virtual. Lab engineering. Where are we in 2013? *International Journal of Online Engineering, IAOE, 9*(6), 12–14.

Tawfik, M., Salzmann, C., Gillet, D., Lowe, D., Saliah-Hassane, H., Sancristobal, E., & Castro, M. (2014). Laboratory as a Service (LaaS): A novel paradigm for developing and implementing modular remote laboratories. *International Journal of Online Engineering, 10*(4), 13–21.

VDI/VDE-GMA. (2013). *Cyber-physical systems: Chances and benefits from the viewpoint of automation* (in German). Theses and action fields.

Reinhard Langmann is the director of the Competence Center Automation Duesseldorf (CCAD) at the Duesseldorf University of Applied Sciences (Germany).

Chapter 13
The Development and Implementation of Instruction and Remote Access Components of Additive Manufacturing

Ismail Fidan, Amy Elliott, Mel Cossette, Thomas Singer, and Ed Tackett

Abstract Additive manufacturing (AM), also known as 3D printing, is one of the latest emerging widespread production technologies. Almost any complex-part geometry is easily made using this technology and is usually used reliably. Many implementations of AM exist from areas as diverse as food industry to biomedical engineering; such a broad-spectrum usage of this technology makes it extremely attractive when combined with its low cost, reliability, color range, and complexity abilities.

Though the cost of buying new AM machines varies greatly depending on the size of the machines (AM equipment ranges from desktop printers to very large production machines), AM equipment is still not affordable for many educational institutions due to limited or low equipment, consumable supplies, physical space, and maintenance budgets. Such issues become even more important for educational organizations in underserved and underdeveloped districts, which typically have inadequate support from their constituents.

I. Fidan (✉)
College of Engineering, Tennessee Technological University, Cookeville, TN, USA
e-mail: ifidan@tntech.edu

A. Elliott
Manufacturing Demonstration Facility, Oak Ridge National Laboratory, Knoxville, TN, USA
e-mail: elliottam@ornl.gov

M. Cossette
National Resource Center for Materials Technology Education, Edmonds Community College, Lynnwood, WA, USA
e-mail: mel.cossette@edcc.edu

T. Singer
The Science, Mathematics and Engineering Division, Sinclair Community College, Dayton, OH, USA
e-mail: thomas.singer@sinclair.edu

E. Tackett
J B Speed School of Engineering, University of Louisville, Louisville, KY, USA
e-mail: ed.tackett@louisville.edu

M. E. Auer et al. (eds.), *Cyber-Physical Laboratories in Engineering and Science Education*, https://doi.org/10.1007/978-3-319-76935-6_13

To address this issue, AM laboratories and their functionalities can easily be made available through the internet. Educational institutions which do not have the capability of AM technologies can easily access and utilize other laboratories' capabilities. In the past, various remotely accessible AM laboratories such as these have been introduced, and their advantages and limitations in various P16 STEM (science, technology, engineering, and mathematics) practices have been reported. In this chapter, the authors introduce a novel concept of accessing external AM laboratories via smartphones and advanced computer technologies.

Keywords Additive manufacturing · Laboratory · STEM · Smartphone · Remote access

13.1 Background

The number of job announcements requiring workers with AM skills increased 1834% in four years and 103% when comparing August 2013 to August 2014 (Columbus 2016). This analysis found that AM skills were the most sought-after skill listings in engineering jobs, representing 35% of all engineering ads posted in the previous 30 days. Few technologies match the promise and potential of AM. Since AM is increasingly used in a variety of industries including aerospace and biomedical, automotive, defense, and materials manufacturing, trained technical experts in AM are needed today (Huang and Leu 2013).

This sharply increasing trend in AM will motivate the creation of innovative, entrepreneurial skills and initiatives at the early stages of higher education so that future generations will likely have more opportunities and capabilities in today's rapidly advancing workplace environment. A highly skilled workforce educated in STEM is able to advance basic scientific knowledge in innovative ways and transform that knowledge into useful products and services. Success stories of many start-up companies and their high-tech undertakings have provided unique and innovative advancements that are gaining power in today's competitive workplace. Unfortunately, very few educational institutions have developed or even have access to instructional guides and other educational materials needed for courses and lab activities in AM.

Tennessee Technological University (TTU) has been developing a number of online courses and remotely accessible AM laboratory environments for almost 10 years. These efforts have been funded by three NSF ATE grants to date (Fidan et al. 2009, 2016; Patton et al. 2008). They are listed below.

With DUE #0536509, remotely accessible laboratory environment has been developed and implemented in junior- and senior-level engineering and technology courses. Along with the laboratory environment, web-based accessible course materials have been developed using WebCT, D2L, and Moodle course management systems. Various workshops have been provided to K-16 STEM teachers so that they learn about and translate these technologies to their classrooms (Fidan et al. 2009).

With DUE #0501527, the Rapid Prototyping Instructional Development website (http://rpids.csc.tntech.edu/) and various workforce development workshops have been accomplished. This project developed a number of instructional materials for the K-16 STEM teachers so that they could use them in STEM courses as supplemental material (Patton et al. 2008).

With DUE #1601587, the goal is to develop an additive manufacturing coalition in which institutions can jointly share and utilize their resources (Fidan et al. 2016). At this time, the target institutions are the University of Louisville, Tennessee Technological University, Edmonds Community College, and Sinclair Community College. Oak Ridge National Laboratory also provides expert support in the development of several MOOCs (massive open online courses) and instructional support materials. Recent work has developed a new web access platform for AM at these institutions. One smartphone application was also utilized to link the access of all laboratories from any smartphone.

In the past, various remotely accessible AM laboratories have been observed, and their best practices, success stories, pros, and cons have been reported in various technical publications (Fidan 2017; Gao et al. 2015; Aziz et al. 2012; Meisel and Williams 2015; Lan 2009). However, there have been no reported studies on the success of utilizing technology such as smartphones and the latest computer technologies in AM. Therefore, the findings of this study are essential in presenting the success story of such high-tech implementation in AM.

Beta test results of the current development show positive feedback from undergraduate engineering and engineering-technology students. Students were able to access the AM laboratory remotely and interact with the laboratory features in various capacities, such as talking to a student assistant, observing the AM operation, and seeing finished pieces. It is expected that all institutions might join the network in the near future and start sharing their AM laboratory capabilities with each other.

This chapter provides the existing framework developed through a smartphone application that links the AM labs to each other and highlights pros and cons of contemporary practices. Also, the AM course that is presently structured for mechanical engineering students will be highlighted with its components. Further, student feedback will be provided on the remote access features of the currently developed system.

13.2 AM Technologies

AM is one of the latest manufacturing processes for making physical objects through digital design files. Design files are created on computers using 3D design software. The design file is converted to a .STL or .AMF file type; these files are digitally sliced into layers, and the AM machines build the final part layer by layer using data from each slice. The final produced piece is the physical form of the digital file. These steps are highlighted in the flowchart in Fig. 13.1.

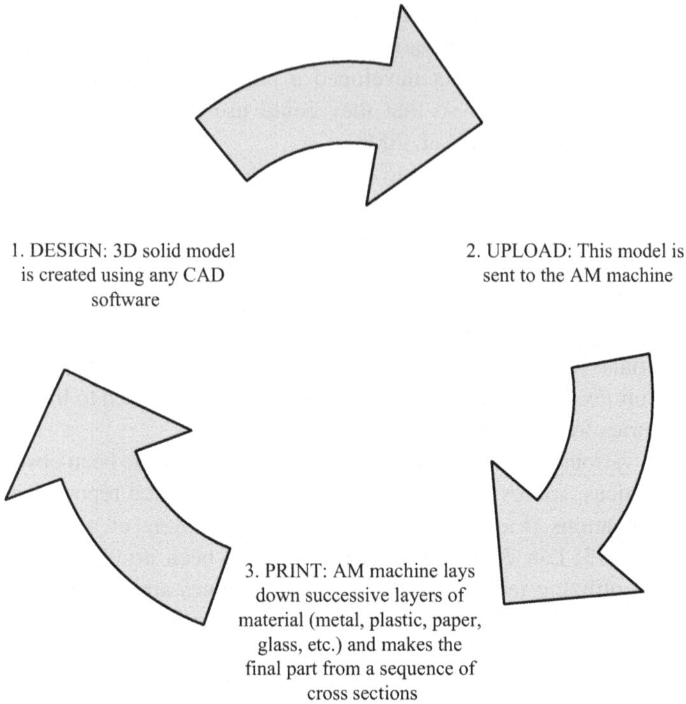

Fig. 13.1 Design, upload, and print flow in AM

3D design stages are accomplished using computer-aided design software like Creo, SolidWorks, SketchUp, and Inventor. Design files are sent to AM machines using a USB cable, Wi-Fi, or an SD card. AM machines have their own software programs to receive the part geometry, digitally slice it into layers, and produce the tool path or "G-code" to recreate each layer.

There are multiple AM technologies that utilize different methods for shaping a feedstock material layer by layer (Gibson et al. 2010). Some methods melt the material to create the layers of a part. Fused deposition modeling is the most commonly used technology in this category. In stereolithography, a part is created in a vat of liquid polymer that is selectively cured using laser technology. In binder jetting technology, small silica particles called plasters and starches are spread in layers over a build cavity, and a binder is inkjetted to stick the layers of the part to each other.

13.3 AM Remote Access Network

In the current AM remote access network, AM laboratories are linked with exceptionally precise network cameras. All network cameras are equipped with two-way communication, infrared night vision, an SD card slot, digital zoom ($\times 10$), pan and tilt abilities, and motion alerts. They also have two-way audio connection, which is a useful feature that lets anyone chat with the laboratory personnel through the remote access. These cameras also let users monitor the part production from start to end and inform the laboratory personnel when there is an issue. The cameras are also instrumental in documenting the production steps and laboratory experiments. Currently, the system does not provide any control on the design software tools but lets the users access the laboratory, watch the production real time, and see the finished product without any delay. Through the AM remote access network, participating institutions target to share their AM resources and capabilities with the latest remote access technologies.

The remote access smartphone application of the AM remote access network can be easily downloaded through any smartphone and gives users the opportunity to watch live video footage anywhere that they have an Internet connection (mydlink 2017). All cameras have the pan and tilt features for custom viewing.

The developed system provides around-the-clock observation with night vision capability, allowing users to see up to 26 ft in complete darkness. Videos of laboratory exercises and snapshots can be recorded up to 32 GB on microSD cards. The system also sends alerts to users' cellular phones when any motion is detected. This type of feature is important for the safety and security of the laboratory and its high-value equipment and tools. Table 13.1 provides the key features of the AM network camera system developed at Tennessee Technological University. Figure 13.2 shows the structure of the developed system.

The following list provides the basic features of the AM network system.

- Allows the user to view, control, and communicate through the built-in microphone and speaker using the free network application available for iOS, Android, and Windows devices.
- Easy access and use of the laboratory with a user-friendly application for any kind of computer and smartphone system.
- Enhanced sound and motion detection with a built-in PIR sensor.
- Sends the user automatic push alerts and triggers clip recording that can be viewed on the network application or web portal.
- Using Internet service, entire access is managed through Wi-Fi or a hard-wired communication system.
- Easy addition of any AM laboratory into the network after a password is provided. Ultra-smooth 340° pan and 120° tilt capability with 10× digital zoom, allowing users to keep an eye on a larger spectrum of the AM laboratory area.
- HD 720p video resolution, giving users clear and detailed live and recorded video day or night with 26-foot night vision.

Table 13.1 Highlights of the AM network camera system

Video capture resolution	*720p HD*
Wi-Fi	*802.11 g/n*
Product dimensions	*4.7 × 0.8 × 5.1 in.*
Battery	*No*
Recharge time	*N/A*
Power supply	*110–240 V AC*
Operating systems	*Microsoft Windows 7, 8, Vista, Mac OS X (v10.6 or higher)*
Storage	*PC or local SD card*
Two-way communication	*Yes*
Night vision	*26 ft in darkness*
Zoom	*Digital zoom 10×*
Field of view	*98° × 52° × 115°*
Pan and tilt	*340° Pan 120° Tilt*
Live feeds/remote monitoring	*Yes, via any PC or mobile device with an Internet connection*
Alert notifications	*Motion alerts can be received on mobile devices*
Connection	*Wi-Fi/Wired*
Password protection	*Yes*
Color	*Black*
Item weight	*0.75 lb*
Video resolution	*Up to 1280 × 720 for 16:9 and 960 × 720 for 4:3*
Frames per second	*30*

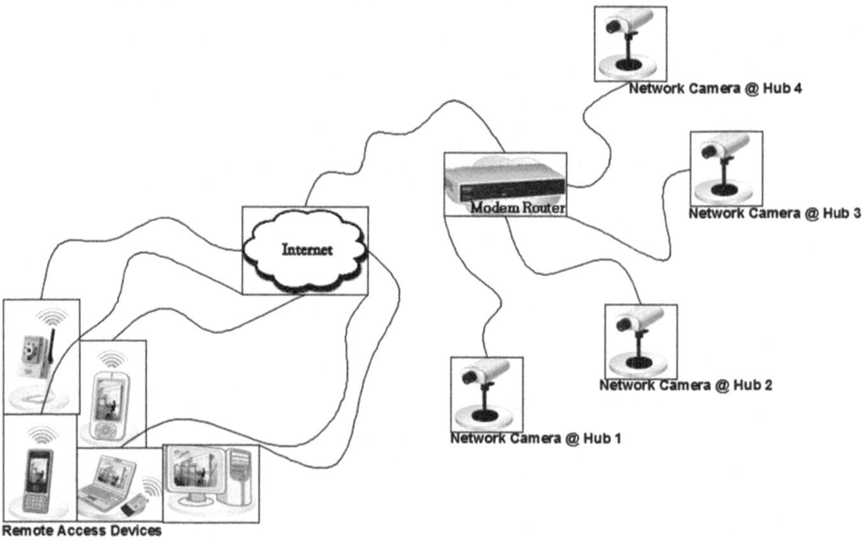

Fig. 13.2 Structure of the remotely accessible AM laboratory network

Fig. 13.3 Accessing TTU AM Laboratory from a Google Chrome Browser

- Built-in microSD card slot that supports 32 GB capacity to record video clip recordings continuously or based on motion triggers or schedules.

Figure 13.3 provides a sample snapshot of the AM laboratory collaboration network from Google Chrome. Figure 13.4 is the access to the same laboratory from a smartphone application.

13.4 Instructional Support Materials

Besides the AM remote access network, instructional AM materials have been developed and placed onto Desire2Learn (D2L), a course management system so that instructors could use them in their engineering and technology courses as a supplement. This system provides several content study tools, video links, virtual lecture series, and Dropbox assignments. A sample screenshot of the Content Study Tools is provided in Fig. 13.5 The reason for selecting this system is the current availability of it through the Tennessee Board of Regents and Tennessee Technological University. Brief summaries of the presently available resources are given below.

The settings provided in the system (at the bottom of Fig.13.4) are explained below.

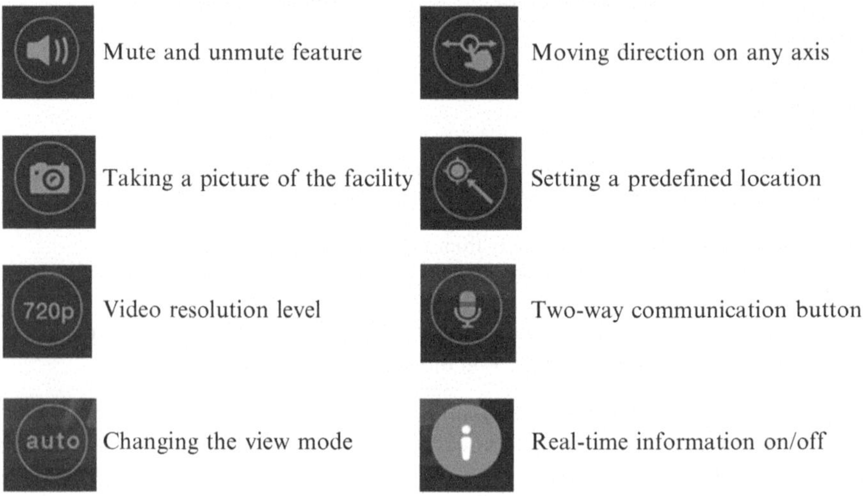

Fig. 13.4 Accessing the TTU AM Laboratory from iPhone application

- Content Study Tools: provides concise information about the AM technologies, processes, and materials.
- Dropbox: offers some assessment exams and quizzes related to AM technologies, processes, and materials.
- Video Links: presents short AM-related video clips from lectures and laboratories.
- Virtual Lecture Series: offers recordings of short AM lectures, which are organized frequently and are publicly available.

One unique practice of the AM collaboration is to provide frequent virtual AM talks, housed at TTU, using the ZOOM webinar tool. Students and remote

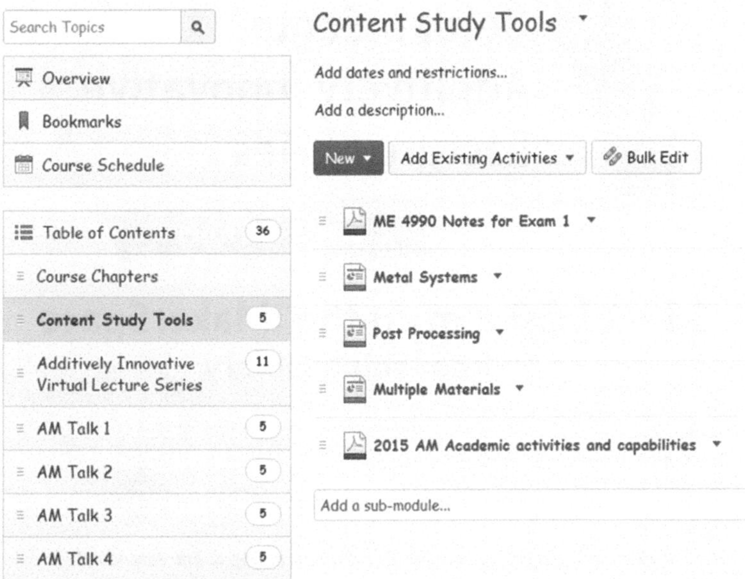

Fig. 13.5 Content study tools of AM

attendees access the provided link and listen to the latest trends, technologies, and advancements in the AM field. Figure 13.6 provides the Fall 2016 lecture series schedule. Their recorded links are provided below.

http://www.cae.tntech.edu/~nguo/private/2016fall/lecture1/
http://www.cae.tntech.edu/~nguo/private/2016fall/lecture2/
http://www.cae.tntech.edu/~nguo/private/2016fall/lecture3/
http://www.cae.tntech.edu/~nguo/private/2016fall/lecture4/

13.5 Feedback Provided by Students

A number of engineering students provided feedback on the pros and cons of the developed systems. The following list provides a short summary of their responses.

Pros:

- Enables students to see how a real AM research lab operates
- Allows students to see items that they have designed as they are being manufactured
- Can be used as a teaching tool without having to leave the classroom
- Cost-effective solution of advanced manufacturing practices
- Can be adapted and implemented for other manufacturing processes

Fig. 13.6 Fall 2016 schedule of the Golden Eagle Additively Innovative Virtual Lecture Series

- Exceptionally intuitive and easy-to-use system
- Offers the feel of an AM laboratory experience without leaving the classroom

Cons:

- Privacy issues with workers in the lab
- No control over how the video is being used
- Requires a number of trial-and-error processes to learn the whole system
- Problems with network traffic, slow Internet, and heavy access to the system
- Unsatisfactory night mode
- While similar, does not provide the same experience as being in the laboratory in person

13.6 Conclusion

Although the utilization of AM technologies is expanding in almost all fields of daily life, it is still not affordable for some underserved and underrepresented districts due to initial cost, budget, maintenance, service, and consumable purchase factors. However, this study proves that additive manufacturing could be practiced and learned with a remotely accessible network environment. In this chapter, four institutions established a remote AM collaboration network in order to utilize their resources in AM teaching and workforce development. Also, the brief details of the AM course management system and additively innovative virtual lecture series have been provided. These types of innovative advanced manufacturing practices will likely be popular in the near future, considering the tight budget issues in purchasing and maintaining the equipment, supplies, and consumables.

Acknowledgments This work is part of a larger project funded by the Advanced Technological Education program of the National Science Foundation, DUE #1601587.

References

Aziz, E. S., Chang, Y., Esche, S. K., & Chassapis, C. (2012). Capturing assembly constraints of experimental setups in a virtual laboratory environment. In *ASME 2012 International Mechanical Engineering Congress and Exposition* (pp. 191–200).

Columbus, L. (2016). *Demand for 3D printing skills is accelerating globally. Forbes*. Retrieved from http://www.forbes.com/sites/louiscolumbus/2014/09/15/demand-for-3d-printing-skills-is-accelerating-globally/

Fidan, I. (2017). Remotely accessible rapid prototyping laboratory: Design and implementation framework. *Rapid Prototyping Journal, 18*(5), 344–352.

Fidan, I., Elkeelany, O., & Ghani, N. (2009). *The development of a remotely accessible rapid prototyping laboratory, National Science Foundation Award*. Retrieved from http://www.nsf.gov/awardsearch/showAward?AWD_ID=0536509&HistoricalAwards=false

Fidan, I., Singer, T., & Cossette, I. (2016). *AM-WATCH: Additive manufacturing – workforce advancement training coalition and hub. National Science Foundation Award*. Retrieved from http://www.nsf.gov/awardsearch/showAward?AWD_ID=1601587&HistoricalAwards=false

Gao, W., Zhang, Y., Ramanujan, D., Ramani, K., Chen, Y., Williams, C. B., Wang, C. C., Shin, Y. C., Zhang, S., & Zavattieri, P. D. (2015). The status, challenges, and future of additive manufacturing in engineering. *Computer-Aided Design, 69*, 65–89.

Gibson, I., Rosen, D., & Stucker, B. (2010). *Additive manufacturing technologies* (Vol. 238). New York: Springer.

Huang, Y., & Leu, M. C. (2013). *Frontiers of additive manufacturing research and education: An NSF additive manufacturing workshop report*. Arlington: National Science Foundation.

Lan, H. (2009). Web-based rapid prototyping and manufacturing systems: A review. *Computers in Industry, 60*(9), 643–656.

Meisel, N. A., & Williams, C. B. (2015). Design and assessment of a 3D printing vending machine. *Rapid Prototyping Journal, 21*(5), 471–481.

mydlink. (2017). *Wifi cameras*. Retrieved from https://www.mydlink.com/entranc

Patton, K., Dostie, D., Leonard, W., White, M., & Fidan, I. (2008). *Rapid prototyping instructional delivery support, National Science Foundation Award*. Retrieved from http://www.nsf.gov/awardsearch/showAward?AWD_ID=0501527&HistoricalAwards=false

Ismail Fidan is a Professor of Engineering and Faculty Fellow in Innovation and Techno-entrepreneurship at Tennessee Technological University.

Amy Elliott is a Researcher at Oak Ridge National Laboratory's Manufacturing Demonstration Facility located in Tennessee.

Mel Cossette is Principal Investigator for the Technician Education in Additive Manufacturing & Materials Project funded by the NSF Advanced Technological Education Directorate and affiliated with the Edmonds Community College located in Washington.

Thomas Singer is Professor of Mechanical Engineering Technology, Coordinator of STEM Guitar Project, Sinclair Community College, Dayton, Ohio.

Ed Tackett is Director of Additive Manufacturing Education and Manager of the Additive Manufacturing Competency Center at the J B Speed School of Engineering of the University of Louisville.

Chapter 14
Design and Implementation of a Remote Laboratory for Heat Transfer Experiments

Ridha Ennetta, Ibrahim Nasri, Soufiene Bouallègue, and Thrasyvoulos Tsiatsos

Abstract The current chapter describes the design and the implementation of a remote laboratory for heat transfer learning purposes. It summarizes the main steps of the work carried out to adapt and redesign the heat exchanger bench to be remotely accessed and controlled. This new device introduced many fundamental aspects of heat transfer, both theoretically and practically. An evaluation procedure was also carried out on this remote lab. This evaluation was focused on some technical and pedagogical aspects. The evaluation results demonstrated that the expected learning outcomes of remote labs seem to be very interesting compared to a conventional lab.

Keywords Remote laboratory · e-Learning · iLab · Virtual instruments · Heat exchanger

14.1 Introduction

As we know, the use of laboratory experiments is a critically important aspect of education. Experience in teaching has shown that a complementary approach combining theoretical and practical exercises is vital for effective learning. According to Hansen (1990), students retain 25% of what they listen to, 45% of what they listen to and see, and 70% when they manipulate, control, and modify experiments, putting into practice what they are learning.

R. Ennetta (✉) · S. Bouallègue
Higher Institute of Industrial Systems of Gabes (ISSIG), University of Gabes, Gabes, Tunisia
e-mail: ridha.ennetta@issig.rnu.tn; soufiene.bouallegue@issig.rnu.tn

I. Nasri
College of Engineering, University of Tabuk, Tabuk, Saudi Arabia
e-mail: inasri@edu.ut.sa

T. Tsiatsos
Aristote University of Thessaloniki, Thessaloniki, Greece
e-mail: tsiatsos@csd.auth.gr

M. E. Auer et al. (eds.), *Cyber-Physical Laboratories in Engineering and Science Education*, https://doi.org/10.1007/978-3-319-76935-6_14

Even in an e-Learning context, learners need to manipulate real systems to be able to assimilate many fundamental aspects. The only realistic solution for them to perform practical experiments could be probably through remote access to distance laboratories.

These remote laboratories use real equipment that can be remotely operated using a computer and Web-based tools (Scanlon et al. 2004). A remote laboratory simplifies the logistics and requirements involved in conventional laboratory work including equipment, lab space, staffing, training, and safety. The students can conduct their experiment from any computer on their own schedule instead of conducting it in a specialized laboratory on the staff's schedule. Remote labs can also enrich science and engineering education by vastly increasing the scope of experiments that students have access to throughout their academic careers. Moreover an online laboratory facilitates efficient sharing of expensive equipment.

In the literature, several architectures of remote laboratories have been proposed for different disciplines such as mechanics (Schauer et al. 2008), automatic control and automation (Coquard et al. 2008), electronics (Tobarra et al. 2015), electrical engineering (Guimarães et al. 2011), and so on (Odeh 2014; Guerra et al. 2007). Overviews of state of the art about technologies and remote laboratory paradigms are given in Gravier et al. (2008) and Gomes and Bogosyan (2009). Recent and interesting examples in several areas of education as well as current trends and challenges in this topic are identified and discussed.

Schauer and his co-authors (2008) proposed an integrated e-Learning-based laboratory for mechanical oscillations. The three constituting components of such an e-Lab, i.e., the remote experiments, e-simulations and e-textbooks, are illustrated. The proposed method of integrated e-Learning was verified at Trnava University in cooperation with Charles University in Prague. A new e-Lab platform was based on AIP-Primeca RAO training network for the Rhône-Alpes French Region (AIP-RAO) (Coquard et al. 2008). The proposed AIP tool sets up new laboratories related to automation as local and remote resources to handle the constraints inherent in using heavy and shared industrial resources. In Tobarra et al. (2015) and Guimarães et al. (2011), two e-Learning laboratories based on OpenSocial and WebLabs concepts have been developed, respectively. They can be easily deployed over different networks such as the public Internet, campus-wide networks, or high-speed private networks. Odeh (2014) presented a Web-based remote lab platform for electronics teaching was built with reusability capability. Such a proposed solution allows the implementation of a variety of electronic experiments and does not necessitate the creation of any kind of software including the user interface. Guerra and his co-authors (2007) developed an experimental platform for electrical machines training and remote control through the Internet. The developed tele-operation-based e-Laboratory allows the use of several computers to distribute the task and help access with IP telephony.

The current chapter presents the work carried out to adapt and redesign a heat exchanger bench to be fully accessed and remotely controlled. This new device enables a mechanical engineering student to apprehend many fundamental and practical aspects of heat exchangers. The objective of this work that was performed in the framework of the "e-Sience" Tempus project was to create an efficient remote labs network in the Maghreb region.

14.2 The Framework of This Study

In the Maghreb region, the strong demand for top technicians and engineers needs an increase of training capacity. Enrollment growth and quality of courses are often incompatible. In order to avoid typical constraints of traditional laboratories, such as scheduling, cost of equipment, and location, the remote operation of real plants can be incorporated into engineering courses (Fabregas et al. 2011).

In this context, an international cooperation project involving 16 partners from 4 European countries (France, Austria, Romania, and Greece) and 3 Maghreb countries (Tunisia, Algeria, and Morocco) was conducted between 2012 and 2015 with the aim of establishing a Maghreb network of remote laboratories.

This project called "e-Sience" offered an innovative pedagogical approach that is complementary to classroom learning using e-Learning tools (Zimmer et al. 2013). The main objective of this project was the creation of an efficient remote labs' network in the Maghreb region for the modernization of higher education in technological sciences. This remote labs network will enable students to conduct real-world experiments at a distance and to assess the educational potential of such a system. So far, this network has been based on four e-Labs (two in Tunisia, one in Algeria, and one in Morocco) constituted of about ten remote labs.

The Higher Institute of Industrial Systems of Gabes (ISSIG) is a partner of the e-Sience project. The ISSIG major task was to build an e-Lab by adapting and developing distant access solutions for two remote labs allowing students to perform practical experiments on heat transfer and mechanical vibration. These two remote labs are directly accessible through the ISSIG e-Lab server. The current chapter presents the most important aspects of the ISSIG e-Lab architecture and summarizes only the work carried out on the heat exchanger (HE) bench to be fully accessed and remotely controlled.

14.3 Architecture of the ISSIG e-Lab

Different architectures can be used to support e-Learning environments (Chandre et al. 2014). Some of them are based on proprietary software solutions such as LabVIEW, while others are supported by open-source software such as PHP, JavaScript, Java, Phyton, etc. The National Instruments' LabVIEW software (2017), which is considered as professional software for analysis, data acquisition, real-time control, and remote laboratories, has had an enormous impact on engineering education. Benefiting from that, the ISSIG has developed an innovative and flexible

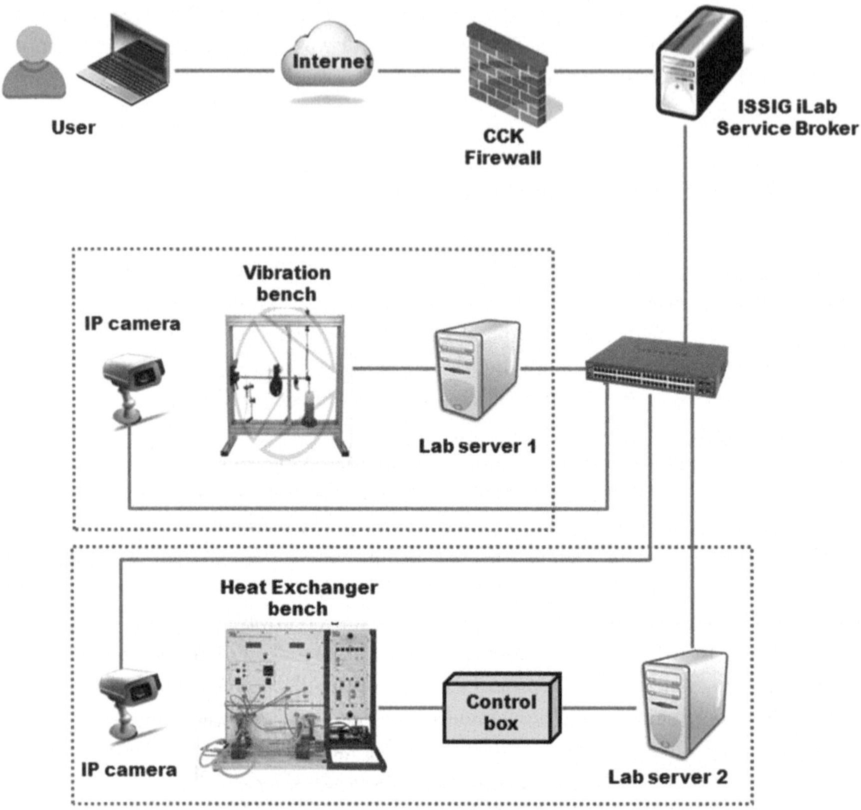

Fig. 14.1 The ISSIG e-Lab architecture

remote laboratory to conduct real-world experiments on some physical phenomena such as mechanical vibration and heat transfer.

The architecture of the ISSIG e-Lab is presented in Fig. 14.1. This e-Lab is composed of two remote labs: the heat exchanger remote lab and the mechanical vibration one (Nasri et al. 2015). More details about this e-Lab will be given throughout the following sections.

There are many software applications used to share online laboratories (Hardison et al. 2008; Shroff et al. 2009). Therefore, to provide access to ISSIG remote labs, we have chosen the iLab Shared Architecture (ISA) of Massachusetts Institute of Technology (MIT) that provides a framework for the development and the deployment (Shroff et al. 2009). The ISA divides an online lab in three distinct parts: the Lab Client, the Service Broker, and the Lab Server, that simplify the development of remote labs around the word by providing reusable components for common lab administration function. In ISA, the Service Broker is the core of the architecture. It provides user authentication, authorization, experiment data storage, and access to scheduling services (Dashboard (2017; Zutin et al. 2011).

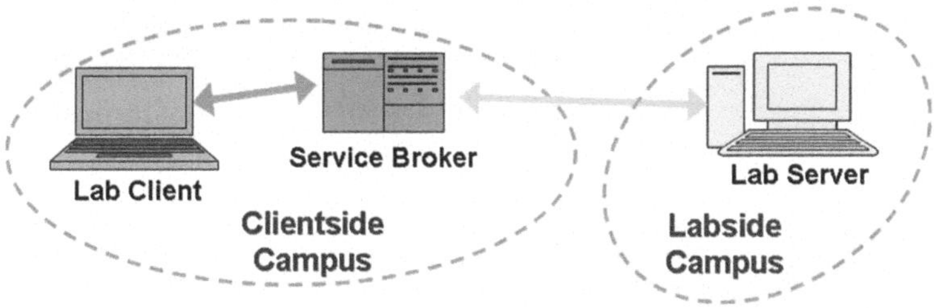

Fig. 14.2 Topology of a batched experiment based on the iLab Shared Architecture

14.3.1 Presentation of the iLab Shared Architecture (ISA)

The ISA tool is an open-source architecture built on Web services that provides a unifying software framework to support access to a wide variety of online laboratories. The ISA provides a set of generic lab services such as user account management, scheduling, and data storage in a middleware system that can be accessed using Web services. There are three different types of architecture available in the framework of online laboratories: batched, sensor, and interactive experiments.

In the first architecture, the batched iLabs are labs where experiments are completely specified prior to submission and run without intervention. Batched iLabs are deployed with Lab Client, Service Brokers, and Lab Server that communicate over the Internet using Web services. In this model, shown in Fig. 14.2, Lab Client and Lab Server communicate with each other exclusively through the iLab Service Broker (Hardison et al. 2008).

In the second architecture, i.e., the sensor experiment, students cannot specify or configure any of the parameters. They can monitor and control real-time data streams without influencing the phenomena being measured (Mao 2007).

In the last architecture, as shown in Fig. 14.3, in addition to the Lab Client, Service Broker, and Lab Server, stand-alone Web services are added to manage experiment storage and lab scheduling. The student must first schedule the time to use the lab. At the scheduled time, the student logs in, and he is able to launch a Lab Client. The student interacts directly with the Lab Server, and, once a lab session begins, the Service Broker steps out of the picture.

In our case, the user needs to control and vary some experimental parameters in order to see the response of the studied system. That is why we adopted the interactive experiments' architecture because it is more suitable to conduct experiments that need some user monitoring. This architecture, as shown in Fig. 14.2, consists of three parts: the Interactive Lab Client Server, the Interactive Service Broker, and the Interactive Lab Server.

The Interactive Lab Client Server is the interface through which students access the iLab. It provides an intuitive representation of the iLab that is being run, allowing users to specify parameters and interact with the lab hardware. The ISA supports

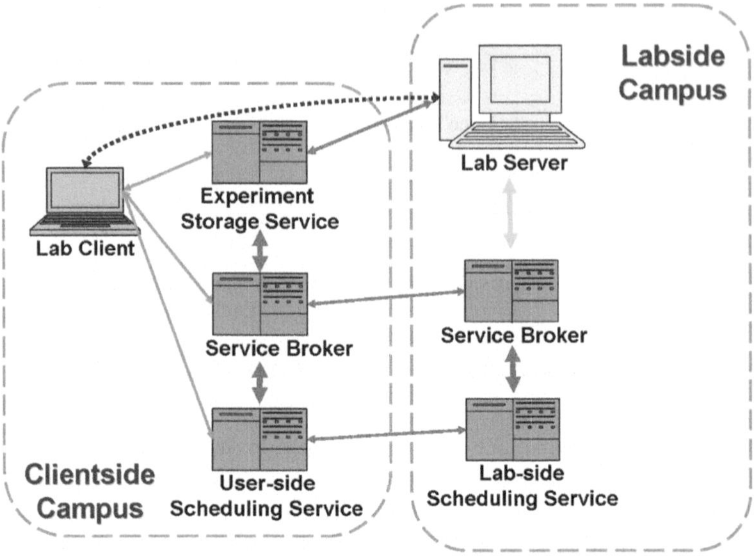

Fig. 14.3 Topology of an interactive experiment based on the iLab Shared Architecture

multiple clients' technology, including Java Apples, JavaServer Pages, Windows Forms Clients, and LabVIEW Front Panels (Naddami et al. 2014). The Interactive Lab Server is directly connected to the lab equipment and deals with the actual operation of the experiment hardware. It is the administrative interface of the lab equipment, and it enables laboratory administrators to set up and configure each experiment independently. The Interactive Service Broker is the heart of the ISA. It provides generic administrative services such as authentication, authorization, user management, scheduling, and data storage. The Service Broker serves as the gateway in an interinstitutional relationship. To support these functions for interactive labs, three tiers have been added to the ISA: the Experiment Storage Service (ESS), the User-side Scheduling Service (USS), and the Lab-side Scheduling Service (LSS).

The ESS component is a stand-alone Web service that allows Service Brokers, Lab Servers, and Lab Clients to store experiment data. It also provides storage of binary data (images, video, or audio) and XML-based text/numeric data. In addition, the students can execute interactive experiments by running the interactive lab client. To support interactive experiments that require scheduling access, the iLab interactive architecture envisions scheduling servers and services that enable students, from different campuses, to reserve time periods to execute experiments. Since the user side and lab side require different scheduling functionalities, the USS and LSS components are introduced to the ISA architecture to manage the reservation (Shroff et al. 2009; Cazacu 2014).

The USS tool is used in conjunction with the LSS to allocate lab time to the users. Using the LSS, a student who wants to schedule time in a given lab must select from a set of available blocks of time. Additionally, the USS is responsible for notifying students if the reservation must be canceled and for considering course/lab requirements when distributing time blocks (Shroff et al. 2009; Cazacu 2014). The LSS tool is responsible for defining the scheduling policy for a particular lab. It is designed to run in conjunction with multiple USS software and may schedule multiple lab servers. The LSS defines the broad lab availability for individual USS/Service Brokers. In turn, a given USS/Service Broker will distribute experiment time to students based on lab requirements, instrument availability, and instructor policy (Shroff et al. 2009; Cazacu 2014).

14.3.2 Access to the ISSIG e-Lab

As previously described, our e-Lab is based on the Interactive Shared Architecture in order to deploy and share the labs. As shown in Fig. 14.4, the platform is accessed through the following Web link: http://onlinelab.issig.rnu.tn/iLabServiceBroker/

The students browse from any place where the Internet connection exists to register and request membership to the group associated with the laboratory. They can access the Service Broker and login page and supply their usernames and passwords. When the students have permission to access to the booking service, they can choose the available experiment to launch.

14.4 Development of the Heat Exchanger Remote Lab

Students studying thermodynamics and heat transfer need to know how well different heat exchangers work. They can use this information to decide the correct heat exchanger for their own designs. Heat exchanger bench shows students how different small-scale heat exchangers work. They mimic the most common heat exchangers used in the industry and compare how well they work for different flow rates and temperatures.

The main tasks in the development of the remote lab were the adaption of an existing heat exchanger bench to be remotely controlled and the design of the user interface (UI) application that enable students to communicate with this new device.

14.4.1 Presentation of the Heat Exchanger Bench

The available heat exchanger bench, as shown in Fig. 14.5, is a compact frame with two water circuits (hot and cold) and instruments to measure and display water flow and temperature. This module can work with various types of heat exchangers (concentric tube heat exchanger, plate heat exchanger, shell and tube heat exchanger,

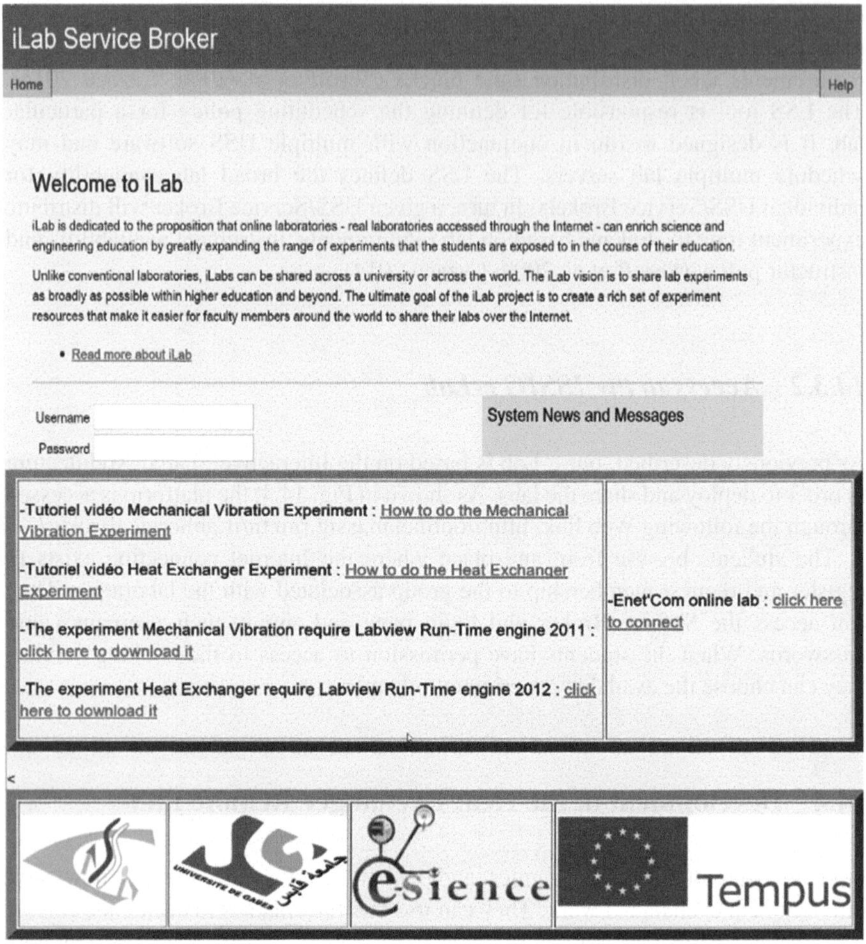

Fig. 14.4 Online iLab ISSIG platform

and jacketed vessel with coil and stirrer). Students test each of the optional heat exchangers and record the flow and temperature changes to see how well the heat exchanger works. If they have one or more of the heat exchangers, students can compare them to see which is the best for any application.

In the present experiment, we use only the concentric tube heat exchanger. This heat exchanger is a simple shell and tube heat exchanger. It has two tubes, one inside the other. The outer tube is the shell. The inner tube carries the water from the hot circuit, whereas the other tube carries the water from the cold circuit. The heat transfers between these two tubes. It is possible to connect the water circuits to give contraflow (counterflow) or parallel flow experiments. This heat exchanger is in two equal parts with extra thermocouples at the midpoint. These experiments

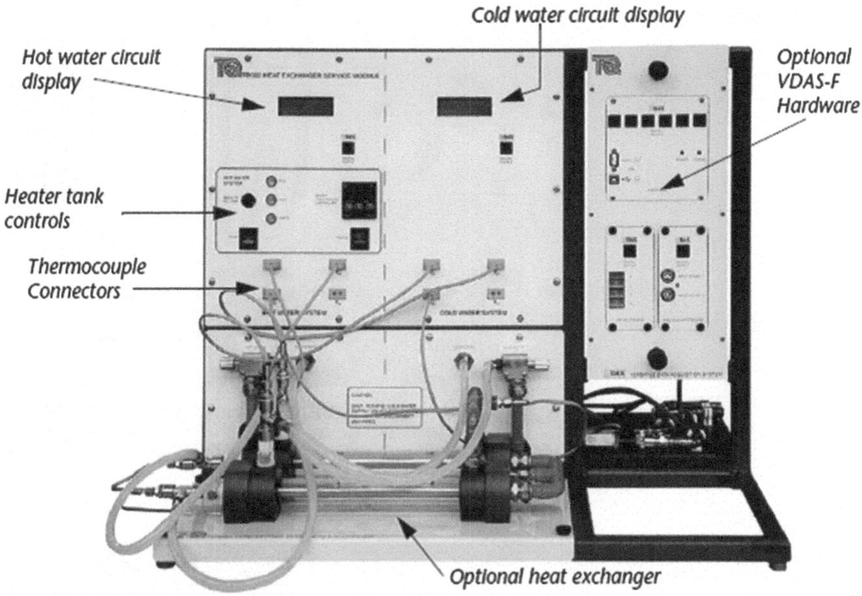

Fig. 14.5 TD360 heat exchanger bench

help students understand more clearly how the temperature changes along the heat exchanger.

14.4.2 Adaptation of the Heat Exchanger Bench

The heat exchanger bench was adapted for remote operations. The adaptations included control of the hot water supply pump, the heater, and the cold as well as the hot water supply flows (see Fig. 14.6). To provide students with an overview of the whole heat exchanger bench, an IP camera was located in the laboratory.

14.4.3 The User Interface Application

The UI application illustrated in Fig. 14.7 was created using the LabVIEW software. This UI was developed to enable students to control the heat exchanger remotely.

Using the LabVIEW Web derver to publish the virtual instruments (VIs) to be remotely controlled via the Internet, the client needs to install the LabVIEW runtime engine (National Instruments 2017). The Web Publishing Tool is a LabVIEW built-in tool to publish the front panel of a VI as a HTML document to the Web. There are

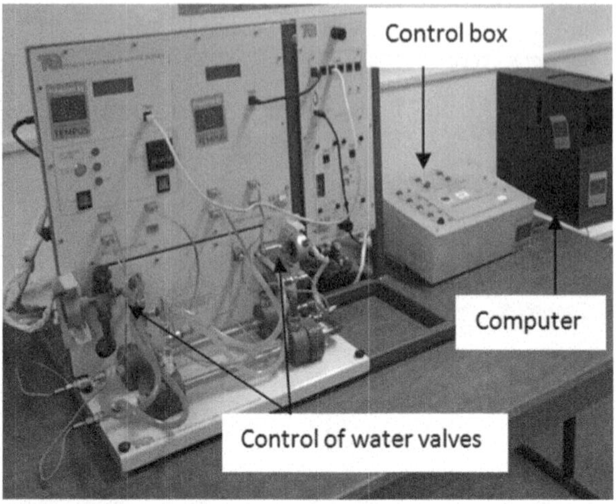

Fig. 14.6 Heat exchanger remote lab

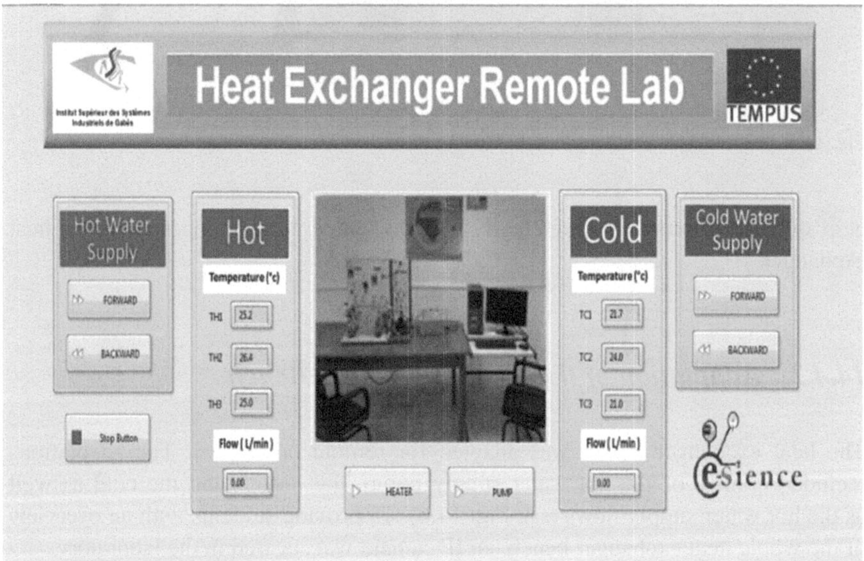

Fig. 14.7 Screenshot of the UI used to control the heat exchanger remote lab

three steps from choosing a VI till saving the HTML file to the disk. In the first step, i.e., “Select VI and Viewing Options,” the VI to publish, which must be in memory, has to be selected. Also the “Viewing Mode” can be changed between “Embedded,” “Snapshot,” and “Monitor,” where Embedded allows clients to view and control the front panel, Snapshot to only display a static image of the front panel, and Monitor to display a snapshot with a configurable updating interval. For the next step “Select HTML Output,” we can type in a title (document title), a text before (header) the front panel, and a text after (footer) the front panel which is going to be displayed in the respective place. The next step is the "Save the New Web Page" where the created HTML file of the VI is going to be saved in a directory with the selected file name and a URL will be created (National Instruments 2017). After saving it, the VI is now ready to be remotely controlled from a client by typing the created URL into the address or URL field of his Web browser window. The URL for the front panel is http://41.229.94.95:8000/TQ.html

This URL must be integrated in the LabApp table on the LabExperiments page. The LabExperiments page is the configuration page for experiments. To complete this page, we must indicate the title, the client guide, the application key, and the path which contains the target application. To access the LabApp table, we must configure also the experiment on Manage Lab Clients in the Service Broker service. Four items must be indicated as we have done for the LabExperiments: the title of the experiment, the client guide, the version, and the loader script. We choose here an interactive redirect client-type experiment, and we put the same URL as the Web Page URL put in the LabExperiments page.

The UI included buttons for controlling the different parts of the heat exchanger bench and displayed the feeds from water supply flow meters, thermocouples, and the IP video camera.

14.4.4 Running Experiment

The heat exchanger remote lab is an interactive experiment that requires users to schedule experiment in advance. To run the experiment, the student should log into the Service Broker, select the heat exchanger experiment group, and redeem the reservation that he/she had already scheduled in advance. The Service Broker checks to make sure that the user has a valid reservation and that he/she is authorized to use the selected experiment. After that, a "Lunch Lab" button appears, and the user is able to start the experiment client as shown in Fig. 14.8.

When the experiment is lunched, the Service Broker facilitates the exchange of credentials so that the Lab Client running in the users' browser can communicate directly with the Lab Server and the experiment hardware. The IP camera, installed in the heat exchanger lab, allows users to watch the experiment while in progress. We noticed that the camera is essential for the student to understand and feel that he is working on and controlling a real hardware.

14.5 Evaluation of the Remote Lab

The evaluation of our remote labs was in the framework of the whole evaluation strategy adopted by all e-Sience project partners (Tsiatsos et al. 2014).

The evaluation strategy of the project focused on five different, but interrelated, directions given as follows: (a) usability of remote labs; (b) learners' attitude toward

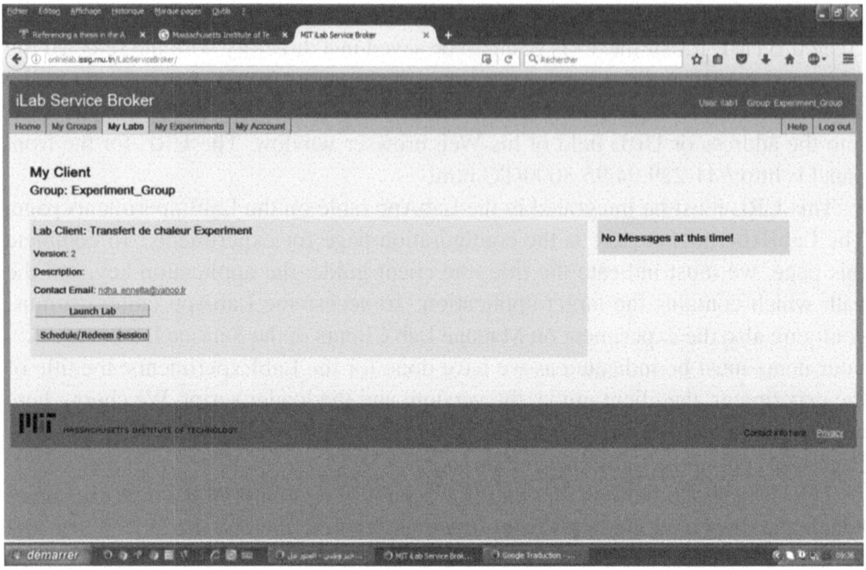

Fig. 14.8 Student view of the heat exchanger experiment page on the Service Broker

remote labs; (c) technical evaluation of remote labs operation; (d) evaluation of the e-Learning content, namely, the teaching units previously described; and (e) learning outcome.

The evaluation was conducted in two sequential phases:

- Phase I: Pilot evaluation
- Phase II: Large-scale evaluation

The first phase was the pilot evaluation of all remote labs in small-scale usage. During this phase, we assessed the remote labs' usability and proper functioning as well as learners' attitude toward remote labs and the learning outcome.

The first evaluation phase of our remote labs was conducted with a population of 30 students that have performed their experiments on every remote lab. According to the results of the first stage evaluation of the remote labs, operation and user interface were improved, and they were deployed in a large-scale usage. After this period of usage, the large-scale evaluation was accomplished. The large-scale evaluation has been focused on usability of remote labs, learners' attitude toward remote labs, evaluation of the e-Learning content, and assessing learning outcome (Tsiatsos et al. 2014).

The evaluation tools which were used in each phase are presented in Table 14.1. As referred by Gomes and Bogosyan (2009), the nature of the learning outcomes arising from laboratory experiences has a complex relationship with the characteristics of the interaction modality. Therefore, a research on the impact of remote labs in

Table 14.1 E-science evaluation instruments

Evaluation goal/phase	Phase I: pilot evaluation	Phase II: large-scale evaluation
(a) Usability of remote labs	USE questionnaire (Lund 2001)	USE questionnaire (Lund 2001)
(b) Learners' attitude toward remote labs	Learners' attitude questionnaire (Douka 2010)	Learners' attitude questionnaire (Douka 2010)
(c) Technical evaluation of remote labs operation	ISO/IEC – SQuaRE functional suitability	
(d) E-learning content evaluation		Checklist for a didactically sound design of eLearning content (Schoor and Körndle 2012)
(e) Learning outcome		Knowledge test adapted in every course (Felder and Solomon 1991)

education, among others, should consider the way in which the technologies which are used affect the nature of the interaction.

14.5.1 Technical Evaluation

The technical evaluation focuses on the functional suitability of these experiments; this is expressed and quantified through measurable parameters as functional correctness, functional completeness, and functional appropriateness. The technical evaluation is based on the specific standard ISO/IEC 25010:2011 Systems and software engineering – Systems and software Quality Requirements and Evaluation (SQuaRE) – system and software quality models. The measurable quality-related properties of a system are called quality properties, with associated quality measures. According to this standard, the product quality model categorizes product quality properties into eight characteristics (functional suitability, reliability, performance efficiency, usability, security, compatibility, maintainability, and portability).

The performed technical evaluation focuses on the functional suitability which means a degree to which a product or system provides functions that meet stated and implied needs when used under specified conditions. The functional suitability has three components: the functional completeness, the functional correctness, and the functional appropriateness. The remote labs were evaluated through the abovementioned parameters. The functional completeness is the degree to which the set of functions covers all the specified tasks and user objectives. The functional correctness means the degree to which a product or system provides the correct results with the needed degree of precision. The functional appropriateness is the degree to which the functions facilitate the accomplishment of specified tasks and objectives.

The technical evaluation was conducted by three experts from Petru Maior University, Romania, during the pilot evaluation phase. Results of this evaluation were quite satisfactory. According to these results, some operation conditions were reviewed, and user interface was redesigned and improved.

14.5.2 Pedagogical Evaluation

The major part of the pedagogical evaluation was conducted in collaboration with experts from Aristotle University of Thessaloniki, Greece. It focused on the (a), (b), (d), and (e) directions cited previously.

In order to evaluate the usability of the remote labs, we utilized the USE Questionnaire presented by Lund (2001) in all evaluation phases. USE stands for usefulness, satisfaction, and ease of use. For many applications, usability appears to consist of usefulness and ease of use. However, usefulness and ease of use are correlated.

Table 14.2 presents the descriptive statistics calculated for the four post-task questionnaire dimensions. The study findings reveal a positive students' opinion toward the usefulness, ease of use, ease of learning, and satisfaction of the remote labs session. A Pearson product-moment correlation coefficient was also computed to investigate the relationship between the results from the USE questionnaire and the grades of the students. The results of the correlation analysis did not yield any statistically significant correlations among the compared variables (see Table 14.3).

The learners' attitude toward remote labs was evaluated using an instrument developed by Douka (2010). This instrument allowed students to rate the remote labs about the following characteristics: comprehensive, sensible, educational, easy, enjoyable, interesting, satisfactory, well done, scientific, serious, well prepared, important, innovative, modern, pedagogic, targeted, and different. The results of the learners'attitude test are listed in Table 14.4.

Table 14.2 Use questionnaire statistics results

Descriptive statistics						
		Usefulness	Ease of use	Ease of learning	Satisfaction	Overall
N		42	42	42	42	42
Min.		1.63	1.73	1.25	1.57	1.54
Max.		5.00	5.00	5.00	5.00	5.00
Mean		3.9345	3.8442	4.0595	3.8571	3.9242
Std. deviation		0.70777	0.70826	0.89000	0.72157	0.71055
Skewness	Statistic	−1.353	−1.179	−1.431	−1.243	−1.597
	Std. error	0.365	0.365	0.365	0.365	0.365
Kurtosis	Statistic	2.877	2.290	1.895	2.698	3.468
	Std. error	0.717	0.717	0.717	0.717	0.717

Table 14.3 Use questionnaire correlations results

Correlations			
		Overall	Grade
Overall	Pearson correlation	1	−0.070
	Sig. (2-tailed)		.474
	N	108	108
Grade	Pearson correlation	−.070	1
	Sig. (2-tailed)	.474	
	N	108	108

Table 14.4 Attitude questionnaire corrections results

Descriptive statistics					
	N	Min.	Max.	Mean	Std. deviation
[Comprehensive]	42	1.0	5.0	3.762	1.2457
[Sensible]	42	1.0	5.0	3.452	1.0170
[Educational]	42	1.0	5.0	3.857	1.0258
[Easy]	42	1.0	5.0	3.667	1.2815
[Enjoyable]	42	1.0	5.0	3.738	1.1699
[Interesting]	42	1.0	5.0	3.929	1.1347
[Satisfactory]	42	1.0	5.0	3.595	1.1489
[Well done]	42	1.0	5.0	3.738	1.1489
[Scientific]	42	1.0	5.0	4.119	1.0407
[Serious]	42	1.0	5.0	3.857	1.1806
[Well prepared]	42	1.0	5.0	3.762	1.1436
[Important]	42	1.0	5.0	3.881	1.1306
[Innovative]	42	1.0	5.0	3.786	1.1590
[Modern]	42	1.0	5.0	4.095	1.3759
[Pedagogic]	42	1.0	5.0	4.024	1.0704
[Targeted]	42	1.0	5.0	3.857	1.0493
[Different]	42	1.0	5.0	4.000	1.1262
Attitude overall	42	1.4	5.00	3.8305	0.86669

The results of the attitude questionnaire correlations is presented in Table 14.5. One can see clearly that there is no statistically significant correlation between the results from the attitude questionnaire and the grades of the students.

The e-Learning content evaluation was conducted by experts from the AUF (Agence Universitaire de la Francophonie) for the purpose of making a formative evaluation. They used a tool called "Checklist for a Didactically Sound Design of eLearning Content" developed by Schoor and Körndle (2012). This tool was designed to verify the content, segmenting, sequencing and navigation, adaptation to target audience, design of text and graphics, learning tasks and feedback, and motivation. Table 14.6 gives an overview on the results of this evaluation.

The evaluation of the learning outcome was carried out by exploiting a knowledge test in two different groups of students: the control group consisting of 66 students and the experimental group consisting of 65 students. The knowledge test

Table 14.5 Attitude questionnaire correlations results

Correlations			
		Attitude overall	Grades
Attitude overall	Pearson correlation	1	−0.098
	Sig. (2-tailed)		0.312
	N	108	108
Grades	Pearson correlation	−0.098	1
	Sig. (2-tailed)	.312	
	N	108	108

Table 14.6 Learning content questionnaire statistics results

Descriptive statistics							
	N	Range	Min.	Max.	Mean	Std. deviation	Variance
Content	42	3.00	2.00	5.00	3.9167	0.66182	0.438
Segmentation	42	3.00	1.67	4.57	3.7143	0.68500	0.469
Adaptation	42	3.00	2.00	5.00	3.9286	0.92110	0.848
Conception	42	3.00	2.00	5.00	4.2202	0.63005	0.397
Learning tasks	42	2.75	2.00	4.75	3.6548	0.65321	0.427
Motivation	42	3.00	2.00	5.00	3.9226	0.72096	0.520
Organization	42	2.50	2.50	5.00	3.9643	0.54452	0.296
Content overall	42	2.79	2.02	4.81	3.9030	0.53235	0.283

was created by the tutors of each course. The experimental group used remote labs and e-Learning content in order to learn the concepts assessed by the knowledge test, whereas the control group learned the same concepts by following the traditional educational process. To ensure that the groups are equally distributed according to the factors affecting the results, the Felder-Soloman Individual Learning Style (ILS) questionnaire (Felder and Solomon 1991) has been performed.

The desired outcome of this case study was, at least, an equal performance between the two groups and not a better learning outcome from the experimental group. Thus, the null hypothesis (H0) of this study is that "There is a statistically significant difference between the performance of the control and the experimental group." The analysis of the data concerning the learning outcome was conducted using an independent sample of Mann-Whitney U test (Mann and Whitney 1947). The level of significance was set to 0.05. The results of this test are presented in Table 14.7. It shows that there was no statistically significant difference between the learning achievement of the control group ($M = 65.31$, $SD = 13.15$) and the experimental group ($M = 65.08$, $SD = 14.71$). Therefore, the null hypothesis (H0) of the study is rejected.

Table 14.7 Learning outcomes statistics results

Group statistics					
Group		N	Mean	Std. deviation	Std. error mean
Tunisia	Control	66	65.3182	13.15330	1.61906
	Treatment	65	65.0846	14.71609	1.82531

14.5.3 Evaluation Outcomes

The study findings reveal a positive students' opinion toward the usefulness, ease of use, ease of learning, and satisfaction of the remote labs session.

Concerning the learning outcome evaluation, results reveal that the control group did not achieve better learning outcome than the experimental group. Thus, we can conclude that in our case, the remote lab can efficiently replace the traditional method of teaching. This replacement seems able to produce the similar learning gains for the students by providing several advantages, such as the opportunity to access special equipment and tools without additional costs and the safety to remotely participate in potentially dangerous experiments.

Concerning the learners' attitude toward remote labs, there is evidence that reflective learners had a more positive attitude toward the usage of remote labs as opposed to the active learners, who reported lower attitude scores.

14.6 Conclusions

The present chapter summarized the important aspects of the work carried out to adapt and redesign a heat exchanger bench to be fully accessed and controlled remotely. This work was performed in the framework of a cooperation project involving 16 partners from Europe and the Maghreb region. The main objective of this project called "e-Sience" was the development of an efficient remote labs' network in the Maghreb region.

As a partner of e-Sience project, we have to adapt and develop distant access solutions for two remote labs allowing students to perform practical experiments on heat transfer and mechanical vibration. The current chapter presented only the work carried out on the heat exchanger bench. This includes control of the water supply pump, the heater, and the cold and hot water supply flows. An IP camera was implemented to provide students with a real-world overview of the whole device. In addition, a user interface application was developed under LabVIEW environment to enable students to remotely control the heat exchanger. It incorporated LabVIEW front panel's controls and indicators for controlling the different parts of the heat exchanger bench and displayed feedbacks from sensors (flow meters and thermocouples) and the IP camera.

To test and validate this remote lab, technical and pedagogical evaluations, which were in the framework of the whole evaluation strategy of e-Sience project, were conducted by experts in the fields. This evaluation demonstrated that the expected learning outcomes of remote labs seem to be similar, even better, compared to conventional laboratory work, and there is clear evidence that the deployment and usage of remote labs should be continued and extended.

References

Cazacu, D. (2014). A remote laboratory for frequency-response analysis of vibrating mechanical system. *Procedia Technology, 12*, 675–680.

Chandre, B. R., Geevarghese, K. P., & Gangadhara, K. V. (2014). Design and implementation of remote mechatronics laboratory for eLearning using LabVIEW and smartphone and cross-platform communication toolkit (*SCCT*). *Procedia Technology, 14*, 108–115.

Coquard, P., Guillemot, M., Leleve, A., Noterman, D., Ben, M., & H. (2008). AIP-Primeca RAO remote laboratories in automation. *International Journal of Online Engineering, 4*(1), 12–18.

Dashboard. (2017). *The iLab project*. Retrieved from http://ilab.mit.edu/wiki

Douka, S. (2010). The teaching of Greek traditional dances in the context of physical education lessons, 11th Panhellenic Conference Sport Management, Recreation & Sport Tourism, Serres, Greece.

Fabregas, E., Farias, G., Dormido-Canto, S., Dormido, S., & Esquembre, F. (2011). Developing a remote laboratory for engineering education. *Computers & Education, 57*, 1686–1697.

Felder, R. M., & Solomon, B. A. (1991). *Index of learning styles*. Retrieved from http://www.ncsu.edu/felder-public/ILSpage.html

Gomes, L., & Bogosyan, S. (2009). Current trends in remote laboratories. *IEEE Transactions on Industrial Electronics, 56*(12), 4744–4756.

Gravier, C., Fayolle, J., Bayard, B., Ates, M., & Lardon, J. (2008). State of the art about remote laboratories paradigms – Foundations of ongoing mutations. *International Journal of Online Engineering, 4*(1), 19–25.

Guerra, T. C., de León Morales, J., Traore, D., Glumineau, A., & Boisliveau, R. (2007). Teleoperation of an experimental platform of electrical machines through the Internet. In *Proceedings of the 2nd*International Conference on*Digital Information Management, ICDIM'07, Lyon, France*.

Guimarães, E. G., Cardozo, E., Moraes, D. H., & Coelho, P. R. (2011). Design and implementation issues for modern remote laboratories. *IEEE Transactions on Learning Technologies, 4*(2), 149–161.

Hansen, E. (1990). The role of interactive video technology in higher education: Case study and proposed framework. *Educational Technology, 30*, 13–21.

Hardison, J. L., DeLong, K., Bailey, P. H., & Harward, V. (2008). *Deploying interactive remote labs using the iLab shared architecture*. The 38th SEE/IEEE Frontiers in Education Conference, Saratoga Springs, NY.

Lund, A. M. (2001). Measuring usability with the USE questionnaire. *Usability Interface, 8*(2), 3–6. Retrieved from www.stcsig.org/usability/newsletter/index.html.

Mann, H. B., & Whitney, D. R. (1947). On a test of whether one of two random variables is stochastically larger than the other. *Annals of Mathematical Statistics, 18*(1), 50–60. https://doi.org/10.1214/aoms/1177730491.

Mao, T. (2007). *Scheduling services and security ticket token services in iLab interactive service*. Master's thesis, Massachusetts Institute of Technology Cambridge, USA).

Naddami, A., Falhi, A., Gourmai, M., & Moussetad, M. (2014). Determination of synchronous machine parameters using iLab shared architecture. *International Journal of Online Engineering, 10*(3), 18–23.

Nasri, I., Ennetta, R., & Bouallègue, S. (2015). Mechanical remote laboratory development using iLab shared architecture. *International Journal of Online Engineering, 5*, 31–37.

National Instruments. (2017). Retrieved from http://www.ni.com

Odeh, S. (2014). A web-based remote lab platform with reusability for electronic experiments in engineering education. *International Journal of Online Engineering, 10*(4), 40–45.

Scanlon, E., Colwell, C., Cooper, M., & Di Paolo, T. (2004). Remote experiments, re-versioning and re-thinking science learning. *Computers & Education, 43*(1–2), 153–163.

Schauer, F., Ozvoldova, M., Lustig, F., & Dekar, M. (2008). Real remote mass-spring laboratory experiments across internet-inherent part of integrated E- learning of oscillations. *International Journal of Online Engineering, 4*(1), 52–55.

Schoor, C., & Körndle, H. (2012) Checklist for a didactically sound design of eLearning content. *eLearning Papers, 29*, 1–6. ISSN: 1887–1542.

Shroff, R., Soumare, H., Hardison, J., del Alamo, J., & Harward, J. (2009). *A versatile internet-accessible electronics workbench with troubleshooting capabilities*. The 6th International Conference on Remote Engineering and Virtual Instrumentation, REV 2009, Bridgeport, CT, USA.

Tobarra, L., Ros, S., Hernández, R., Marcos-Barreiro, A., Robles-Gómez, A., Caminero, A. C., Pastor, R., & Castro, M. (2015). Creation of customized remote laboratories using deconstruction. *IEEE Revista Iberoamericana De Tecnologias Del Aprendizaje, 10*(2), 69–76.

Tsiatsos, T., Douka, S., Mavridis, A., Tegos, S., Naddami, A., Zimmer, T., & Geoffroy, D. (2014). Evaluation plan and preliminary evaluation of a network of remote labs in the Maghrebian countries. *International Journal of Online Engineering, 10*(5), 15–20.

Zimmer, T., Geoffroy, D., Pester, A., Oros, R., Tsiatsos, Th., & Douka, S. (2013, July). *eSience: Setting up a network of remote labs in the Maghrebian countries*. International Conference on Engineering Education and Research (iCEER 2013), Marrakesh, Morocco.

Zutin, D. G., Auer, M. E., & Gustavsson, I. (2011). *A VISIR lab server for the iLab shared architecture*. 2011 IEEE Global Engineering Education Conference (EDUCON): Learning Environments and Ecosystems in Engineering Education, 30–33, Amman, Jordan.

Ridha Ennetta is assistant professor of mechanical engineering at the Higher Institute of Industrial Systems, Gabes University, Tunisia.

Ibrahim Nasri was (2011–2014) director of the Higher Institute of Industrial Systems, Gabes University, Tunisia. He is currently assistant professor of mechanical engineering at the college of engineering of Tabuk University, Saudi Arabia.

Soufiene Bouallègue is assistant professor of electrical engineering at the Higher Institute of Industrial Systems, Gabes University, Tunisia.

Thrasyvoulos Tsiatsos is currently assistant professor in the Department of Informatics and member of the Multimedia Lab of Aristotle University of Thessaloniki, Greece.

Chapter 15
Collaborative Virtual Laboratory Environments with Hardware in the Loop

Z. Zhang, M. Zhang, Y. Chang, E.-S. Aziz, S. K. Esche, and C. Chassapis

Abstract Over the last decade, the research community has expanded substantial efforts aiming at designing, agreeing on, and rolling out technical standards and powerful universal development tools that allow the rapid and cost-effective integration of specific experimental devices into standardized remote laboratory platforms. In this chapter, a virtual laboratory system with experimental hardware in the loop is described.

Keywords Virtual laboratories · Remote laboratories · Mixed reality environments · Virtual engineering environments · Human-computer interface

15.1 Introduction

15.1.1 Definition

Distance education has become an increasingly popular form of education which has been deployed at different levels of engineering and science education. In order to enable this modern form of education, remotely accessible laboratory systems – or

Z. Zhang (✉)
Department of Mechanical Engineering Technology, CUNY New York City College of Technology, Brooklyn, NJ, USA
e-mail: ZhZhang@citytech.cuny.edu

E.-S. Aziz · S. K. Esche · C. Chassapis
Department of Mechanical Engineering, Stevens Institute of Technology, Hoboken, NJ, USA
e-mail: EAziz@stevens.edu; SEsche@stevens.edu; CChassap@stevens.edu

M. Zhang
Department of Mechanical and Industrial Engineering, Southern Illinois University Edwardsville, Edwardsville, IL, USA
e-mail: MZhang@siue.edu

Y. Chang
Department of Computer Science, Stevens Institute of Technology, Hoboken, NJ, USA
e-mail: YChang1@stevens.edu

M. E. Auer et al. (eds.), *Cyber-Physical Laboratories in Engineering and Science Education*, https://doi.org/10.1007/978-3-319-76935-6_15

remote laboratories for short – have been developed for over two decades as one of the most important support tools for distance education. Remote laboratories (RLs) are also often referred to as online laboratories or remote workbenches.

More recently, with the continued rapid advances in virtual reality (VR) technologies, the concept of remote laboratories has been expanded to also include collaborative virtual laboratory (VL) environments, and as a result, RLs are gradually replaced by VLs, since they offer all the functions and characteristics of RLs and in addition allow for simulations, cooperation, and immersion. VLs are usually defined as laboratories that are operated remotely in virtual or augmented form through telecommunication. As the name suggests, virtual means that the environment in VLs is created based on the visualization of physical laboratories by means of computer graphics techniques. After that, physics engines are deployed to fulfill the function of simulation. In order to realize remote control of physical experimental devices, connections and communications are established between physical hands-on laboratories and VLs. In order to improve the feel of immersion of the users, VLs are augmented with a variety of data acquisition (DAQ) systems.

15.1.2 Evolution and Current Status

The development of VLs and RLs had progressed in parallel for a period of time before VLs started dominating in recent years. The history of the concept of VLs can be traced back to the late 1970s. The original VLs were simulators, for example, the early military training VL presented in Kocian (1977). Furness and Kocian (1986) presented a more advanced model of a virtual flight simulator during the late 1980s. These simulators were designed through local sensors, mechanical mechanisms, controllers, and motors, and hence, they were also referred to as hardware-intensive VLs (see Fig. 15.1) (Karim 1992). In the 1990s, simulations based on computer-aided platforms were investigated (Griffith et al. 1992; Adam 1993; Vosniakos et al. 1997), and they were referred to as software-intensive VLs. An example of the CAD workflow is shown in Fig. 15.2. In these implementations, modeling, simulation, and optimization were the main features. Both hardware-intensive simulators and software-intensive CAD simulation platforms are operated locally without the support of telecommunication techniques. In the 1990s, with the development of telecommunications based on Ethernet standards, which enable stable data transmission remotely, web-based VLs appeared (Gertz et al. 1994; Sears and Watkins 1996; Harasim et al. 1996; Rzepa and Tonge 1998). The objectives of this kind of VLs are to provide remote access to physical devices, enable distance communications between instructors and learners, share costly equipment and resources, and help the learners in understanding the experiments. In fact, web-based VLs have all the functions provided by traditional RLs. From the late 1990s and early 2000s, with PCs being equipped with high-performance central processing units (CPUs) and graphics processing units (GPUs), VL platforms taking advantage of computer graphics techniques were created (Obeysekare et al. 1997; Avradinis et al. 2000; Witmer and Singer 1998; Kfir 2001). With the support of advanced

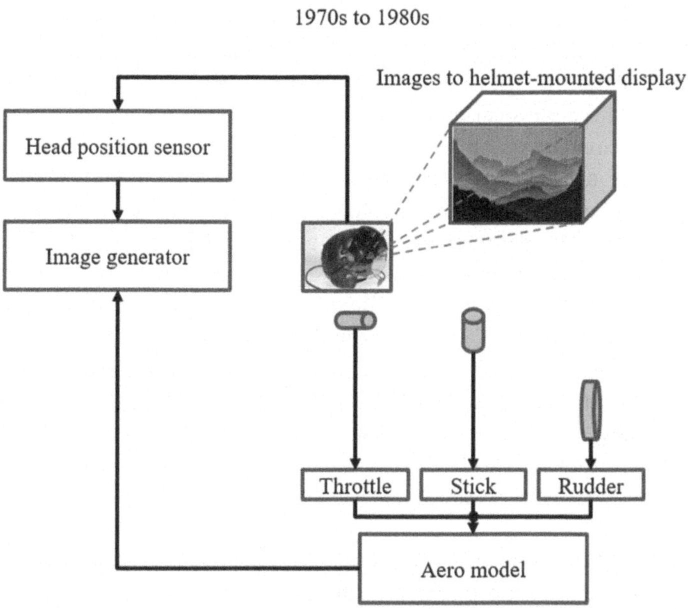

Fig. 15.1 Hardware-intensive simulator: diagram of visually coupled airborne system simulator

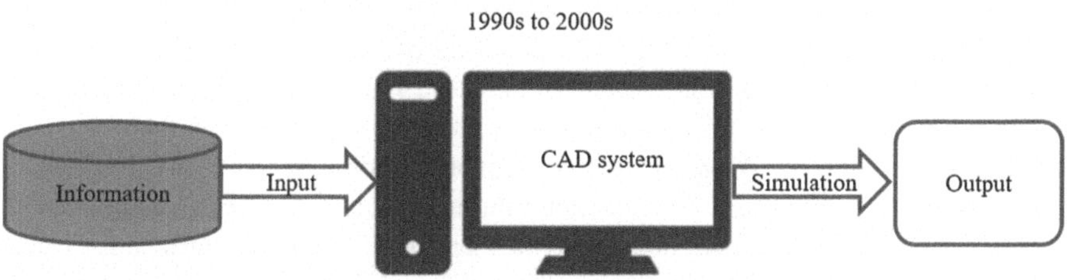

Fig. 15.2 Workflow of a software-intensive CAD system

CPUs and GPUs, the virtual environments (VEs) used in VLs were optimized significantly, but 2D VEs were still prevalent. At the same time, VLs developed based on computer graphics were still used as demonstration tools for experiments during this period.

From the mid-2000s, network techniques became much more advanced than before, which enabled real-time processing and remote transmission of massive amounts of data, while 3D graphics were improving dramatically. As a result, VLs based on 3D graphics and networks were becoming popular (Alexiou et al. 2004, 2005; Familia 2005; Ramasundaram et al. 2005; Adamo-Villani et al. 2006). The main advantage over earlier ones was that these VLs looked more realistic, but they could only let users familiarize themselves with the procedures of an experiment. Upon entering the late 2000s, physics engines were becoming mature, benefiting from a boom in CAD software and video games. At that time, VLs with physics engines were presented in many publications (Jacobson and Lewis 2005; Hummel et al. 2012; Song et al. 2008; Howard and Vance 2007; Jia 2006; Aziz et al. 2006a, b). These VLs could realize simulations and interactions in accordance with the physical properties of the virtual models, while maintaining a realistic virtual mapping of the real world. In order to further improve the users' interest in VLs and facilitate the creation of VLs, game-based VLs (GBVLs) employing game engines were developed (Chang et al. 2006a, b, 2007; Trenholme and Smith 2008). The game engines were endowed with various basic ready-to-use functions such as graphics rendering, sound generation, networking, physics modeling, game logics, artificial intelligence, and user interactions (Zhang et al. 2014). Therefore, developers of VLs could devote most of their effort to the design of the experiments themselves. Certainly, to provide users a better feel of immersion, VLs were augmented and mixed with various kinds of feedback sensors and physical experimental devices (Zhang et al. 2013a, b; Borghetti et al. 2013; Dorozhkin et al. 2012). The nature of these VLs was a seamless integration of hardware-intensive simulators and VEs of software-intensive VLs. This integration also remains the current trend in VLs.

RLs were introduced to the world after the appearance of VLs. The underlying concepts of RLs can be traced back to the late 1980s and early 1990s, when the communication methods had been expanded greatly from the telephone to video conferencing with the development of advanced computers and the Internet (Verma and Lin 1989; Aburdene et al. 1991; Arpaia et al. 1997; De Meyer 1991; Bohus et al. 1995; Taylor and Trevelyan 1995). The function of the early VLs was to acquire data from devices located at a distance. In the late 1990s and during the 2000s, with further improvements in telecommunication techniques, many RLs were used in education (Esche 2005; Aziz et al. 2006; Hahn and Spong 2000; Ma and Nickerson 2006; Deniz et al. 2003) and professional training (Leleve et al. 2003; Lustigova and Lustig 2009). The RLs mentioned above only provided a remote interface for users to conduct physical experiments. Nowadays, physical devices have been integrated into VLs, which is often referred to as VLs with hardware in the loop. Then, VLs not only enable remote access, real-time simulation, and immersive experiences (Zhang et al. 2013a; Balamuralithara and Woods 2009; Jara et al. 2011; Andujar et al. 2011), but they also provide the functions for remotely controlling physical devices. In addition, the basic configuration of current VLs has been nearly the same since the 1990s (see Fig. 15.3). A more advanced form of VLs with integrated physical devices is depicted in Fig. 15.4 (Zhang et al. 2013a).

Based on the above introduction, the following discussion will focus on current developments in VLs and their corresponding components, characteristics, categories, and functions.

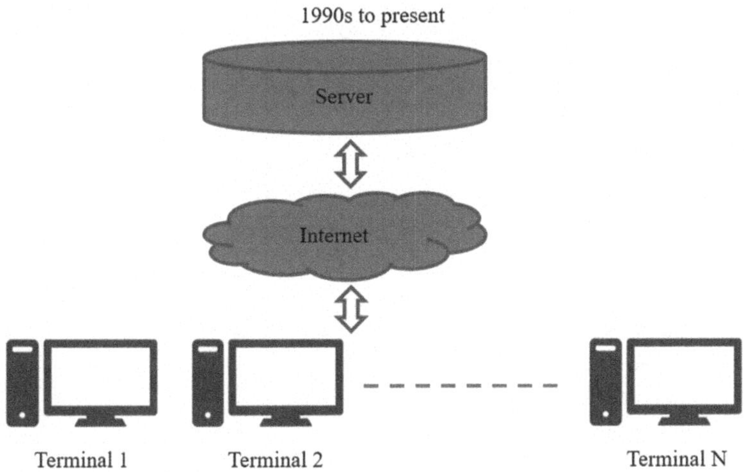

Fig. 15.3 Common configurations of network-based VL

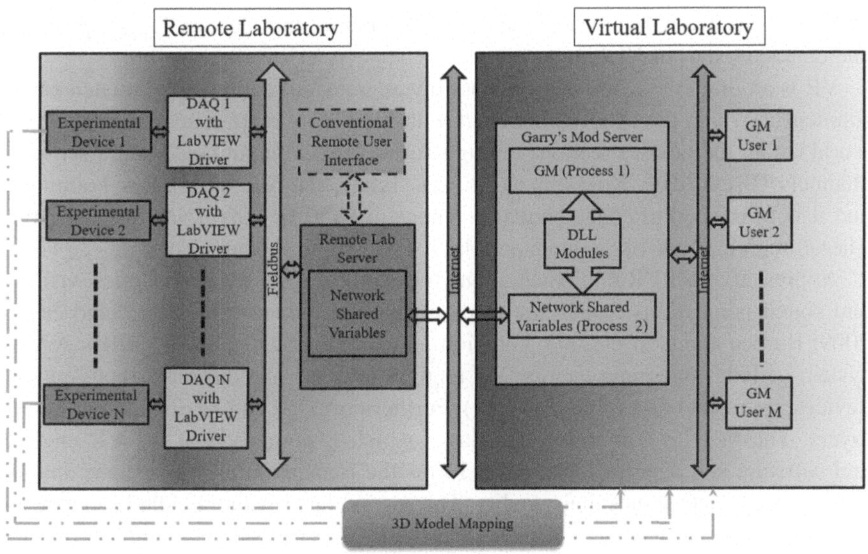

Fig. 15.4 Game-based VL with integrated physical devices

15.1.3 Components and Characteristics

Present VLs are one of the implementations of VR and have the same architecture as VR systems. Therefore, they simultaneously have all of the common characteristics of VR systems, including both their advantages and disadvantages. Below,

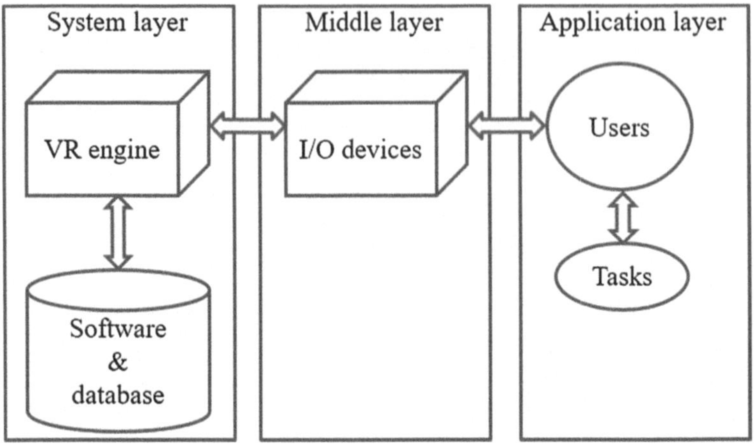

Fig. 15.5 Architecture of virtual reality systems

additional details of VR systems are provided before introducing the components and characteristics of VLs.

VR is an immersive, interactive (possibly augmented) computer-simulated environment in which the users can interact with the virtual representations of the real world (or an imaginary creation) through various input/output devices and sensory channels (Brey 2014; Kozak et al. 2014). Today, the VR industry is booming, and it is forecasted to create annual revenues of $150 by 2020 (Nandwana 2016). Therefore, VR is one of the most popular research topics today.

At present, most VR implementations focus on the generation of haptics, vision, and sound perceptions, which are three of the five human senses (Luciano et al. 2009; Burdea and Coiffet 2003; Gaggioli and Breinin 2001; Dilwort 2010). A VR system includes five main factors: VR engine, software and database, input/output devices, users, and tasks (see Fig. 15.5). Furthermore, VR can be divided into three layers. The first layer is the system layer which is composed of the VR engine and software and database. This layer forms the foundation of VR systems and to a large extent determines their quality. The second layer is the middle layer which includes input/output devices. This layer provides the necessary communication and interaction capabilities. The third layer is the application layer which is formed by the users and the tasks performed by them.

The VR engine is used to perform virtual object modeling and simulation, which includes geometry, texture, intelligent behavior (Luck and Aylett 2000), and modeling of physical characteristics (e.g., hardness, inertia, surface plasticity). The VR engine is the core of any VR system, which reads its input devices, accesses task-dependent databases, performs the real-time computations required to update the state of the virtual world, and feeds the results to the output devices.

Software is used to create the components of the VR engine and to provide powerful programming packages (e.g., IDE, framework, toolkit, API or SDK,

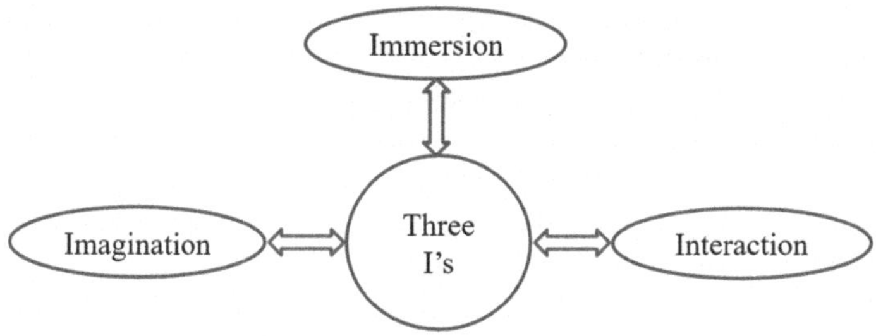

Fig. 15.6 Characteristics of VR

etc.) that help the VR application developers, for example, to create the models manipulated by the VR engine.

The input/output devices include user input devices (such as trackers, gloves, keyboard, or mice) and output devices (such as head-mounted displays (HMDs), 3D large-volume displays, force feedback robotic arms, etc.) and other physical devices which are used to augment the VLs.

The function of the users is to make a detailed plan for performing the given tasks.

The characteristics of VR can be summarized by three words: immersion, interaction, and imagination, and they are the so-called three I's of VR (see Fig. 15.6).

Immersion means to use sensor technology that brings the human senses to digital content, creating experiences that are engaging, impactful, and ultimately more real. Examples include the application of feedback gloves (Merians et al. 2002), haptic sensors (Luciano et al. 2005), fabric gloves, and head-mounted displays (Barfield 2015).

Interaction means to use interfaces of the system to realize the communications between the system and its users. Common interfaces include keyboards and mouse, while advanced interfaces include cameras (Chang et al. 2014) and data gloves (Lu et al. 2012).

Imagination means to use computer graphics techniques to provide the corresponding virtual representations of real environments. In fact, although immersion and interaction are close to our hearts, imagination realized by many sophistical techniques is very important to any VR. It is not just a graphical representation of the world, but it also has applications that involve solutions to real problems in engineering (Zyda 2005), medicine (Ayache 1995), military (Livingston et al. 2002), etc. The extent to which an application is able to solve a particular problem depends very much on the human imagination, the third "I" of VR. Therefore, VR represents the integration of immersion, interaction, and imagination. The imaginative part of VR refers also to the mind's ability to perceive nonexistent things (Luciano et al. 2009).

As mentioned above, VLs are VR systems. Therefore, VLs are composed of the five components introduced above, and their characteristics can be summarized by the three terms of immersion, interaction, and imagination. Based on this point, VLs provide several important benefits.

Firstly, VL systems can reduce the costs and resource consumption. The creation of virtual systems instead of physical ones can make the creation of multiple copies of physical devices unnecessary, thus reducing the consumption of natural resources. In addition, the virtual systems can also alleviate the requirement for human resources during training.

Secondly, VL systems are inherently safer and less failure prone than physical ones. What happens in the virtual world will not threaten the users physically. For example, firefighters cannot get hurt in a virtual firefighter training system.

Thirdly, VL systems can be shared locally and remotely by multiple users simultaneously. Users can visit the same VL server and collaborate to finish the same tasks through the Internet, which can provide the users with flexibility and convenience.

Fourthly, augmented VL systems can provide their users with a feel of immersion.

15.1.4 Categories Based on Different Criteria

15.1.4.1 Non-immersion vs. Immersion

At present, there are no strict criteria for categorizing VLs yet, despite their history of over three decades. Therefore, there are many categorizing methods. To our knowledge, VLs can be categorized based on the three characteristics of immersion, interaction, and imagination.

According to the users' feel of immersion, there are immersive and non-immersive VLs. Immersive VLs use feedback sensors and/or integrate physical devices of experiments into the VEs to give the users the perception of being physically present in a nonphysical world. The sensors include haptic feedback sensors (e.g., fiber-optic-wired gloves (Noor and Wasfy 2001), vision sensors (e.g., shuttle glasses (Lin et al. 2002), head-mounted displays (Azuma et al. 2001), acoustic sensor, etc. The integration of physical devices into VLs can provide the users with real-time data from the experimental devices and give them the perception of in-person participation (Zhang et al. 2013a, b; Luciano et al. 2009).

Compared with immersive VLs, non-immersive VLs are simpler systems. They provide the users with a VE in which they can manipulate the models and perform simulations. These VLs use PCs (Li et al. 2003) or mobile devices (Bottentuit Junior and Coutinho 2007) as the implementation platform of the system and employ keyboards, mice, trackballs, or touch screens as communication interfaces (Smedley and Higgins 2005).

15.1.4.2 Categories with Interaction

VLs can also be divided into single-player and multiplayer VLs.

Single-user VLs provide an environment that only permits one user to interact with it at any given time and lets this user perform preprogrammed or AI-controlled tasks. In addition, single-user VLs are commonly found in locally operated simulators (Valera et al. 2005). The most widely known simulators are virtual flight simulators (Furness and Kocian 1986). These simulators can only be played by one person at a time. Remotely controlled VLs have been reported in Esche and Chassapis (1998), Rohrig and Jochheim (1999), and Casini et al. (2001).

Multiuser VLs are those in which multiple users can perform experiments in the same environment simultaneously. These VLs allow user interaction with other individuals in partnership or competition while providing functions for communication among these users. In multiuser VLs, the users may work in single-user mode or work cooperatively with partners to achieve a common goal. The users of multiuser VLs usually share the common resources remotely with the support of a network. Sample multiuser VLs were described in Macedonia et al. (1995), Nelson et al. (2005), and Chang (2016).

15.1.4.3 2D and 3D VLs Based on Imagination

According to the employment of different graphics techniques, VLs can further be categorized into 2D and 3D.

2D VLs use 2D computer graphics to create digital images. They provide graphical user interfaces that enable the interaction with computer-based VEs. The major benefit of 2D VLs is that they take full advantage of most common input devices, such as mouse, keyboard, and trackball because these input devices are constrained to two dimensions of movement. At the same time, 2D VLs also provide geometric primitives and support procedural models which are necessary for the illustration of experiments (Cockburn and McKenzie 2002). Some examples can be found in Faulkner and Krauss (1996), Hirose (1997), Abe and Cardoso (1999), Nah et al. (2011), and Sharma et al. (2011).

3D VLs use three-dimensional representations of the real components to create VEs. Most of the visualization of the real world is performed by 3D modeling software. This includes the procedures of 3D computer graphics, namely, modeling, rendering, layout, and animation (Hughes et al. 2014). 3D VLs are more similar to the real world than to 2D VLs, and therefore they provide a lifelike experience for users. At present, 3D VLs form the majority, and some examples can be found in Pukhov (1999), James et al. (2004), Arango et al. (2008), and Chang et al. (2013).

15.1.4.4 Other Methods for Categorizing VLs

There are other well-known terms for different forms of VLs, including desktop, distributed, and game-based VLs.

Desktop VLs use a computer monitor to provide a graphical interface for users (McLellan 2001).

In distributed VL systems, which are implemented on several computers instead of a single one, the users can interact with each other in real time through a network (Griffith et al. 1992; Freund and Roßmann 2003).

Game-based VL systems are implemented based on a game engine. Generally, game engines provide the developers with various basic functions such as graphics rendering, sound generation, physics modeling, game logics, artificial intelligence, user interactions, and networking (Baba et al. 2007; Thorn 2010). The resulting GBVL implementations are immersive, distributive, and collaborative. At the same time, because computer game engines provide a suite of development tools, they make the common components of VLs reusable and adaptable. An example can be found in Chang et al. (2012).

15.2 Prevalent Techniques Used to Create and Operate Virtual Laboratories

15.2.1 Techniques to Create a Virtual Environment by Visualizing the Real World

15.2.1.1 Virtual Environment

Visualization of the world is the process of mapping real objects into a VE. In order to clarify this process, the concept of VE is first introduced. VEs include models, a virtual space (VS), avatars, and plots.

Models are virtual mappings of real objects. The main functions of the models in VLs are to lay out the VS and to be used for simulations. The models are usually stored in a model library and ready to be selected. While both models and other virtual components in 3D VEs are created using 3D modeling techniques, they have different requirements according to their functions in VEs.

As part of the creation of a model, a bounding box, collision detection, interactive topology, and constraints have to be defined, which represent the basic requirements for simulations. 3D models can be in the form of wireframes, surfaces, or solids. They use primitives (such as points, lines, polygons, curved surfaces, boxes, cones, cylinders, spheres, wedges, and tori) to represent various geometric entities. Solid models can have physical properties, and therefore, they enable some advanced applications such as virtual assembly, additive manufacturing, etc.

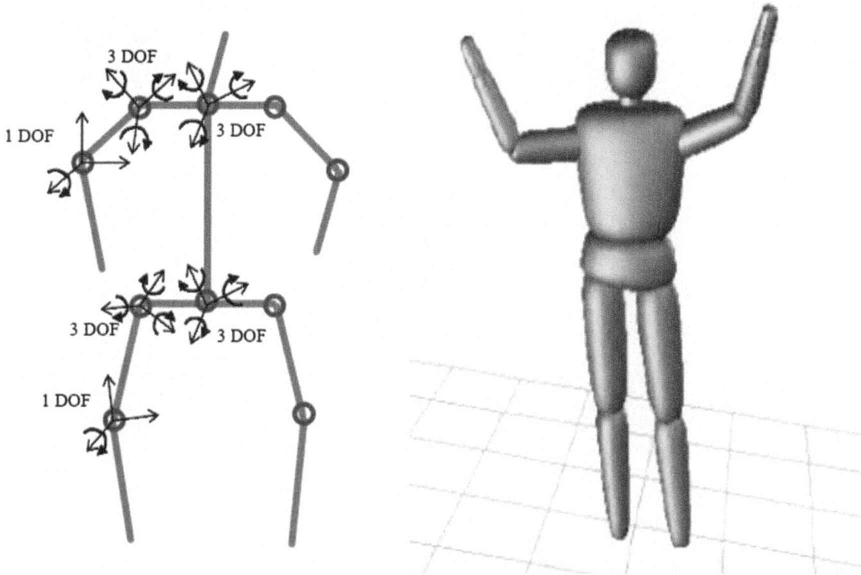

Fig. 15.7 Skeleton and model

Avatars are graphical representations of the users. They may be described by either two-dimensional or three-dimensional forms. The 2D form of avatars appears as an icon in online communities, for example, the profile photo of one's Facebook account. They can also be animated images as being represented in 2D RGB games, for example, the famous game of Super Mario (Classic Games 2017). The most common avatars in modern VLs are 3D models of human bodies, for example, the avatars in Dota (Wikipedia 2017) and Steam (2017). The creation of 3D avatars is a complicated process. It includes tracking, segmentation, model fitting, motion prediction, kinematics, dynamics of articulated structures, etc. Avatars must have a skeleton which represents the human body in accordance with the anatomical structure of the human body. Following that, a mathematical model has to be created to control the degrees of freedom and movements of the avatar. In Fig. 15.7, a skeleton with 22 degrees of freedom is presented (Sarris and Strintzis 2005).

A VS is a map in which all of the interactions between models and avatars can be implemented. A VS is modeled by a single complex surface. It is a 3D representation of a physical space. At the same time, a VS should be a penetrable solid, while at the same time enabling collision detection.

15.2.1.2 VE Creation with CAD Software

As introduced above, this section is only focusing on the creation of 3D VEs. The most popular approach for creating 3D VEs is to use CAD software (e.g., Maya

Fig. 15.8 Typical CAD/CAM system

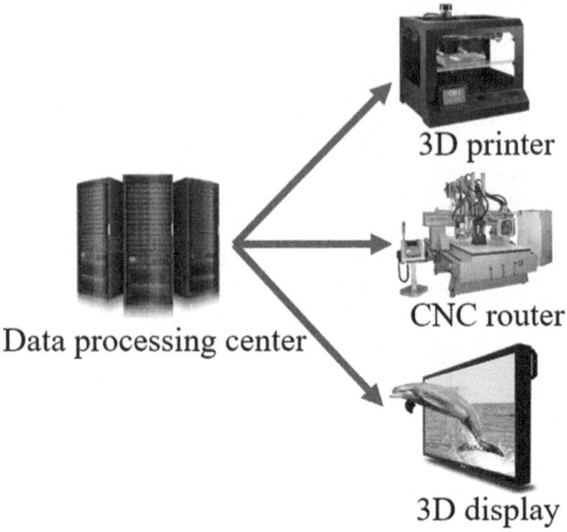

(Autodesk 2017a), 3D Studio Max (Autodesk 2017b)). The process of VE creation includes modeling and animation.

A typical CAD system includes hardware, software, and corresponding peripherals. The primary concern for the 3D modeling is the software part. The software uses graphics techniques to represent real objects in digital form, stores the basic primitives and the reusable models into a database to facilitate the design process, and provides a rendering mechanism, middleware, and drives for displaying the models with various peripherals (monitor, 3D printer, computer numerical control router, etc.) (see Fig. 15.8).

For rigid models, the process of 3D model creation usually starts with a sketch of the wireframe of an object. After that, color and textures are applied to the wireframe. Finally, the 3D models are rendered and ready for sending to the peripherals. As an instance of rigid model creation, Fig. 15.9 illustrates the procedures used to create the map of a GBVL. For the models used in VEs, the pipeline of creation is more complicated than the creation of rigid models because the animation and simulation must be taken into account. The first step is character modeling and prop modeling (including the physical properties and roles of this model) (Patnode 2012). Then, the basic skeleton of this model is created. Subsequently, the parts of this skeleton are textured and rigged (e.g., by adding joints and defining the degrees of freedom among the bones of the skeleton) (Autodesk 2017c). Finally, the models are sent to the output peripherals for manipulation. Figure 15.10 depicts the workflow used to create an animated octopus model with the Maya software (Davidphillips 2017). This model is a typical example for complicated models which may be used in the VE creation.

The common method used to create VEs can be generalized into the following steps:

Fig. 15.9 Flow chart of map creation in CAD software as an instance of rigid models

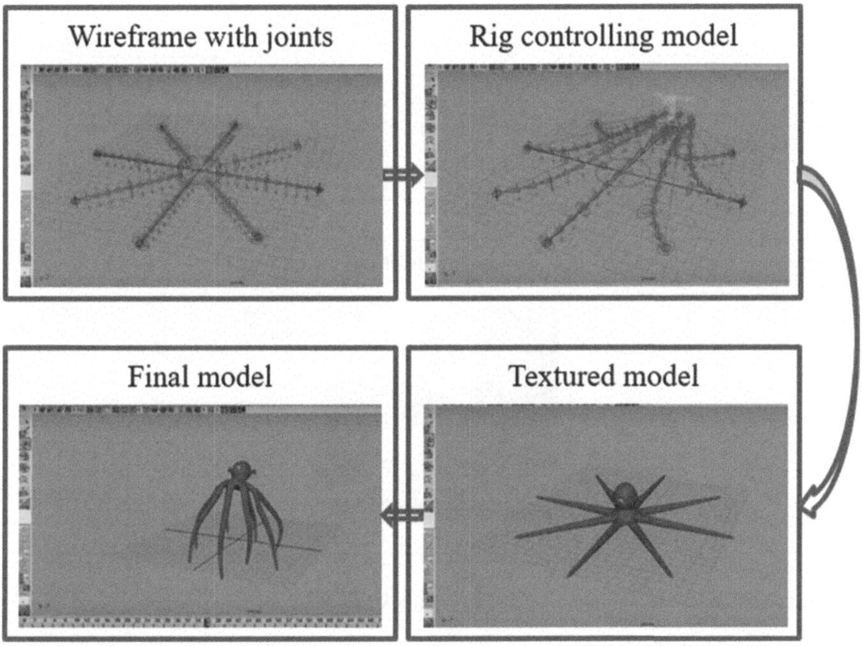

Fig. 15.10 Workflow of octopus modeling with animation

- Obtaining the geometric information of real objects with various contact and/or noncontact measurement tools. The common contact measurement tools include tape measure, speed square, protractor, micrometers, and vernier calipers. The non-contact measurement tools include laser measuring tools CCD camera and digital camera.
- Off-line processing the acquired data from the measurement tools and plotting the models of the real objects with built-in tools such as "SDK of VL" or third-party software such as Solidworks, Maya, and 3D Studio Max.
- Converting the models into the format of the VL.

In order to illustrate the common modeling process, a pipeline used to create the VEs of GBVLs based on Garry's Mod is presented in Fig. 15.11.

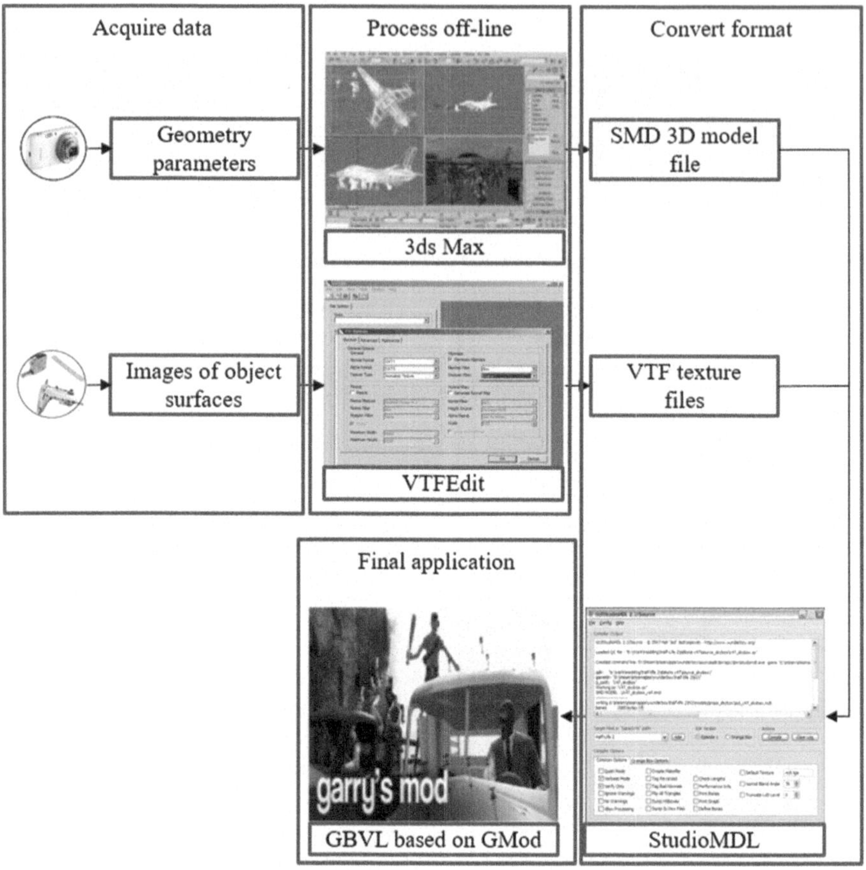

Fig. 15.11 Example for common method of creating VEs

15.2.2 *Simulation in VEs*

15.2.2.1 Physics Engines

For a 3D game, in order to enhance users' feel of immersion, physics simulations of gravity effects, rigid-body dynamics, collision phenomena, etc. are essential. However, the programming of these effects is complicated. For VL developers in small teams, it is a heavy burden to yield satisfactory physics simulation effects. In addition, the generated simulation codes are hard to be reused as they are usually application specific.

Physics engines represent software that is specifically designed for simulating physical systems in other computer applications. They usually combine various physics simulations and offer multiple levels of approximation. Application developers can easily use physics engines for physical simulations rather than create these simulations by themselves. They can also flexibly enable or disable different simulation effects and choose the desired approximation level.

Many commercial physics engines are available. The most frequently applied ones include Havok (2017), NVidia PhysX (2017), Amazon Lumberyard (2017), and True Axis (2017). Many of them are free for non-commercial use. Open-source physics engines include bullet (2017), open dynamics engine (2017), and Tokamak physics (2017). By using an open-source engine, developers have the option to modify the source code if some simulations are not satisfactory (however, this is not easy).

Generally, there are two approaches for simulating dynamics: mass-aggregate physics simulation and rigid-body physics simulation. In the former case, objects are modeled as a linked graph with a mass node (usually in an object all nodes have the same mass) and massless rigid links. For instance, a box can be modeled as 8 mass nodes at its 8 vertices being connected by 12 links at its edges. Such a model greatly simplifies the physics simulation. For some applications, such as beam bridge simulations, this method is very suitable. For other applications, such as a rolling wheel, drawbacks are evident as the simulation cannot be continuous due to the discrete mass distribution.

Another method, which is more complicated but nevertheless used by most commercial engines, is rigid-body physics simulation. In this method, the dynamics of several rigid-body primitives, including box, cylinder, cone, sphere, etc., are preprogrammed. An object, regardless of its real shape, is fitted into another model, called "physics model," which consists of one or several of the primitive geometries. Game designers or VE authors must properly assign a physics model for every geometric model if a rigid-body physics simulation is desired. Some CAD software (e.g., 3D Studio Max) or game development kits (e.g., Source SDK 2017) support automatically generated physics models.

The purpose of collision detection is to determine whether two objects in a VE are overlapping or not and to find the contact point if they are. Collision detection is still a very active research topic because of its complexity. It necessitates significant computational power, which most personal computers do not provide. Therefore, many methods were developed to simplify the collision detection. The most common method is applying a bounding volume for an object.

Bounding volumes, like the previously mentioned physics models for rigid-body physics simulation, are lumped models, which are composed of simplified geometries for collision detection. In some engines, the physics model serves as the bounding volume.

Due to its simplicity, the most commonly encountered bounding volume is a bounding box. A bounding box, by its name, is a cuboid that encompasses the geometry model. There are two types of bounding boxes, namely, axis-aligned bounding boxes (AABBs) and object-oriented bounding boxes (OBBBs). In the former case, the bounding box is always aligned with the coordinate system. Therefore, an AABB bounding box can be represented by its two diagonal vertices (usually the lower-left corner and upper-right corner). However, for models with some special geometries, this bounding box may greatly distort the collision detection as the bounding box may contain significant free space. OBBBs can better approximate the occupied volume of an object as they are not necessarily aligned with the axes. However, along with the coordinates of the two corners, an additional direction vector must be included into the OBBBs in order to represent the bounding volume, and hence the collision computation is more complicated.

15.2.2.2 Modeling with Constraints

When two rigid-body objects are associated with each other (e.g., two bars are connected by a hinge), a dynamical constraint is formed such that the movement of one object affects the movement of the other. Many physics engines allow developers to describe an object with multiple associated rigid bodies and simulate the behaviors of these objects accordingly. Developers only need to specify the constraint position and type, letting the physics engine compute the detailed movement of the object. Table 15.1 lists the common constraints usually included in game engines.

15.2.3 Interface of VLs

Many activities can be performed in a VL, for instance, navigating through the VL, manipulating objects, creating assemblies, and disassembling them. Although novel human-computer interfaces have become available recently, the most frequently used human-computer interaction is still through keyboard and mouse.

One of the most commonly performed actions in laboratory VLs is picking up an object. Through mouse and keyboard, game engines usually define a standard operation for users to define object picking. In Garry's Mod, for example, a "physics gun" is provided by the game engine. An avatar shoots a "laser beam" from the

Table 15.1 List of common constraints

Constraint type	Description	Number of DOF removed
Revolute	Joins two parts with an axis about which they can spin freely; their relative positions in axial direction are fixed	5
Spherical	Joins two parts with same center point about which they can rotate freely in all directions	3
Elastic	Connects two parts with a spring-like rope that, when compressed or stretched, tries to resume its original length	5
Prismatic	Creates a path along a straight line on a part that a matching part can travel along	5
Fixed	Joins two parts such that afterward they can no longer be moved relative to each other	6

Table 15.2 List of common operations in 3D games

Operation	By mouse and keyboard
Navigate forward/backward/left/right	Press key "w"/"s"/"a"/"d"
Turn left/right	Move mouse left/right
Select a mechanical part	Press key "e"
Talk to teammate	Press key "u"
Carry a part and walk	Hold left mouse button and press navigation keys
Pick up a part	Aim part by mouse, hold left key
Manipulate a picked-up part	Hold key "e," hold left mouse button, move mouse
Create assembly	Combine operation of navigation keys, hold left mouse button, use roller on mouse when necessary

physics gun. The first object that the laser beam encounters can then be manipulated by mouse and keyboard. Although this method is not realistic, it is very easy for users to understand and use.

In addition, most game engines provide similar keyboard and mouse operations for other actions. Table 15.2 lists these common operations.

15.2.4 Augmentation of VLs

Augmentation of VLs is a common approach for improving the users' feel of immersion. A VL created based on augmentation is sometimes called an augmented laboratory (AL). In an AL, physical devices are integrated into a VL, thus allowing real objects to interact with the virtual ones in real time. In addition, the ALs discussed here refer to ones created with 3D modeling techniques. Usually, platforms for realizing augmentation of VLs are equipped with some wearable devices (Barfield 2015) (e.g., HMDs, data gloves, or computational clothing) or integrated with physical devices (e.g., smart phones, handheld computers, or experimental devices) (Zhang et al. 2013a; Barfield 2015). For the augmentation of VLs, there are three critical problems to be solved:

- How can experimental data be acquired with appropriate sensors? This leads to a discussion of DAQ systems.
- How can the real and virtual worlds interoperate in real time? This involves the tracking of the real objects, the real-time mapping of the real world, and the synchronization between the real and virtual worlds.
- How can the communication between the virtual and real worlds be realized efficiently and even in real time? This requires the rethinking of the approaches for passing data between the real and virtual worlds.

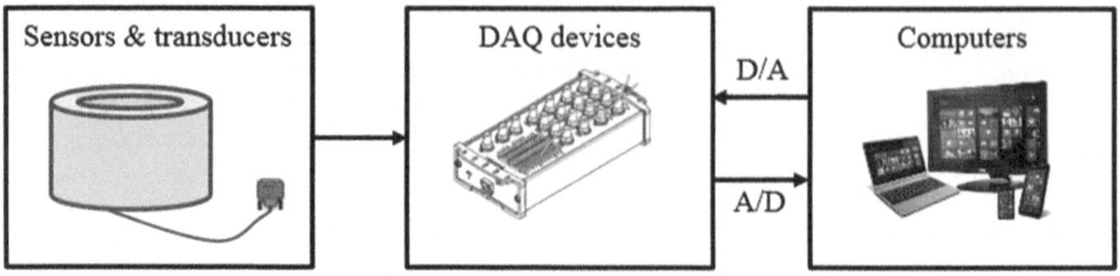

Fig. 15.12 A typical DAQ system

15.2.4.1 DAQ Systems

DAQ systems are used to acquire and process various data. They are composed of sensors, DAQ devices, and computers as shown in Fig. 15.12 (National Instruments 2017).

- Sensors are used to acquire the data on specific physical phenomena (location, temperature, force, etc.) and then represent the corresponding phenomena by appropriate signals (light, voltage, etc.). Sensors are employed in diverse areas such as medical care, communication, automobiles, manufacturing, etc.
- DAQ devices serve as the interfaces between sensors and computers. Their primary functions are to condition the signals and convert them from analog into digital forms. DAQ devices are commonly composed of signal conditioning components (signal filtering, amplifying, and isolation), AD/DA converters, and data buses. Advanced functions of DAQ devices include automatic measurement and intelligent data processing.
- Computers are generalized as the platforms that are used to control the DAQ devices, process the acquired data, visualize the analysis results, and store the measured and processed data. The popular forms of computers include desktops, laptops, single-board computers, and control centers used in a manufacturing.

15.2.4.2 Tracking Module

The identification of the objects' positions is a crucial part of implementing ALs. According to the tracking technique employed, the hardware interface can be categorized into (a) marker and markerless vision-based, (b) sensor-based, and (c) hybrid tracking-based.

- Marker and markerless vision-based tracking systems. In the former case, ID-encoded markers are added to the physical system beforehand, and the objects are recognized and tracked by decoding the visual markers. In this case, the entire system is invasive and requires an additional setup step. Markerless tracking systems which use the natural features of the object and the environment are much more promising when building AL systems (Zhang et al. 2015a, b, c).

- Sensor-based tracking systems. These tracking systems are based on sensors such as ultrasonic, optical, GPS, and inertial sensors (Rolland et al. 2001). Compared to vision-based tracking, sensor-based tracking techniques tend to be faster and more robust. They can be used for motion prediction when fast changes occur, but most systems are not as accurate as vision-based techniques.
- Hybrid tracking systems. In some applications, the sole usage of vision-based or senor-based tracking systems cannot provide robust tracking results. In such cases, a hybrid tracking technique such as combining an inertial sensor and a vision system (Aron et al. 2007) could produce more reliable results.

15.2.4.3 Data Passing Method

The passing of data between a VL and physical devices is realized by middleware. At present, there are three popular communication mechanisms in middleware. They are message-oriented communication mechanisms (Mahmoud 2004), shared memory-based communication mechanisms (Mazzucco et al. 2009), and communication mechanisms based on hybrid message passing and shared memory (Henty 2000). Message-oriented communication mechanisms employ asynchronous interactions by sending/receiving messages among communication participants (callers, receivers). With asynchronous interaction, the caller can retain processing control and thus does not need to block and wait for the called code to return. This model allows the caller to continue the processing regardless of the processing state of the called procedure/function/method. The disadvantage is that, with synchronous interaction, the called code may not be executed straight away. This interaction model requires an intermediary to handle the exchange of requests. Normally, this intermediary is a message queue. Since all participants can retain processing independently, they can continue the processing regardless of other participants' state. The schematic of asynchronous interaction is depicted in Fig. 15.13. The most significant advantage of message-oriented middleware is to allow the users to continue the processing of other messages once a message has been sent. One simple application of a message-oriented middleware infrastructure is depicted schematically in Fig. 15.14. All participants can be callers or receivers, and the messages are sent to the middleware instead of to the target receivers. This mechanism makes the participants independent of each other. Shared memory communication protocols permit users to access the shared memory simultaneously. For inter-process communication with shared memory, the shared memory is only used on a single machine (see Fig. 15.15). For distributed system communication with distributed shared memory, the shared memory allows users to access shared resources by a series of transmissions, namely, private memory which is partitioned in distributed shared memory, a message-passing channel which is created according to message-based communication protocols, and shared resources which are stored in distributed shared memory (see Fig. 15.16). The hybrid communication takes advantage of both shared memory and message-oriented mechanisms. The shared

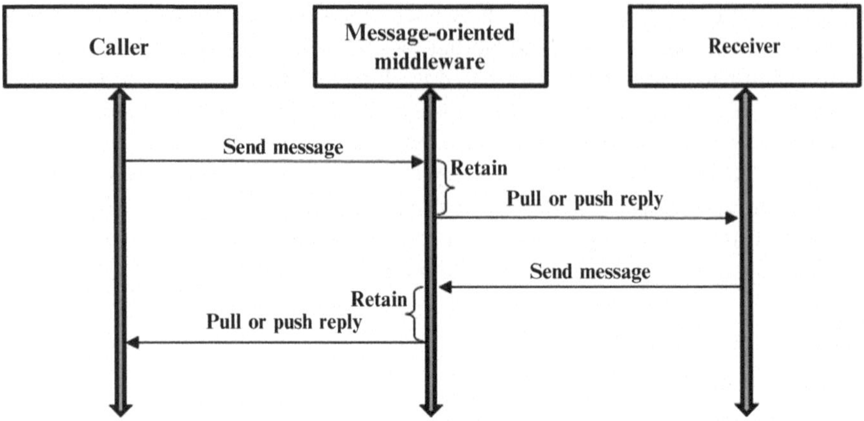

Fig. 15.13 Asynchronous interaction model

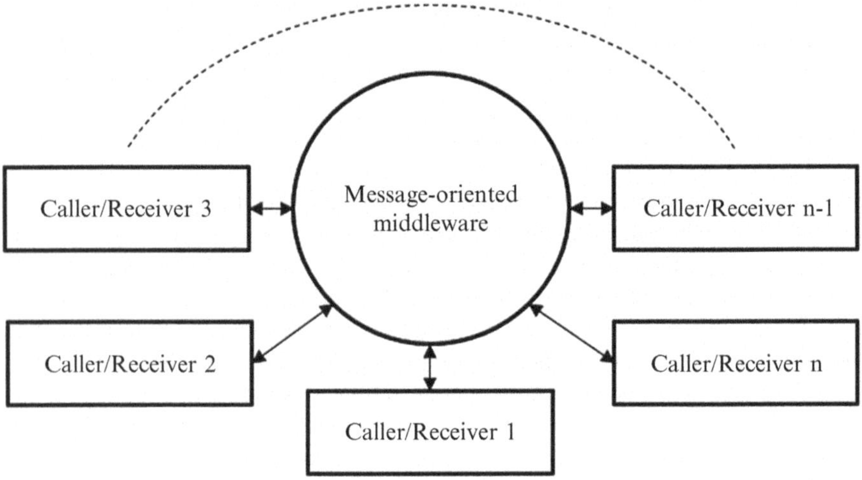

Fig. 15.14 Message-oriented middleware infrastructure

Fig. 15.15 Shared memory on a machine with two processes

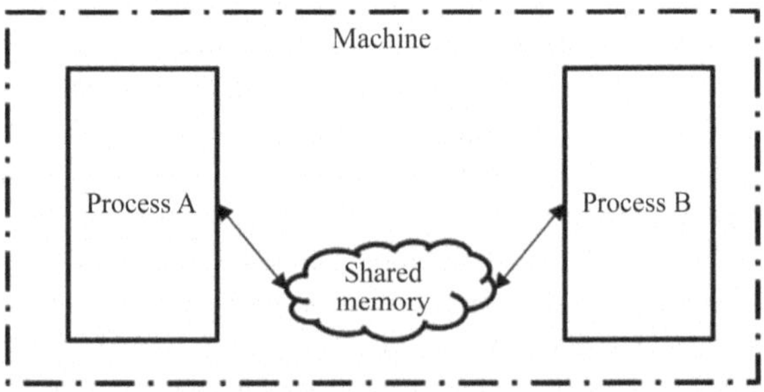

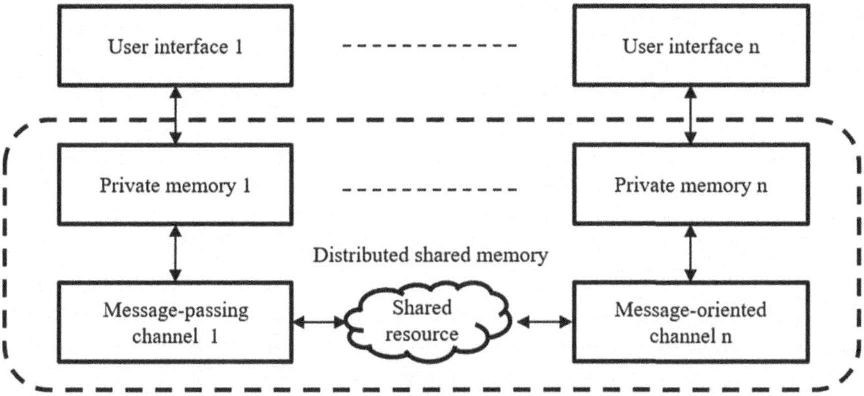

Fig. 15.16 Distributed shared memory architecture

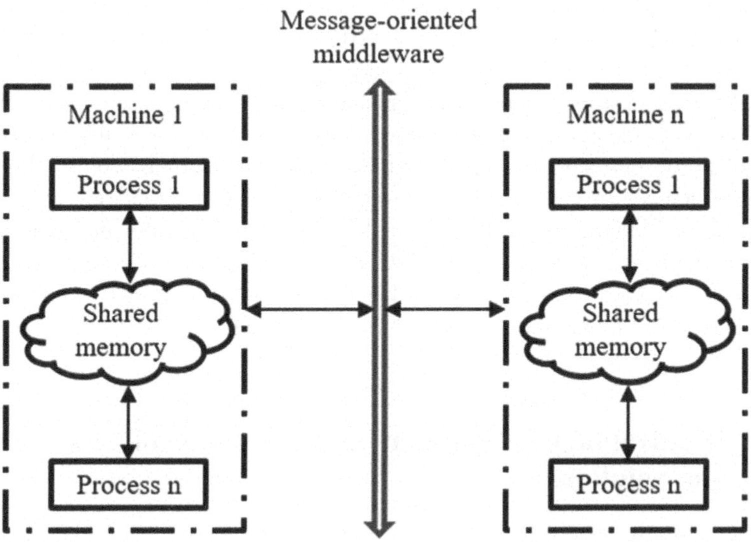

Fig. 15.17 Hybrid communication architecture with shared memory and message-oriented protocol

memory is used to realize inter-process communication on a device, while message-oriented protocols are used to realize communication among distributed devices in a distributed network (see Fig. 15.17).

Although there are three methods used for the communication between the virtual and real worlds, the message-oriented protocol has limitations with respect to licensing, platform dependence, communication latency, scalability, reliability, user-friendliness, and data-type compatibility. In addition, pure shared memory methods cannot realize remote communication. Therefore, the best solution is to use a hybrid communication method of shared memory and a message-oriented protocol.

15.2.5 Operation

For students, “hands-on” experiments are more about interaction, interpretation, and revelation than virtual experiments (Pyatt and Sims 2012). Experiments implemented on VL platforms offer efficiencies over physical hands-on ones because they require less setup time and provide results to students instantaneously (Zacharia et al. 2008). Compared to hands-on experiments, the students are able to perform more experiments and gather more data in the same amount of time. In addition, the students can have flexible schedules by accessing the VL at anywhere and anytime.

ALs, by integrating physical devices into VLs, take full advantage of both traditional hands-on laboratories and VLs. For example, a study of first-year undergraduate students learning optics under three conditions (only virtual, only physical, and a combination) showed that the students in the combined condition outperformed those in the physical only and virtual only conditions (Olympiou and Zacharia 2012). Similar results have been found in multiple studies (Jaakkola et al. 2011; Verschaffel et al. 2010; Chini et al. 2012), which suggests that well-designed combinations of virtual and physical experiments compared with either one alone enable the students to gain a better understanding of the subjects.

Moreover, in the traditional operation of VLs with real proctors, the students are required to log into a VL system with their usernames and passwords. The most common method for supervising an experiment is that proctors (i.e., laboratory administrators, laboratory instructors, or hired proctors) monitor the participants with a surveillance camera system (Toledo 2017; SIS 2017). These proctors are located at the server side of the virtual laboratory and supervise the entire process of the experiment by monitoring video feeds on a screen.

15.3 Disadvantages of Popular Implementation of VLs and Solutions

15.3.1 Disadvantages

Present VL systems have several shortcomings which keep them from gaining further popularity.

Long design and modification period. There are no VL systems that support real-time modeling for customized objects. The process of visualization in VR is often created manually by CAD software. It is complicated and tedious, and thus it is time-consuming and requires patience.

Feel of immersion of the users is limited. VLs are capable of coordinating the users’ interactions both spatially and temporally (McLellan 2001) but fail to induce the feel of presence. Note that the essence of the above shortcomings of VLs follows from the word “virtual.” Despite the fact that one of the main objectives of VLs is to mimic the real world as realistically as possible, it is impossible – at least today – to make the users feel totally immersed.

The capability to simulate real phenomena needs to be developed further. The more vivid VLs are, the more meaningful they become. However, even though in most instances VLs are designed to simulate the real world in a realistic fashion, they still have some shortcomings to be overcome. Firstly, while VR eliminates the physical distances between the users and/or objects in the VE, the users' real identities are lost, thus decreasing the credibility of the users and the VL system itself. Furthermore, the representation of the virtual world is not authentic or objective. Models, figures, and environments are too artificial. The elements of current VL platforms are not an accurate representation of the real world, and therefore they do not facilitate accurate simulations.

Last, the traditional proctoring approaches have two disadvantages. One shortcoming is the cost for setting up the surveillance system, and the other disadvantage is the laborious nature proctoring.

15.3.2 Solutions

15.3.2.1 Real-Time Creation of VEs

Since the creation of VEs is complicated and tedious, new techniques should be developed in order to get a solution that supports the real-time creation of VEs.

Currently, the ability to create VEs is impaired by (i) the conventional methods for acquiring the geometric parameters of real-world artifacts, (ii) the algorithms used to process the acquired data, and (iii) the procedures for creating the virtual models used in VLs.

If sensors can replace the traditional measuring devices, the laborious work of surveying the real world can be completed very quickly. Moreover, some middleware that can speed up the data processing of the acquired raw data is desired. Finally, an automatic workflow which can generate ready-to-use VEs should be provided. A promising method to meet these requirements is a real-time 3D reconstruction technique. Especially during 3D reconstruction, noncontact 3D scanners which can obtain the depth information are more desirable than other measurement tools.

3D reconstruction is the process of acquiring shape information and creating the models of real objects. The created models can be rendered directly in a specific rendering system or they can be exported into other CAD software (e.g., 3D Studio Max, Pro/Engineer, CATIA, etc.) for further post-processing. The most important aspect of 3D reconstruction is the acquisition of the information of the scanned object surfaces and the processing of the obtained data. Although 3D reconstruction techniques are still not mature, their potential for implementation in various fields of application (e.g., archaeological research (Fiz and Orengo 2007; De Reu 2014)), medical applications (Hibbard et al. 1993; McInerney and Terzopoulos 1996), reverse engineering (Werghi et al. 1999), and recovery of 3D shapes of deformable surfaces (Salzmann et al. 2008; Varol et al. 2012) makes them the subject of ongoing research.

In addition, real-time reconstruction is much more challenging than the common 3D reconstruction methods. There are two methods for realizing real-time reconstruction of VEs, namely, improving the hardware processing speed or optimizing the reconstruction algorithms. The data processing capabilities and the storage capacity of the hardware (i.e., CPU, GPU, storage devices, and memory) are limited. In addition, the limitation is that once transistors can be created as small as atomic particles, then there will be no more room for growth in the CPU performance as far as their speed is concerned. This means that the limitation may be overcome only by employing multiple devices (e.g., multiple GPUs) if the improvement of the hardware capabilities is preferred (Moore's Law 2017). Unfortunately, the development of hardware is typically much more difficult than initially anticipated. Therefore, a feasible solution for increasing the speed of the data processing is to design or optimize the data processing algorithms.

In order to compensate for the hardware limitations, parallel computing has been introduced as an alternative to traditional serial computation. Serial computation breaks a problem into a discrete series of instructions, and then these instructions are executed sequentially. Serial computation is executed on a single processor, and only one instruction may be executed at any moment in time as illustrated in Fig. 15.18. Parallel computing is the simultaneous use of multiple computing resources to solve a computational problem. When using parallel computing, a problem is broken into discrete parts, so-called blocks, that can be processed concurrently. Each block is further broken down into a series of instructions. Instructions from the different parts are executed simultaneously on different processors. After that, an overall control/coordination mechanism is employed. In Fig. 15.19, a parallel computing example is illustrated (Parallel Computing 2017).

The proposed method for creating VEs involving real-time 3D reconstruction can be described as follows: (1) obtaining the surface data of real objects using depth scanners, (2) processing the raw surface data employing parallel computing techniques, (3) generating an improved point cloud, and (4) creating the final model. Models for VLs created through 3D reconstruction methods need to be stored in valid 3D model files. Then, these models are imported into and rendered in the specific VL platforms. Figure 15.20 demonstrates a typical VE creation workflow used in a game-based VL (refer to Zhang et al. (2013a, b, 2014, 2015a, b, c) for more details).

Figure 15.21 illustrates a multiuser game-based VE built with the innovative method. In this VE, a flow-development experiment was designed as a pilot application. The experiment was used to measure the characteristics of air distribution systems and teach the basic principles of fluid mechanics with a focus on the flows in ducts and jets. The experimental setup demonstrated how viscous effects permeated the entire flow. In the pilot implementation, Pitot tubes were employed to measure the pressures from which velocity distributions at various cross sections of the pipe were then determined. The Pitot tubes were controlled using stepper

Fig. 15.18 Serial computing example

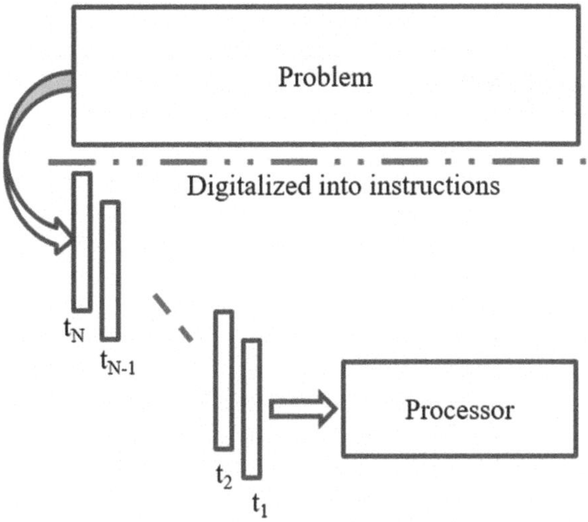

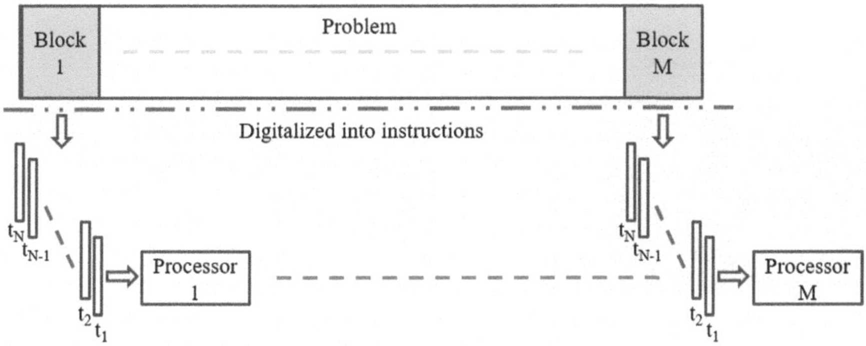

Fig. 15.19 Parallel computing example

motors according to the user commands received from inside the game server. The students, the instructor, and the teaching assistant were represented as avatars. They interacted with a virtual flow rig apparatus that represented the real physical device. The virtual apparatus consisted of a test tube on a base with blower, diffuser, Pitot tubes, stepper motors, pressure reading taps, and an orifice.

15.3.2.2 Real-Time Stereo Tracking, Mapping, and Simulating

The performance of many augmented VR applications is limited due to either failing to track targets in real time or unreliable tracking results. Tracking objects using 3D data has been a challenging task for many years.

Tracking techniques can be rendered computationally efficient using approaches such as representing objects by their centroid (Kaestner et al. 2012) or by the center

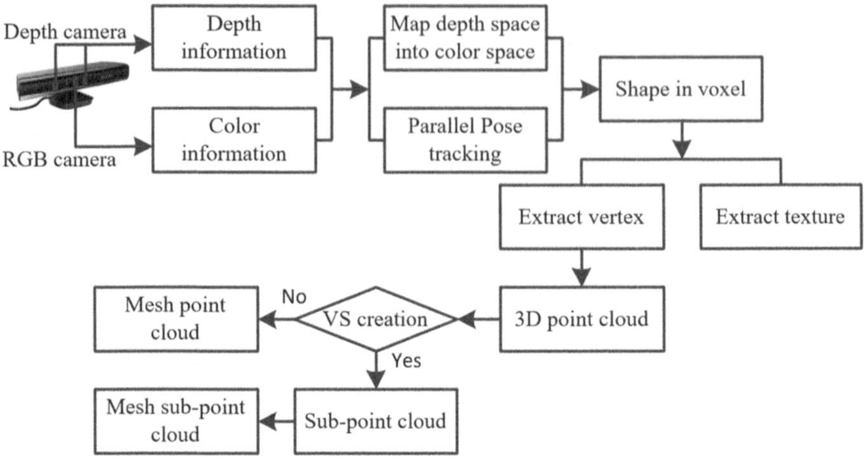

Fig. 15.20 VE creation procedures

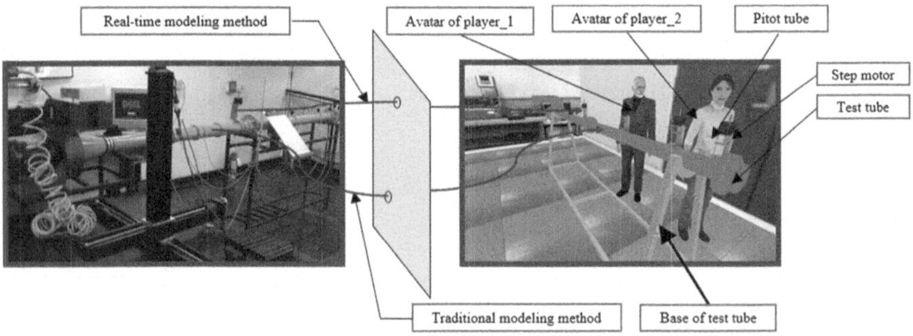

Fig. 15.21 GBVR with fluid experiment setup

of their bounding boxes wrapped in a Kalman filter (Azim and Aycard 2012). These techniques discard almost the entire available depth information in the 3D data. As a result, they are not accurate. Another approach can achieve accuracy and computational efficiency at the same time by fitting models composed of primitive geometric shapes to the point cloud of the tracked object. For instance, rectangular shapes can be used to track the motion of moving vehicles (Petrovskaya and Thrun 2008) or line, and corner features enable the tracking of partially visible vehicles. These techniques start by assuming the shape of target objects, and thus they are not suitable for tracking objects of arbitrary shape.

Aligning the objects' point clouds with the 3D scene data using the iterative closest point method and its variants (Feldman et al. 2012; Moosmann and Stiller 2013) is a popular approach because they are comparatively efficient and they use the full 3D information of the target objects. These techniques start by assuming an initial position and improve the alignment iteratively using approaches such as the hill-climbing method (Thornton and Boulay 1998). The quality of the tracking results relies heavily on a good estimation of the initial alignment.

Simultaneous localization and mapping (SLAM) is a category of grid-based methods that can be used for developing tracking techniques with an acceptable accuracy. However, higher accuracy requires fine grids, which make SLAM infeasible for real-time applications. Several approaches have been reported for overcoming this disadvantage, such as multi-resolution grid-based SLAM systems (Ryde and Hu 2010) and coarse-to-fine sampling with annealed dynamic histograms (Held et al. 2016).

In order to achieve computational efficiency and tracking reliability at the same time, a three-stage approach for capturing general motions of objects with a low-cost 3D scanner was designed for applications in ALs (Zhang et al. 2015a, b, c). This approach consists of point cloud preprocessing with a focus on computational efficiency, object tracking employing recognition, and post-processing including motion analysis. This approach can be tailored to special cases. Specifically, the algorithms focus more on computational efficiency when the objects of interest have simple shapes and the same colors while they focus more on reliability for objects with complex geometries or textures. The three-stage approach was proved to have acceptable results; however, it requires the model of the target objects to be built beforehand. In order to improve the usability of this system, a technique for the simultaneous tracking and reconstruction of objects was developed (Zhang et al. 2015a, b, c). This tracking scheme was accomplished without generating the model of the target objects a priori. After recording the positions and orientations of the target objects as tracking results, the object model is generated as well. The tracking scheme is believed to be feasible and suitable to be used in vision-based object tracking applications.

Figure 15.22 provides an example in which the Microsoft Kinect was used to track a spherical bob (which was taken as a Foucault pendulum). Only those points that represented the spherical part were extracted from a series of point clouds and the bob positions (x, y, and z coordinates of the sphere's centroid) were computed and recorded. More details are described elsewhere (Zhang et al. 2015a, b, c).

15.3.2.3 Advanced Interfaces

15.3.2.3.1 Vision-Based Human-Computer Interaction

Vision-based human-computer interaction employs cameras to realize the interaction between the users and the VEs. One such implementation is depicted in Fig. 15.23. The Microsoft Kinect was used to track the human gestures, and then, the tracking results were sent to the VE to control the avatars.

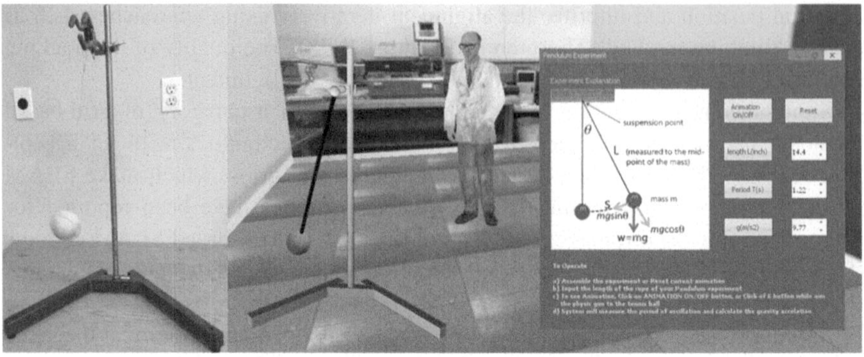

Fig. 15.22 Experimental setup of Foucault pendulum (left); Foucault pendulum experiment implementation in GBVL (right)

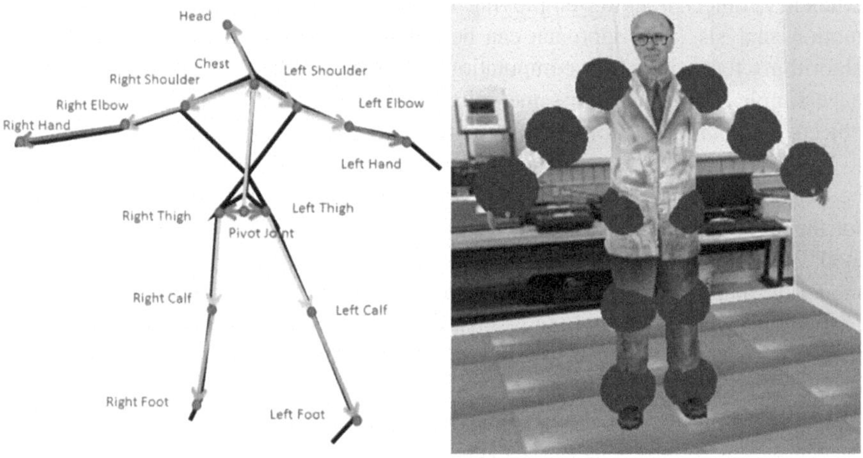

Fig. 15.23 Skeleton recognized by Kinect (left), corresponding skeleton in GMod (right)

15.3.2.3.2 Speech Input

The usage of voice commands can greatly reduce the students' difficulties in getting started with a VL platform. The speech recognition can be accomplished at the same time as the gesture tracking. In order to make the commanding process convenient and easy to use for students, a command usually includes several speeches, the semantics of which are close to the meaning of the command in natural language. For instance, the words "accelerate" and "faster" are both interpreted as "move faster." Thus, these two words are chosen for the speech of "accelerate." Since there are four categories of voice commands, the method to process these commands is more complicated than that for processing the gesture tracking. A simplified flow chart of the voice command processing is shown in Fig. 15.24.

A prototype of a GBVL for mechanical assembly training with an advanced interfaced is shown in Fig. 15.25. More details can be found in Chang et al. (2014).

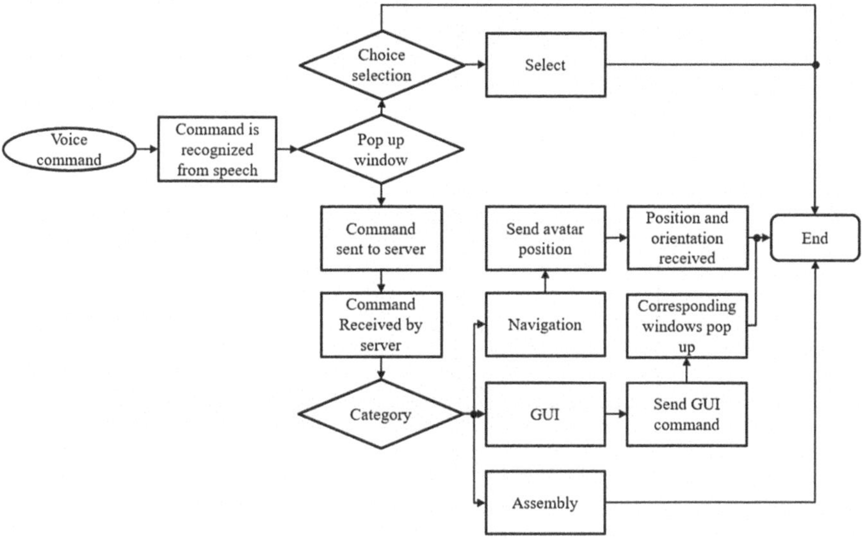

Fig. 15.24 Flow chart for voice command processing

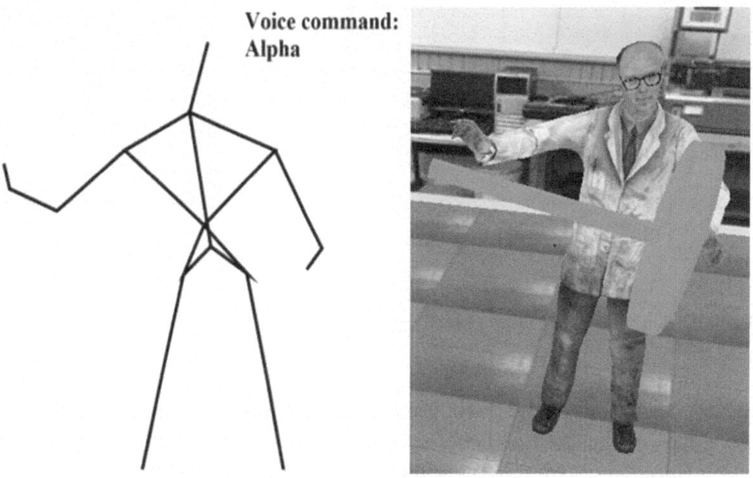

Fig. 15.25 Skeleton captured by Kinect (left), Avatar picking up a part with both hands (right)

15.3.2.4 Biometric Authentication and Remote Proctoring

Virtual proctors employ biometric technologies in order to identify the learners, to monitor their actions, and to validate the test results without a need for real people. Among biometrics methods, facial recognition is one of the preferred choices for authenticating and tracking the users. The flow chart of a virtual proctor based on biometric technology is shown in Fig. 15.26. The frames that are used to

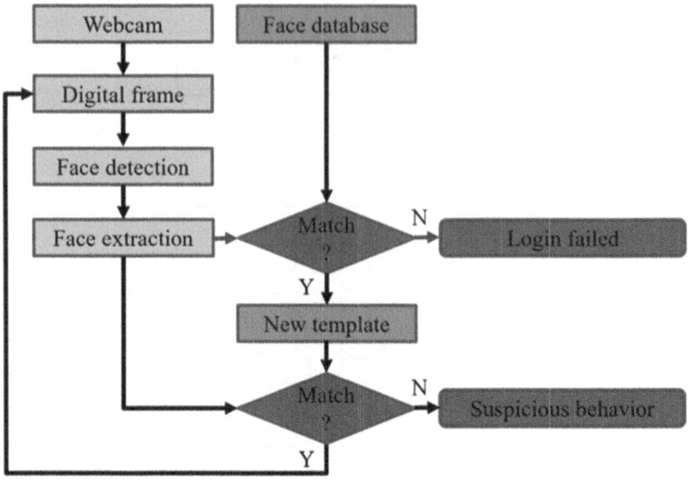

Fig. 15.26 Flow chart for virtual proctor

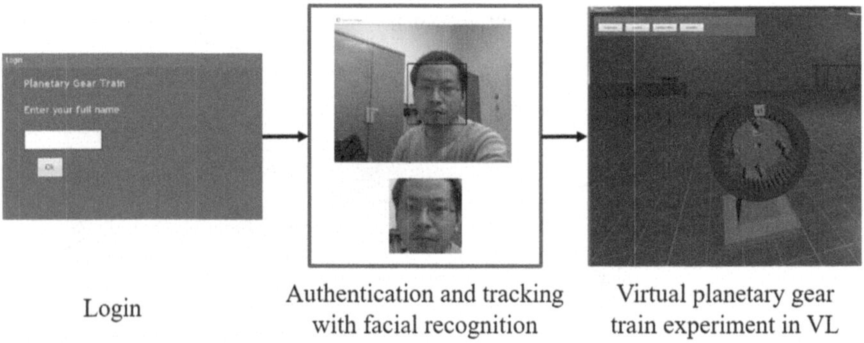

Fig. 15.27 Implementation of virtual proctor in VL

authenticate and monitor the users are created by webcams at the users' location. The users' face is detected and extracted repeatedly from a single sample frame at a specified frequency. Upon passing through the authentication, the user either can log into the VL or is forced to exit the VL. After the authentication, a loop follows. If a suspicious behavior is identified, a video clip is recorded, which can then be used by the instructors of the experiments to verify whether there was any cheating attempt. Figure 15.27 shows the implementation of a virtual proctor. A more detailed discussion can be found elsewhere (Zhang et al. 2016).

15.4 Pilot Implementation

In order to evaluate the performance of a VL with the solutions described above, a simple gear train experiment (see Fig. 15.28) and a planetary gear train experiment (see Fig. 15.29) were implemented in the VL and evaluated in an undergraduate mechanical engineering course (see Aziz et al. 2012, 2014).

A total of 94 students participated in the evaluation. Most students worked in groups of 2, and the remaining students worked individually. While the students working in groups were prevented from physically contacting their partners, they could communicate within the VE.

An analysis of the prior gaming experience of the students was conducted (see Table 15.3). Then, the students involved in the study were grouped based on their previous exposure to games. Then, the number of actions taken by each student was compared for experienced vs. inexperienced game players (see Fig. 15.30). The figure indicates that those students who had little or no prior exposure to

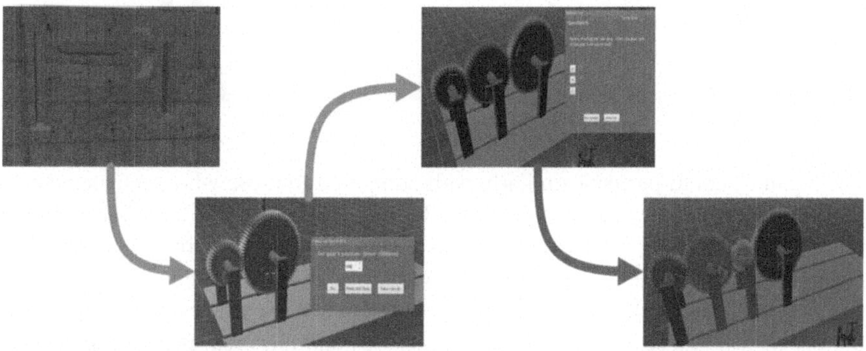

Fig. 15.28 Simple gear train experiment

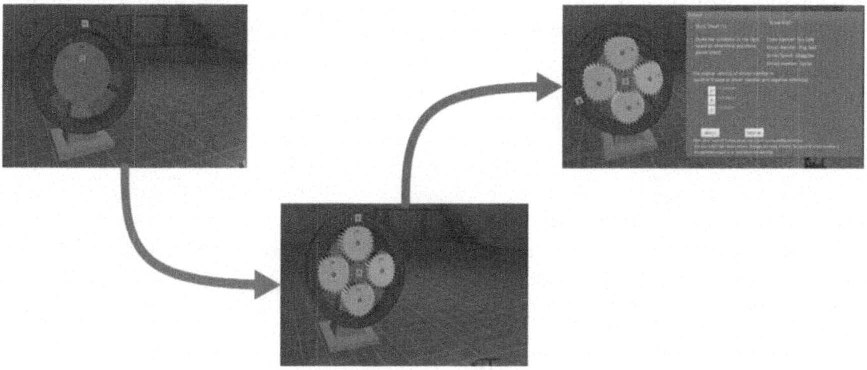

Fig. 15.29 Planetary gear train experiment

Table 15.3 Frequency of video game playing of students

Gender	>2 h/week	>20 min/week; <2 h/week	A few times only	Never or almost never
Male	36	22	18	2
% male	46.15%	28.21%	23.07%	2.56%
Female	1	2	8	5
% female	6.25%	12.50%	50.00%	31.25%
% Total	39.36%	25.53%	27.66%	7.45%

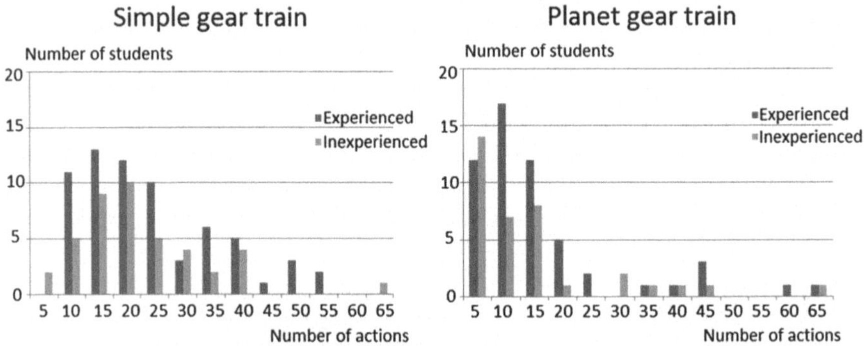

Fig. 15.30 Comparison of number of actions taken by student vs. their prior gaming experience

video games could perform similarly well compared to those who are experienced players. Therefore, it can be concluded that the laboratory exercises have strong potential for learning as the inexperienced players were also able to acquire the basic skills necessary for operating the virtual laboratory system without major struggles. Moreover, the virtual proctor was used to test the proctoring quality. After the experiments, a survey with six questions was administered. It was designed to determine the students' attitudes toward the designed virtual experiment. Figure 15.31 summarizes the results of the student survey which indicates that their satisfaction with this virtual laboratory was very high. More details about the evaluation of the VL can be found in Chang et al. (2016a, b).

15.5 Summary

This book chapter is composed of four main parts.

The first part introduces the basic concept, history, and current status of VLs. In this part, the fundamental structure of VLs is described, and the development and characteristics of the VLs are discussed. The history of VLs reflects the different stages of VL development from the introduction of the very concept. The current status mainly provides a survey of VLs published in the literature so far.

The second part describes the techniques used to create and operate VLs based on the review of both mature and pilot implementations of VLs currently in existence.

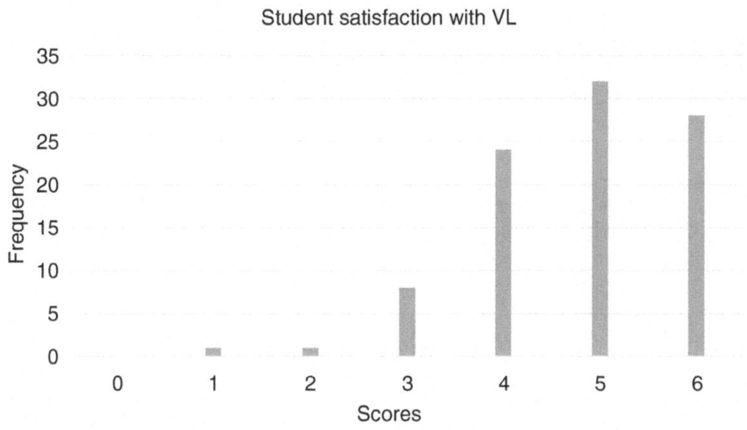

Fig. 15.31 Histogram of student satisfaction with VLs

The content of this part ranges from DAQ development, user interface innovation, and simulation optimization to VE creation.

The third part discusses the disadvantages associated with the current VLs. Then, promising approaches for overcoming these limitations are introduced. VLs have been developed and used for quite some time, but they fail to fully replace hands-on laboratories in practice because there are limitations and shortcomings with respect to the feel of immersion, cost of system creation, and data acquisition methods. Therefore, some of these constraints and their causes are discussed in detail.

In the fourth part, a pilot implementation and some assessment results were presented. There are two experiments, namely, a simple gear train experiment and a planetary gear train experiment. Ninety-four students participated in two evaluation studies. The results indicated that the VL built with the techniques discussed here was well received by the students.

In the future, the research on VLs will focus on the improvement of the users' feel of immersion, the efficiency of creation and operation of VLs, and the advanced peripheral hardware of VLs. With the further development of modern technologies in many disciplines, VLs are expected to become more suitable for benefitting a broader variety of learners.

References

Abe, N. M., & Cardoso, J. R. (1999). A virtual lab for electric motors and drives. *IEEE Transactions on Magnetics, 35*(3), 1674–1677.

Aburdene, M. F., Mastascusa, E. J., & Massengale, R. (1991). A proposal for a remotely shared control systems laboratory. In *Proceeding of frontiers in education conference*, West Lafayette, IN, September 21–24.

Adam, J. A. (1993). Virtual reality is for real. *IEEE Spectrum, 30*(10), 22–29.

Adamo-Villani, N., Richardson, J., Carpenter, E., & Moore, G. (2006). A photorealistic 3D virtual laboratory for undergraduate instruction in microcontroller technology. In *Proceeding of ACM SIGGRAPH 2006 Educators program*, Boston, MA, US, July 30–August 3.

Alexiou, A., Bouras, C., Giannaka, E., Kapoulas, V., Nani, M., & Tsiatsos, T. (2004). Using VR technology to support e-learning: The 3D virtual radiopharmacy laboratory. In *Proceedings of the 24th international conference on distributed computing systems workshops*, Tokyo, Japan, March 23–26.

Alexiou, A., Bouras, C., & Giannaka, E. (2005). Virtual laboratories in education. In J. P. Courtiat, C. Davarakis, & T. Villemur (Eds.), *Technology enhanced learning. IFIP International Federation for Information Processing* (Vol. 171). Boston: Springer.

Amazon Lumberyard. (2017). *AAA game engine integrated with AWS*. Retrieved from https://aws.amazon.com/lumberyard/

Andujar, J. M., Mejías, A., & Marquez, M. A. (2011). Augmented reality for the improvement of remote laboratories: An augmented remote laboratory. *IEEE Transactions on Education, 54*(3), 492–500.

Arango, F., Altuger, G., Aziz, E.-S., Esche, S. K., & Chassapis, C. (2008). Piloting a game-based virtual learning environment. *Computers in Education Journal, 18*(4), 82–91.

Aron, M., Simon, G., & Berger, M. O. (2007). Use of inertial sensors to support video tracking. *Computer, 18*, 57–68.

Arpaia, P., Baccigalupi, A., Cennamo, F., & Daponte, P. (1997). A remote measurement laboratory for educational experiments. *Measurement, 21*(4), 157–169.

Autodesk. (2017a). *Make it with Maya computer animation software*. Retrieved from http://www.autodesk.com/products/maya/overview

Autodesk. (2017b). *3D modeling, animation, and rendering software*. Retrieved from http://www.autodesk.com/products/3ds-max/overview

Autodesk. (2017c). *3DS Max2104*. Retrieved from http://docs.autodesk.com/3DSMAX/16/ENU/3ds-Max-Tutorials/index.html

Avradinis, N., Vosinakis, S., & Panayiotopoulos, T. (2000). Using virtual reality techniques for the simulation of physics experiments. In *Proceeding of 4th systemics, cybernetics and informatics international conference*, Orlando, Florida, US.

Ayache, N. (1995). Medical computer vision, virtual reality and robotics. *Image and Vision Computing, 13*(4), 295–313.

Azim, A., & Aycard, O. (2012). Detection, classification and tracking of moving objects in a 3D environment. In *Proceeding of the IEEE symposium on intelligent vehicles*, Alcal de Henares,

Aziz, E.-S., Esche, S. K., & Chassapis, C. (2006a). An architecture for virtual laboratory experimentation. In *Proceedings of the 2006 ASEE annual conference and exposition*, Chicago, Illinois, USA, June 18–21.

Aziz, E.-S., Esche, S. K., & Chassapis, C. (2006b). A scalable platform for remote and virtual laboratories. *World Transactions on Engineering and Technology Education, 5*(3), 445–448.

Aziz, E., Corter, J., Chang, Y., Esche, S., & Chassapis, C. (2012). Evaluation of the learning effectiveness of game-based and hands-on gear train laboratories. In *Proceeding of frontiers in education conference (FIE)*, Seattle, WA, US, October 3–6.

Aziz, E. S., Chang, Y., Esche, S. K., & Chassapis, C. (2014). A multi-user virtual laboratory environment virtual laboratory environment for gear train design. *Computer Applications in Engineering Education, 22*(4), 788–802.

Azuma, R., Baillot, Y., Behringer, R., Feiner, S., Julier, S., & MacIntyre, B. (2001). Recent advances in augmented reality. *IEEE Computer Graphics and Applications, 21*(6), 34–47.

Baba, S. A., Hussain, H., & Embi, Z. C. (2007). An overview of parameters of game engine. *IEEE Multidisciplinary Engineering Education Magazine, 2*(3), 10–12.

Balamuralithara, B., & Woods, P. C. (2009). Virtual laboratories in engineering education: The simulation lab and remote lab. *Computer Applications in Engineering Education, 17*(1), 108–118.

Barfield, W. (Ed.). (2015). *Fundamentals of wearable computers and augmented reality*. London: CRC Press.

Bohus, C., Aktan, B., Shor, M. H., & Crowl, L. A. (1995). *Running control engineering experiments over the Internet*. Oregon State University, Department of Computer Science, Technical Report 95-60-07.

Borghetti, M., Sardini, E., & Serpelloni, M. (2013). Sensorized glove for measuring hand finger flexion for rehabilitation purposes. *IEEE Transactions on Instrumentation and Measurement, 62*(12), 3308–3314.

Bottentuit Junior, J. B., & Coutinho, C. P. (2007). Virtual laboratories and M-learning: learning with mobile devices. In *Proceedings of the international multi-conference on society, cybernetics, and informatics*, Orlando, Florida, US, July 12–15.

Brey, P. (2014). *Virtual reality and computer simulation. In ethics and emerging technologies* (pp. 315–332). Basingstoke: Palgrave Macmillan.

Bullet Engine. (2017). *Real-time physics simulation*. Retrieved from http://bulletphysics.org/wordpress/

Burdea, G., & Coiffet, P. (2003). *Virtual reality technology*. US: Wiley.

Casini, M., Prattichizzo, D., & Vicino, A. (2001). The automatic control telelab: a remote control engineering laboratory. In *Proceedings of the 40th IEEE conference on decision and control*, Orlando, Florida, US, December 4–7.

Chang, C., Kodman, D., Esche, S. K., & Chassapis, C. (2006a). Immersive collaborative laboratory simulations using a game engine. In *Proceedings of the 2006 ASEE annual conference and exposition*, Chicago, Illinois, US, Jun 18–21.

Chang, C., Arango, F., Kodman, D., Esche, S. K., & Chassapis, C. (2006b). Utilization of immersive collaborative student laboratory simulations developed using a game engine. In *Proceedings of the ASME international mechanical engineering congress and exposition IMECE'06*, Chicago, Illinois, US, November 5–10.

Chang, C., Arango, F., Esche, S. K., & Chassapis, C. (2007). On the assembly of experimental setups in virtual laboratory environments. In *Proceedings of the ASME international mechanical engineering congress and exposition IMECE'07*, Seattle, Washington, US, November 10–16.

Chang, Y., Aziz, E.-S., Esche, S. K., & Chassapis, C. (2012). A game-based laboratory for gear design. *Computers in Education Journal, 22*(1), 21–31.

Chang, Y., Aziz, E.-S., Esche, S. K., & Chassapis, C. (2013). A multi-user virtual laboratory environment for gear train design. *Computer Applications in Engineering Education, 22*(4), 788–802.

Chang, Y., Aziz, E.-S., Zhang, Z., Zhang, M., Esche, S. K., & Chassapis, C. (2014). A platform for mechanical assembly education using the Microsoft Kinect. In *Proceedings of the ASME international mechanical engineering congress & exposition*, Quebec, Canada, November 14–20.

Chang, Y., Aziz, E.-S., Zhang, Z., Zhang, M., & Esche, S. K. (2016a). Evaluation of a video game adaptation for mechanical engineering educational laboratories. In *Proceedings of 46th ASEE/IEEE frontiers in education conference*, Erie, US.

Chang, Y., Aziz, E.-S., Zhang, Z., Zhang, M., Esche, S. K., & Chassapis, C. (2016b). Usability evaluation of a virtual educational laboratory platform. *Computers in Education Journal, 7*(1), 24–26.

Chini, J. J., Madsen, A., Gire, E., Rebello, N. S., & Puntambekar, S. (2012). Exploration of factors that affect the comparative effectiveness of physical and virtual manipulatives in an undergraduate laboratory. *Physical Review Special Topics – Physics Education Research, 8*(1), 010113-1- 010113-12.

Classic Games. (2017). *Meet with confidence*. Retrieved from https://www.classicgames.me/super-mario-2d-land.html

Cockburn, A., & McKenzie, B. (2002). Evaluating the effectiveness of spatial memory in 2D and 3D physical and virtual environments. In *Proceedings of the SIGCHI conference on human factors in computing systems*, Minneapolis, Minnesota, US, April 20–25.

Davidphillips. (2017). http://www.davidphillips.me/3d-art-and-animation.php. Accessed in Feb 2017.

De Meyer, A. (1991). Tech talk: How managers are stimulating global R&D communication. *MIT Sloan Management Review, 32*(3), 49.

De Reu, J., De Smedt, P., Herremans, D., Van Meirvenne, M., Laloo, P., & De Clercq, W. (2014). On introducing an image-based 3D reconstruction method in archaeological excavation practice. *Journal of Archaeological Science, 41*, 251–262.

Deniz, D. Z., Bulancak, A., & Ozcan, G. (2003). A novel approach to remote laboratories. In *Proceeding of 33rd annual frontiers in education conference*, Westminster, Colorado, November 5–8.

Dilwort, J. (2010). Realistic virtual reality and perception. *Philosophical Psychology, 23*, 23–42.

Dorozhkin, D. V., Vance, J. M., Rehn, G. D., & Lemessi, M. (2012). Coupling of interactive manufacturing operations simulation and immersive virtual reality. *Virtual Reality, 16*(1), 15–23.

Esche, S. K. (2005). On the integration of remote experimentation into undergraduate laboratories – pedagogical approach. *International Journal of Instructional Media, 32*(4), 397–407.

Esche, S. K., & Chassapis, C. (1998). An Internet-based remote-access approach to undergraduate laboratory education. In *Proceedings of the fall regional conference of the middle atlantic section of ASEE*, Washington, DC, US, November 6–7.

Familia, R. (2005). A virtual laboratory for cooperative learning of robotics and mechatronics. In *Proceedings of the 6th international conference on information technology based higher education and training*, Juan Dolio, Dominican Republic, July 7–9.

Faulkner, G., & Krauss, M. (1996). Guidelines for establishing a virtual reality lab [medical applications]. *IEEE Engineering in Medicine and Biology Magazine, 15*(2), 86–93.

Feldman, A., Hybinette, M., & Balch, T. (2012). The multi-iterative closest point tracker: An online algorithm for tracking multiple interacting targets. *Journal of Field Robotics, 29*(2), 258–276.

Fiz, I., & Orengo, H. A. (2007). The application of 3D reconstruction techniques in the analysis of ancient Tarraco's urban topography. In *Proceedings of 35th international conference on computer applications and quantitative methods in archaeology*, Berlin, Germany, April 2–6.

Freund, E., & Roßmann, J. (2003). Distributed virtual reality: System concepts for cooperative training and commanding in virtual worlds. *Journal of Systemics, Cybernetics and Informatics, 1*(1), 47–54.

Furness, T. A., & Kocian, D. F. (1986). Putting humans into virtual space. In *Proceedings of the 16th conference on aerospace simulation*, San Diego, California, US.

Gaggioli, A., & Breinin, R. (2001). In G. Riva & F. Davide (Eds.), *Communications through virtual technology, identity community and technology in the Internet age*. Amsterdam: IOS Press.

Gertz, M. W., Stewart, D. B., & Khosla, P. K. (1994). A human machine interface for distributed virtual laboratories. *IEEE Robotics & Automation Magazine, 1*(4), 5–13.

Griffith, R., Chiprout, E., Zhang, Q. J., & Nakhla, M. (1992). A CAD framework for simulation and optimization of high-speed VLSI interconnections. *IEEE Transactions on Circuits and Systems I: Fundamental Theory and Applications, 39*(11), 893–906.

Hahn, H. H., & Spong, M. W. (2000). Remote laboratories for control education. In *Proceedings of the 39th IEEE conference on decision and control*, Sydney, NSW, Australia; December 12–15.

Harasim, L., Calvert, T., & Groeneboer, C. (1996). Virtual-U™: A web-based environment customized to support collaborative learning and knowledge building in post secondary courses. In *Proceedings of the 1996 international conference on learning sciences*, Evanston, IL, US.

Havok engine. (2017). Retrieved from http://www.havok.com/

Held, D., Levinson, J., Thrun, S., & Savarese, S. (2016). Robust real-time tracking combining 3D shape, color, and motion. *The International Journal of Robotics Research, 35*(1–3), 30–49.

Henty, D. S. (2000). Performance of hybrid message-passing and shared-memory parallelism for discrete element modeling. In *Proceedings of the 2000 ACM/IEEE conference on supercomputing*, Dallas, Texas, US, November 5–10.

Hibbard, L. S., Grothe, R. A., Arnicar-Sulze, T. L., Dovey-Hartman, B. J., & Page, R. B. (1993). Computed three-dimensional reconstruction of median eminence capillary modules. *Journal of Microscopy, 171*, 39–56.

Hirose, M. (1997). Image-based virtual world generation. *IEEE Multi Media, 4*(1), 27–33.

Howard, B. M., & Vance, J. M. (2007). Desktop haptic virtual assembly using physically based modelling. *Virtual Reality, 11*(4), 207–215.

Hughes, J. F., Van Dam, A., Foley, J. D., & Feiner, S. K. (2014). *Computer graphics: Principles and practice*. US: Pearson Education.

Hummel, J., Wolff, R., Stein, T., Gerndt, A., & Kuhlen, T. (2012). An evaluation of open source physics engines for use in virtual reality assembly simulations. In *Proceeding of international symposium on visual computing*, Rethymnon, Crete, Greece, July 16–18.

Jacobson, J., & Lewis, M. (2005). Game engine virtual reality with CaveUT. *Computer, 38*(4), 79–82.

Jara, C. A., Candelas, F. A., Puente, S. T., & Torres, F. (2011). Hands-on experiences of undergraduate students in automatics and robotics using a virtual and remote laboratory. *Computers & Education, 57*(4), 2451–2461.

Karim, M. A. (Ed.). (1992). *Electro-optical displays*. Boca Raton: CRC Press.

Kfir, R. E. (2001). Virtual laboratories in education. In *Proceedings of the 1st international conference on computer graphics, virtual reality and visualisation*. Cape Town, South Africa, November 05–07.

Kocian, D. F. (1997, May 17–18). A visually-coupled airborne systems simulator (VCASS) – An approach to visual simulation. In *Proceeding of the IMAGE conference sponsored by air force human resources laboratory*. Phoenix: Williams AFB.

Kozak, I., Banerjee, P., Luo, J., & Luciano, C. (2014). Virtual reality simulator for vitreoretinal surgery using integrated OCT data. *Clinical Ophthalmology, 2014*(8), 669–672.

Leleve, A., Benmohamed, H., Prevot, P., & Meyer, C. (2003). Remote laboratory-towards an integrated training system. In *Proceeding of 4th international conference on information technology based higher education and training*, Marrakech, Morocco, July 7–9.

Li, J. R., Khoo, L. P., & Tor, S. B. (2003). Desktop virtual reality for maintenance training: An object oriented prototype system (V-REALISM). *Computers in Industry, 52*(2), 109–125.

Lin, F., Ye, L., Duffy, V. G., & Su, C. J. (2002). Developing virtual environments for industrial training. *Information Sciences, 140*(1), 153–170.

Livingston, M. A., Rosenblum, L. J., Julier, S. J., Brown, D., Baillot, Y., Swan, I. I., Gabbard, J. L., & Hix, D. (2002). An augmented reality system for military operations in urban terrain. In *Proceedings of interservice/industry training, simulation & education conference*, Orlando, Florida, December 2–5.

Lu, G., Shark, L. K., Hall, G., & Zeshan, U. (2012). Immersive manipulation of virtual objects through glove-based hand gesture interaction. *Virtual Reality, 16*(3), 243–252.

Luciano, C., Banerjee, P., Florea, L., & Dawe, G. (2005). Design of the ImmersiveTouch™: A high-performance haptic augmented virtual reality system. In *Proceedings of the 11th international conference on human-computer interaction*, Las Vegas, Nevada, USA, Jul 22–27.

Luciano, C., Banerjee, P., & DeFanti, T. (2009). Haptics-based virtual reality periodontal training simulator. *Virtual Reality, 13*(2), 69–85.

Luck, M., & Aylett, R. (2000). Applying artificial intelligence to virtual reality: Intelligent virtual environments. *Applied Artificial Intelligence, 14*(1), 3–32.

Lustigova, Z., & Lustig, F. (2009). Remote and open laboratory in science education – technological, educational and psychological issues. In *Proceeding of the ITI 2009 31st international conference on information technology interfaces, Cavtat/Dubrovnik*, Croatia, June 22–25.

Ma, J., & Nickerson, J. V. (2006). Hands-on, simulated, and remote laboratories: A comparative literature review. *ACM Computing Surveys, 38*(3), 7–32.

Macedonia, M. R., Brutzman, D. P., Zyda, M. J., Pratt, D. R., Barham, P. T., Falby, J., & Locke, J. (1995). NPSNET: a multi-player 3D virtual environment over the Internet. In *Proceedings of the symposium on interactive 3D graphics*, Monterey, CA, US, April 9–12.

Mahmoud, Q. (2004). *Middleware for communications*. New York: Wiley.

Mazzucco, M., Morgan, G., Panzieri, F., & Sharp, C. (2009). Engineering distributed shared memory middleware for java. In *On the move to meaningful internet systems*. Berlin: Springer.

McInerney, T., & Terzopoulos, D. (1996). Deformable models in medical image analysis: A survey. *Medical Image Analysis, 1*(2), 91–108.

McLellan, H. (2001). *Virtual realities*. McLellan Wyatt Digital.
Merians, A. S., Jack, D., Boian, R., Tremaine, M., Burdea, G. C., Adamovich, S. V., Recce, M., & Poizner, H. (2002). Virtual reality-augmented rehabilitation for patients following stroke. *Physical Therapy, 82*(9), 898–915.
Moore's Law. (2017). Retrieved from http://www.mooreslaw.org/
Moosmann, F., & Stiller, C. (2013). Joint self-localization and tracking of generic objects in 3D range data. In *Proceeding of the IEEE international conference on robotics and automation*, Karlsruhe, Germany, May 6–10.
Nah, F. F. H., Eschenbrenner, B., & DeWester, D. (2011). Enhancing brand equity through flow and telepresence: A comparison of 2D and 3D virtual worlds. *MIS Quarterly, 35*(3), 731–747.
Nandwana, T. P. (2016). *Virtual reality and augmented reality: The next best thing to being there*. Retrieved from http://www.carriermanagement.com/features/2016/05/25/154815.htm
National Instruments. (2017). *What is data acquisition?* Retrieved from http://www.ni.com/data-acquisition/what-is/
Nelson, B., Ketelhut, D. J., Clarke-Midura, J., Bowman, C., & Dede, C. (2005). Design-based research strategies for developing a scientific inquiry curriculum in a multi-user virtual environment. *Educational Technology, 45*(1), 21–27.
Noor, A. K., & Wasfy, T. M. (2001). Simulation of physical experiments in immersive virtual environments. *Engineering Computations, 18*(3/4), 515–538.
NVidia. (2017). *PhysX games*. Retrieved from http://www.geforce.com/hardware/technology/physx
Obeysekare, U., Grinstein, F. F., & Patnaik, G. (1997). The visual interactive desktop laboratory. *IEEE Computational Science and Engineering, 4*(1), 63–71.
ODE. (2017). *Open dynamics engine*. Retrieved from http://www.ode.org/
Olympiou, G., & Zacharia, Z. C. (2012). Blending physical and virtual manipulatives: An effort to improve. *Science Education, 96*(1), 21–47.
Parallel Computing. (2017). *Introduction to parallel computing*. Retrieved from https://computing.llnl.gov/tutorials/parallel_comp/#Whatis
Patnode, J. (2012). *Character modeling with Maya and ZBrush: Professional polygonal modeling techniques*. Oxford: CRC Press.
Petrovskaya, A., & Thrun, S. (2008). Model based vehicle tracking for autonomous driving in urban environments. In *Proceedings of robotics: Science and systems IV*, Zurich, Switzerland, June 25–28.
Pukhov, A. (1999). Three-dimensional electromagnetic relativistic particle-in-cell code VLPL (Virtual Laser Plasma Lab). *Journal of Plasma Physics, 61*(3), 425–433.
Pyatt, K., & Sims, R. (2012). Virtual and physical experimentation in inquiry-based science labs: Attitudes, performance and access. *Journal of Science Education and Technology, 21*(1), 33–147.
Ramasundaram, V., Grunwald, S., Mangeot, A., Comerford, N. B., & Bliss, C. M. (2005). Development of an environmental virtual field laboratory. *Computers & Education, 45*(1), 21–34.
Rohrig, C., & Jochheim, A. (1999). The virtual lab for controlling real experiments via internet. In *Proceedings of the 1999 IEEE international symposium on computer aided control system design*, Hawaii, USA, August 22–27.
Rolland, J. P., Davis, L., & Baillot, Y. (2001). A survey of tracking technology for virtual environments. *Fundamentals of Wearable Computers and Augmented Reality, 1*(1), 67–112.
Ryde, J., & Hu, H. (2010). 3D mapping with multi-resolution occupied voxel lists. *Autonomous Robots, 28*(2), 169–185.
Rzepa, H. S., & Tonge, A. P. (1998). VChemLab: A virtual chemistry laboratory. The storage, retrieval, and display of chemical information using standard internet tools. *Journal of Chemical Information and Computer Sciences, 38*(6), 1048–1053.
Salzmann, M., Urtasun, R., & Fua, P. (2008). Local deformation models for monocular 3D shape recovery. In *Proceeding of the IEEE conference on computer vision and pattern recognition*, Anchorage, AK, US, June 23–28.

Sarris, N., & Strintzis, M. G. (Eds.). (2005). *3D modeling and animation: Synthesis and analysis techniques for the human body*. USA: IGI Global.

Sears, A. L., & Watkins, S. E. (1996). A multimedia manual on the World Wide Web for telecommunications equipment. *IEEE Transactions on Education, 39*(3), 342–348.

Sharma, S., Azeemuddin, S., & Anwar, M. (2011). A self learning VLSI lab along with web-based platform to design schematics and layouts. In *Proceeding of IEEE international conference on technology for education*, Chennai, India, July 14–16.

SIS. (2017). *The scholars' international school*. Retrieved from http://www.scholarsqatar.com/virtual-lab/

Smedley, T. M., & Higgins, K. (2005). Virtual technology: Bringing the world into the special education classroom. *Intervention in School and Clinic, 41*(2), 114–119.

Song, P., Yu, H., & Winkler, S. (2008). Vision-based 3D finger interactions for mixed reality games with physics simulation. In *Proceedings of the 7th ACM SIGGRAPH international conference on virtual-reality continuum and its applications in industry*, Fusionopolis, Singapore, December 8–9.

Steam. (2017). *Steam support*. Retrieved from https://support.steampowered.com/kb_article.php?ref=7285-QDGB-1502

Taylor, K., & Trevelyan, J. (1995). Australia's telerobot on the web. In *Proceedings of 26th international symposium on industrial robots*, Singapore, October 4–6.

Thorn, A. (2010). *Game engine design and implementation. Chapter 1*. Sudbury: Jones & Bartlett Publishers.

Thornton, C., & Boulay, B. (1998). *Artificial intelligence: Strategies, applications, and models through search* (2nd ed.). London: Routledge.

Tokamak. (2017). *Tokamak physics*. Retrieved from http://www.tokamakphysics.com/

Toledo. (2017). *College of business and innovation*. Retrieved from https://www.utoledo.edu/business/InfoTech/ITLabs.html

Trenholme, D., & Smith, S. P. (2008). Computer game engines for developing first-person virtual environments. *Virtual Reality, 12*(3), 181–187.

True Axis. (2017). *Authentic skateboarding game*. Retrieved from http://www.trueaxis.com/

Valera, A., Díez, J. L., Vallés, M., & Albertos, P. (2005). Virtual and remote control laboratory development. *IEEE Control Systems, 25*(1), 35–39.

Valve Developer Community. (2017). *SKD installation*. Retrieved from https://developer.valvesoftware.com/wiki/SDK_Installation

Varol, A., Shaji, A., Salzmann, M., & Fua, P. (2012). Monocular 3D reconstruction of locally textured surfaces. *IEEE Transactions of Pattern Analysis and Machine Intelligence, 34*(6), 1118–1130.

Verma, S. P., & Lin, K. S. (1989). System for automatically reading utility meters from a remote location. U.S. Patent 4,833,618.

Verschaffel, L., de Corte, E., de Jong, T., & Elen, J. (Eds.). (2010). *Use of representations in reasoning and problem solving: Analysis and improvement*. London: Routledge.

Vosniakos, G. C., Ziaaie-Moayyed, M., & Mamalis, A. G. (1997). Design of a system for computer-aided engineering of manufacturing facilities. *Computer Integrated Manufacturing Systems, 10*(1), 1–7.

Werghi, N., Fisher, R., Robertson, C., & Ashbrook, A. (1999). Object reconstruction by incorporating geometric constraints in reverse engineering. *Computer-Aided Design, 31*(6), 363–399.

Wikipedia. (2017). *Dota 2*. Retrieved from https://en.wikipedia.org/wiki/Dota_2

Witmer, B. G., & Singer, M. J. (1998). Measuring presence in virtual environments: A presence questionnaire. *Presence: Teleoperators and Virtual Environments, 7*(3), 225–240.

Zacharia, Z. C., Olympiou, G., & Papaevripidou, M. (2008). Effects of experimenting with physical and virtual manipulatives on students' conceptual understanding in heat and temperature. *Journal of Research in Science Teaching, 45*(9), 021–1035.

Zhang, Z., Zhang, M., Tumkor, S., Chang, Y., Esche, S. K., & Chassapis, C. (2013a). Integration of physical devices into game-based virtual reality. *International Journal of Online Engineering, 9*(5), 25–38.

Zhang, Z., Zhang, M., Chang, Y., Aziz, E.-S., Esche, S. K., & Chassapis, C. (2013b). Real-time 3D model reconstruction and interaction using Kinect for a game-based virtual laboratory. In *Proceedings of ASME international mechanical engineering congress & exposition*, San Diego, CA, US, November 15–21.

Zhang, Z., Zhang, M., Chang, Y., Esche, S. K., & Chassapis, C. (2014). An efficient method for creating virtual spaces for virtual reality. In *Proceedings of ASME 2014 international mechanical engineering congress and exposition*, November 14–20, 2014, Montreal, QC, Canada.

Zhang, M., Zhang, Z., Chang, Y., Esche, S. K., & Chassapis, C. (2015a). Kinect-based universal range sensor and its application in educational laboratories. *International Journal of Online Engineering, 11*(2), 26–35.

Zhang, Z., Zhang, M., Chang, Y., Esche, S. K., & Chassapis, C. (2015b). A smart method for developing game-based virtual laboratories. In *Proceedings of ASME international mechanical engineering congress and exposition*, Houston, Texas, US, November 13–19.

Zhang, M., Zhang, Z., Chang, Y., & Esche, S. K. (2015c). Simultaneous tracking and reconstruction of objects and its application in educational robotics laboratories. In *Proceedings of the ASEE annual conference & exposition*, Seattle, Washington, US, June 14–17.

Zhang, Z., Zhang, M., Chang, Y., Esche, S. K., & Chassapis, C. (2016). A virtual laboratory system with biometric authentication and remote proctoring based on facial recognition. In *Proceedings of the ASEE annual conference & exposition*, New Orleans, LA, US, June 26–29.

Zyda, M. (2005). From visual simulation to virtual reality to games. *Computer, 38*(9), 25–32.

Dr. Mingshao Zhang is an Assistant Professor in the Department of Mechanical and Industrial Engineering at Southern Illinois University, Illinois, USA.

Dr. Zhou Zhang is an Assistant Professor in the Department of Mechanical Engineering Technology at the CUNY New York City College of Technology, New York, USA.

Dr. Chang Yizhe Chang is a Post-doctoral Researcher in the Department of Computer Science at Stevens Institute of Technology, New Jersey, USA.

Dr. El-Sayed Aziz is a Teaching Assistant Professor in the Department of Mechanical Engineering at Stevens Institute of Technology, New Jersey, USA.

Dr. Sven Esche is an Associate Professor, Associate Department Chair, and Director of Graduate Programs in the Department of Mechanical Engineering at Stevens Institute of Technology, New Jersey, USA.

Dr. Constantin Chassapis is the Vice Provost for Academics and a Professor in the Department of Mechanical Engineering at Stevens Institute of Technology, New Jersey, USA.

Chapter 16
Mobile Cyber-Physical Labs: Integration of Mobile Devices with System and Control Laboratories

Jared A. Frank, Anthony Brill, and Vikram Kapila

Abstract Recent years have witnessed the adoption of mobile devices to deliver valuable interactive learning experiences to students. Although prior efforts have led to the development of mobile applications that enhance access to virtual and remote laboratories, research has not yet explored the comprehensive integration of mobile technologies into traditional laboratory activities. In this chapter, we present the development of mobile cyber-physical laboratories (MCPLs) in which hardware and software of mobile devices are leveraged in measurement, control, monitoring, and interaction with physical test-beds in the laboratory. Two separate approaches for realizing cost-effective and portable educational test-beds are proposed that utilize the sensing, storage, computation, and communication (SSCC) capabilities of mobile devices to facilitate inquiry-based educational experiences. In the first approach, smartphones are mounted directly to test-beds to allow inertial- and/or vision-based measurement and control of the test-bed. In the second approach, tablets are held such that their rear-facing cameras allow vision-based measurement and control of the test-bed. By developing mobile applications that incorporate interactive plots and augmented reality visualizations, unique and engaging learning experiences are provided from learners' personal mobile devices. The implementation and evaluation of each approach is discussed with a motor test-bed used to teach concepts of dynamic systems and control. Results of investigations indicate that by intimately linking concrete physical and cyber representations of phenomena through interactive, visually engaging interfaces, the MCPLs allow learners to make connections necessary for deep conceptual understanding and to engage in activities that hone their design skills.

Keywords Augmented reality · Dynamic systems · Mobile learning · Mixed-reality learning · Virtual reality

J. A. Frank · A. Brill · V. Kapila (✉)
Mechatronics, Controls, and Robotics Laboratory, Department of Mechanical and Aerospace Engineering, NYU Tandon School of Engineering, Brooklyn, NY, USA
e-mail: jared.alan@nyu.edu; brill.anthony@nyu.edu; vkapila@nyu.edu

M. E. Auer et al. (eds.), *Cyber-Physical Laboratories in Engineering and Science Education*, https://doi.org/10.1007/978-3-319-76935-6_16

16.1 Introduction

Since the mid-nineteenth century, educational laboratories have been environments that enable learning through hands-on experiences and have been essential in providing training in scientific inquiry and problem-solving, fostering deeper understanding of abstract concepts and theories introduced in the classroom, and arousing learners' interest in science and engineering (Blosser 1983). These roles of hands-on laboratories have been well-established, remaining relatively unaltered over the years even as laboratories underwent transformations in the delivery of experiences in response to changes in the economic, technological, and cultural landscapes. With the adoption of new technologies, educational institutions, especially those with limited budgets, have had to face the challenge of providing quality, engaging hands-on laboratory experiences despite the increased cost and complexity of equipment. The technology creating the most profound impact on laboratory instruction has been the personal computer (PC), whose integration into the laboratory has enhanced the laboratory experience, from data acquisition and data analysis to design assistance and simulations (Feisel and Rosa 2005).

With the advent of the Internet, virtual and remote laboratories have been developed to address issues of accessibility, affordability, and learner engagement in laboratory experiences. Interest in these novel laboratory paradigms stemmed from the notion that engaging laboratory experiences can be provided to learners, including those separated by a distance, without requiring a physical presence with laboratory-grade equipment. By interacting with either simulated equipment or real equipment located at a remote site, using a graphical user interface on a PC, learners can readily access learning experiences with reduced restrictions on time and equipment *vis-à-vis* traditional laboratory settings. Under the virtual or remote laboratory paradigms, learners often perform experiments through manipulation of virtual or augmented reality graphics (Andujar et al. 2011). In fact, such interactions have been shown to provide learners highly engaging experiences with scientific phenomena that are impossible to achieve in the real world (Klopfer and Squire 2008; Liu et al. 2007), such as the ability to visualize electron flow (Finkelstein et al. 2005) or the effects that physical parameters have on mechanical vibrations (Aziz et al. 2007). Although virtual and remote laboratories have allowed for reduced costs, ubiquitous learning, and engaging interactive visualizations, they have been accompanied by increases in system complexity, requiring well-trained personnel in the automation of existing test-beds, installation of networking infrastructure, and development of user interfaces. Furthermore, these laboratory paradigms have sometimes experienced difficulty meeting educational objectives due to limitations in mathematical models and the lack of physical presence of learners with equipment and collaborators (Corter et al. 2007).

As researchers attempt to address the technological and pedagogical challenges associated with virtual and remote laboratories, mobile devices such as smartphones and tablets offer several key features that may make them uniquely suitable in the development of novel solutions (Fig. 16.1a). For example, mobile devices

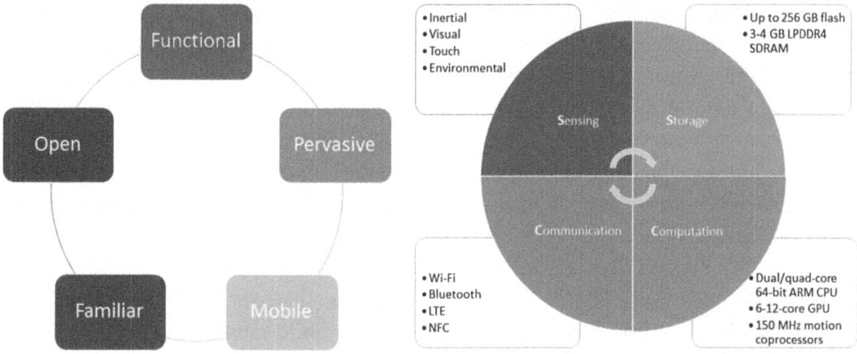

(a) (b)

Fig. 16.1 Characteristics of mobile devices that make them suitable platforms for (**a**) learning and (**b**) cyber-physical systems

have become highly pervasive and familiar, replacing laptops and desktops as the primary computational devices used by society (Bonnington 2015; Gillett 2012). In other words, learners already own these devices, carry them along wherever they go, and already know how to use them effectively. Thus, mobile devices can yield more affordable, portable, and natural laboratory experiences for today's learners *versus* solutions designed for laptop or desktop computers. Moreover, since manufacturers have opened their mobile platforms to third-party development and distribution of applications, educationally valuable applications can have immediate and widespread impact. These qualities have led mobile devices to be considered as platforms for transforming the delivery of education and training in a variety of subject areas (Ally 2009). Most importantly, mobile devices have achieved an unprecedented level of functionality that may be leveraged to enhance both on- and off-site laboratory instruction. To meet the needs of today's learners, who have become accustomed to mobile, high-quality experiences with interactive media, virtual and remote laboratories have already begun to be made accessible from interfaces on mobile devices. With mobile interfaces, learners can interact with virtual or remote laboratories from anywhere and at any time (Maiti and Tripathy 2012). In addition to increased accessibility and mobility, researchers believe that mobile interfaces to virtual and remote experiments will better engage and motivate learners (da Silva et al. 2013). However, developing mobile interfaces for the same virtual and remote laboratories may lead to many of the same technological and pedagogical issues that have been encountered with earlier implementations, while creating new issues, such as increases in time delays in the video streams over mobile networks, which can reduce responsiveness and thus disrupt sensations of presence that promote learning. Alternatively, a more comprehensive integration of the hardware and software components of mobile devices and of the conventional test-beds used in laboratory instruction can result in a cyber-physical laboratory format whose characteristics and affordances are distinct from what has been previously developed.

Research and development in mobile technologies have led to accelerating advances in the SSCC capabilities of smartphones and tablets (Fig. 16.1b). These advances offer new ways of integrating the hardware and software of mobile devices with that of laboratory test-beds, such that the SSCC tasks performed by the mobile devices are intimately coupled with the physical behavior of the test-beds. The complex interactions between computation, control, and communication associated with these cyber-physical systems introduce technical challenges that may push the limits of the mobile technologies and compromise the stability and performance of the test-beds' physical dynamics. Despite these challenges, a MCPL paradigm (one wherein test-beds must be treated as cyber-physical systems due to their extensive utilization of the onboard SSCC capabilities of mobile devices) is proposed that can deliver to learners affordable, portable, and engaging laboratory experiences in which their personal devices are responsible for measurement, control, and intuitive user interaction with the test-beds.

The MCPL approaches of this work leverage the aforementioned affordances of mobile devices to deliver educational experiences that allow learners to obtain deep understanding of concepts and build critical thinking and problem-solving skills in dynamic systems and control. The development of the MCPL approaches, their associated interfaces, and educational activities have been informed by various constructs of pedagogy and learning theories that facilitate effective learning, namely, minimalist theories of instructional design, cognitive load theory, constructivist theory, adult learning theories, and dual coding theory. The minimalism theories (Carroll 1990) emphasize the role that learners' prior experiences ought to play in designing instructional tools and activities. The cognitive load theory (Sweller 1988) suggests that instructional design should reduce anxieties and cognitive load associated with performing educational activities. The constructivist theory (Honebein et al. 1993; Vygotsky 1978) promotes self-creation of knowledge through meaningful experiences. The principles of adult learning theories (Knowles 1975) recommend inclusion of self-directed, problem-based learning activities, wherein learners discover concepts for themselves and instructors provide minimal guidance. Finally, the dual coding theory (Clark and Paivio 1991) advocates for the delivery of both verbal and visual representations of knowledge so that learners can form associations between them.

This chapter is organized as follows. In Sect. 16.2, the state of the art regarding the utilization of mobile devices to deliver laboratory instruction in science and engineering is reviewed briefly. Section 16.3 describes the proposed MCPL paradigm, including its distinct characteristics, alignment with pedagogical and learning theories, and potential educational affordances. Then, Sects. 16.4 and 16.5 present two MCPL formats in which mobile devices and laboratory test-beds are integrated by mounting mobile devices directly to the test-beds and through mobile vision-based mixed-reality interfaces, respectively. The measurement, control, and user interaction techniques utilized by implementations of each format are summarized, as well as some of the benefits and drawbacks that have been discovered in their developments. Section 16.6 presents an illustrative example, a motor test-bed used in the instruction of the concepts of dynamic systems and control, i.e., stability, damping, and pole placement, as well as the learning objectives and the original laboratory setup used to conduct activities. Then, the development and evaluation of solutions according to each of the two proposed MCPL formats are outlined. Finally, Sect. 16.7 offers concluding remarks and future directions for the proposed MCPL paradigm.

16.2 State of the Art

Equipping a laboratory with sufficient experimental resources can be a serious burden for many educational institutions with limited means. In developing countries around the world, the lack of material, financial, and human resources significantly constrains the development of modern laboratory facilities. As the power and pervasiveness of mobile devices have expanded, with most students now bringing these devices with them to class, so has their potential to enhance the delivery of educational experiences (Martin and Ertzberger 2013). Interactive mobile applications have been developed on mobile devices to teach everything from a new language (Godwin-Jones 2011) to algebra (Franklin and Peng 2008). Moreover, these applications have begun to be employed as effective tools to aid in laboratory instruction in science and engineering fields, providing domain-specific services like scientific calculators, unit converters, and scientific and engineering data tables (Libman and Huang 2013; Thilmany 2014; Williams and Pence 2011), as well as more general laboratory-related tasks, such as note-taking, data logging, photo and video documentation, report preparation and submission, and recording, analyzing, and sharing of data with peers (Nguyen et al. 2015). Furthermore, many of these mobile applications, which can be downloaded at little to no cost, can lower environmental impact, improve efficiency, increase productivity, enhance data quality, and provide immediate access to data (Hesser and Schwartz 2013).

Of the mobile applications developed thus far, some of the most useful have arguably been those that provide elegant alternatives to the services performed by more sophisticated laboratory equipment, reducing the cost and complexity associated with delivering laboratory experiences. Specifically, recent advances have included applications that leverage the computational power of mobile processors to provide visually stimulating and virtual laboratory experiences (as discussed in Sect. 16.2.1), the embedded sensing and onboard storage of devices to enable portable experimentation (as discussed in Sect. 16.2.2), and the communication capabilities of devices to extend access to remote laboratories via mobile interfaces (as discussed in Sect. 16.2.3).

16.2.1 Computation: Interactive, Visual, and Virtual Labs on Mobile Devices

Virtual, augmented, and mixed-reality technologies provide engaging interactive visualizations that have the ability to transform the delivery of science education (Cheng and Tsai 2013). With young learners, these interactive media have been used in teaching concepts that rely on spatial relations (Kerawalla et al. 2006), such as the 3D structure of compounds and materials (Núñez et al. 2008) and the rotation and revolution of planetary orbits (Woods et al. 2004). In higher education, they have been integrated to enable learners to visualize the 3D geometry associated with engineering graphics and advanced mathematics and sciences (Lee 2012). In industry, recent efforts are exploring how the power of head-mounted displays can be employed in providing interactive and immersive learning environments that can teach complex processes, such as the assembly of engines, through realistic 3D holographic visualizations (Noor 2016). One current issue with these head-mounted solutions is that they are not yet affordable and pervasive devices in society. Alternatively, the central and graphic processing units onboard mobile devices have become powerful enough to support similar learning experiences and present such experiences to learners on platforms that they already own and bring into the laboratory (Hürst and Van Wezel 2011; Liu et al. 2012a).

The ability to display engaging interactive visualizations of abstract concepts is not the only affordance provided by the powerful mobile processors. By combining such visualizations with simulations of physical processes, new opportunities exist for the implementation of virtual laboratories that run natively on mobile devices. Such laboratories have been developed to provide simulations that teach engineering students the concepts of digital signal processing, such as convolution, Fourier analysis, and filter design (Liu et al. 2012b; Ranganath et al. 2012). An important consideration in these developments is the design of the user interfaces, which feature multi-touch gestures (e.g., taps, double-taps, drag-and-drop, etc.), interactive graphical block-based programming, and plots that further aid learners in visualizing the procedures of algorithms. This is because graphically rich user interfaces, with stimulating visualizations and high levels of interactivity, have consistently been shown to provide engaging learning experiences that ultimately stimulate student interest and deeper conceptual understanding in a variety of subject areas (Naps et al. 2002; Schweitzer and Brown 2007; Venkataraman 2009).

16.2.2 Sensing and Storage: Mobile Experimentation

In addition to their powerful computational hardware, mobile devices' diverse sets of embedded sensors have begun to be leveraged in the development of portable measurement systems, impacting areas such as instrumentation, healthcare, transportation, and environmental monitoring (Alexander 2015; Lane et al. 2010).

Moreover, these sensing capabilities have allowed instructors to use smartphones and tablets as experimental tools to aid learners in performing authentic hands-on explorations of physics concepts (Kuhn and Vogt 2013). Applications are developed that analyze data recorded by the cameras and motion sensors of learners' personal devices to study phenomena in projectile motion, acoustics, optics, and radiation (Klein et al. 2014a,b; Kuhn et al. 2014; Vogt and Kuhn 2013). These studies demonstrate that the ability of mobile devices to collect physical measurements can simplify the scientific process and allow learners to use their devices for performing experimentation outside of the classroom. Moreover, the ability to conduct mobile experimentation with their personal devices may not only change learners' perspectives of their devices from consumer products to technological tools, but it may also stimulate their interest in doing science and work toward bridging the gap between informal and formal learning (Cook et al. 2008).

16.2.3 Communication: Accessing Remote Labs from Mobile Devices

In engineering laboratories, the development of user interfaces that leverage augmented graphics may allow sophisticated laboratory equipment, such as atomic force microscopes, to be effectively operated by untrained users (Vogl et al. 2006). Moreover, the effective integration of network communication capabilities and interfaces with augmented graphics can enhance the experiences of remotely conducting experiments with laboratory equipment (Andujar et al. 2011). In an effort to connect learners to laboratory experiments from their mobile devices, the necessary system architectures (Frank and Kapila 2014), software platforms (Maiti and Tripathy 2012), and interface development strategies (Orduña et al. 2011) have been investigated. With this type of extended access, educators can expand the boundaries of their classrooms and provide the combined benefits of both mobile learning and remote experimentation, namely, effective interactions with real data from truly anywhere at any time (da Silva et al. 2013), including in creative and spontaneous moments or as part of collaborations between distant learners (May et al. 2012). Mobile remote experimentation may expose learners to investigative experiences with real data that arouse interest in the field that they may not otherwise be able to obtain (de Lima et al. 2014). However, recent implementations have often been approached simply as adaptations of the traditional desktop interfaces on smaller screens (Orduña et al. 2011) and developed as web applications that lack support for several of the most powerful and attractive mobile capabilities (e.g., motion sensors, cameras, interactive 3D graphics, and computational resources) (Maiti and Tripathy 2012). Thus, current implementations fail to address the disadvantages commonly encountered with remote experimentation resulting from removed physical presence (Ma and Nickerson 2006).

16.3 Mobile Cyber-Physical Laboratories

Since the beginning of modern science and engineering education, laboratories have had to continuously adapt their methods of delivery to the release of new technologies. The most recent and revolutionary changes to laboratory instruction occurred with the adoption of PCs and the Internet (Feisel and Rosa 2005). However, just as PCs and the Internet transformed the way people interact with digital information and with one another over a distance, cyber-physical systems are expected to transform the ways in which they interact with the physical world around them (Sha et al. 2009). Research in the area of cyber-physical systems involves looking at new ways in which communication, control, and computational elements can be managed and interfaced with physical elements and can bring about significant impacts in areas such as transportation, healthcare, manufacturing, agriculture, energy, and defense (Rajkumar et al. 2010). When the cyber-physical system under investigation has mobility, it is characterized as a mobile cyber-physical system (Hu et al. 2013). Smartphones and tablets have become convenient and economic platforms for implementing mobile cyber-physical systems for a number of reasons, particularly due to the characteristics that have been illustrated in Fig. 16.1. Examples of mobile cyber-physical systems include applications to track and analyze CO_2 emissions (Froehlich et al. 2009), monitor cardiac patients (Leijdekkers and Gay 2006), and measure traffic (Rose 2006).

Applying mobile cyber-physical systems to address the needs of modern laboratories and learners can lead to the creation of novel test-beds that operate by interfacing the SSCC elements of mobile devices with physical hardware from test-beds of conventional laboratories. This is because the accuracy and precision of mobile sensors, the storage space available with mobile memory, the computational power of mobile processors, and the speed of mobile communication modules have reached levels that allow these embedded technologies to be utilized in place of conventional laboratory-grade equipment in a variety of experiments. Thus, educational institutions will be able to develop, with reduced cost and complexity, experiments in science and engineering that rely on the capture, storage, processing, and communication of the data from the sensors of mobile devices. For example, in engineering fields such as dynamic systems and control, laboratory test-beds often undergo planar motion (Apkarian 1995). Fortunately, several different technologies embedded in mobile devices, such as its inertial sensors and cameras, can be used to measure planar motion. Laboratory test-beds that display rotational or translational motion in a plane are thus amenable to being integrated with mobile devices that serve as platforms for contactless sensing and wireless control of the test-beds. This integration can produce laboratory test-beds with reduced wiring, cost, and form factor.

By allowing learners to utilize their own mobile devices in performing lab activities, MCPLs exploit intrinsic motivational aspects of the devices, since learners have been shown to be drawn to and engaged with the devices' user-friendly features (Kim et al. 2013). Moreover, by leveraging intuitive and engaging metaphors of interaction that are available on mobile devices, user interfaces can deliver educationally effective learning activities with MCPLs. For example, in accordance with the minimalist theories of instructional design (Carroll 1990), interfaces offer interaction modalities that learners are already familiar with from their day-to-day activities, such as gestures on the touchscreen for starting and stopping data collection, altering the values of experimental parameters, and zooming in and out of plots. For application navigation and operation of laboratory equipment, familiar techniques are utilized that capitalize on learners' experience with their personal devices. In this manner, the MCPL application interfaces limit the amount of training required and direct learners' attentions toward learning through the performance of engaging and interactive laboratory activities. In addition, by rendering simple but intuitive metaphors for interaction, MCPL application interfaces can reduce learners' anxieties and cognitive load, which according to cognitive load theory (Sweller 1988) is known to promote motivation in learning. This allows learners to conduct almost immediately meaningful and self-directed project-based activities, which align with the constructivist learning theories (Honebein et al. 1993; Vygotsky 1978).

To reinforce learners' visualization of concepts, MCPL interfaces render stimulating digital graphics that are relevant to the instructional material associated with activities. These graphics provide learners with feedback that facilitates the recognition of cause-effect relationships, as well as allow learners to recover from errors made in activities. The interactive graphics, such as dynamic plots and 3D augmented graphics, are included to assist learners in visualizing representations of phenomena of interest. By making use of both cyber and physical representations of phenomena that are linked and interact with each other, MCPLs appeal to learners with varied learning styles and preferences through the activation of deeper levels of information processing, which has been shown to improve the retention and recall of information (Wu et al. 2001). Moreover, MCPLs allow learners to form connections between concepts and the dynamic imagery observed with the interface. To achieve these connections with laboratory test-beds, augmented reality provides a useful technique. Wu et al. (2013) identifies five affordances of incorporating augmented reality for educational purposes: (1) learning content in 3D perspectives; (2) ubiquitous, collaborative, and situated learning; (3) learners' senses of presence, immediacy, and immersion; (4) visualizing the invisible; and (5) bridging formal and informal learning. With MCPLs, these affordances are exploited through the development of mobile applications that allow learners to interact with experimental hardware in mixed-reality.

MCPLs can not only provide economic and educational improvements for on-site activities, but they can also enable new forms of off-site activities. By eliminating the need for several laboratory-grade equipment, mobile cyber-physical systems can be made compact and thus portable. To reap further reductions in size and cost, science and engineering faculty, who nowadays have access to and experience with 3D printing, can design and fabricate their own physical hardware for research and instruction in a variety of different subjects (Irwin et al. 2014; Pearce 2012). Moreover, with the ability to print customizable equipment, instructors can rapidly prototype test-beds that are amenable to being readily integrated with mobile devices, as well as tailored to the educational objectives of the test-bed, such as the specific concepts to be illustrated.

To date, two distinct MCPL formats have been proposed whose feasibility have been validated and whose educational benefits have begun to be explored in teaching topics in dynamic systems and control. The two formats, whose implementations consist of the physical integration of the mobile device and test-bed hardware and development of a mobile application, differ mainly in how the test-bed is modified to be integrated with a mobile device. In the first format, smartphones are mounted directly to test-beds such that their embedded sensors, e.g., inertial sensors and cameras, can be readily exploited in the measurement and control of the test-bed's physical dynamics. As the back end of a mobile application executes the necessary processes to perform the feedback control of the test-bed, the front end can be utilized to provide a user interface that sits attached to the test-bed, providing learners with one central location for performing all stages of their educational activities, from experimental operation and design to the acquisition, analysis, interpretation, storage, and sharing of experimental data.

In the second format, laboratory test-beds are fitted with visual markers and learners interact with test-beds by holding their mobile devices in their hands such that the devices' rear-facing cameras are pointed at the test-beds. In the back end of a mobile application, image processing routines are employed to detect the visual markers and obtain vision-based measurements that are used in the estimation of the test-bed's physical state. These measurements are also used in the development of a mixed-reality interface, provided by the front end of the application, that supplies a live view of the system augmented with 3D interactive visualizations. Not only can this interactive augmented media aid in the visualization of concepts, but it can also be manipulated by learners to command the system to a desired set point.

Due to the differences between how mobile devices are integrated with laboratory test-beds in each of the two formats, they may render slightly different characteristics and educational affordances. Despite their differences, the two formats share several fundamental benefits regarding the cost and complexity associated with delivering engaging laboratory experiences. In both formats, the utilization of mobile device eliminates the need for traditional laboratory workstations that rely on desktop PCs and suites of sophisticated data acquisition and control equipment. User interfaces provide interactive plots that allow learners to observe measurements as they are collected in real time and to adjust the values of parameters to alter the system's behavior on the fly. These plots enable learners to make important connections between visually observed responses of system behavior, corresponding response plots of measured behavior, and the associated values of parameters so that they may understand the effects that parameters have on system behavior and the presence of certain physical phenomena.

16.4 Smartphone-Mounted Laboratory Test-Beds

The technologies embedded in mobile devices have been advanced to the point where they can be used in the measurement, estimation, and control of the state of physical systems to which the device has been rigidly attached. This concept has been demonstrated with the development of robots and unmanned vehicles with mounted smartphones that are not only able to be integrated with the vehicles' hardware but can also handle the large computational loads associated with executing the positioning and navigation algorithms for the vehicle (Aroca et al. 2012). Mounted smartphones have also been used in the stabilization and control of unmanned aerial vehicles (Desai et al. 2013) and marine vehicles (El-Gaaly et al. 2013), eliminating the need for several conventional hardware elements and thus reducing the cost, weight, and complexity of the vehicles. In these applications, measurements from the embedded inertial and visual sensors of smartphones are leveraged to detect obstacles in the environment, sense collisions, estimate the poses of vehicles, and compute vehicle velocities.

In the laboratory, the same smartphone-mounted approach can be applied. A majority of learners and educators own and bring their smartphones with them to the laboratory. Since a diverse array of laboratory test-beds can be readily mounted with such smartphones, smartphone-mounted laboratory test-beds (SMLTBs) can be employed whose states are measured and controlled by the back ends of mobile applications. Meanwhile, the front ends of applications can host user interfaces that present experimental data and provide interactive controls for learners to quickly and seamlessly perform experiments with test-beds. Although SMLTBs may have the potential to increase the portability and reduce the cost of laboratory test-beds, the development of SMLTBs presents several challenges related to mounting the smartphones, modeling and designing controllers, and programming mobile applications.

16.4.1 Measurement, Modeling, and Control

Results from experiments and simulations have validated the feasibility of mounted smartphones as measurement and control platforms for a variety of laboratory test-beds. In integrating smartphones with test-beds according to the SMLTB format, the smartphones must be rigidly mounted to the test-beds such that measurements from the embedded sensors can be used in the state estimation and control of the test-beds. To achieve such an integration successfully, three principal considerations must be addressed: (1) what smartphone sensors will be used to measure the state of the test-bed; (2) where and how the smartphone will be mounted to effectively capture these measurements; and (3) how the placement of the smartphone will affect other stages of development, including the modeling of the system and design of controllers.

A wide variety of embedded sensors have been utilized on smartphones to take physical measurements (Khan et al. 2013). Of these sensors, the inertial sensors and digital cameras present the most promising opportunities for measuring the planar motions of test-beds. For example, inertial measurement units (IMUs), which have become a standard sensor on board smartphones, consist of 3-axis gyroscopes, accelerometers, and magnetometers, whose raw data can be used directly or processed to estimate the attitude of the smartphone. By rigidly mounting the smartphone to a test-bed, the attitude readings of the smartphone indicate the attitude of the test-bed. When using smartphone IMUs in the estimation and control of the test-bed's state, the accuracy and speed of the attitude readings can have a significant impact on the stability and performance of the test-bed. Since IMUs give readings that are relative and tend to drift with time, important consideration will need to be given to the sampling rate and calibration of the IMUs. Nevertheless, research has shown that the IMUs of smartphones are capable of producing attitude readings accurate enough to stabilize a variety of motorized laboratory platforms.

In recent years, digital cameras have provided a particularly attractive modality for sensing in a variety of control applications due to their affordable, data-rich, and inherently contactless operation. Practically all smartphones now contain two digital cameras that can capture high-resolution video at impressive frame rates: one on the front surface of the device with the screen, facing toward the user (front-facing), and one on the rear surface of the device, facing away from the user (rear-facing). When leveraged alongside powerful processors, smartphone cameras can be used in the vision-based state estimation and control of a physical system in planar motion. Two main vision-based approaches have been explored to control physical systems (Sanderson and Weiss 1983). The first approach, known as the image-based approach, utilizes relations between the pixel coordinates of visual features detected in video frames as a feedback to indirectly control system state. In the second approach, called the pose-based approach, a calibrated camera is used to estimate the pose between the camera's coordinate frame and a fixed real-world coordinate frame so that the image locations of visual features in pixel coordinates can be mapped to representations of the physical locations of physical features in real-world coordinates, which can be used with traditional feedback algorithms to control the system. To keep the screen accessible to learners and visual features of interest in view of one of the smartphone's cameras, the smartphone can be mounted on the test-bed in two principal configurations: one in which the visual features are fitted to the test-bed itself and the state of the test-bed is directly observed by the camera (eye-to-hand), and another in which the state of the test-bed is indirectly estimated through observations of visual features located off of the test-bed (eye-in-hand) (Hutchinson et al. 1996). Research has explored the stabilization of an inverted pendulum on cart test-bed using a mounted smartphone in the eye-in-hand configuration (Brill et al. 2016a) and the control of a ball and beam test-bed using a mounted smartphone in the eye-to-hand configuration (Brill et al. 2016c). Due to undesirable features of many test-beds, such as inherent nonlinearities, underactuation, large bandwidth, and open-loop instability, the vision-based control poses challenging demands on processing time, frame rate, and video quality.

However, one important advantage of mounting the smartphone to the test-bed is that data from both the smartphone's inertial sensors, which are collected at faster rates but contain more noise, can be fused with that of its vision system, which are obtained at much slower rates but are less noisy, to improve the stability and performance of test-beds (Frank et al. 2016b).

Since mounting a smartphone to a test-bed can add a significant load on the actuators of the test-bed, one key disadvantage of the SMLTB format is that the dynamics of the test-bed must often be remodeled to take into account the presence of the smartphone. For example, it has been shown that farther a smartphone is placed from driven axes, the larger the effects of nonlinear terms in the dynamic model and the higher the loads on the actuators, which may not only contribute destabilizing disturbances on the system but may also cause the system analysis and control design to become more complex (Frank et al. 2016b). However, depending on the structure of the test-bed, there may be mounting configurations of the smartphone that reduce the new model to a relatively simple form. Thus, careful consideration of the consequences of smartphone placement must be made early on in the development of test-beds that invoke the SMTLB format.

16.4.2 User Interaction

Mounting a smartphone to a laboratory test-bed not only allows for its onboard hardware to monitor and control the test-bed but can also give learners access to interactive user interfaces that are mounted directly to the test-bed. Such interfaces can provide learners graphical feedback of collected data and control signals that enhance their performance of experiments. Thus, when mounting smartphones to test-beds, consideration must be taken not to block the buttons or screen of the smartphone so that a user interface remains accessible to learners. Since the interface becomes the means by which learners conduct experiments, its design is critical to the quality of the learning experience with the MCPL. Thus, through the interface, instructors can provide learners with educational content, instructions for performing the experimental procedure, and controls for interacting with the test-bed. With dynamic plots, learners have access to experimental data as it is collected in real time and can tap on touchscreen buttons to make adjustments to system parameters, to command the test-bed to a desired state, and to start or stop data collection. During the experiment, optional feedback of the data generated by the application and of the status of the application's back-end algorithms can be used by more advanced users to easily calibrate and troubleshoot the experiment. At the end of the experiment, learners can email their experimental data for sharing with others, post-processing in standard software, or incorporating their results into lab reports and presentations.

Over the course of the mid-twentieth century, several research efforts investigated the manipulation of physical models as a means of making abstract concepts more accessible to concrete thinkers (Gabel and Sherwood 1980). Copolo and Hounshell

(1995) found that students in a high school chemistry class that used both computer and physical ball-and-stick models to learn molecular structures scored significantly higher on retention test than those using either one of the models alone. Moreover, Wu et al. (2001) supports the use of multiple linked representations of scientific phenomena to account for different learning styles and preferences of individual learners. User interfaces for SMLTBs provide the benefit of access to both a physical and cyber representation of a phenomenon. By linking the representations and providing them to learners simultaneously at the same place, the mobile device creates multimodal learning experiences wherein learners can make connections to form deeper understanding of phenomena.

16.5 Mobile Mixed-Reality Test-Beds

Although SMLTBs illustrate how state-of-the-art mobile technologies may be integrated with laboratory test-beds in the development of educational platforms, they are feasible with smaller, lighter smartphones and not particularly amenable with larger, heavier tablets. As an alternative to SMLTBs, this section offers an approach that allows for the integration of mobile devices of varied form factors with laboratory test-beds in the development of a new class of educational platforms that leverage mobile mixed-reality to enhance laboratory instruction. Rather than utilizing an eye-in-hand configuration, learners may hold mobile devices in an eye-to-hand configuration such that their rear-facing cameras are pointed toward the test-beds to capture and display video from an arbitrary perspective. In this configuration, computer vision techniques can be used to obtain accurate spatial measurements that are used not only in augmenting the video with interactive visualizations but also in the feedback control of the test-bed. By having the laboratory test-bed and augmented visualizations coexist and interact in real time, mobile mixed-reality test-beds (MMRTBs) can promote immersive learning experiences in which learners command test-beds by manipulating augmented graphics through touchscreen gestures.

16.5.1 Measurement, Modeling, and Control

The MMRTB format has been utilized in three different architectures, distinguished by the information transmitted from the mobile device to the test-bed in each architecture: commands from user interactions, sensor data, or control actions (Frank and Kapila 2016b). In the first architecture, the role of the mobile device is to capture user interactions, such as gestures from moving the device or from tapping on the touchscreen, while all measurement and control of the test-bed is performed by hardware connected to the test-bed. This architecture has been demonstrated using a 2D robotic mechanism that learners can interact with by manipulating a virtual representation of the mechanism on a tablet (Frank and Kapila 2016a). Because the behavior of the virtual mechanism on the interface is programmed using the forward and inverse kinematic models of the actual mechanism, and because it responds immediately to user interactions before commands are issued to the test-bed, learners can leverage both the simulated and actual mechanisms in reinforcing their conceptual understanding of spatial relationships and robot kinematics. In the second architecture, the role of the mobile device is to capture both user interactions as well as measurements of the test-bed. This architecture, which can reduce cost introduced by laboratory-grade sensors, has been demonstrated to interact with a ball and beam system (Frank et al. 2015). This chapter discusses the third and most powerful architecture in which learners' devices are responsible for not only providing the user interface to learners and for performing sensing but also for implementing the state estimation and feedback control of the test-bed, resulting in the highest reductions in size, cost, and complexity in the process.

As the learner points the mobile device at the test-bed, video of the test-bed is captured from its rear-facing camera and displayed on the device screen. By attaching bright visual markers to critical components of the test-bed, these components can be detected in real time by processing the video using a computationally efficient and simple-to-implement color segmentation approach. Since the test-bed exhibits planar motion, the visual markers are arranged in the plane with a known model, in which their locations are represented with respect to some specified coordinate frame in the plane. By using an algorithm that exploits some combination of the colors of the markers, the distances between them, or the shapes they form, the marker association problem can be solved to uniquely identify each of the markers. Once all the markers have been detected, identified, and localized in the video, their 2D image coordinates can be fit to their corresponding 3D real-world coordinates, and the pose of the specified coordinate frame relative to the camera coordinate frame can be solved in a least-squares sense (Haralick et al. 1989). Once this relative pose is estimated from the 2D-3D point correspondences, positions and orientations of system components in the plane can be accurately measured in real-world coordinates with respect to the specified coordinate frame established by the model, using only video frames from the mobile device camera. These spatial relations estimated using computer vision techniques are used not only to render realistic augmented graphics in the scene of the video but also in the feedback control of the test-bed.

The MMRTB format holds several significant differences from the SMLTB format with regard to measurement, modeling, and control. First, MMRTBs do not require the attachment of a significant load to the test-bed. Thus, the mathematical models of MMRTBs remain relatively unchanged compared to the models of the original test-beds. This significantly simplifies efforts in modeling, analysis, and control design for test-beds and allows more time to be focused on user interface design and the development of educational activities. One disadvantage of MMRTBs is that by not being mounted to test-beds, mobile devices will not be able to use their inertial sensors in measuring the planar motion of test-beds. Thus, mobile devices will only have access to measurements collected using computer vision techniques, which may make the measurements less robust to test-beds with unmodeled nonlinearities, large bandwidth, or open-loop instability. Thus, when developing MMRTBs with more complex test-beds, it is critical to examine the effects of parameters like camera frame rate, algorithm efficiency, and video resolution on the stability and performance of the test-bed (Frank and Kapila 2016b). However, one fundamental advantage of MMRTBs is that, since they exploit an eye-to-hand configuration, they more readily support the use of mixed-reality interactions that have been proven effective in education.

16.5.2 User Interaction

As seen in Sect. 16.4.2, SMLTBs provide learners access to both a physical and cyber representation of a phenomenon. In designing user interfaces for MMRTBs, additional benefits may be obtained by having the physical and cyber representations coexist in an interactive mixed-reality environment. As the interface projects augmented visualizations of physical phenomena onto the scene of the test-bed, learners manipulate the visualizations to command the test-bed or alter its behavior. This is accomplished by executing a series of transformations that accurately map the 2D taps that learners make on the touchscreen to real-world locations in the plane of the test-bed. First, the location on the screen tapped by the learner is detected and represented in terms of screen coordinates. Then, these screen coordinates are converted to image coordinates through conversion factors between the resolutions of the screen and the video. Finally, the relative pose obtained between the plane of the device camera and the plane of the test-bed is used to map the image coordinates of the tapped location to coordinates in the frame established in the plane of the test-bed. Once known, these real-world coordinates can be used to send control actions to the test-bed, to drive the augmented graphics, or to adjust values of parameters relevant to the laboratory experiment.

By simulating the system's governing equations on the mobile application, the augmented graphics can provide responsive and predictive visualizations of system behavior before commands are issued to the test-bed. This combination of low-latency video from the device camera, extrasensory visualizations afforded by augmented graphics, and fluid interactivity provided by the touchscreen can yield mixed-reality environments that provide stimulating visual feedback to enhance the monitoring and operation of test-beds. By providing not only the actual physical response of the test-bed and the augmented graphics projected on the live video but also conventional 2D time plots of the test-bed behavior as measured using vision techniques, learners can make important connections between linked representations of the system. This allows learners to be engaged in new immersive and interactive experiences with laboratory hardware and to more effectively visualize and understand relationships between parameter values and physical phenomena.

16.6 Example: Control and Interaction with a DC Motor Test-Bed to Teach Dynamic Systems and Control Concepts

To illustrate how MCPLs that integrate mobile devices with test-beds can be developed and implemented with students, a DC motor test-bed is considered to teach concepts of dynamic systems and control. In this section, we discuss how the laboratory has traditionally been conducted, as well as the results of evaluations with both a SMLTB and MMRTB implementation of the laboratory.

16.6.1 System Description

The test-bed under consideration is a geared DC motor with a 6-in. (0.15-m)-long rectangular metal arm attached to its output shaft. Traditionally, this test-bed is fitted with an incremental optical encoder and multi-turn potentiometer to measure the motor orientation, and a tachometer to measure its angular rate. The motor is driven by an amplifier that receives control signals from a PC via a data acquisition and control board (DACB) (see Fig. 16.2). With the PC, learners have access to desktop applications, developed with engineering software such as MATLAB/Simulink or LabView, where they can enter the values of parameters, select the voltage with which to drive the motor in an open-loop control, or choose the values of control gains to use in a closed-loop proportional-plus-derivative (PD) control law. These applications typically provide knobs, sliders, or text fields for inputting this information and can also provide plots of collected data and video feedback to learners.

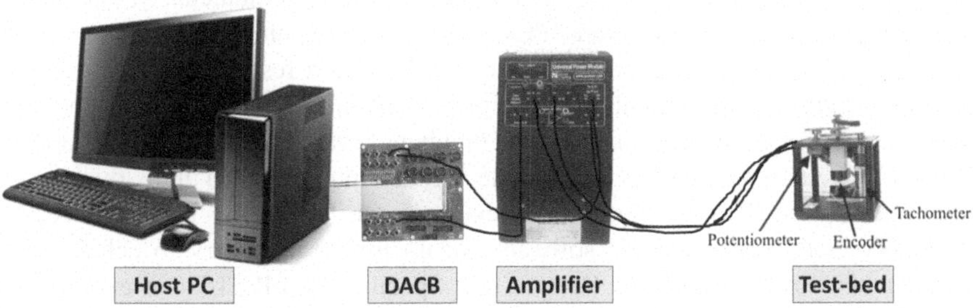

Fig. 16.2 Conventional laboratory setup for experimentation in dynamic systems and control using a motor test-bed

16.6.2 Educational Activity and Learning Objectives

A popular approach to full-state feedback control design is the pole placement technique. In a closed-loop full-state feedback control system, the locations of the poles in the s-plane directly determine the system's response to standard inputs such as unit step. If the system model is linear and controllable, the Ackermann formula (Dorf and Bishop 2008) is used to calculate the values of the control gains that yield the desired locations of the poles. However, a common problem facing learners is not the application of the formula, but rather developing an understanding of the relationship between the pole locations and resulting system behavior.

Traditionally, learners use software applications to design and run controllers with different gains to examine the effect that the values of the gains have on characteristics of the system's response, such as its stability and damping. By monitoring both the response of the actual test-bed and plots of the sensor data on the computer screen, learners can investigate phenomena such as overshoot, oscillations, and steady-state error. Finally, design challenges are assigned in which learners must place the closed-loop poles of the system in locations that result in a system response that meets specified performance criteria. Learning objectives include the ability to make associations between directly observed phenomena and plotted data, to effectively analyze data, and to hone skills designing controllers that yield a desired performance.

16.6.3 Mobile Cyber-Physical Lab

Figure 16.2 depicts a common laboratory setup for conducting the activities in dynamic systems and control. However, for institutions that must install enough experimental stations to cater to several dozen students simultaneously, equipping each station with laboratory-grade PCs, DACBs, power amplifiers, sensors, and motors can become a significant financial burden. As a result, many institutions can only afford a limited number of equipment and then cannot afford to upgrade their facilities for many years. The limited number of available experimental stations often forces learners to work in large teams that prevent each individual from having educationally meaningful experiences in the laboratory. However, costly laboratory-grade hardware and software are no longer necessary to achieve the learning objectives outlined in Sect. 16.6.2. Specifically, MCPLs can exploit the SSCC capabilities of mobile devices to produce performance data that is as central to authentic learning experiences as that produced by some of the more expensive components in a traditional laboratory station. Moreover, the integration of mobile devices allows for the development of interfaces that introduce more intuitive interactive techniques for designing controllers, commanding the test-bed, and monitoring changes in the test-bed's behavior. The proposed MCPLs manage to economize in-lab activities without sacrificing the benefits of hands-on experiences with concrete physical platforms.

The proposed design of MCPL activities leverages the affordances of mobile devices in ways that have previously been shown to support effective learning. For example, by conducting a pre-assessment prior to initiating the MCPL activities, learners' prior knowledge is activated as they review the concepts to be explored in the activity. Next, learners begin activities by navigating through a series of screens that consist of brief written introductions and illustrations of concepts. This approach allows for instructional material to be delivered in both verbal and visual forms in alignment with the dual coding theory (Clark and Paivio 1991), as well as in small digestible parts, which has been known to yield higher retention of knowledge (Miller 1956). By having learners interactively adjust parameter values and command the test-bed until a desired closed-loop performance is achieved, the MCPL activities are aligned with the constructivist theories of learning (Honebein et al. 1993; Vygotsky 1978). By specifying the target performance characteristics for the controlled dynamic system, activities create a meaningful context that extrinsically motivates learners in exploring and analyzing the system behavior. As learners interactively adjust the system's parameter values to meet the desired performance, they observe incremental changes in system behavior that allows them to perform sensemaking and form mental models of the effects of system parameters on its behavior. Finally, through the use of self-directed problem-based learning and exploration, wherein learners discover concepts for themselves and instructors provide minimal guidance, the MCPL activities are aligned with the principles of adult learning (Knowles 1975).

16.6.4 Smartphone-Mounted Approach

To control the orientation of the arm of the motor test-bed, a smartphone is rigidly mounted so that its inertial sensors or camera can be used to estimate the arm's orientation and angular velocity. Three distinct approaches have been employed to obtain these estimates. In the first approach, the smartphone's inertial sensors are used to measure both the orientation and angular velocity of the motor arm. The gyroscope provides raw measurements of the angular velocity, while fusing measurements from the gyroscope and accelerometer produces accurate estimates of the arm's orientation. In the second approach, the front-facing camera of the smartphone is used to collect vision-based measurements of orientation. This is accomplished in an eye-in-hand configuration by fitting a platform with colored markers, whose image locations can be easily detected as long as the platform remains in view of the smartphone camera. As the smartphone rotates with the motor arm, changes in its orientation are determined from the resulting changes in the location of each marker in the image (Brill et al. 2016b). To obtain accurate estimates of the arm's angular velocity, the mobile application implements a Kalman filter that uses a discretized linear model of the arm's dynamics. Finally, in the third approach, a multimodal sensing technique is used wherein inertial- and vision-based measurements are fused to produce reliable estimates of the motor arm's motion.

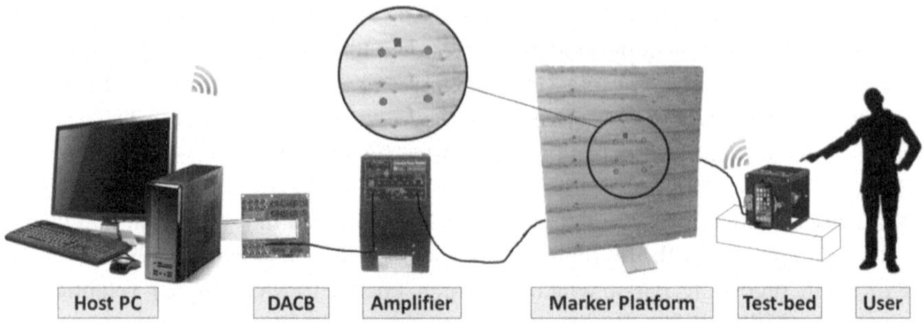

Fig. 16.3 Laboratory setup for experimentation with the SMLTB

The difference between the variance of each measurement is considered in the data fusion technique implemented directly on the mounted smartphone.

Once estimates of the arm's orientation and angular velocity are obtained, they are fed into a feedback control algorithm running in the back end of the mobile application. This algorithm computes the control action for the motor, which is wirelessly transmitted to a PC that drives the test-bed (see Fig. 16.3). Now, with the necessary SSCC tasks offloaded onto the mounted smartphone, the PC's only responsibility in this implementation is to relay control signals received from the smartphone to the motor test-bed.

The user interface developed for the proposed SMLTB is designed in two parts, which correspond to the training and design phases of the instructional activity conducted with learners. In the training phase of the activity, learners are guided through several screens of the interface that interactively introduce them to the fundamental concepts behind the lesson: the different damping and stability conditions of system responses (see Fig. A1). First, learners navigate through screens that present written descriptions and illustrative 2D plots of four fundamental classes of system responses: underdamped, overdamped, critically damped, and unstable responses. These screens help learners to begin to characterize the system responses. To reinforce this content, the next set of screens begin to use the mobile application to control the smartphone-mounted test-bed. Specifically, each of the four characteristic responses is again illustrated one at a time by redesigning a controller such that the motor test-bed demonstrates the response under investigation. Each screen is split into two segments: a top segment where an s-plane is displayed with the locations of the poles marked and bottom segment where real-time plots of the motor arm's orientation and angular velocity are shown as they are measured by the smartphone.

While the training phase provides learners with a guided learning activity to gain an understanding of the relationships between pole locations and system response, the design phase attempts to reinforce this understanding and hone design skills by allowing learners to explore the s-plane with freedom, as shown in Fig. A2. In this phase of the activity, learners are first introduced to the design challenge, which asks them to place the poles of the system such that the system exhibits no more than 10% overshoot and settles in less than 1.5 s. Then, the top segment of the screen is used to provide an interactive pole-zero plot with which learners can intuitively place the poles in the s-plane by tapping at their desired locations on the touchscreen, triggering a new controller to be designed on the fly. By pressing a button in the center of the screen, the motor is commanded to rotate by 90° using the controller redesigned with the pole locations chosen by the learner. With access to the same real-time plots at the bottom segment of the screen, learners can guide their own learning through trial and error, observing trends in the responses to make connections between the locations of the poles in the s-plane, the observed response of the actual motor arm, and plots of the measured response of the system. Thus, using this mobile tool, learners can gain valuable experience designing controllers in the s-plane and gain a deeper understanding of the changes in system response, including the appearance of phenomena like oscillations and steady-state error that can result from their designs.

As a preliminary validation of the proposed system, a group of 17 graduate-level mechanical engineering students with experience in the material covered by the activity were asked to perform it and provide an expert analysis of the potential educational value of the SMLTB. The results of the expert analysis suggest that the students found the activity enjoyable and useful in demonstrating the content (Brill et al. 2016b). Specifically, they indicated that the characteristic responses investigated in the activity were demonstrated well using the SMLTB. The interactive pole-zero plot was also deemed an effective tool for investigating relationships between the closed-loop pole locations and response of the system. The results suggest not only that the activity with the SMLTB adequately addresses concepts of dynamic systems and control but also that the students support the incorporation of SMLTBs into the formal curriculum.

To evaluate the content learning outcomes of the proposed system, a cohort of 38 undergraduate mechanical engineering students performed the activity using the SMLTB and responded to a pre- and post-assessment of their content knowledge. The results of the evaluation indicate that learners' understanding of closed-loop poles and their effects on the system response significantly improved after using the SMLTB (Brill et al. 2016b). In fact, learners' average score nearly doubled from pre- to post-assessment, suggesting that the SMLTB was successful in teaching the students the damping, stability, and pole-location concepts presented in the experiment. Moreover, all of the students in the evaluation were able to successfully complete the design challenge to produce a desirable system response. However, a majority of the students required an iterative approach toward finding a solution and thus made extensive use of the interactive and visual features of the activity. The responses to a survey conducted with the students using a five-point scale indicate that the activity was found to be useful in demonstrating the concepts presented. Additionally, the majority of students would like to see similar applications that make use of smartphones developed and applied to other laboratory experiments in the future.

16.6.5 Mixed-Reality Approach

To illustrate the activity using the MMRTB approach, the same motor test-bed is fitted with visual markers so that an eye-to-hand vision-based sensing approach can be used to measure the motor arm's angular position (Frank et al. 2016a; Frank and Kapila 2017). To demonstrate that laboratory-grade PCs and DACBs are not necessary in the MCPLs discussed in this study, a low-cost microcontroller with an external digital-to-analog converter (DAC) is used to relay control actions to the amplifier that drives the motor (see Fig. 16.4). These control actions are received over Wi-Fi from the mobile application executing on a tablet held by the learner. Once the colored markers affixed to the test-bed are visually detected by the application, their locations in the image are used to establish the relative pose between the plane of the device camera and the plane of the test-bed. This relative pose is used, along with a recursive state estimation algorithm, to obtain accurate vision-based estimates of the orientation and angular velocity of the motor arm. These estimates are then fed into the feedback controller running on the mobile application to compute the control actions sent to the microcontroller.

Due to its larger screen real estate, a tablet is used in the development of the user interface for the proposed MMRTB. Using this interface, learners can transition between the training and design phases of the activity without switching between multiple different screens. In fact, the interface developed for the MMRTB is split into three main views which appear on the screen simultaneously (see Fig. B1). In the bottom left-hand view of the interface are the same real-time plots of the arm's orientation and angular velocity as were used in the interface for the SMLTB. Above these plots in the top left-hand portion of the screen is a view whose display changes depending on the phase of the activity, which the learner can switch between by pressing buttons located at the bottom-right corner of the interface. During the training phase of the activity, the top-left view presents the same written content used by the interface for the SMLTB to introduce the four fundamental classes of system responses. During the design phase, the top-left view shows the same interactive pole-zero plot as used in the interface for the SMLTB.

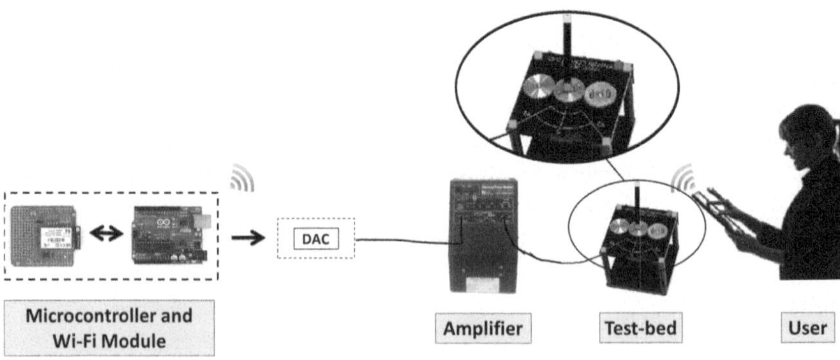

Fig. 16.4 Laboratory setup for experimentation with the MMRTB

Regardless of the phase of the activity, the large view on the right displays the live video from the rear-facing camera of the tablet. Projected onto the view is a purple semitransparent virtual arm, which lies in the plane of the actual arm and represents the set point for the test-bed. As the virtual arm is tapped and dragged, it pivots under the user's finger about its fixed end, resembling the actual motor arm rotating about its axis. Not only can this interactive augmented media aid in the visualization of concepts, but its manipulation by learners commands the system to a desired set point. Thus, by presenting interactive content in 3D perspectives that is spatially connected to the actual test-bed and updates in real time, the interface allows learners to maintain a sense of presence, immediacy, and immersion while practicing critical thinking and problem-solving skills associated with controller design in a practical and situated context. Moreover, augmented graphics of the interface enhance learners' understanding of dynamic systems and control concepts such as oscillation and steady-state error by enabling them to visualize normally invisible deviations between desired and actual system behavior. By having all three views on the screen at the same time, the user interface makes it easy for learners to make important connections between visually observed behavior of the motor arm, the real-time plots of the motor arm's behavior, and the associated locations of the system poles. That is, the interface integrates what are normally informal learning techniques (e.g., trial and error, observation, etc.), into a formal learning structure. In this manner, the MMRTB user interface integrates the previously discussed (see Sect. 16.3 and Wu et al. 2013) affordances of augmented reality for educational purposes.

To assess the educational effectiveness and user experience associated with the proposed MMRTB, an evaluation was conducted with 75 undergraduate students from two different years in mechanical and aerospace engineering using the same pre- and post-assessments that were used to evaluate the SMLTB (Frank and Kapila 2017). The results of the evaluation show that the students scored significantly higher on their post-assessment than on their pre-assessment, both in their overall scores as well as their performance in each of the three specific topic areas (damping, stability, and poles). Moreover, the results indicate that both classes benefited almost equally from their experiences with the MMRTB, which allowed learners to refresh previously seen material and gain understanding of concepts that they did not recall seeing in the classroom.

16.6.6 3D-Printed Portable Test-Beds

The results discussed above show that MCPLs can provide learners the benefits of working with physical platforms while eliminating the high costs of laboratory-grade equipment. Moreover, the test-beds of MCPLs can be developed according to two principal approaches: one in which smartphones are mounted to the test-beds and one in which tablets are held in the user's hands. In fact, the mobile devices may eliminate the need for so much of the traditional laboratory station that new,

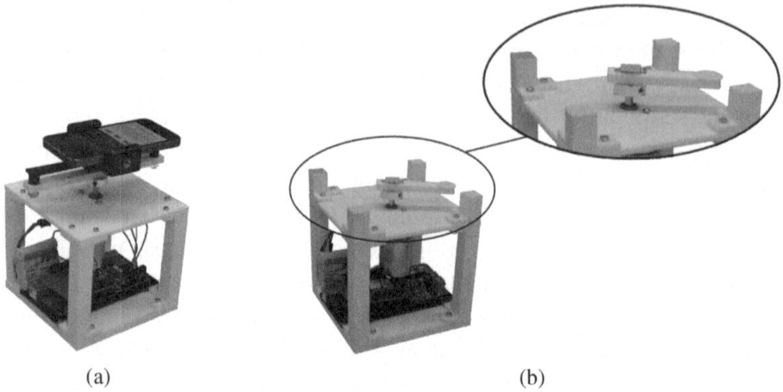

Fig. 16.5 3D-Printed portable motor test-beds developed according to the (**a**) SMLTB and (**b**) MMRTB formats

more portable platforms may be feasible using low-cost commercial-of-the-shelf (COTS) electronics to perform necessary functions (i.e., wireless communication, data acquisition, and power amplification). Figure 16.5 shows two such platforms that have been designed to support each of the two MCPL formats. With a 3D-printed base, these platforms are replicas of the motor test-beds used in the research explored in this chapter. Since the platform is designed to leverage the SSCC capabilities of learners' mobile devices, no sensors are contained in the platform, and a low-cost microcontroller with Wi-Fi module is embedded in the platform to relay the control actions wirelessly received from the mobile device to a motor controller. A multicell nickel-metal hydride battery pack powers the motor and onboard electronics. Thus, not only do the platforms in Fig. 16.5a, b provide the same interactive learning experiences as those presented in Figs 16.3 and 16.4, they do so for less than 10% of the price and are completely portable. One benefit of developing reduced-cost platforms whose base is 3D-printed and whose electronics are all COTS is that many more platforms can be built and provided to learners, creating the opportunity for individualized learning experiences, in which each learner is involved in every aspect of the activity and at his/her own pace. Since the platforms are portable, learners can even bring them home to complete assignments and to use them as study aids in preparation for exams.

16.7 Discussions and Conclusions

With mobile devices such as smartphones and tablets becoming the primary personal computers in people's lives, this chapter investigated how such devices might be integrated with laboratory test-beds to engage learners in new and interactive ways. This investigation was motivated because, despite their benefits, virtual and remote experiments lack the benefit of presence in the laboratory with actual equipment. We have found that the use of MCPLs, which intimately couple the SSCC capabilities of mobile devices with the laboratory platform's physical dynamics, creates opportunities to deliver portable, economic, and engaging hands-on experiences with both interactive visualizations and physical laboratory equipment. By enabling natural and intuitive interactions with laboratory equipment using the devices that learners already own and bring to the laboratory, MCPLs can facilitate readily accessible inquiry-based learning. To implement MCPLs, this chapter proposed two principal approaches to integrate mobile devices with test-beds that exhibit planar motion. In the first approach, smartphones are mounted directly to test-beds to obtain the benefits of both physical equipment and interactive plots provided by a user interface on the smartphone. In the second approach, mobile devices are held in learners' hands to obtain the benefits of the physical equipment, interactive plots, as well as mixed-reality graphics. By linking these disparate physical and cyber representations of phenomena, the user interfaces of MCPLs can provide learners with multimodal learning experiences that allow them to obtain deeper understanding of concepts.

To investigate the learning effectiveness and learners' experiences with MCPLs, two MCPLs were developed according to each of the two approaches, based on a motor test-bed to demonstrate concepts of dynamic systems and control to engineering students. Results of evaluations with students indicate that students' understanding of dynamic systems and controls concepts (e.g., stability, damping, and poles) improved from pre- to post-assessment by having conducted a learning activity with the MCPLs consisting of a training phase and a design phase that introduce and then reinforce the concepts, respectively. Moreover, students reported having successful, enjoyable experiences with the MCPLs, requiring little time or assistance before becoming comfortable with the user interfaces, and recommending that more such systems be developed and formally introduced into the curriculum. As a step in that direction, we briefly introduced a novel way to utilize 3D-printing and COTS electronics to support the production of low-cost MCPLs that institutions can afford and that can deliver individualized mobile learning experiences to students.

With the release of mobile devices with even more powerful processing, higher-resolution sensing, and faster communication, we expect to see wider opportunities for their integration with physical equipment in the development of MCPLs. Thus, future work will consider the use of MCPLs with other test-beds for teaching dynamic systems and control concepts, as well as in the teaching of other engineering topics. Finally, with the growing popularity of smart wearable devices, in particular of augmented reality headsets, future research will explore the educational effectiveness and user experiences associated with extending MCPLs to this class of mobile devices.

Acknowledgements This work is supported in part by the National Science Foundation awards RET Site EEC-1542286 and EEC-1132482, ITEST DRL: 1614085, DRK-12 DRL: 1417769, and GK-12 Fellows DGE: 0741714, and NY Space Grant Consortium grant 76156-10488.

Appendix A: SMLTB User Interface

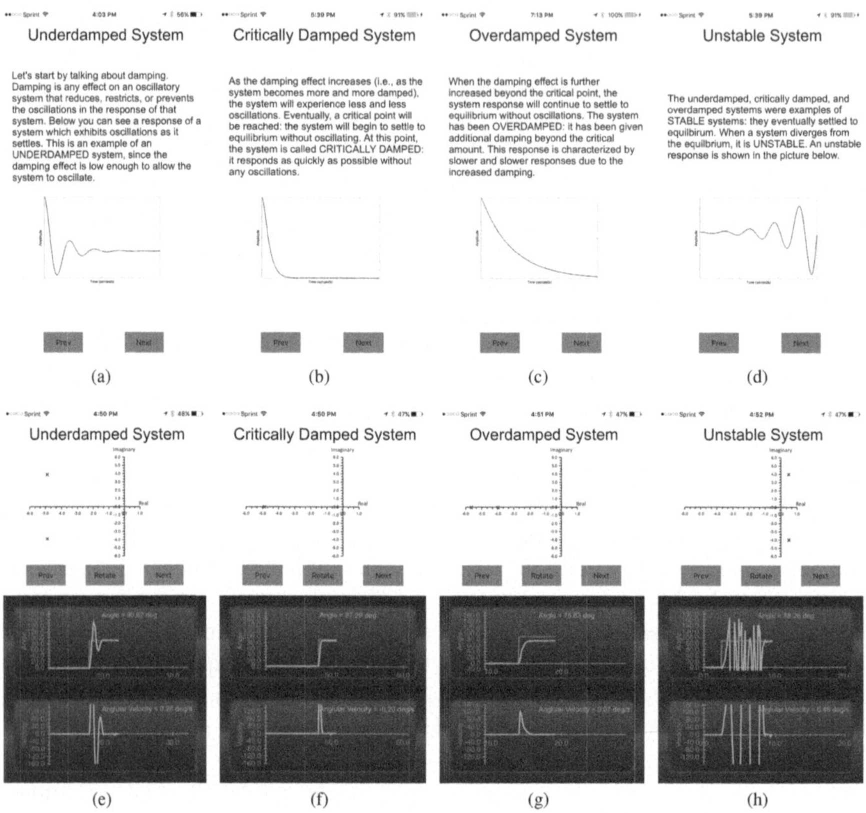

Fig. A1 Screenshots of the interface developed for training phase of activities with the SMLTB

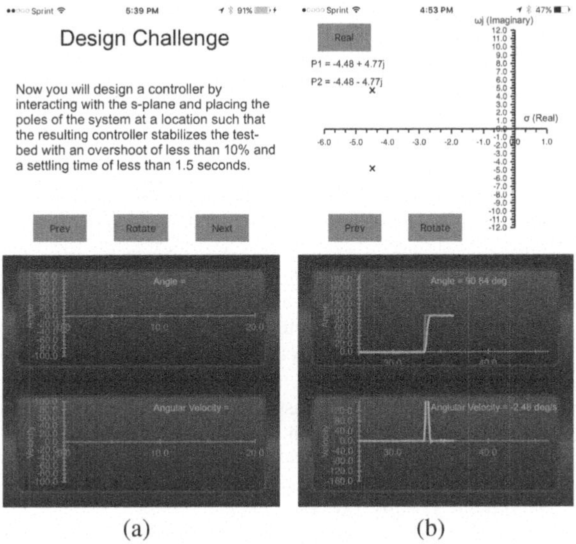

Fig. A2 Screenshots of the interface developed for design phase of activities with the SMLTB

Appendix B: MMRTB User Interface

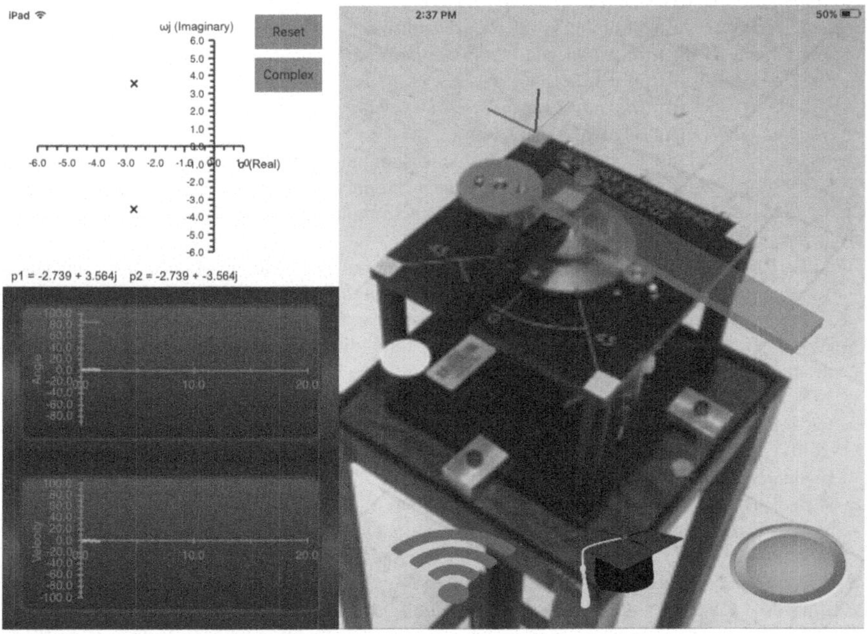

Fig. B1 Screenshot of the interface developed for the MMRTB

References

Alexander, K. (2015). Instruments in your pocket. *Mechanical Engineering Magazine, 137*(9), 43–46.

Ally, M. (2009). *Mobile learning: Transforming the delivery of education and training*. Edmonton: Athabasca University Press.

Andujar, J., Mejías, A., & Marquez, M. (2011). Augmented reality for the improvement of remote laboratories: An augmented remote laboratory. *IEEE Transactions on Education, 54*(3), 492–500.

Apkarian, J. (1995). *A comprehensive and modular laboratory for control systems design and implementation*. Toronto: Quanser Consulting.

Aroca, R. V., Péricles, A., de Oliveira, B. S., & Marcos, L. (2012). Towards smarter robots with smartphones. In *Proceedings of the 5th Workshop in Applied Robotics and Automation*, Bauru (pp. 1–6).

Aziz, E.-S., Esche, S. K., & Chassapis, C. (2007). On the design of a virtual learning environment for mechanical vibrations. In *Proceedings of the ASEE/IEEE Frontiers in Education Conference*, Milwaukee (pp. F2H-7–F2H-12).

Blosser, P. E. (1983). The role of the laboratory in science teaching. *School Science and Mathematics, 83*(2), 165–169.

Bonnington, C. (2015). In less than two years, a smartphone could be your only computer. Wired. http://www.wired.com/2015/02/smartphone-only-computer/. Accessed 10 Feb 2015.

Brill, A., Frank, J. A., & Kapila, V. (2016a). Visual servoing of an inverted pendulum on cart using a mounted smartphone. In *Proceedings of the American Control Conference*, Boston (pp. 1323–1328).

Brill, A., Frank, J. A., & Kapila, V. (2016b). Using mounted smartphones as a platform for laboratory education in engineering. In *American Society for Engineering Education Annual Conference*, New Orleans (10.18260/p. 27153).

Brill, A., Frank, J. A., & Kapila, V. (2016c). Using inertial and visual sensing from a mounted smartphone to stabilize a ball and beam test-bed. In *Proceedings of the American Control Conference*, Boston (pp. 1335–1340).

Carroll, J. M. (1990). *The nurnberg funnel: Designing minimalist instruction for practical computer skill*. Cambridge, MA: MIT Press.

Cheng, K.-H., & Tsai, C.-C. (2013). Affordances of augmented reality in science learning: Suggestions for future research. *Journal of Science Education and Technology, 22*(4), 449–462.

Clark, J. M., & Paivio, A. (1991). Dual coding theory and education. *Educational Psychology Review, 3*(3), 149–170.

Cook, J., Pachler, N., & Bradley, C. (2008). Bridging the gap? Mobile phones at the interface between informal and formal learning. *Journal of the Research Center for Educational Technology, 4*(1), 3–18.

Copolo, C. E., & Hounshell, P. B. (1995). Using three-dimensional models to teach molecular structures in high school chemistry. *Journal of Science Education and Technology, 4*(4), 295–305.

Corter, J. E., Nickerson, J. V., Esche, S. K., Chassapis, C., Im, S., & Ma, J. (2007). Constructing reality: A study of remote, hands-on, and simulated laboratories. *ACM Transactions on Computer-Human Interaction, 14*(2), 7.

da Silva, J. B., Rochadel, W., Marcelino, R., Gruber, V., & Bilessimo, S. M. S. (2013). Mobile remote experimentation applied to education. In O. Dziabenko, & J. García-Zubia (Eds.), *IT innovative practices in secondary schools: Remote experiments* (pp. 281–302). Bilbao: University of Deusto.

de Lima, J. P. C., Rochadel, W., Silva, A. M., Simão, J. P. S., da Silva, J. B., & Alves, J. B. M. (2014). Application of remote experiments in basic education through mobile devices. In *IEEE Global Engineering Education Conference*, Istanbul (pp. 1093–1096).

Desai, A., Lee, D. J., Moore, J., & Chang, Y. P. (2013). Stabilization and control of quadrotor helicopter using a smartphone device. In *Proceedings of SPIE Conference* (Vol. 8662); *Intelligent Robots and Computer Vision XXX: Algorithms and Techniques*, Burlingame (pp. 1–9).
Dorf, R., & Bishop, R. (2008). *Modern control systems*. Upper Saddle River: Pearson Education.
El-Gaaly, T., Tomaszewski, C., Valada, A., Velagapudi, P., Kannan, B., & Scerri, P. (2013). Visual obstacle avoidance for autonomous watercraft using smartphones. In *Proceedings of the Autonomous Robots and Multirobot Systems Workshop*, Saint Paul (pp. 1–15).
Feisel, L., & Rosa, A. (2005). The role of the laboratory in undergraduate engineering education. *Journal of Engineering Education, 94*(1), 121–130.
Finkelstein, N. D., Adams, W. K., Keller, C. J., Kohl, P. B., Perkins, K. K., Podolefsky, N. S., et al. (2005). When learning about the real world is better done virtually: A study of substituting computer simulations for laboratory equipment. *Physical Review Special Topics-Physics Education Research, 1*, 010103-1–010103-8.
Frank, J. A., & Kapila, V. (2014). Development of mobile interfaces to interact with automatic control experiments [Focus on education]. *IEEE Control Systems, 34*(5), 78–98.
Frank, J. A., & Kapila, V. (2016a). Towards teleoperation-based interactive learning of robot kinematics using a mobile augmented reality interface on a tablet. In *Indian Control Conference*, Hyderabad (pp. 385–392).
Frank, J. A., & Kapila, V. (2016b). Using mobile devices for mixed-reality interactions with educational laboratory test-beds. *Mechanical Engineering, 138*(6), 52–56.
Frank, J. A., & Kapila, V. (2017). Mixed-reality learning environments: Integrating mobile interfaces with laboratory test-beds. *Computers & Education, 110*, 88–104.
Frank, J. A., Gómez, J. A. D., & Kapila, V. (2015). Using tablets in the vision-based control of a ball and beam test-bed. In *Proceedings of the International Conference on Informatics in Control, Automation and Robotics*, Colmar (pp. 92–102).
Frank, J. A., Brill, A., & Kapila, V. (2016a). Interactive mobile interface with augmented reality for learning digital control concepts. In *Indian Control Conference*, Hyderabad (pp. 85–92).
Frank, J. A., Brill, A., & Kapila, V. (2016b). Mounted smartphones as measurement and control platforms for motor-based laboratory test-beds. *Sensors, 16*(8), 1331, 1–21.
Franklin, T., & Peng, L. W. (2008). Mobile math: Math educators and students engage in mobile learning. *Journal of Computing in Higher Education, 20*(2), 69–80.
Froehlich, J., Dillahunt, T., Klasnja, P., Mankoff, J., Consolvo, S., Harrison, B., et al. (2009). UbiGreen: Investigating a mobile tool for tracking and supporting green transportation habits. In *Proceedings of the SIGCHI Conference on Human Factors in Computing Systems*, Boston (pp. 1043–1052).
Gabel, D., & Sherwood, R. (1980). The effect of student manipulation of molecular models on chemistry achievement according to Piagetian level. *Journal of Research in Science Teaching, 17*(1), 75–81.
Gillett, F. (2012). Why tablets will become our primary computing device. Forrester. http://blogs.forrester.com/frank_gillett/12-04-23-why_tablets_will_become_our_primary_computing_device/. Accessed 27 Aug 2016.
Godwin-Jones, R. (2011). Emerging technologies: Mobile apps for language learning. *Language Learning & Technology, 15*(2), 2–11.
Haralick, R. M., Joo, H., Lee, C. N., Zhuang, X., Vaidya, V. G., & Kim, M. B. (1989). Pose estimation from corresponding point data. *IEEE Transactions on Systems, Man, and Cybernetics, 19*(6), 1426–1446.
Hesser, T., & Schwartz, P. (2013). iPads in the science laboratory: Experience in designing and implementing a paperless chemistry laboratory course. *Journal of STEM Education: Innovations & Research, 14*(2), 5–9.
Honebein, P. C., Duffy, T. M., & Fishman, B. J. (1993). Constructivism and the design of learning environments: Context and authentic activities for learning. In: Duffy, T.M., Lowyck, J., Jonassen, D.H., and Welsh, T.M. (eds.) *Designing environments for constructive learning*. Berlin/Heidelberg: Springer.

Hu, X., Chu, T. H. S., Chan, H. C. B., & Leung, V. C. M. (2013). Vita: A crowdsensing-oriented mobile cyber-physical system. *IEEE Transactions on Emerging Topics in Computing, 1*(1), 148–165.

Hürst, W., & Van Wezel, C. (2011). Multimodal interaction concepts for mobile augmented reality applications. In *Proceedings of International Multimedia Modeling Conference*, Taipei (pp. 157–167).

Hutchinson, S., Hager, G. D., & Corke, P. I. (1996). A tutorial on visual servo control. *IEEE Transactions on Robotics and Automation, 12*(5), 651–670.

Irwin, J. L., Pearce, J. M., Anzolone, G., & Oppliger, D. E. (2014). The RepRap 3-D printer revolution in STEM education. In *ASEE Annual Conference & Exposition*, Indianapolis. https://peer.asee.org/23175.

Kerawalla, L., Luckin, R., Seljeflot, S., & Woolard, A. (2006). Making it real: Exploring the potential of augmented reality for teaching primary school science. *Virtual Reality, 10*(3–4), 163–174.

Khan, W. Z., Xiang, Y., Aalsalem, M. Y., & Arshad, Q. (2013). Mobile phone sensing systems: A survey. *IEEE Communications Surveys & Tutorials, 15*(1), 402–427.

Kim, Y. H., Kim, D. J., & Wachter, K. (2013). A study of mobile user engagement (MoEN): Engagement motivations, perceived value, satisfaction, and continued engagement intention. *Decision Support Systems, 56*, 361–370.

Klein, P., Gröber, S., Kuhn, J., & Müller, A. (2014a). Video analysis of projectile motion using tablet computers as experimental tools. *Physics Education, 49*(1), 37–40.

Klein, P., Hirth, M., Gröber, S., Kuhn, J., & Müller, A. (2014b). Classical experiments revisited: Smartphones and tablet PCs as experimental tools in acoustics and optics. *Physics Education, 49*(4), 412.

Klopfer, E., & Squire, K. (2008). Environmental detectives – the development of an augmented reality platform for environmental simulations. *Educational Technology Research and Development, 56*(2), 203–228.

Knowles, M. S. (1975). *Self-directed learning: A guide for learners and teachers*. New York: Associated Press.

Kuhn, J., & Vogt, P. (2013). Applications and examples of experiments with mobile phones and smartphones in physics lessons. *Frontiers in Sensors, 1*(4), 67–73.

Kuhn, J., Molz, A., Gröber, S., & Früebis, J. (2014). iRadioactivity: Possibilities and limitations for using smartphones and tablet PCs as radioactive counters. *The Physics Teacher, 52*(6), 351–356.

Lane, N. D., Miluzzo, E., Lu, H., Peebles, D., Choudhury, T., & Campbell, A. T. (2010). A survey of mobile phone sensing. *IEEE Communications Magazine, 48*(9), 140–150.

Lee, K. (2012). Augmented reality in education and training. *TechTrends, 56*(2), 13–21.

Leijdekkers, P., & Gay, V. (2006). Personal heart monitoring and rehabilitation system using smart phones. In *International Conference on Mobile Business*, Copenhagen (pp. 1–7).

Libman, D., & Huang, L. (2013). Chemistry on the go: Review of chemistry apps on smartphones. *Journal of Chemical Education, 90*(3), 320–325.

Liu, W., Cheok, A. D., Mei-Ling, C. L., & Theng, Y.-L. (2007). Mixed reality classroom: Learning from entertainment. In *International Conference on Digital Interactive Media in Entertainment and Arts*, Perth (pp. 65–72).

Liu, C., Huot, S., Diehl, J., Mackay, W., & Beaudouin-Lafon, M. (2012a). Evaluating the benefits of real-time feedback in mobile augmented reality with hand-held devices. In *Proceedings of the SIGCHI Conference on Human Factors in Computing Systems*, Austin (pp. 2973–2976).

Liu, J., Hu, S., Thiagarajan, J. J., Zhang, X., Ranganath, S., Banavar, M. K., et al. (2012b). Interactive DSP laboratories on mobile phones and tablets. In *IEEE International Conference on Acoustics, Speech and Signal Processing*, Kyoto (pp. 2761–2764).

Ma, J., & Nickerson, J. V. (2006). Hands-on, simulated, and remote laboratories: A comparative literature review. *ACM Computing Surveys, 38*(3), 7, 1–24.

Maiti, A., & Tripathy, B. (2012). Different platforms for remote laboratories in mobile devices. *International Journal of Modern Education and Computer Science, 4*(5), 38–45.

Martin, F., & Ertzberger, J. (2013). Here and now mobile learning: An experimental study on the use of mobile technology. *Computers & Education, 68*, 76–85.

May, D., Terkowsky, C., Haertel, T., & Pleul, C. (2012). Using e-portfolios to support experiential learning and open the use of tele-operated laboratories for mobile devices. In *International Conference on Remote Engineering and Virtual Instrumentation*, Bilbao.

Miller, G. A. (1956). The magical number seven, plus or minus two: Some limits on our capacity for processing information. *Psychological Review, 63*, 81–97.

Naps, T. L., Rößling, G., Almstrum, V., Dann, W., Fleischer, R., Hundhausen, C., et al. (2002). Exploring the role of visualization and engagement in computer science education. In *ACM SIGCSE Bulletin, 35*, 131–152.

Nguyen, L., Barton, S., & Nguyen, L. (2015). iPads in higher education: Hype and hope. *British Journal of Educational Technology, 46*(1), 190–203.

Noor, A. K. (2016). The HoloLens revolution. *Mechanical Engineering Magazine, 138*(10), 32–37.

Núñez, M., Quirós, R., Núñez, I., Carda, J. B., & Camahort, E. (2008). Collaborative augmented reality for inorganic chemistry education. In *Proceedings of the WSEAS/IASME International Conference on Engineering Education*, Heraklion (pp. 271–277).

Orduña, P., García-Zubia, J., Irurzun, J., López-de-Ipiña, D., & Rodriguez-Gil, L. (2011). Enabling mobile access to remote laboratories. In *Global Engineering Education Conference*, Amman (pp. 312–318).

Pearce, J. M. (2012). Building research equipment with free, open-source hardware. *Science, 337*(6100), 1303–1304.

Rajkumar, R., Lee, I., Sha, L., & Stankovic, J. (2010). Cyber-physical systems: The next computing revolution. In *Proceedings of the 47th Design Automation Conference*, Seattle (pp. 731–736).

Ranganath, S., Thiagarajan, J. J., Ramamurthy, K. N., Hu, S., Banavar, M., & Spanias, A. (2012). Work in progress: Performing signal analysis laboratories using Android devices. In *Frontiers in Education Conference*, Seattle (pp. 1–2).

Rose, G. (2006). Mobile phones as traffic probes: Practices, prospects and issues. *Transport Reviews, 26*(3), 275–291.

Sanderson, A. C., & Weiss, L. E. (1983). Adaptive visual servo control of robots. In: Pugh, A. (ed.) *Robot vision* (pp. 107–116). Berlin/Heidelberg: Springer.

Schweitzer, D., & Brown, W. (2007). Interactive visualization for the active learning classroom. *ACM SIGCSE Bulletin, 39*, 208–212.

Sha, L., Gopalakrishnan, S., Liu, X., & Wang, Q. (2009). Cyber-physical systems: A new frontier. In: Yu, P.S. and Tsai, J.J.P. (eds.) *Machine learning in cyber trust* (pp. 3–13). Boston, MA: Springer.

Sweller, J. (1988). Cognitive load during problem solving: Effects on learning. *Cognitive Science, 12*, 257–285.

Thilmany, J. (2014). Working away. *Mechanical Engineering Magazine, 136*(1), 40–43.

Venkataraman, B. (2009). Visualization and interactivity in the teaching of chemistry to science and non-science students. *Chemistry Education Research and Practice, 10*(1), 62–69.

Vogt, P., & Kuhn, J. (2013). Analyzing radial acceleration with a smartphone acceleration sensor. *The Physics Teacher, 51*(3), 182–183.

Vogl, W., Ma, B. K.-L., & Sitti, M. (2006). Augmented reality user interface for an atomic force microscope-based nanorobotic system. *IEEE Transactions on Nanotechnology, 5*(4), 397–406.

Vygotsky, L. S. (1978). Mind in society: The development of higher mental processes. Cambridge, MA: Harvard University Press.

Williams, A., & Pence, H. (2011). Smart phones, a powerful tool in the chemistry classroom. *Journal of Chemical Education, 88*(6), 683–686.

Woods, E., Billinghurst, M., Looser, J., Aldridge, G., Brown, D., Garrie, B., et al. (2004). Augmenting the science centre and museum experience. In *Proceedings of the International Conference on Computer Graphics and Interactive Techniques in Australasia and South East Asia*, Suntec City (pp. 230–236).

Wu, H.-K., Krajcik, J. S., & Soloway, E. (2001). Promoting understanding of chemical representations: Students' use of a visualization tool in the classroom. *Journal of Research in Science Teaching, 38*(7), 821–842.

Wu, H. K., Lee, S. W. Y., Chang, H. Y., & Liang, J. C. (2013). Current status, opportunities and challenges of augmented reality in education. *Computers & Education, 62*, 41–49.

Jared A. Frank was a doctoral degree student at the NYU Tandon School of Engineering (2013–2017) and now serves as an R&D Sensor System Engineer at Siemens Gamesa Renewable Energy.

Anthony Brill was a masters degree student at the NYU Tandon School of Engineering (2014–2016) and now serves as an R&D Control Engineer at Siemens Gamesa Renewable Energy.

Vikram Kapila is a Professor in the Department of Mechanical and Aerospace Engineering at the NYU Tandon School of Engineering.

The manufacturer's authorised representative in the EU is Springer Nature Customer Service Centre GmbH, Europaplatz 3, 69115 Heidelberg, Germany. If you have any concerns regarding our products, please contact ProductSafety@springernature.com

Printed and bound by CPI Group (UK) Ltd, Croydon, CR0 4YY

25/09/2026

02222831-0007